Water Resource Management Issues

Water Resource Management Issues

Basic Principles and Applications

Louis Theodore
R. Ryan Dupont

CRC Press
Taylor & Francis Group
Boca Raton London New York

CRC Press is an imprint of the
Taylor & Francis Group, an **informa** business

CRC Press
Taylor & Francis Group
6000 Broken Sound Parkway NW, Suite 300
Boca Raton, FL 33487-2742

Printed on acid-free paper

International Standard Book Number-13: 978-0-367-18385-1 (Hardback)

Visit the Taylor & Francis Web site at
http://www.taylorandfrancis.com

and the CRC Press Web site at
http://www.crcpress.com

Contents

SECTION III Water Treatment Technologies

SECTION IV The Future

Preface

In the last five decades, people have become aware of a wide range of environmental issues. All sources of air, land, and water pollution are under constant public scrutiny. In particular, increasing numbers of professionals are being confronted with problems related to drinking water issues. Because some of these are of relatively new concern, individuals must develop a proficiency and an improved understanding of technical and scientific, as well as regulatory, issues regarding resource management issues water and related topics to cope with these challenges.

Water is as important a resource as energy. It is an integral part of virtually all societal activities, including food consumption, energy production and distribution, transportation, environmental management activities, industrial development, habitat for fish species, animal and human health, etc. Yet, water resources are not only unevenly but also, irregularly, distributed with some parts of the world experiencing extreme shortages of water. Addressing this problem will become a major undertaking of the technical community in this century. Dealing with these problems and potential solutions is perhaps the main objection of this book.

Although this is not the first professional book to treat this particular subject, it is one of the few books that attempts to highlight all aspects of the spectrum of water resource management issues. This book is intended primarily for engineers, industrial hygienists, health and safety officers, and plant engineers and managers. Lawyers, news media personnel, and regulatory officials can also benefit from this text. The authors' aim is to offer the reader a perspective on water resource management issues and solutions and to provide an introduction to the specialized literature in this and related areas. The readers are encouraged, through the reference lists at the end of each chapter, to continue their own development beyond the scope of this book.

As is the case in preparing a book, the problem of what to include and what to omit has been particularly difficult. However, every attempt has been made to offer material to individuals with a limited technical background at a level that should enable them to better cope with some of the complex problems encountered in water management today.

The book is divided into four parts. Following a detailed Glossary, Section I provides an introduction to background issues such as water properties and water chemistry, regulatory approaches, international concerns, etc. Section II covers issues related to water resources. Separate chapters are provided on the United States and global sectors. The relatively new topic of water in space is also described. Section III is devoted to water treatment technologies and processes, including chapters on evaporation, membrane processes, crystallization, and nanotechnology. The future is addressed in Section IV. The general topics of terrorism, pollution prevention, and sustainability, plus a separate chapter on desalination that highlights some recent activity in this area are addressed in Section IV, which includes a chapter on optimization and concludes with a chapter on ethical considerations and future U.S. water security. Also note that Illustrative Examples complement the presentation in many of the chapters. The Appendix includes separate sections on units and common abbreviations and conversion constants.

During the preparation of this book, the authors were ably assisted by a number of individuals. These people devoted much time and energy researching and writing parts of various sections of this book. Their invaluable assistance is gratefully acknowledged; the names of these individuals are listed under the titles of the chapters to which they contributed. The authors are also particularly indebted to Drs. Johnny Jeris and Wally Matyistik for their technical support and to the contributing authors that provided invaluable assistance in the preparation of various chapters throughout the text. The authors' sincere gratitude is due to all those who patiently assisted with the typing and proofreading of this manuscript, especially Mary K. Theodore and Sarah Ruane, and to Ivonne Harris of the Utah Water Research Laboratory that prepared all final copies of figures and illustrations.

Louis Theodore
East Williston, NY

R. Ryan Dupont
Smithfield, UT

Authors

Raised in Hell's Kitchen, **Louis Theodore** received the degrees of M.Ch.E and Eng.Sc.D from New York University and B.Ch.E. from The Cooper Union. Over the past 50 years, Dr. Theodore has been a successful educator at Manhattan College (holding the rank of Full Professor of Chemical Engineering), Graduate Program Director, researcher, professional innovator, and technical communicator. During this period, he was primarily responsible for his program achieving a No. 2 ranking by the *U.S. News & World Report* and particularly successful in placing students in internships, jobs, and graduate schools.

Theodore is an internationally recognized lecturer who has provided more than 200 courses to industry, government, and technical associations. Theodore developed and served as the principal moderator/lecturer for U.S. Environmental Protection Agency (EPA) courses on hazardous waste incineration and air pollution control equipment, consulted for several industrial companies in the field of pollution prevention and environmental management, and served as a consultant/expert witness for the U.S. EPA and U.S. Department of Justice.

He is the author of more than 110 text/reference books ranging from pollution prevention to air pollution control to hazardous waste incineration and engineering and environmental ethics.

Theodore is the recipient of the Air and Waste Management Association's (AWMA) prestigious Ripperton award that is "presented to an outstanding educator who through example, dedication and innovation has so inspired students to achieve excellence in their professional endeavors." He was also the recipient of the American Society for Engineering Education (ASEE) AT&T Foundation award for "excellence in the instruction of engineering students."

He currently serves as a part-time consultant to Theodore Tutorials. Theodore is a member of Phi Lambda Upsilon, Sigma Xi, Tau Beta Pi, American Chemical Society, American Society of Engineering Education, Royal Hellenic Society, and a Fellow of the Air & Waste Management Association.

R. Ryan Dupont has more than 35 years of experience teaching and conducting applied and basic research in environmental engineering at the Utah Water Research Laboratory at Utah State University (USU). His main research areas have addressed soil and groundwater bioremediation, stormwater management via green infrastructure, field remediation technology demonstration and treatment system performance verification, and water reuse technology performance and risks. He received a BS degree in Civil Engineering and MS and PhD degrees in Environmental Health Engineering from the University of Kansas–Lawrence. He has been a Full Professor of Civil and Environmental Engineering at USU since 1995, served as the Head of the Environmental Engineering Division for 10 years, was instrumental in establishing an undergraduate degree in Environmental Engineering at USU, and has been responsible for attracting more than $6 million in extramural funding through the Water Research Lab since joining the faculty in 1982. Dupont is a member of Sigma Xi, Tau Beta Pi, Chi Epsilon, the American Society of Civil Engineers, the American Society of Engineering Educators, the Water Environment Federation, the Solid Waste Association of North America, Engineers without Borders, and the Air and Waste Management Association. He was recognized as an Outstanding Young Engineering Educator by the American Society of Engineering Education and was a 2015 recipient of the Richard I. Stessel Waste Management Award for "distinguished achievement as an educator in the field of waste management" from the Air and Waste Management Association.

Authors

Section I

Overview

1 Glossary of Terms

1.1 INTRODUCTION

This chapter provides definitions for many, but not all, of the terms that the reader will encounter in this book. It should be noted that some of the definitions specifically refer to the United States and that a good number of definitions pertain to wastewater and wastewater-related terms. It should also be noted that many of the terms have come to mean different things to different people, and this will become evident as one delves deeper into the applicable literature. Finally, the bulk of the material in this chapter was drawn from the previous work of Theodore et al. (1997).

1.2 GLOSSARY

absolute humidity: the amount of water vapor present in a unit mass of air, which is usually expressed as kilograms of water vapor per kilogram of dry air or pounds of water vapor per pound of dry air.

absolute pressure: the actual pressure exerted on a surface that is measured relative to zero pressure; it equals the gauge pressure plus the atmospheric pressure.

absolute pressure gauge: a device that measures the pressure exerted by a fluid relative to a perfect vacuum.

absolute scale: a temperature scale that is based on absolute zero and that uses units of measurement equivalent to centigrade degrees on the Kelvin scale or to Fahrenheit degrees on the Rankine scale.

absolute temperature: the temperature expressed in degrees Kelvin (K) or degrees Rankine (°R).

absolute temperature scale: a scale (e.g., Kelvin, Rankine) in which temperatures are measured relative to absolute zero.

absolute vacuum: a void that is completely empty of matter.

absolute zero temperature: the temperature of zero degrees on either the Kelvin or Rankine scale at which molecular motion is thought to cease.

absorbate: a substance that is taken up and retained by an absorbent.

absorbent: any substance that takes in or absorbs other substances.

absorber: a device in which a gas is absorbed by contact with a liquid.

absorption: the process in which one material (the absorbent) takes up and retains another (the absorbate) to form a homogenous solution; it often involves the use of a liquid to remove certain gas components from a gaseous mixture.

absorption tower: a vertical tube or pipe in which a rising gas is partially absorbed by a liquid, usually in the form of falling droplets.

absorption trench: a part of a subsurface sewage disposal system that consists of a trench, aggregate, soil, and a distribution pipe.

acid: a material containing hydrogen that produces at least one hydrogen ion when dissolved in a water solution; it can react with and neutralize a base to form a salt.

Contributing Author: Bridget Forster

acid deposition: a complex chemical and atmospheric phenomenon that occurs when emissions of sulfur and nitrogen compounds plus other substances are transformed by chemical processes in the atmosphere (often far from the original source) and then are deposited on surfaces in either a wet or dry form; the wet form, properly called "acid rain," can fall as rain, snow, ice, or fog.

acid dew point: the dew point of flue gases that contain any significant quantity of sulfur trioxide, SO_3; this temperature is approximately 300°F.

acid rain: precipitation contaminated with sulfur dioxide, nitrous oxide, and other chemicals from power plants and industrial sites.

acre-foot: a measure of volume; 43,560 cubic feet, 325,900 gallons, or the volume of water covering 1 acre, 1 foot deep.

action level: the level of a contaminant that, if exceeded, triggers treatment or other requirements that a water system must follow.

activated sludge: a suspension of microbial biomass grown within an aeration tank in an activated sludge process; the biomass is used to remove biodegradable organic material in wastewater.

activated sludge process: a biological process employed in secondary wastewater treatment in which a mixture of wastewater and microorganisms is agitated and oxygenated in an aeration tank to encourage microorganisms to biodegrade organic material in a wastewater stream; the sludge is settled then recirculated to provide additional contact time between the incoming wastewater and the recycled biomass (return activated sludge).

acute health effect: an immediate effect (i.e., within seconds, minutes, hours or days) that may result from exposure to certain drinking water contaminants (e.g., pathogens).

adiabatic: a term used to describe a system in which no gain or loss of heat occurs.

adjudication: a process performed by the courts that determines ownership of groundwater and assigns a Watermaster to manage or enforce pumping rights and sometimes water quality; this is often a multiyear process.

advanced wastewater treatment: any process that is employed for the treatment of wastewater that follows secondary treatment and serves to improve the quality of effluent prior to reuse or discharge; this may include the removal of phosphorus, nitrogen, suspended solids, dissolved organic compounds, dissolved solids, etc.

aerated lagoon (aerated pond): an engineered wastewater treatment pond or earthen basin in which mechanical or diffused aeration is employed to provide an oxygen supply for biological wastewater treatment.

aerated pond: see aerated lagoon.

aeration: a process in which water or another fluid is brought into intimate contact with air so that oxygen may be absorbed by the fluid; this is accomplished by diffused or mechanical surface agitation, and is employed in biological wastewater treatment processes.

aeration tank: a tank in which wastewater and biological solids (activated sludge) are brought into contact with oxygen or air by aeration.

aerobic: a term used to describe a system that requires oxygen to sustain itself.

aerobic bacteria: bacteria that require dissolved oxygen for metabolism and growth.

aerobic biological oxidation: any process involving the use of aerobic organisms.

aerobic digestion: the stabilization of microbial solids wasted from an activated sludge process (Waste Activated Sludge) through aeration.

aerobic respiration: a process in which organisms, using dissolved oxygen as a terminal electron acceptor, generate energy for growth and metabolism from the breakdown of organic material or waste substances.

aerobic treatment: a process by which microbes decompose complex organic compounds in the presence of oxygen and use the liberated energy for growth and metabolism.

agricultural irrigation: water distribution systems and practices in agriculture.

air pollutant: any substance in air that, at a sufficiently high concentration, could harm humans, animals, vegetation, or materials of construction.

air stripping tower: a tower used to remove volatile organic chemicals (such as solvents) from contaminated water by causing them to evaporate; polluted water is sprayed downward through a tower filled with packing materials while air is blown upward through the tower.

algae: a large group of aquatic plants that contain chlorophyll; they vary from single cell to multicellular organisms; algae exist in salt water and freshwater and can adversely affect water quality when excessive growth and death occurs (eutrophication) resulting in the lowering of the oxygen content in water.

algae bloom: a sudden proliferation of algae in water bodies, usually stimulated by the excessive input of nutrients (e.g., nitrogen and phosphorous); this can adversely affect water quality by causing a lowering of the oxygen content in the water.

algae harvesting: an advanced wastewater treatment method employed to remove nutrients by their uptake and assimilation into algal cells.

ambient: a term used to describe the surrounding area or environment.

amoeba: a small protozoa that moves and ingests food by changing the shape of its cell body.

anaerobic: a chemical reaction, process, or microorganism that occurs in the absence of oxygen.

anaerobic bacteria: bacteria that live and grow in the absence of free oxygen.

anaerobic biological treatment: any treatment using anaerobic organisms to reduce the organic matter in wastes.

anaerobic contact process: an anaerobic wastewater treatment process where microorganisms are removed from the effluent stream by sedimentation and recycled to the process to provide additional treatment; an anaerobic activated sludge process.

anaerobic digestion: the process of the anaerobic stabilization of wastewater solids settled in a primary clarifier; the solids are anaerobically converted to soluble organic materials then to carbon dioxide and methane by a consortium of anaerobic organisms.

anaerobic lagoon: a waste stabilization pond that is devoid of dissolved oxygen and employed to stabilize high organic content wastes; these lagoons are deep with a small surface area to minimize oxygen diffusion into the liquid.

anaerobic waste treatment: wastewater treatment processes that use anaerobic microorganisms in the absence of free oxygen to stabilize biodegradable wastewater or solids.

appropriative water rights: holders can use available water that is not taken by anyone else.

aquatic: a term used to describe any organism growing in, living in, or frequenting water.

aquatic growth: any floating, drifting, or attached organism in a body of water.

aqueduct: a conduit or channel employed to convey water from one location to another.

aqueous solution: a solution with water as the liquid phase.

aquifer: a geologic formation, group of formations, or part of a formation that is capable of yielding a significant amount of groundwater to wells or springs.

artesian well: a well tapping a confined aquifer in which the static water level is above the bottom of the upper confining unit; a flowing artesian well is a well in which the water level is above the land surface.

audit: the examination of something with intent to check, verify, or inspect.

available oxygen: the quantity of dissolved oxygen available for oxidation of organic matter in a water body.

back wash: water used in a filtration system to separate clogging material from a filter media to clean the filter so it may be put back into service.

backflow: a flow condition induced by a differential in pressure that causes the flow of water or other fluid into the distribution pipes of a reservoir from any source other than its intended primary source.

base: any compound that dissociates in aqueous solution to yield hydroxyl ions; it is employed to neutralize acids.

basin plan: a planning document produced and updated usually every 5 years by regional water boards that establishes the beneficial use for each water body and water quality improvements necessary to maintain or achieve water quality standards for these uses.
batch process: an unsteady-state process that is not continuous; its operations are carried out with discrete quantities of material.
beach erosion: the deterioration of the shoreline by wave action, shore currents, or other natural occurrences.
bedrock: the solid rock mass located below the loose material on the Earth's crust such as soil, glacial drift, or alluvium.
bio-oxidation unit: a piece of equipment that consists of an aeration tank and a clarification chamber with an adjustable overflow weir.
biochemical oxygen demand (BOD): the amount of dissolved oxygen required for the microbial decomposition of biodegradable organic matter in a wastewater; it is a standardized means (5-day, at 20°C, in the dark) of estimating the level of biodegradable organic contamination in a wastewater sample.
biodegradability: the susceptibility to decompose by the action of microorganisms.
biological wastewater treatment: a method of wastewater treatment in which bacterial or biochemical processes are employed to stabilize biodegradable organic materials in wastewater.
biomass: all of the living material in a given area, and the solid fuels that are composed of crop, plant, and tree residues, as well as cattle manure.
biota: all of the living organisms that exist in an area.
blackwater: a water that contains human wastes.
blowdown: the cyclic or constant removal of a portion of any process flow to maintain the constituents of the flow at a desired level.
boiler feedwater: the water fed into a boiler to replace that evaporated in the generation of steam.
boiling point: the temperature of a liquid at which its vapor pressure is equal to that of the atmospheric pressure of the environment; it is 212°F for water at sea level.
bound water: the water molecules that are tightly held by various chemical groups in a larger molecule.
brackish water: a water with a salt content in the range between that of freshwater and seawater.
brine: a concentrated solution of salt and water that remains after the removal of a distilled product.
by-pass: the avoiding of a particular portion of a process or system.
by-product: a material that is not one of the primary products and is not solely or separately produced by the production process.
calibration: the determination, checking, or adjustment of the accuracy of any instrument that gives quantitative measurements.
carcinogen: any substance that can cause cancer.
carryover: the entrainment of liquid or solid particles in the vapor evolved by a boiling liquid or from a process unit.
catch basin: a chamber or well, usually built at the curbline of a street, that admits surface water for discharge into a stormwater drain.
CBOD5: the amount of dissolved oxygen consumed in 5 days from the carbonaceous portion of biodegradable materials in a wastewater.
cesspool: a lined underground basin to which raw household wastewater is sent and from which the water leaks into the surrounding soil.
chemical analysis: the analysis by chemical methods that provides the composition and concentration of a substance.
chemical oxygen demand (COD): a measure of the oxygen that is needed for the chemical oxidation of organic matter present in wastewater.
chemical sludge: the sludge obtained by treatment of wastewater with chemicals.

chemical treatment: any one of a variety of technologies that use chemicals or a variety of chemical processes to treat a system or waste.
chlorine demand: the amount of chlorine necessary to produce a free chlorine residual in a water sample.
chlorine water: a clear, yellowish liquid that deteriorates on exposure to air and light; it is employed as a deodorizer, disinfectant, and antiseptic.
clarification: the removal of suspended solids from wastewater by gravity settling; this process is often accelerated through coagulation of and flocculation of small solids with chemicals.
closed loop: a term used to describe an enclosed process.
closed loop cooling tower: water-conserving cooling tower system in which water used for cooling is recycled through a piping system that cools the water.
coastal waters: the waters of the coastal zone, except for the Great Lakes and specified ports and harbors on inland rivers.
coastal zone: the lands and water adjacent to the coast that exert an influence on the uses of the sea and its ecology, or whose uses and ecology are affected by the sea.
coastline: the line separating the land surface and the water surface of the sea.
coliform: a group of related bacteria whose presence in water may indicate contamination by disease-causing microorganisms; indicators organisms of recent fecal contamination.
combined sewer: a single sewer system that carries stormwater runoff and sewage to a wastewater treatment plant.
communicable disease: an illness that is caused by a specific infectious agent or its toxic products and that arises through transmission to a susceptible host.
community water system: a public water system that serves at least 15 service connections employed by year-round residents or regularly serves at least 25 year-round residents.
condenser: any device that cools gases or vapors to liquid.
conduit: any artificial or natural duct, either opened or closed, employed for conveying fluids.
confined aquifer: an aquifer in which groundwater is confined under pressure.
confined groundwater: water in an aquifer that is bounded by confining beds and is under pressure significantly greater than atmospheric pressure.
confining bed: a layer or mass of rock having very low hydraulic conductivity that hampers the movement of water into and out of an adjoining aquifer.
connate water: water entrapped in the interstices of sedimentary rock at the time of its deposition.
conservative pollutant: a pollutant that does not decay, does not react, is persistent, and is not biodegradable.
contaminant: any physical, chemical, biological, or radiological substance that has a harmful effect on human health or the environment when contained in air, water, or soil.
continental shelf: the comparatively shallow area surrounding the continents and falling steeply to the deep ocean floor.
continuous sampling: the continuous withdrawal for analysis of a sample from some larger quantity of liquid, air or solid.
cooling tower: a hollow, vertical structure, perhaps with internal baffles, to disperse water so it is cooled by flowing air and by evaporation at ambient temperature.
cooling tower makeup: water added to a recirculating cooling tower water stream to compensate for water evaporation losses.
cooling water: water typically used to cool heat-generating equipment or to condense gases in a thermodynamic cycle.
cooling water blowdown: the procedure used to reduce total dissolved solids by removing a portion of poor-quality recirculating water.
cooling water drift: unevaporated water carried out of a cooling tower by the airflow; it has the same composition as the recirculating water.

cooling water evaporation: cooling water recycling approach in which water loses heat when a portion of it is evaporated.

creep: the movement of water under or around a structure built on permeable foundations.

crust: the outermost layer of the Earth consisting of felsic and mafic rocks, which are less dense than the rocks of the mantle below.

cryogenics: the production and utilization of extremely low temperatures.

crystallization: the change of state of a substance from a liquid to a solid by the phenomenon of crystal formation by nucleation and accretion (e.g., the freezing of water into ice).

dam: a barrier formed across a waterway that is employed to create a reservoir, or to divert water into a specific conduit or channel.

deaeration: a process by which dissolved air and oxygen are removed from water.

decantation: the separation of a liquid from a solid or a higher density liquid with which it is immiscible by drawing off the fluid.

dechlorination: the removal of chlorine from a substance by chemically replacing it with hydrogen or hydroxide ions to detoxify the substances.

decontamination/detoxification: a process that converts toxic wastes into nontoxic compounds.

deep-well injection: a method of ultimate disposal that involves depositing liquid waste into a deep well beneath the surface of the earth for permanent storage.

dehumidifier: a device incorporated into many air conditioning systems to dry incoming air by passing it across a bed of a hygroscopic substance or through a spray of very cold water.

dehydration: the chemical process where water in a chemical or material is removed.

deionized water: common industrial water devoid of dissolved salts and organics used to remove contaminants from products and equipment.

demineralization: the process of removing dissolved minerals from water by physical, chemical, or biological means.

demister: a device composed of plastic threads, wire mesh, or glass fibers employed to remove liquid droplets entrained in a gas stream.

denitrification: the use of nitrate by soil and wastewater bacteria as an electron acceptor in the degradation of organic material resulting in the production of free nitrogen gas.

desalination: the extraction of freshwater from sea or other salt water by the removal of salts, usually by evaporation, reverse osmosis, or crystallization.

desert: a terrestrial environment where evaporation exceeds precipitation, with consequent lack of vegetation.

detention basin: a man-made facility to hold stormwater temporarily until such time as there is room in the storm drain system to release it safely.

detoxification: the destruction of the toxic aspects of a substance.

dew point: the temperature at which the first droplet of water forms on the progressive cooling of a mixture of air and water vapor; at the dew point, the air becomes saturated with water.

dialysis: the separation of smaller molecules from larger ones in a solution by means of the diffusion from a concentrated solution to a dilute solution across a semipermeable membrane.

digested sludge: sludge that has been stabilized under either aerobic or anaerobic conditions to remove biodegradable organic material and active microbes so the sludge can be disposed of without cause nuisance conditions or negative impacts to the environment.

dike: an embankment or ridge of either natural or manmade materials employed to prevent the movement or overflow of liquids, sludge, solids, or other materials.

discharge: the volume of water that flows past a given area in a given time.

dissolved solids (TDS): total dissolved solids, minerals, or salts in water.

direct discharger: a municipal or industrial facility that introduces wastewater to surface water bodies through a defined conveyance or system.

disinfectant: any substance that destroys harmful microorganisms or inhibits their activity.

disposal well: a well employed for the disposal of waste into a subsurface stratum.

dissolved oxygen (DO): the oxygen freely available in water; it is one of the most important indicators of the quality of a water supply because oxygen is necessary for the life of aquatic organisms.

distillation: a process of separating the constituents of a liquid mixture by means of partial vaporization of the mixture and separate recovery of vapor and residue as a result of a difference of vapor pressure.

distilled water: a water of high purity, prepared by repeated distillation.

domestic wastewater: wastewater generated from household activities that include sanitary waste, liquid waste from food preparation and laundering, bathing and showering waste, and other liquid cleaning wastes generated from households.

downstream: the regions of a river system located in a hydraulically lower location than a given position along a stream or river. A section of the river system is hydraulically lower if gravity transports the water in the direction of or nearer to the mouth of the stream or river.

downtime: a term used to describe the periods when a system is unavailable or not operating.

drain: any channel that carries off surface water.

drawdown: the difference between the water level in a well before pumping and the water level in the well during pumping.

drilled well: a well constructed by either percussion or rotary hydraulic drilling.

drinking water supply: any raw water source that is or may be employed by a public water system or as drinking water by one or more individuals.

dry cooling: cooling system using air instead of water as the cooling fluid to eliminate the use and evaporative loss of water.

dry well: a shallow well used in stormwater systems for the infiltration of collected stormwater for shallow groundwater recharge; the well is dry during dry weather conditions.

duct: a round or rectangular conduit, usually metal or fiberglass, employed to transport fluids, usually air.

dystrophic lake: an acidic, shallow body of water that contains high concentrations of humic substances and organic acids that are often brown in color; it contains many plants but few fish.

ebb tide: the tide occurring at the ebb period of tidal flow; it is sometimes referred to as *falling tide* to describe the direction of the current.

ecosphere: the global sum of all ecosystems on Earth.

ecosystem: the interacting system of a biological community and its nonliving surroundings.

eddy: a circular movement occurring in flowing water caused by currents in the water induced by obstructions or changes and irregularities in the banks or bottom of the channel or by differences in temperature.

effluent: any fluid emerging from a pipe or similar outlet that enters the environment; it usually refers to treated wastewater from municipal or industrial treatment plants.

electrolysis: the use of a direct electric current to drive an otherwise nonspontaneous chemical reaction.

elution: the process of moving a substance through a bed by means of a slow-moving stream.

elutriation: a process for separating particles based on their size, shape, and density using a stream of gas or liquid flowing in a direction usually opposite to the direction of sedimentation. This method is mainly used for particles smaller than 1 μm.

embankment: a ridge of earth or stone employed to prevent water from passing beyond desired limits.

empirical: anything derived from experimentation or observation and not from fundamentals or theory.

endothermic: a term used to describe a process or change that occurs with absorption of heat.

enrichment: the addition of constituents, generally nutrients, from wastewater treatment plant effluent, or agricultural runoff to surface water.

environment: the system of external conditions affecting the existence and development of an individual or organism.

environmental audit: an independent assessment of the current status of a company's compliance with applicable environmental requirements.

epidemiology: the study of diseases as they affect population, including the distribution of disease, or other health-related states and events in human populations, the factors that influence this distribution, and the application of this study to control health problems.

erosion: the breakdown of solid rock into smaller particles and its removal by wind, water, or ice; the removal of surface soil by wind water or ice.

estuary: a region of interaction between rivers and near-shore ocean water where tidal action and river flow create a mixing of freshwater and salt water.

euphotic zone: the upper zone of a sea or lake into which sufficient light can penetrate for active photosynthesis to take place.

eutrophic lake: a body of water that is characterized by an abundant accumulation of nutrients that support a dense growth of algae and other organisms, the decay of which depletes the shallow waters of oxygen in summer.

eutrophication: the process by which a body of water becomes enriched in dissolved nutrients (such as phosphates) that stimulate the growth of aquatic plant life usually resulting in the depletion of dissolved oxygen.

evaporation: the conversion of a liquid into a vapor.

evaporation pond: a shallow pond or impoundment with a large surface area that is designed to contain wastewater and allow it to evaporate with no discharge to the environment.

excursion: an unintentional occurrence, such as a discharge of pollutants above the permitted amount, for unplanned reasons.

exempted aquifer: an underground body of water defined in the underground injection control program as an aquifer that is "not reasonably expected" to be a source of drinking water and that is exempted from regulations barring underground injection activities.

exfiltration: the quantity of wastewater that leaks into the surrounding ground through leaks in the sewer system.

exothermic: a term used to describe a reaction or process during which heat is released.

faucet aerator: device that can be installed in a sink to reduce water use.

feed: the material supplied to a processing unit for treatment or processing.

feedforward control system: a system in which changes are detected at the process input and an anticipated correction signal is applied before process output is affected.

feedstock: the raw materials supplied to manufacturing or processing plants for use in the production of goods or materials.

film-type condensation: the process by which a saturated vapor comes into contact with a surface whose temperature is below the saturation temperature and uniformly condenses on the surface.

filter membrane: a thin film containing many fine pores that is employed to filter a liquid or solid stream.

filter press: a mechanical device that forces sludge, usually conditioned with coagulant addition, between moving filter membranes to squeeze water out of it and produce a dry sludge cake of between 15% and 25% solids.

finished water: water that has been treated and is ready to be delivered to customers.

five-day BOD: a standard test that measures the amount of oxygen consumed during the first 5 days of biological oxidation of biodegradable organic material; see also biochemical oxygen demand.

flash drum: a unit in which volatile components are vaporized and separated from a liquid stream for further fractionation.

flash evaporator: a distillation device in which saline water is injected in a superheated state into a vessel under vacuum and in which boiling occurs without the usual heat source.

flocculation: the process by which solids in water or sewage are made to increase in size by biological or chemical action so that they can be more easily separated from the water by sedimentation.

fluid: any material or substance that changes shape or direction uniformly in response to an external force imposed on it; the term applies to liquids and gases.

fluidization: a technique in which a finely divided solid is caused to behave like a fluid by suspending it in a moving gas or liquid.

fluoridation: the addition of fluoride to public water supplies to prevent or delay the onset of dental decay.

fractional distillation: a distillation process in which countercurrent distillation is employed to obtain a product as nearly pure as possible; it is also any distillation process in which the product collected may be a series of separate components of similar boiling range.

freeze-drying: a method of dehydration or separating water from materials; the material is first frozen and placed in a high vacuum so that the water (ice) vaporizes (sublimes) in the vacuum without melting and the nonwatery components are left behind in an undamaged state.

freshwater: the water that generally contains less than 1,000 mg/L of dissolved solids.

froth: a foamy mass of bubbles that exists on a body of water.

gas permeation: the movement of gas from the high pressure side of a membrane to the low pressure side.

geohydrology: the branch of hydrology involving the study of groundwater and its physical and chemical interactions with the physical environment.

geology: the science that deals with the origin, history, and structure of the earth, as recorded in rocks, along with forces and processes that modify rocks.

geothermal energy: the energy derived from the superheated water and steam trapped in underground reservoirs.

geothermal gradient: the change of temperature in the earth with depth, usually expressed in degrees per unit of depth.

glacial drift: the rock material transported and deposited directly by glaciers or indirectly by ice or water emanating from a glacier.

glacier: a large body of ice originating on land by the compaction and recrystallization of snow and showing evidence of present or past movement.

grab sample: a sample that is collected at such a time and place so that it is ideally most representative of a total discharge.

gray water: domestic wastewater composed of water from kitchen sinks, bathroom sinks, and tubs, clothes washers, and laundry tubs.

greenhouse effect: the gradual rise in the average global temperature as a result of the absorption of infrared radiation reflected by the Earth's surface by increasing amounts of carbon dioxide, methane, nitrous oxide, and other greenhouse gases in the air.

grit: the dense, suspended inorganic material present in wastewater that poses an abrasion hazard to pumps and that can accumulate in reactors and clarifiers if not removed from the incoming wastewater.

grit chamber: a chamber or basin in a wastewater treatment plant designed to reduce the velocity of flow of the wastewater so that inorganic solids (grit) will settle out of the water before subsequent treatment steps.

groundwater: water that exists underground in saturated zones beneath the land surface.

groundwater basin: a porous formation with sides and bottom of relatively impervious material in which groundwater is held or retained.

groundwater hydrology: the branch of hydrology dealing with groundwater sources, movement, recharge, depletion, etc.

groundwater reservoir: an area below the ground surface in which groundwater is stored; also called an aquifer.

groundwater recharge: the input of surface water to resupply groundwater reservoirs through infiltration at the soil surface or from the bottom of rivers and streams; the use of reclaimed wastewater, by surface spreading or direct injection, to replenish freshwater aquifers; to resupply freshwater aquifers to prevent saltwater intrusion, to control or prevent ground subsidence, and to augment non-potable or potable groundwater aquifers.

groundwater system: a groundwater reservoir and its contained water; also, the collective hydrodynamic and geochemical processes at work in the reservoir.

habitat: a dwelling place of a species or community, providing a particular set of environmental conditions (e.g., forest floor, sea shore, etc.).

hard water: water containing high levels dissolved divalent cations, primarily calcium and magnesium, that interfere with some industrial processes and prevent soap from lathering.

hardness (water): a property of water causing formation of an insoluble residue when the water is used with soap and forming a scale in vessels in which water has been allowed to evaporate; it is due primarily to the presence of divalent ions of calcium and magnesium.

heat exchanger: a unit or vessel in which a hot fluid stream transfers part of its energy to a cooler fluid stream or vice versa.

heat sink: a structure designed to absorb heat.

heavy rain: rain that is falling with an intensity in excess of 0.03 inches/hour during an interval of 6 minutes.

heterogenous: a term used to describe a mixture of different phases (e.g., liquid-vapor, liquid-vapor-solid).

high dam: a dam that is taller than 165 feet; the first high dam was Hoover Dam in the United States; four other notable high dams are Grand Coulee Dam in the United States, Aswan High Dam in Egypt, Three Gorges Dam in China, and the Sardar Sarovar Dam in India.

holding pond: a pond or reservoir, usually made of earth, that is built to store polluted runoff.

holdup: a volume of material held or contained in a process vessel or line.

homogenous: a term used to describe a mixture or solution comprised of two or more compounds or elements that are uniformly dispersed in each other.

hot brine: a slightly salty subterranean water, the temperature of which is markedly higher than that dictated by the normal geothermal gradient; it can be employed as a source of geothermal energy.

hot rock: a subterranean rock, the temperature of which is higher than would be dictated by the normal geothermal gradient; it can be employed as a source of geothermal energy.

humidifier: a device for increasing the water content of air; it is usually incorporated into an air conditioning system.

hydration: the chemical process of combination or union of water with other substances.

hydraulic fill: an earth structure or grading operation in which the fill material is transported and deposited by means of water being pumped through a flexible or rigid pipe.

hydrogeology: the geology of groundwater, with particular emphasis on the chemistry and movement of water.

hydrological cycle: the constant movement and cycling of water by evaporation, precipitation, and condensation in the Earth-atmosphere system.

hydrology: the science dealing with the properties, distribution, and circulation of water in relation to the land surface.

hydrolysis: the reaction of a salt with water to form an acid and a base.

hydrophilic: a term used to describe a substance with an affinity for water.

hydrophobic: a term used to describe a substance that separates from water or surfaces that repel water.

hydrosphere: the part of the earth that is composed of water, including oceans, seas, lakes, rivers, icecaps, etc.

hydrothermal: a term used to describe any geological process involving heated or superheated water.

ice: the allotropic, crystalline form of water.

imhoff tank: a unit providing for both sedimentation and anaerobic sludge digestion in a single tank.

imported water: water supply that is conveyed from one watershed to be used in another.

impurity: the presence of one substance in another, often in such low concentration that it cannot be measured quantitatively by ordinary analytical methods.

in situ: a term used to describe any reaction occurring in place, and a term used to describe a fossil, mineral, or rock found in its original place of deposition, growth, or formation.

indirect discharger: an industry that sends waste to a publicly owned treatment works (POTW); see also POTW.

induced draft: the negative pressure created by the action of a fan, blower, or other gas-moving device.

industrial wastewater: the wastewater generated from industrial processes.

industrial water: the water that is withdrawn from a source for sole use in an industrial process.

infiltration: the penetration of water through the ground surface into sub-surface soil or the penetration of water from the soil into sewer, or other pipes through defective joints, connections, or other leaks in a system, or a land application technique where large volumes of wastewater are applied to land, allowed to penetrate the surface, and percolate through the underlying soil.

inflow: an entry of extraneous stormwater into a sewer system from sources above the ground surface, typically through manhole covers or direct pumping (i.e., via sump pumps).

influent: any untreated wastewater stream flowing into a wastewater treatment plant.

injection well: a well into which fluids are injected for purposes such as waste disposal, remediation, hydraulic control, etc.

injection zone: a geological formation, group of formations, or part of a formation receiving fluids through a well.

inland waters: the waters of the United States in the inland zone, waters of the Great Lakes, and specified ports and harbors on inland rivers.

inland zone: land inside the coastal zone, excluding the Great Lakes and specified ports and harbors on inland rivers.

insoluble: a term used to describe a substance that is incapable of being dissolved in a liquid.

instantaneous sampling: the collecting of a sample in a very short period of time so that the sampling time is insignificant in comparison with the duration of the operation or the period being studied.

instream use: the water use taking place within a stream channel, hydroelectric power generation, navigation, water quality improvement, fish propagation, or recreation.

interstitial water: the water contained in the interstices of rocks, where the origin of the water is unknown or unspecified.

ion exchange: a mass transfer process that involves the interchange of ions between a liquid and a solid material; it can be employed to concentrate and recover desired materials or to remove undesired ions from a water supply.

irrigation: a technique for applying water or wastewater to land areas to supply the water and nutrient needs of plants.

irrigation districts: special units of local government that control the bulk of surface water supplies, primarily in the Western states in the United States.

irrigation field practices: techniques that keep water in the field, more efficiently distribute water across the field, or encourage the retention of soil moisture.

irrigation withdrawals: withdrawal of water for application on land to assist in the growing of crops and pastures.

isothermal: a term used to describe a process that exhibits no change in temperature (i.e., constant temperature).

lagoon: a large pond, sometimes called a *stabilization* or *oxidation pond*, into which sewage and industrial wastes may be pumped for decomposition by bacterial action; algal growth from sunlight generates oxygen to support bacterial degradation of organics in the wastewater.

lagooning: the placement of solid or fluid material in a basin, reservoir, or artificial impoundment for purposes of treatment, storage, or disposal.

lake: an inland body of freshwater or salt water of considerable size, occupying a basin on the Earth's surface.

land application: the discharge of wastewater onto the ground surface for treatment or disposal.

leach: a process by which something is removed by percolating, or trickling liquid; it usually refers to the removal of components in soil by water.

leachate: any liquid, including any suspended components in the liquid, that has percolated through or drained from a solid, such as liquid generated from waste in a landfill.

leakage: an undesired and gradual escape or entry of a fluid into or out of a tank or vessel.

levee: a dike or ridge at the side of a river, intended to protect the land side from flood waters or to confine the stream flow to its regular channel.

limnetic zone: the well-lit, open-water surface region of a lake away from shore.

limnology: the scientific study of the physical, chemical, and biological components of freshwater with emphasis on plants and animal life.

liquid: an amorphous, noncrystalline state of matter; the molecules are much more highly concentrated than in gases and usually less concentrated than in solids.

liquid permeation: the movement of liquid components from one side of a membrane to the other.

liquid-liquid extraction: the separation of a solute based on its relative solubility in two immiscible liquids, usually water and an organic solvent.

low-flow showerhead: a showerhead that produces 2.5 gallons per minute or less, as compared to the 4.5 gallons per minute produced by most older standard showerheads.

low-flush toilet: a toilet that requires 1.6 gallons of water per flush or less, as compared to the 3.5 to 5 gallons of water required to flush most older, standard toilets.

magma: the molten material within the Earth's crust that is composed of silicates and volatiles (water and gases) in complex solution.

main sewer: the principal sewer to which the branch sewers and submains are tributaries.

make-up water: the water that is employed to replenish a system that loses water through leakage, evaporation, etc.

manifold: a pipe fitting with numerous branches to convey fluids between a large pipe and several smaller pipes or to permit the choice of diverting flow from one of the several sources or to one of the many discharge points.

mantle: the part of the interior of the Earth between the crust and the core.

maximum contaminant level (MCL): a maximum allowable level of a compound within a treated drinking water deemed safe for human consumption; concentrations above this level are deemed unsafe and unacceptable for human consumption.

maximum daily discharge limit: the highest allowable daily discharge of pollutants in a wastewater effluent.

mechanical aeration: the mixing by mechanical means of wastewater in an aeration tank to generate liquid droplets that come into contact with the atmosphere and carry oxygen into the bulk liquid in the tank to supply oxygen for the aerobic stabilization of organics in the wastewater.

membrane: a thin sheet of material through which a gaseous or liquid solution may pass.

membrane hydrolysis: a hydrolysis process that occurs when a colloidal electrolyte is separated from pure water by a membrane.

membrane selectivity: the ability of a membrane to allow passage of only anions or cations.

microorganism: an organism of microscopic size generally considered to include bacteria, algae, protozoa, fungi, and rotifers but excluding viruses; they (primarily bacteria) are employed in biological wastewater treatment processes to remove biodegradable organic matter and suspended solids.

mine water: the water encountered in mining operations that, if discharged into surface streams or other bodies of water, often contaminates them and makes them unfit for use.

mineral spring: a spring that contains high levels of mineral salts.
miscibility: the ability of a liquid or gas to dissolve uniformly in another liquid or gas.
mixed liquor: a mixture of microbial biomass and wastewater containing organic matter undergoing activated sludge treatment in an aeration tank.
monitoring: a periodic or continuous surveillance or testing to determine the level of compliance with statutory requirements and pollutant levels in various media, or in living things.
monitoring well: a well drilled to collect groundwater samples for analysis to determine the amounts, types, and distribution of contaminants in the groundwater.
mother liquor: a concentrated solution that is substantially freed from undissolved matter by filtration, centrifuging, or decantation; the product can be obtained by evaporation or crystallization.
municipality: a city, town, borough, county, parish, district, or other public body.
natural resources: the land, fish, wildlife, biota, air, water, groundwater, drinking water supplies, and other such resources belonging to, managed by, held in trust by, pertaining to, or otherwise controlled by the United States.
navigable water: any stream, lake, arm of the sea, or other natural body of water that is navigable and of sufficient capacity to float watercraft for the purposes of commerce, trade, transportation, or recreation, as indicated under the Clean Water Act.
new source: any stationary source that is built or modified after publication of final or proposed regulations which prescribe a standard of performance that is intended to apply to that type of emission source.
nitrates: oxidized nitrogen that forms a monovalent anion; is the form of nitrogen that is produced from the aerobic oxidation of ammonia (nitrification); and that serves as an electron acceptor for the anaerobic oxidation of organic matter (denitrification); it is a nutrient that can result in eutrophication of lakes and ponds and is the cause of "blue baby" syndrome in humans if in high concentrations in drinking water: its MCL is 10 mg/L in drinking water.
nonpoint source: a source of pollutant discharge that is not traceable to a discrete location.
nonpotable water: water considered unsafe and/or unpalatable for drinking (see also potable water).
nutrient: any element or compound that is essential to the life and growth of plants or animals, either as such or as transformed by chemical or enzymatic reactions; nitrogen and phosphorous are the major nutrients of concern in polluted waters.
ocean water (seawater): a uniform solution contained approximately 96.5% water and 3.5% ionized salts; ingestion of substantial amounts will create a bodily chloride imbalance with harmful effects.
oceanic: a term used to describe the parts of the oceans deeper than 200 meters.
oceanography: the science dealing with oceans, including their form, physical and chemical features, and related phenomena.
open channel: any natural or artificial waterway or conduit in which a fluid flows with a free surface exposed to atmospheric pressure.
osmosis: the passage of a pure liquid (usually water) through a semipermeable membrane from a solution of low concentration into a solution of a higher concentration (e.g., the flow of pure water into a solution of salt and water); *see also* reverse osmosis.
osmotic pressure: the pressure that results from osmosis.
outfall: the place where an effluent is discharged into a receiving water.
outfall sewer: a pipe or conduit that transports wastewater effluent, to a final point of discharge.
oxidation pond: an engineered pond that uses mechanical or diffused aeration to provide oxygen used by aerobic bacteria to remove biodegradable organics in a wastewater.
oxygen demand (OD): the quantity of oxygen used in the biochemical oxidation of biodegradable organic matter.
oxygenation: the increase of the dissolved oxygen content within a wastewater stream through aeration.

ozonation: the addition of ozone to a water supply to reduce taste and odor problems and for disinfection.

ozonator: a device that creates ozone from oxygen and adds it to water.

parts per billion (ppb): the fraction (ppb_m for mass fraction and ppb_v for volume fraction) multiplied by 10^9; it is a unit used to measure extremely small concentrations of a substance; equivalent to units of µg/L in water.

parts per million (ppm): the fraction (ppm_m for mass fraction and ppm_v for volume fraction) multiplied by 10^6; it is a unit used to measure small concentrations of a substance; equivalent to units of mg/L in water.

pathogenic waste: a discarded waste that contains organisms capable of causing disease.

percolation: the flow of a liquid downward through a filtering medium or soil layer.

permafrost: the portion of the earth which is permanently frozen, such as the Artic regions and portions of Alaska.

permeability: the degree to which a liquid can move freely through soils.

physical quality: the physical characteristics possessed by a material; it includes temperature, color, odor, and turbidity.

physical treatment: a water or wastewater treatment process that uses physical means for pollutant removal; processes include screening, grinding, settling, filtration, and centrifugation.

physicochemical: a term used to describe processes that involve both physical and chemical characteristics; adsorption is one such process that involves both physical and chemical attraction of an adsorbate to an adsorption site.

point source of pollution: pollution originating from a discrete source, such as the outflow from a pipe, ditch, tunnel, concentrated animal-feeding operation, or floating craft.

pollutant: any harmful substance added to the air, water, or soil.

pollution: the direct or indirect alteration of the physical, thermal, biological, or radioactive properties of any part of the environment in such a way as to create a hazard or potential hazard to the health, safety, or welfare of any living species.

potable water: water that is safe for human consumption.

POTW: publicly-owned treatment works, or municipal wastewater treatment plants.

pre-aeration: a wastewater treatment process in which the removal of dissolved gases and the addition of oxygen are performed, typically in aerated grit chambers.

precipitate: a solid that separates out from a liquid because of some physical or chemical change in the liquid.

pressure sewers: a system of pressurized pipes in which wastewater is transported to a higher elevation by the use of pumping force.

pretreatment: any process employed to partially remove pollutants from a waste stream prior to any subsequent treatment process; required by industries that discharge into municipal wastewater treatment plants for any industrial pollutants not removed by or that can harm the proper operation of the municipal plant.

primary clarifiers: the settling basins that receive wastewater after preliminary treatment and prior to biological treatment. Also called *primary sedimentation tanks.*

primary sewage sludge: a semisolid of from 4% to 6% solids resulting from primary clarification; these solids are settled, raw organic solids and must be stabilized via anaerobic digestion. Also known as *primary sludge* or *primary solids.*

psychrometric chart: a chart employed to determine the properties of moist air that provides humidity as a function of temperature.

public water system (PWS): any system that provides piped water for human consumption to at least 15 service connections or regularly serves 25 individuals.

purging: a cleansing or removal of impurities, foreign matter, or undesirable contaminants from a process through periodic withdrawals of liquids or solids.

purification: the removal of undesirable constituents from a substance by one or more separation techniques.

rank: the stage reached by coal in the course of its carbonation; the chief ranks, in order of increasing carbon content, are lignite, subbituminous coal, bituminous coal, and anthracite.

raw sewage: the sewage that enters a wastewater treatment plant that has not been treated.

raw water: the untreated water that enters the first treatment unit of a water treatment plant.

receiving water: any body of water (e.g., river, lake, ocean, stream, etc.) into which treated wastewater is discharged to.

recirculating cooling water: the recycling of cooling water to greatly reduce water use by reusing the water to perform several cooling operations.

reclaimed water: treated wastewater that is reused for generally nonpotable uses to supplement or replace other raw water supplies.

recycled water: wastewater that has been treated for reuse and is recycled, generally for nonpotable uses within a home or industrial facility.

red tide: a proliferation and accumulation of certain microscopic algae, predominantly dinoflagellates, in coastal waters; some species produce toxins that are labeled harmful algae blooms, or HABs, that pose a serious and recurring threat to human health, wildlife, marine ecosystems, fisheries, and coastal aesthetics.

red water: a rust-colored water, usually resulting from the presence of precipitated ferric iron salts.

release: any spilling, leaking, pumping, pouring, emitting, emptying, discharging, injecting, escaping, leaching, dumping, or disposing into the environment.

renewable water supply: the rate of supply of water (volume per unit time) potentially or theoretically available for use in a region on an essentially permanent basis.

reproducibility: the ability to repeat an experiment, reaction, measurement, or process and produce the same results.

reservoir: any body of water employed for the storage, control, or regulation of water.

respiration: a process in which oxygen is taken into an organism for the generation of energy during metabolism, with the production of oxidized end products such as water and carbon dioxide.

reverse osmosis: a water treatment process employed to separate water from pollutants by the application of pressure to force the water through a semipermeable membrane.

rinse: the removal of foreign materials from a surface by using a flow of liquid.

rinse water: water used to remove debris and contaminants from products and equipment.

riptide: a strong surface current of short duration flowing outward from the shore.

river basin: the land area drained by a river and its tributaries.

river bed: the bottom of a river.

runoff: the water from precipitation that exceeds an areas infiltration and storage that flows over the ground into a surface water body.

rural area: the area outside the limits of any city, town, village, or other designated residential or commercial area.

saline water: water that generally is considered unsuitable for human consumption or for irrigation because of its high content of dissolved solids, generally greater than 10,000 mg/L of dissolved solids; with 35,000 mg/L dissolved solids is normally assigned to seawater.

salinity: the amount of salts or minerals dissolved in water.

salinization: a process in which a soluble salt accumulates in soils.

salt: a chemical compound formed when the hydrogen ion of an acid is replaced by a metal, or when an acid reacts with a base in an aqueous solution.

saltwater intrusion: the displacement of fresh groundwater by higher density salt water near coastal regions.

salting out: a reduction in the water solubility of certain molecules in a solution of very high ionic strength.

sample: a representative specimen of a liquid, solid, or gas collected for the purpose of determining its composition.

sampling: a method employed to obtain representative test samples; it consists of the collection, isolation, and the possible concentration of a small fractional part of a larger volume of a media.

sand bar: a ridge of sand built up by deposition to the surface or near the surface of a river or along a beach.

sand dune: a mound or ridge of loose sand blown by prevailing winds.

sanitary sewer: a pipe network that carries wastewater from residences, commercial buildings, industrial plants, and institutions, together with minor quantities of groundwater, stormwater, and surface waters that unintentionally enter the system to a wastewater treatment plant for treatment prior to discharge to the environment.

sanitary survey: an on-site review of water sources, facilities, equipment, operation, and maintenance of a public drinking water system to evaluate the adequacy of those components for producing and distributing safe drinking water.

saturated rock: a rock that has all of its void spaces filled with fluid.

saturated soil: a soil that has all of its void spaces filled with fluid.

saturated zone: a subsurface soil or rock zone in which all the interstices or voids are filled with water.

saturation temperature: the minimum temperature at which air is saturated with water vapor; the boiling point of water.

scour: the action of a flowing liquid as it erodes and carries away material on the sides or bottom of a channel.

screening: the use of screens to separate and remove coarse floating and suspended solids from sewage.

sea: a large body of salt water, second in rank to an ocean, that is generally part of, or connected to an ocean at some point.

sea level: the surface of the sea that is employed as a reference for elevation.

seawall: a coastal wall built to provide protection against erosion or flooding from the ocean.

seawater intrusion: the movement of seawater into freshwater aquifers near the coast when these freshwater aquifers are over pumped.

secondary clarifiers: the settling basins that receive wastewater after biological treatment in a wastewater treatment plant. Also called *secondary sedimentation tank*.

secondary drinking water regulation: a regulation that sets a maximum acceptable level for contaminants that adversely affect the taste, odor, or appearance of water or otherwise adversely affect the public welfare.

secondary treatment: a wastewater treatment process used to remove organic matter and suspended solids in wastewater to meet secondary treatment standards of 30 mg/L BOD_5 and 30 mg/L TSS; a treatment standard that represents biological wastewater treatment.

secondary wastewater treatment plant: a facility that produces an effluent that meets secondary treatment standards; a range of biological treatment processes ranging from lagoons and activated sludge to trickling filters and rotating biological contactors.

sediment: the solid material or deposits that have settled from a fluid.

sedimentation tank: a tank in which water or wastewater containing settleable solids is retained for a period of hours to allow these solids to move to the bottom of the tank by gravity; the settled solids are removed from the bottom and the floating solids are skimmed off the top for further treatment and disposal; also called *clarifiers*. *See* settling tank.

semiconfined aquifer: an aquifer that is partially confined by a layer (or layers) of low permeability soil through which recharge and discharge nevertheless may occur.

semipermeable membrane: a membrane that allows substances of a certain size to pass through it while preventing the passage of larger ones.

separate sewer: a sewer intended to receive only wastewater, as opposed to a combined sewer that coveys both wastewater and stormwater.

septic tank: an underground, watertight sedimentation tank that receives domestic wastewater and in which solids settle and are decomposed anaerobically.

septic wastewater: wastewater devoid of oxygen and held under anaerobic conditions.

settleable solids: the materials that are of sufficient size and density to sink to the bottom of a wastewater sedimentation tank.

settling tank: a tank used in water and wastewater treatment to hold water for a period of hours, during which heavier particles sink to the bottom for removal, treatment and disposal; also called a *clarifier. See* sedimentation tank.

sewage: the wastewater produced by residential, commercial, institutional, and industrial facilities.

sewer: the system of pipes or conduits employed to collect and deliver wastewater to treatment plants or stormwater surface water bodies.

shore: the land bordering any body of water.

slow sand filtration: a treatment process that involves the passage of raw water through a bed of sand at low velocity, which results in the substantial removal of chemical and biological contaminants through the development of a biolayer (schmutzdecke) at the sand surface.

sludge: the thick, semisolid waste that accumulates as a result of the chemical coagulation, flocculation, and settling which occurs during drinking water treatment; the thick, semisolid biomass that is produced in the biological treatment of wastewater.

sludge cake: the dewatered sludge from a treatment plant that has a solids content of 18% to 30% solids.

sludge dewatering: the process of removing water from sludge using methods such as air drying, pressure filtration, vacuum filtration, centrifugation, or belt presses.

sludge digestion: the process by which raw organic matter in primary sludge or excess microbial biomass in secondary sludge is liquified, gasified, and converted to more stable end products through the activity of anaerobic and aerobic microorganisms, respectively.

sludge dryer: a mechanical device for the removal of a large percentage of moisture from sludge by heat.

sludge filter: mechanical devices in which wet sludge, usually conditioned by a coagulant, is dewatered by means of vacuum (vacuum filter) or elevated pressure (pressure filter).

slurry: a high solids content mixture of particulate matter and liquid.

sodium chloride (NaCl): a colorless, transparent, crystalline solid or white, crystalline powder; it is noncombustible.

soft water: a water with a low concentration of calcium and magnesium ions.

softening: the chemical precipitation of divalent cations which cause the hardness of water; chemical coagulants, alum or ferric chloride, are normally used as coagulants, with flocculation and sedimentation following coagulant addition.

soil drainage: the removal of excess water from a soil by gravity.

soil moisture: water content in the soil, generally given as volume or weight percent.

solubility: the ability of one substance to be dissolved by another.

spray chamber: a chamber equipped with water sprays that cool and clean the fluids passing through it.

spray irrigation: the application of water to a land surface via spray droplet application; a method for disposing of some wastewaters by spraying them on land.

stabilization pond: a large, shallow basin for purifying many types of municipal and industrial wastewater by allowing bacteria and algae to convert organic materials into stabilized end products.

steam drum: a vessel in a boiler in which the saturated steam is separated from the steam-water mixture and into which the feedwater is introduced.

still: an apparatus for purify liquids through heating to selectively boil and then cool to condense the vapor (e.g., to prepare alcoholic beverages, distilled water, etc.).

storm drain: a drain employed for conveying stormwater runoff to a sewer.

storm sewer: a piping system employed exclusively for the transport of stormwater from streets, building, and surface runoff.

stormwater runoff: the portion of the volume a rainfall event that exceeds the infiltration and storage capacity of a watershed.

subsurface sewage disposal system: a system for the treatment and disposal of domestic sewage by means of a septic tank in combination with a soil absorption field.

subterranean water: the water that occurs in open spaces within rock materials of the Earth's crust.

superheated steam: steam at a temperature above its boiling point at a given pressure.

supersaturation: an unstable condition in which a solvent contains more dissolved matter or gas than is present in a saturated solution of the same components at the same temperature.

surface water: all water that is above the surface of the ground and is naturally open to the atmosphere.

surge irrigation: the intermittent application of water to irrigation pathways; this method pulses water down the furrow and creates more uniform irrigation water distribution.

suspension: a system in which very small particles are uniformly dispersed in a liquid or gaseous medium.

tank: a stationary device that is essentially a container (e.g., designed to contain an accumulation of waste) that is constructed primarily of non-earthen materials which provide structural support.

temperature gradient: the change in temperature with distance or position.

tertiary treatment: the advanced treatment of wastewater beyond secondary treatment; it may involve combinations of physical, chemical, and biological treatment processes to remove solids, nutrients, metals, salts, nonbiodegradable organics, etc., to prepare the waste for disposal into highly sensitive environments or for reuse.

thickener: a small circular or rectangular sedimentation tank, designed to increase the concentration of solids in a suspension.

thickening agent: any of a variety of substances employed to increase the viscosity of liquid mixtures and solutions without changing its other properties.

tidal wave: an exceptionally large wave, tsunami, or increase in the water level along a shore as a result of strong winds, volcanic eruption, or earthquake.

tide: the periodic rising and falling of water that results from the gravitational attraction of the moon and sun acting on the rotating Earth.

total organic carbon (TOC): the total amount of organic carbon present in water as organic compounds (e.g., amino acids, hydrocarbons, proteins, etc.).

toilet displacement device: object placed in a toilet tank to reduce the amount of water used per flush; for example, weighted plastic jugs filled with water or toilet dams that hold back a reservoir of water when the toilet is flushing.

toxic: a term used to describe a poisonous substance that has a harmful effect on an organism by ingestion, inhalation, or skin absorption.

trace: a very small quantity of a constituent, the amount of which cannot often be determined precisely because of its low concentration.

transpiration: the process by which water passes through living organisms, primarily plants, and into the atmosphere.

treatment: any method, technique, or process that is designed to change the physical, chemical, or biological composition of a waste so as to neutralize it, recover energy or material resources from it, render it nonhazardous or less hazardous, or make it safer to transport, store, or dispose of.

tributary: a stream or river that flows into a larger stream or main stem (or parent) river or a lake

trough: a structure employed to hold or transport fluids.

tsunami: a sea wave caused by an underwater seismic disturbance such as sudden faulting, a landslide, or volcanic activity.

turbid water: water that is cloudy due to fine particles in suspension.

turbidity: a measure of the fine particles suspended in a fluid; measured as the proportion of light passed through a sample that is refracted by suspended particles in a water column.

ultimate oxygen demand (UOD): the quantity of oxygen consumed by bacteria during the degradation of biodegradable organic compounds in a water sample over an extended period of time, normally defined as 20 days.

ultrafiltration: the separation of a solute with a specific molecular size and shape from a solution by applying pressure to force the solvent to flow through a membrane.

underground sources of drinking water: the aquifers that are currently being employed as a source of drinking water, and those that are capable of supplying a public water system.

unconfined aquifer: an aquifer whose upper surface is free of a confining layer and thus is able to fluctuate under atmospheric pressure.

upstream: the regions of a river system located in the direction opposite to the flow of a stream from a given position; a section of a river system that is hydraulically higher, if gravity transports the water away from the given location; the direction opposite to the flow in a process.

uptake: the act of taking up, drawing up, or absorbing.

urban runoff: the stormwater from city streets and adjacent domestic or commercial properties.

utility: public water service provider.

vapor pressure: the pressure exerted by a vapor in equilibrium with a liquid at a given temperature.

virgin material: a raw, unused material.

wake: the visible trail of turbulence left behind a moving stream.

wash solvent: a liquid added to a liquid-liquid extraction to wash or enrich the purity of the solute used in the extraction process.

wash water: the water employed to wash equipment.

waste load allocation: the maximum load of pollutants each discharger is allowed to release into a particular waterway based on surface water pollutant transport and degradation modeling.

wastewater: the water used to carry liquid waste material, consisting of dissolved and suspended solids, organics, and nutrients, from homes, businesses, institutions, and industries to wastewater treatment plants for contaminant removal prior to release to surface water.

wastewater operations and maintenance: the actions taken after construction of wastewater treatment facilities to assure that the facilities will be properly operated and maintained.

wastewater treatment: a series of processes in which wastewater is treated to remove or alter its objectional constituents to a degree that renders it less harmful or dangerous.

wastewater treatment plant: a series of unit operations including screening, sedimentation, digestion, stabilization, dewatering, disinfection, and other processes for removing pollutants from wastewater before it discharges into the environment.

wastewater treatment unit: a device that is part of a wastewater treatment facility that is subject to regulation.

water: a colorless, odorless, tasteless liquid composed of the elements hydrogen and oxygen.

water audit: program involving sending trained water auditors to participating family homes, free of charge, to identify water conservation opportunities such as repairing leaks and installing low-flow plumbing and to recommend changes in water use practices to reduce home water use.

water conditioning: the treatments, excluding disinfection, that are intended to produce a water which is free of taste, odor, and undesirable contaminants.

water conservation: activities designed to reduce the demand for water, improve efficiency in use, and reduce losses and waste of water in a potable water system.

water consumption: the quantity of water supplied in a municipality or district for a variety of uses during a given period.

water main: the water pipe, located in the street, from which domestic water supply is delivered to specific premises.

water pollution: the contamination of fresh or salt water with materials that are toxic, noxious, or otherwise harmful to fish, man, or other animals.

water purification: any process in which water is treated in such a way as to remove or reduce undesirable impurities.

water quality: the chemical, physical, or biological characteristics of water with respect to its suitability for a particular purpose.

water quality criteria: the specific levels of water quality that, if reached, are expected to render a body of water suitable for its designated beneficial use.

water quality standards: the state-adopted, Environmental Protection Agency (EPA)-approved, allowable numeric ambient water quality concentrations for surface water bodies defined for each beneficial use.

water recycling: reuse of water for the same application for which it was originally used.

water reuse: using treated wastewater (reclaimed water) for some beneficial purpose rather than discharging it to surface water; the deliberate use of reclaimed wastewater must be in compliance with applicable rules for a beneficial purpose (landscape irrigation, agricultural irrigation, aesthetic uses, groundwater recharge, industrial uses, or fire protection).

water rights: the rights acquired under the law to use surface or groundwater for a specified purpose, in a given manner, and usually within the limits of a given period.

water softening: a treatment process designed to completely or partially remove hardness-producing ions, Ca^{2+} or Mg^{2+}, for a potable water.

water solubility: a measure of the maximum concentration of a chemical compound that can result when it is dissolved in water.

water supplier: a person who owns or operates a water supply system; they can be public or private.

water supply system: the collection, treatment, storage, and distribution of potable water from source to consumer.

water surcharge: imposition of an increased cost because of excessive water use.

water table: the upper level of the groundwater below which the ground is saturated with water.

water treatment: the purification of water to make it suitable for drinking or other beneficial uses.

waterlog: occurs when water is added to land faster than it can drain.

watershed: the area surrounding a stream that supplies it with runoff.

waterway: any body of water, other than the open sea, that is or can be employed by boats as a means of travel.

well: a bored, drilled, or driven shaft, or a dug hole, whose depth is greater than the largest surface dimension and whose purpose is to reach underground water supplies, or to store or bury fluids below ground.

wetland: an area covered or saturated permanently, occasionally, or periodically by freshwater or salt water.

WHO: World Health Organization of the United Nations, based in Geneva (www.who.int).

REFERENCE

Theodore, L., J. Reynolds, and K. Morris. 1997. *Concise Dictionary of Environmental Terms*. Amsterdam, the Netherlands: Gordon and Breach Science Publishers.

2 Historical Perspective

2.1 INTRODUCTION

The Big Bang. In 1948, physicist G. Gamow proposed the Big Bang Theory of the origin of the universe. He believed that the universe was created in a gigantic explosion as all mass and energy were created in an instant of time. Estimates on the age of the universe at the present time range between 7 and 20 billion years, with 13.5 billion years often mentioned as the age of the planet Earth (Theodore and Theodore 1996).

The bang occurred in a split second and within a minute the universe was approximately a trillion miles wide and expanding at an unbelievable rate. Several minutes later, all the matter known to humanity had been produced. The universe as it is known today was in place.

Gamow further believed that the various elements present today were produced within the first few minutes after the Big Bang, when near infinitely high temperatures fused subatomic particles into the chemical elements that now comprise the universe. More recent studies suggest that hydrogen and helium would have been the primary products of the Big Bang, with heavier elements being produced later within the stars. The extremely high density within the primeval atom caused the universe to expand rapidly. As it expanded, the hydrogen and helium cooled and condensed into stars and galaxies. This perhaps explains the expansion of the universe and the physical basis of Earth and our galaxy.

This galaxy is a massive ensemble of hundreds of millions of stars, all gravitationally interacting and orbiting about a common center. All the stars visible to the unaided eye from Earth belong to the Earth's galaxy defined as the Milky Way. The Sun with its associated planets is just one star in this galaxy. Besides stars and planets, galaxies contain clusters of stars that in turn consist of atomic hydrogen gas, molecular hydrogen, complex molecules composed of hydrogen, nitrogen, carbon, and silicon. These galaxies are generally not isolated in space but are often members of small or moderate-sized groups, which in turn form large clusters of galaxies. The Milky Way is one of a small group of about 20 galaxies that is referred to as the Local Group. The Milky Way and the Andromeda galaxy are the two largest and members, each with approximately 10^{12} stars.

Water on Earth is believed to have been brought to the surface by comets that collided with Earth early in its history. This water is a prerequisite for all life on Earth, and it is liquid water that is the prerequisite not steam or ice. Thus, water must exist at temperatures constrained by the boiling and freezing points of this unique material for life to flourish.

Over the course of history, people who learned from water and water-related problems survived to learn again and reproduced, whereas those who did not disappeared. Environmental concerns took hold at about 3,000 B.C.E. Urban areas on the Indian continent developed sanitation programs such as underground drains and public baths. Aspects of health were integrated with daily activities including personal hygiene, health education, dietary practices plus food, and environmental sanitation.

This chapter introduces the reader to reflections on the generation and utilization of water on the planet through topics that include: the Earth and Moon, the hydrologic cycle, early humans, the development of agriculture, colonization of the New World, the Industrial Revolution, and the environmental movement and the Environmental Protection Agency. An application section provides six Illustrative Examples related to the general subject of water on the Earth.

2.2 THE EARTH AND MOON

Once the dust settled following the Big Bang, the Earth had become an integral part of the Milky Way. Liquid water and ice covered approximately 70% and 5% of the Earth's surface, respectively. Interestingly, all three phases of water exist on Earth, something no other planet in the Milky Way can lay claim to. This original amount of water on the planet has essentially remained constant over several billion years and, thus, may be viewed as both a renewable and sustainable resource if managed properly.

From a molecular perspective, all matter on Earth consists of atoms. Each atom in simple terms can be considered to be made of three classes of particles, referred to as *subatomic particles*. These three particles are the electron, proton, and neutron. The electron (negative) and the proton (positive) have equal but opposite charges. Because an atom is neutral in charge, the number of electrons must equal the number of protons. Further, these particles are the same in all atoms. The difference between atoms of distinct elements, for example, hydrogen and lead, is due entirely to the difference in the number of subatomic particles in each atom. Thus, an atom can be viewed as the smallest form of a unique element because an atom loses its identity when reduced to these basic subatomic particles.

At the turn of this century, 112 elements were known. These elements vary widely in location and abundance on Earth. For example, more than 75% of the Earth's crust consists of oxygen and silicon. Interestingly, approximately 65% by mass of the human body is oxygen. As the number and information on elements increased, chemists attempted to find similarities in elemental as well as chemical behavior. These efforts ultimately resulted in the development of the Periodic Table, an arrangement of elements in order of increasing atomic number, with elements having similar properties placed in vertical columns. Elements may be viewed as substances that cannot be decomposed, or broken into more elementary substances, by ordinary chemical means.

Water is in motion on Earth and is part of a closed cycle, the *hydrologic cycle*, that is impacted by both the Earth's Moon and the energy of the Sun. This results in the tides, evaporation of water, and its condensation to form precipitation in various forms. In comparison to the total water on Earth, only a mere 1% participates in the short-term water cycle that is relevant for human consumption. Most water is stored in the oceans, ice caps, and deep groundwater aquifers and moves on time scales of several thousand years or more, clearly outside the realm of human generation.

The Moon plays an interesting role in the movement of water on Earth, namely the tides. The Moon is a satellite that orbits around the Earth every 27 days. There are several theories as to how the Moon was formed: some argue that it was created many years ago when a body about the size of Mars collided with the Earth and debris from the collision came together to form the Moon. The tides are caused primarily by the Moon's gravity. The Sun's gravity also plays a role, but the Moon is more important because it exerts more than twice as much gravitational influence as the Sun because the Moon is so much closer. The Moon is 239,000 miles from the Earth, whereas the Sun is 92 million miles from Earth. The Moon's gravity pulls water toward the side of the Earth closest to the Moon, causing the water level to rise. The Moon's gravity is exerted on everything but because water is a fluid, it is able to move. On the opposite side of the Earth, where the Moon's gravity is felt least, there is also a bulge in the water level caused by inertia. On the sides of the Earth perpendicular to the moon, the water level is lower because the water is pulled to the sides parallel to the Moon. As the Earth rotates, it moves through and then away from these bulges. When it moves through a bulge, high tide occurs, and then 6 hours later there is a low tide, and then another high tide occurs as the Earth rotates through the bulge on the opposite side of the globe. The sides perpendicular to the Moon have low tide at the time that the side facing the Moon is at high tide.

Interestingly, there are different kinds of tides. In addition to regular high and low tides there are spring and neap tides. Spring tides have nothing to do with the season of spring. Both spring and neap tides are dependent on the phases of the Moon. Spring tides occur during a full or new Moon when the gravitational pull of the Moon and the sun are aligned. Neap tides occur when the Moon is in one of its quarters and the gravitational forces of the Sun and the Moon are perpendicular to each other relative to the Earth, minimizing the overall gravitational influence on the oceans. When this happens there is less of a difference between high and low tide. As one might expect, spring and neap tides happen approximately twice a month.

The tides are the most visible way the Moon affects the Earth, but there are many other ways. For example, without tides, there would be no tide pools along the ocean shores. Also, if the Moon did not exist, the Earth would be much darker at night because no light would be reflected by the Moon. Without the Moon there is a chance that we could not exist on Earth. The Earth would be less stable in its rotation because the gravity of the Moon stabilizes our rotation. Without the Moon, the tilt of the Earth's axis would vary more, which would change the length of days and seasons, and climate changes might be more extreme.

There has been significant research focused on studying the relationship between the Moon and the Earth's tides. People have wondered for centuries about the relationship between the Moon and the Earth. Scientists like Johannes Kepler, in the late sixteenth and early seventeenth centuries, were researching and learning about the Moon. Kepler's laws of planetary motion explained that gravity was responsible for the tides. Over the years; theories were formed and discoveries made by figures such as Isaac Newton and George Darwin (the son of Charles Darwin). Newton argued that the ocean tides were created not only by a gravitational force from the Moon and the Sun but that the size of the Earth, Moon, and Sun, and their distance from each other, all played a role. Darwin also hypothesized that the Moon is moving further away from the Earth, a little bit each year; a hypothesis proven by astronauts in 1969. In the future, global warming, the heating of the Earth accompanied by the rising of the oceans, will make the tides more pronounced, especially during full and new moons.

The Earth's atmosphere also plays an important role in sheltering living matter from cosmic rays, and radiation in the form of protons and atomic nuclei from the Sun. The Earth's surface temperature can range from –90°F to 150°F, but the atmosphere and the water in it help significantly moderate these extremes. Water evaporation provides a cooling effect if the air is not fully saturated because water evaporation draws the heat necessary for water vaporization from the surrounding environment.

2.3 THE HYDROLOGIC CYCLE

Water is the original renewable resource. Although the total amount of water on the surface of the Earth remains fairly constant over time, individual water molecules carry with them a rich history. The water molecules contained in the fruit eaten yesterday may have fallen as rain last year in a distant place or could have been used decades, centuries, or even millennia ago by one's ancestors.

Water is always in motion, and the hydrologic cycle describes this movement from place to place. The vast majority (96.5%) of water on the surface of the Earth is contained in the oceans. Solar energy heats the water at the ocean surface and some of it evaporates to form water vapor. Air currents take the vapor up into the atmosphere along with water transpired from plants and evaporated from soil. The cooler temperatures in the atmosphere cause the vapor to condense into clouds. Clouds move around the world until the moisture capacity of the cloud is exceeded and the water falls as precipitation. Most precipitation in warm climates falls back into the oceans or onto land, where the water flows over the ground as surface runoff. Runoff can enter rivers and streams, which transport the water back to the oceans; it can accumulate and be stored as freshwater in lakes; or it can soak into the ground as infiltration. Some of this water may infiltrate deep into the ground and replenish aquifers that store huge amounts of freshwater for long periods of time.

In cold climates, precipitation falls as snow and can accumulate as ice caps and glaciers which can store water for thousands of years. Three important processes are involved in the hydrologic cycle:

1. Physical processes. Evaporation and condensation are physical processes. The energy that is needed in the evaporation of water on the Earth's surface is stored in the water vapor in the air and can be transported over great distances and then released through condensation elsewhere. Two-thirds of the planet's entire energy transport occurs via this evaporation and condensation mechanism.
2. Chemical processes. Crystallization, dissolution, and chemical reactions are chemical processes. Water is an extremely good solvent. Water wears away mountains, forms soils from bedrock, and sets minerals free, which in turn flow with the water through the soil and then to the plants. Minerals serve as natural fertilizers for vegetation.
3. Biological processes. Photosynthesis transforms carbon dioxide and water into sugar and oxygen. Both of these are the necessary basic products for building all further complex molecules for life on Earth.

Humans take water out of the hydrologic cycle to use it for a wide range of different purposes. Only a small part of this water is used. On its way through the cycle, the purity and composition, the aggregate state, or the temperature of the water is changed. The benefits of use from this change can be numerous. It can be used for drinking, for industrial manufacturing, for industrial cooling processes and thermal electricity plants, for providing electricity from hydroelectric power, or even for producing artificial snow for winter tourism.

2.3.1 Rivers and Streams

From a water quality engineering point of view rivers have been studied over time more extensively and longer than other bodies of water, probably reflecting the fact that many early peoples lived close to or interacted with streams and rivers.

Hydrologically, interest in rivers begins with the analysis of river flows. The magnitude and duration of flows, coupled the chemical quality of the waters, determine (to a considerable degree) the biological characteristics of a stream. A river is an extremely rich and diverse ecosystem, and any water quality analysis must recognize this diversity.

The principle physical characteristics of rivers that are of interest include:

1. Geometry: width, depth
2. River slope, bed roughness, "tortuosity"
3. Velocity
4. Flow rate
5. Mixing characteristics (dispersion in the river)
6. Water temperature
7. Suspended solids and sediment transport

For river water quality management, the important chemical characteristics are:

1. Dissolved oxygen (DO) variations, including associated effects of oxidizable nitrogen on the DO regime
2. pH, acidity, alkalinity relationships in areas subjected to such discharges (i.e., drainage from abandoned mines)
3. Total dissolved solids and chlorides in certain river systems (i.e., natural salt springs in the Arkansas-White-Red River basins)
4. Chemicals that are potentially toxic

Biological characteristics of river systems that are of special significance in water quality studies are:

1. Bacteria and viruses
2. Fish populations
3. Rooted aquatic plants
4. Biological slimes (i.e., *Sphaerotilus*)

As with all water quality analyses, the objective in river water quality engineering is to recognize and quantify, as much as possible, the various interactions between river hydrology, chemistry, and biology. The study of river hydrology includes many factors of water movement in river systems, including precipitation, stream flow, droughts and floods, groundwater, and sediment transport (Linsley et al. 1982). The most important aspects of river hydrology are the river flow, velocity, and geometry. Each of the characteristics are used in various ways in the water quality modeling of rivers. Measurements of river flows focus on those times when the flow is "low" because of dilution. If a discharge is running into a stream, then conditions will probably be most critical during the times when there is less water in the channel. The flow at a given point in a river will depend on:

1. Watershed characteristics such as the drainage area of the river or stream basin up to the given location
2. Geographical location of the basin
3. Slope of the river
4. Dams, reservoirs, or locks which may regulate flow
5. Flow diversions into or out of the river basin

In recent years significant effort has been devoted to the modeling of rivers and streams. The flow in the river can be obtained by several methods. A direct measurement of river velocity and cross-sectional area at a specific location can give an estimate of the flow at that location and time. River velocities are measured either directly by current meters or indirectly by tracking the time for objects in the water to travel a given distance. Because the velocity of a river varies with width and depth due to frictional effects, the mean vertical velocity must be estimated.

With an estimate of the velocity at hand, a first approximation can be made to the time of travel between various points on the river. This relationship ignores dispersion or mixing in the river and any effects of "dead" zones such as deep holes or side channel coves. With the flow and hydraulic properties of a river system defined and the estimates of these properties at hand, some first approaches to describing the discharge of residual substances into rivers and streams can be examined. Such residuals may include discharges from waste treatment plants, from combined sewer overflows, or from agricultural and urban runoff.

2.3.2 Estuaries, Bays, and Harbors

The region between a free-flowing river and the ocean is a fascinating, diverse, and complex water system: the coastal regime of estuaries, bays, and harbors. Since the beginning of time, the ebb and flow of the tides, the incursion of salinity from the ocean, and the influx of nutrients from upstream drainages all contribute to the generation of a unique aquatic ecosystem. The estuarine and wetland regions are considered to be crucial to the maintenance of major fish stocks, such as the striped bass and blue fish, which to varying degrees use the estuarine areas as spawning and nursery grounds.

The movement of the tides into and out of estuaries and the associated density effects created by the incursion of salinity, are of particular importance in describing the water quality of such bodies of water. Many major cities are located along estuaries primarily as a result of the historical need for ready access to national and international commerce routes. For many years, such cities discharged large quantities of untreated waste, but, because of the large volumes of water moving through estuaries, effects were not immediately felt. Later, however, especially in, the last century, the pollutant load on estuaries became excessive, quality deteriorated rapidly, and great interest centered on the analysis and improvement of water quality in estuaries.

Several distinct zones of a river can be defined. The *tidal river* is that region of a river where there is some current reversal but sea salts have not penetrated to this region so that the tidal river is still "fresh." The *estuary* is the "drowned" part of a river system due to incursion of the ocean landward with marked current reversal and brackishness because of the saline water. It should be note that if a river discharges to a large lake such as to one of the Great Lakes in the United States, a condition similar to that in an estuary can be created through the incursion of lake water up into the mouth of the river.

Early dwellers by their shores have always been fascinated by the movement of water into and out of estuaries and bays along coastal regions. No coastline is without tides, and over the many centuries of observation, a great degree of regularity in the vertical and horizontal motion of water along the coast has been noticed. Tides are the movement of water above and below a datum plane, usually mean sea level. Tidal currents are the associated horizontal movement of the water into and out of an estuary. As indicated previously, tidal motions occur on a more or less regular cyclical basis reflecting the regularity of the lunar and solar cycles (Defant 1958; Neumann and Pierson 1966; Ippen 1966). Tides are also generated in lakes and inland seas, produced principally by winds blowing across the lake surface and "piling up" the water, which, in turn, sets the lake into an oscillatory motion or seiche. The approximately regular motion of the lake results in a motion in lake tributaries similar to estuarine tides.

Tidal excursion is the approximate distance a unit of volume will travel along the main axis of an estuary in going from low to high water, or vice versa. The *tidal flow* is the total volume of water passing a given point in the estuary over time.

The tidal currents in open offshore waters behave in an interesting fashion due to the lack of physical boundaries. The tidal current tends to move about a point in a rotary-type current. This type of current, therefore, will tend to move any wastes discharged offshore in an elliptical motion on which may be superimposed a net current drift. The current structure in offshore waters is, therefore, quite complex and is of particular significance in the transport of wastes discharged at sea.

An important characteristic of estuarine hydrology is the net flow through the estuary over a tidal cycle or a given number of cycles. This is the flow that, over a period of several days or weeks, flushes material out of the estuary and is a significant parameter in the estimation of the distribution of estuarine water quality. If the estuary is well mixed from top to bottom and from side to side, then the net flow at any location in the estuary is approximately equal to the sum of the upstream external flow inputs to the estuary, assuming no other significant net hydrologic inputs or losses. This is so because it is known that the estuary is not overflowing due to the flow inputs. Therefore, this flow must, on balance, be leaving the estuary at any cross section.

Estimating the time and spatial behavior of water quality in estuaries is complicated by the effects of tidal motion. The upstream and downstream currents produce substantial variations of water quality at certain points in the estuary, and the calculation of such variation is indeed a complicated problem. Some simplifications can, however, be made which provide some remarkably useful results in estimating the distribution of estuarine water quality. The simplifications can be summarized through the following assumptions (Thomann and Mueller 1987):

1. Estuary is one-dimensional (i.e., it is subject to reversals in direction of the water velocity, and only the longitudinal gradient of a particular water quality parameter is dominant).
2. Water quality is described as a type of average condition over a number of tidal cycles.
3. Area, flow, and reaction rates are constant with distance.
4. Estuary is in a steady-state condition.

2.3.3 Lakes

A major portion of water-based recreational activities centers on the thousands of lakes, reservoirs, and other small, relatively quiescent bodies of water located throughout the United States. In addition, these lakes and reservoirs serve as a source of water for municipal and industrial use, including water for agricultural purposes, water quality control, and fisheries management. The ecosystems and quality of lakes throughout the world have long been a primary concern of water quality managers.

Lakes and reservoirs vary from small ponds and dams to the magnificent and monumental large lakes of the world such as Lake Superior, one of the Great Lakes, and Lake Baikal in the Soviet Union, the deepest lake in the world (5,310 ft). The ecosystems supported by this broad range of water bodies vary from the very attractive local sport fishes such as bass and perch to the large top predators of both sport and commercial value such as lake trout and landlocked and migratory salmon.

Limnology is the study of the physical, chemical, and biological behavior of lakes. Recreation, sport fishing (and for the larger lakes, commercial fishing), and water supply for municipal and industrial uses are all intimately related to the quality of these water bodies. The distinguishing physical features of lakes include relatively low flow-through velocities and development of significant vertical gradients in temperature and other water quality variables. Lakes, therefore, often become sinks for nutrients, toxicants, and other substances in incoming rivers. As a result, eutrophication is one of the more significant water quality problems of lakes.

The principal physical features of a lake are length, depth, area (both of the water surface and of the drainage area), and volume. The overall physical relationships for a lake can be summarized in area-depth and volume-depth curves. The relationship between the flow out of a lake or reservoir and the volume is also an important characteristic. The ratio of the volume to the flow represents the hydraulic retention time (i.e., the average residence time of water within a lake assuming the lake contents are completely mixed). The hydraulic retention times, as a function of the ratio of lake drainage area to surface area, for northern U.S. lakes and reservoirs range from 1 day to about 6,000 days, or 16 years (Bartsch and Gakstatter 1978). A long detention time does not necessarily indicate a large lake; a small lake with a small flow may still have a long detention time.

As with rivers and estuaries, an understanding of the water balance and circulation of lakes is of considerable importance in water quality analysis and engineering (Linsley et al. 1982). A general and simple hydrologic balance equation for a given body of water is:

The net flows into and out of the lake due to river and/or groundwater flow + precipitation directly on the lake + lake evaporation = The change in the lake volume over a period of time (2.1)

Inflows may include surface inflow, subsurface inflow, and water imported into the lake. Outflows may include surface and subsurface outflow from the reservoir and exported water. The change in storage in the lake or reservoir may also include subsurface storage or "bank" storage of water. In determining the hydrologic balance of a lake, the change in volume and surface inflow and outflow can usually be easily measured. Precipitation can also be measured without difficulty except for large lakes, where it must be estimated for the open water. The remaining unknowns include subsurface water movements and evaporation.

2.3.4 Oceans

The location of oceans has slowly changed since the beginning of time. Today's oceans contain the bulk of water available to humans; however, because of its high salinity, it is not available as potable water without significant treatment. Other factors as described have also come to affect ocean water quality, and for many years, oceans have been explored for their varied energy production potential.

In parts of the world where oil was discovered near coastlines, explorers often found that the deposits extended far out beneath the sea. Offshore oil-fields, at first, were reached by building piers as far out as possible to support drilling equipment. However, it soon became clear that other techniques would be needed to drill for oil lying under deeper water. During the last half-century, new techniques were developed to tackle oil exploration at sea.

A drilling rig was specially designed for exploration in deeper water. This has huge buoyancy tanks that enable it to float out to the drilling site. There, the tanks are partly filled with water, making the rig sink lower in the sea and giving it more stability in stormy weather. Anchors are also used to keep it in place above the well. This type of rig, known as a *semisubmersible rig*, allows exploration in water depths of more than a thousand feet. For exploration in even deeper water, oil companies use a specially equipped drillship. In addition to the usual rear-mounted propeller, this type of vessel has thruster units installed in the hull, enabling it to move in any horizontal direction. These prevent the ship from being moved out of position by tides and currents as it drills. The action of the thrusters is controlled by a computer installed on board. Signals from a beacon fixed on the seabed warn the computer as soon as the ship starts to change position. The computer then turns on one or more thrusters to produce a force that counteracts the movement and keeps the vessel directly above the oil well.

The discovery of vast oilfields beneath the frozen areas of Alaska and the Arctic Circle confronted technicians and engineers with a spectacular challenge. Oil could hardly have been found in a more remote and inhospitable place. The supplies are thousands of miles from the industrial centers where fuel is needed, and because the Arctic seas are frozen for much of the year, it is impossible to carry cargoes of oil by sea. Because pack ice forms on the surface of the sea around the Arctic Circle most of the year, exploration with conventional offshore drilling rigs is difficult. Oil companies have solved this problem by building artificial islands in the comparatively shallow water during the short ice-free periods in the summer. These islands, made of gravel dredged from the surrounding seabed, are able to support the weight of a normal land-based drilling unit, which can be transported in sections across the surface of the frozen ocean in winter.

Ocean tidal energy, thermal energy, and wave energy have existed for billions of years. It is only recently, however, that the technical community has come to realize and attempt to harness these forms of energy from the sea. Skipka and Theodore (2014) provide some details as will be discussed.

2.3.4.1 Tidal Energy

The tides offer a virtually inexhaustible natural source of energy that is essentially unused. The idea of harnessing the rise and fall of the oceans has received the attention of engineers, scientists, and inventors in the past. The maximum tidal ranges do not occur on a daily basis. Tides are caused by the gravitational attraction of the moon and the sun on the waters of the Earth and on the Earth itself. The moon has the greater effect. It "pulls" the water away from the Earth, increasing on the side toward the moon. It also draws the Earth away from the water, increasing the water height on the other side, creating two low tides and two high tides each day. The sun acts in a similar manner but has a reduced effect on tide. When the sun and moon are "pulling" together (or opposite), their tides are in phases and result in a high tidal range. About a week later, when the sun and moon are "pulling" at right angles (half-moon), their tides tend to cancel each other, and the tidal range is smaller.

Tidal energy is the most promising source of ocean energy. A dam called a *barrage* is built across an inlet. The barrage has one-way gates that allow the incoming flood tide to pass into the inlet. When the tide turns, the water flows out of the inlet through turbines built into the barrage, producing electricity. Unfortunately, generating electricity in the middle of the ocean is simply extremely difficult because of transport problems; there is no one there to use it. One can only use the energy near shore, where people need it. For the technical community, tidal energy is considered a renewable energy source. The power generated is free and nonpolluting and the plants are easy to maintain. The plants can affect the ecology and there can be aesthetic objections as well. Unfortunately, the United States has no tidal plants at this time and only a few sites where tidal energy could be economically produced.

2.3.4.2 Thermal Energy

Ocean thermal energy is also referred to as ocean thermal energy conversion (OTEC). Vast amounts of energy are available from ocean thermal gradients. When two extensive currents of water, one warm and one cold, exist in close proximity to one another, it is possible to operate a power plant using this temperature differential (Theodore 2014). Theoretically, this energy may be extracted wherever a temperature difference driving force exists. The extraction of such energy becomes (as one might suppose) more difficult, more costly, and less efficient, as the temperature difference between the high- and low-temperature reservoirs decreases (Theodore, Ricci, and VanVliet 2007). The economics are not currently competitive where the technology for this idea has been shown to work.

If this energy source were developed, however, there would be a number of environmental impacts to be considered. The large-scale mixing of warm and cold water could have significant impacts on the ocean, biota, and climate. The large surface areas in the heat exchangers would be continually subjected to the flow of corrosive seawater, and metallic elements will therefore be introduced into the seawater. Loss of working fluid (typically ammonia) might also be a problem if leaks in the system are significant, or if there are unexpected spills. Other problems include the impacts of techniques used to inhibit biofouling and corrosion, the impacts of coastal zone facilities associated with the operation of the offshore plants, and the installation and operation of the electrical distribution systems (U.S. DoE 1978).

2.3.4.3 Wave Energy

There is also tremendous energy in waves. Waves are caused by the wind blowing over the surface of the ocean. In some areas of the world, the wind blows with enough intensity and force to produce large waves. The west coasts of the United States and Europe and the coasts of Japan, Australia, and New Zealand are excellent candidates for harnessing wave energy. There are no large, commercial wave-energy plants, but there are a few small ones in Australia (Carnegie Clean Energy 2019). This resource might produce enough energy to power local communities. Interestingly, Japan, which must import almost all of its fuel, has an active wave-energy program (Marine Energy.biz 2018).

2.4 THE FIRST HUMANS

Environmental problems have bedeviled humanity since the first person discovered fire. The earliest humans appear to have inhabited a variety of locales within a tropical and semitropical belt stretching from Ethiopia to southern Africa about 1.9 million years ago. These first humans provided for themselves by a combination of gathering food and hunting animals. Humans, for the majority of their two million years existence, lived in this manner. The steady development and dispersion of these early humans was largely due to an increase in their brain size. This led to the ability to think abstractly, which was vital in the development of technology and the ability to speak. This in turn led to cooperation and more elaborate social organization (Ponting 1991). The ability to use and communicate the developed technology to overcome the hostile environment ultimately led to the expansion of these first human settlements.

With the use of primitive tools and skins of animals for clothes, the first humans moved outside Africa about one and a half million years ago. The migration led them into the frost free zones of the Middle East, India, southern China, and parts of Indonesia. The humans at this time could only adapt to those ecosystems found in the semitropical areas that contained a wide variety of vegetation and small, easily hunted animals to supplement their diet. Despite relatively easy access, Europe was not settled for a long period of time because of the deficient ecosystem, which was later overcome by an increase in technology. The first evidence of human settlement in Europe is dated to about 730,000 years ago. The settlement of the Americas was almost the last stage in the movement of humans across the globe about 20,000 years ago. This was made possible by crossing to Alaska in the last glaciation when the reduced sea levels turned the Bering Strait into a land bridge. Once the first human settlers were able to move south through the passes, they found an enormously rich environment that supplied an abundance of food. The human population multiplied rapidly and within a few thousand years had spread to the tip of South America. By about 10,000 years ago humans had spread over every continent, living in small mobile groups. A minority of these groups lived in close harmony with the environment and did minimal damage. Evidence has been found where groups tried to conserve resources in an attempt to maintain subsistence for long periods of time, but many examples of waste of resources by early man are also evident in the archeological record.

2.5 THE DEVELOPMENT OF AGRICULTURE

A major shift in human evolution took place between 10,000 and 12,000 years ago. Humans learned how to domesticate animals and cultivate plants and in doing so made a transition from nomadic hunter gatherer to rooted agriculturalist. The global population at this time was about 4 million people, which was about the maximum that could readily be supported by a gathering and hunting way of life (Ponting 1991). The increasing difficulty in obtaining food is believed to be a major contributor to this sudden change. The farmer changed the landscape of the planet and was far more destructive than the hunter. Although farming fostered the rise of cities and civilizations, it also led to practices that denuded the land of its nutritional and water holding capacity. Great civilizations flourished and then disappeared as once fertile land, after generations of over farming and erosion, was transformed into barren wasteland.

The adoption of agriculture, combined with its two major consequences, settled communities and a steadily rising population, placed an increasing strain on the environment. The strain was localized at first, but as agriculture spread so did its effects. Agriculture involved removing the natural habitat to create an artificial habitat where humans could grow the plants and animals they would need.

The natural balance and inherent stability of the original ecosystem were thereby destroyed. Instead of a variety of plants and permanent natural ground cover, a small number of crops made only part-time use of the space available. The soil was exposed to wind and rain to a far greater extent than before, particularly where fields were left barren for part of the year, leading to a higher rate of soil erosion than under natural ecosystem conditions. Nutrient cycling processes were also disrupted and extra inputs in the form of manures and fertilizers were therefore required if soil fertility was to be maintained. The adoption of irrigation was even more disruptive because it created an environment that was even more artificial. Adding large amounts of water to a poor soil would allow the farmer to grow their preferred crop, but it would have catastrophic long-term effects. The extra water would drain into the underlying water table, sometimes leading to rising water levels, which caused the soil to become waterlogged. This additional water not only altered the mineral content of the soil but also increased the amount of salt and would eventually, especially in hot areas with high evaporation rates, produce a thick layer of salt on the surface that made agriculture impossible. The emergence of villages and towns meant that the demand for resources was now more concentrated.

2.6 COLONIZATION OF THE NEW WORLD

Only a little more than 500 years, a mere second on the geological clock, have passed since Christopher Columbus's discovery opened a fresh and verdant new world to the Europeans: a land with few indications of human occupation except for a few thin plumes of smoke rising from cooking fires in small clearings in the woods. These clearings belonged to the Native Americans, which numbered about 4 million at this time. Over the centuries these people had created their own complex culture. Their means of sustaining themselves did not rely on scarring or subduing the Earth, but on using what it offered. Native American society was not separate from nature but part of it. Geography, as well as history, began to change when Columbus anchored his little fleet off the island of San Salvador. Like most of those who followed, Columbus and his company risked the voyage to the New World for what they could take from it. They came for gold, a trade route to the spices of India and other riches of Asia: land, goods to sell, glory, adventure, religious, and personal freedom, and in some cases, to convert the heathen to Christianity (Shabecoff 2003). Although the explorers, adventurers, and settlers came to seize whatever riches and opportunities the land had to offer, it was what they brought with them, far from what they took, that changed the face of the continent forever. What they brought was Europe's 2,000 or more years of western history, customs, prejudices, and methodology. They brought European technology, philosophy, religion, aesthetics, a market economy, and a talent for political organization. They brought European diseases that decimated the native people. They also brought with them European ideas of what the New World was and visions of what it should be.

In the beginning, the explorers and first settlers were faced with a dark forbidding line of forest behind which was a vast, unmapped continent, inhabited, they thought, by savages, and filled with ferocious wild beasts. Mere survival meant conquering the wilderness. The forest had to be cleared to make living space and to provide wood for shelters and fires (Shabecoff 2003). Behind the trees lurked the Indians, ready, the settlers suspected, to commit unspeakable atrocities. The forest was filled with wolves, bears, and panthers that would pounce on their children and domestic animals, or so they feared. The greater the destruction of the forest, the greater the safety for the tiny communities clinging to the edge of the hostile continent. Removing the trees also opened land for crops and cattle. Killing the wild animals not only filled the pot with meat but also eliminated the deer and other grazing animals that stole the settlers' corn (Shabecoff 2003).

The European population quickly grew beyond the carrying capacity of the land. Cropland was frequently exhausted by permanent cultivation; cattle, swine, and sheep introduced by immigrants made far heavier demands on field and forest than wild animals. As each new field was harvested, the chemical, mineral, and biological nature of the soil itself was depleted. The Europeans also brought technology that contributed to the heavy impact they had on the land. Horses and oxen enabled the settlers to open and cultivate much more acreage. Plows could dig deeply into the soil, exposing far more loam. With draft animals, the Europeans could harvest heavier loads and transport them to markets. Sailing ships could then transport those loads along the coast or across the ocean.

Whereas the Native Americans would take from the land only what they could consume, the colonist and their successors sought to grow surplus that they could sell for cash or trade for manufactured goods and other commodities. The production of surplus led to the accumulation of capital and the creation of wealth, largely in the towns that served as marketplaces. That meant clearing more land, cutting more timber, planting more crops, and raising more cattle, all at a rate that could be sustained only at a cost of permanent damage to the land. The deforestation of New England and the disappearance of the beaver in the East are but two dramatic examples of how the demands of the market could deplete abundant resources in short order. By the time of the American Revolution, the wilderness along the eastern seaboard had been tamed. Although some pockets of forest remained, the thirteen colonies were largely covered with farms, dotted with villages, and punctuated by a few substantial cities, notably Boston, New York, Philadelphia, and Charleston.

2.7 THE INDUSTRIAL REVOLUTION AND BEYOND

Early in the nineteenth century, an awesome new force was gathering strength in Europe. The term "industrial revolution" was coined by the French as a metaphor of the affinity between technology and the great political revolution of modern times. Soon exported to the United States, the industrial revolution swept away any visions of America being an agrarian society. The steam engine, the railroad, the mechanical thresher, and hundreds of other ingenious artifacts that increased man's ability to transform the natural world and put it to use would soon be puffing and clattering and roaring in all comers of the land. The new machines swiftly accelerated the consumption of raw materials from the nation's farms, forests, and mines.

Lumbering became the nation's most important industry in the late eighteenth century. Wood was the most widely used raw material for heating and building; the same can be said for ships, furniture, railroad ties, factories and paper-making. The supply seemed inexhaustible since the forest still darkened huge parts of the country. The forest melted away before the axes of the advancing Americans. The settlers never thought of their axe work as deforestation but as the progress of civilization. Soon after the tree cover was removed, the forest soil began to lose nutrients. The soil began washing away, turning clear streams into slow, muddy ditches, filling lakes, and killing fish.

Meanwhile, the big cities and growing wealth of the East were creating a more rapidly expanding market for wheat, corn, beef, and other cash crops. New roads and canals, the steamboat and the locomotive, made domestic and foreign markets increasingly accessible to farms in the center of the continent. Eli Whitney' s cotton gin, Cyrus McCormick's reaper, Benjamin Holt's combine, and other ingenious inventions encouraged the development of a highly productive and efficient agriculture that sharply reduced the biological diversity of the land. Mining both preceded and quickly followed settlement of the interior, and left deep and permanent scars on the continent's land and waters. Gold in California, lead in Illinois, coal and oil in Pennsylvania, iron ore in Minnesota, and copper in Montana attracted fortune hunters and job seekers. Reports of a strike would draw thousands of prospectors and workers as well as those who lived off them. Mines were operated without care for the surrounding countryside. The picks and shovels, the hoses and dredges, and the smaller fires of the miners created the nation's first widespread pollution and environmental health problems. Mining left behind gutted mountains, dredged-out streams, despoiled vegetation, open pits, polluted creeks, barren hillsides and meadows, a littered landscape, and abandoned camps. Mining contributed to deforestation of the countryside. Woodlands were often cleared for mining operations; enormous amounts of timber were needed for the posts and beams that supported the mine shafts and fueled smelter operations (Shabecoff 2003).

Steam shovels came into use in the 1880s, enabling the coal operators of Pennsylvania and the iron ore producers of Minnesota to peel away the very crust of the earth to extract raw materials for industry and wealth for themselves. Spoils from the coal started to turn streams more acidic. The discovery of oil in Pennsylvania in 1859 brought drilling rigs that poked into the skyline: large areas of soil were soaked with black ooze (Shabecoff 2003). It was in the cities that environmental pollution and its effects were most pervasive, however. Garbage and filth of every kind were thrown into the streets, covering the surface, filling the gutters, obscuring the sewer culverts that sent forth perennial emanations. In the winter, the filth and garbage would accumulate in the streets to the depth of sometimes 2 or 3 feet. Most cities were nightmares of primitive sanitation and waste disposal systems. Privies for sewage and private wells for water were still widely used in metropolitan areas until the end of the nineteenth century.

Perhaps the national government could have done more to protect the land and its resources as well as public health. But, for most of the nineteenth century, the national government was still a weak presence in most areas of the country. There was, moreover, no body of laws with which the government could assert its authority. Laissez-faire was the order of the day. By the end of the century there was a growing body of information about the harm being done and some new ideas on how to set things straight. Yet, there was no acceptable ethic that would compel people to treat the land, air, and water with wisdom and care. To a large extent the people did not know what they were doing to the environment (Shabecoff 2003).

The federal government ultimately entered into the environmental and conservation business in a significant fashion when Teddy Roosevelt's second cousin Franklin Delano entered the White House in 1933. It was his political ideology as much as his love of nature that led Roosevelt to include major conservation projects in his New Deal reforms. The Civilian Conservation Corps, the Soil Conservation Service, and the Tennessee Valley Authority were among the many New Deal programs created to serve both the land and the people.

2.8 THE ENVIRONMENTAL MOVEMENT AND THE ENVIRONMENTAL PROTECTION AGENCY

At this point in time, muscle and animal power were replaced by electricity and internal combustion engines. At the same time, industry was consuming natural resources at an incredible rate. All of these events began to escalate at a dangerous rate after World War II. Soon after, in the late summer of 1962, a marine biologist named Rachel Carson, author of *Silent Spring*, the best-selling book about pesticide impacts on the ecosystem, opened the eyes of the world to the dangers of attacking the environment. It was perhaps at this point that America began calling in earnest for reform of the destruction of nature and constraints on environmental degradation.

It is now 1970, a cornerstone year for modern environmental policy. The National Environmental Policy Act (NEPA), enacted on January 1, 1970, was considered a "political anomaly" by some. NEPA was not based on specific legislation; instead it referred in a general manner to environmental and quality of life concerns. The Council for Environmental Quality (CEQ), established by NEPA, was one of the councils mandated to implement legislation. April 22, 1970, brought Earth Day, where thousands of demonstrators gathered all around the nation. NEPA and Earth Day were the beginning of a long, seemingly never-ending debate over environmental issues.

The Nixon Administration at that time became preoccupied with not only trying to pass more extensive environmental legislation but also implementing the laws. Nixon's White House Commission on Executive Reorganization proposed in the Reorganizational Plan # 3 of 1970 that a single, independent agency be established, separate from the CEQ. The plan was sent to Congress by President Nixon on July 9, 1970, and this new U.S. Environmental Protection Agency (EPA) began operation on December 2, 1970. The EPA was officially born.

For additional literature regarding the early history and the environmental movement, the interested reader is referred to the book by Philip Shabecoff, *A Fierce Green Fire* (Shabecoff 2003). This outstanding book, as well as Ponting's *A Green History of the World* (Ponting 1991) are "musts" for anyone who works in or has interests in the environmental arena.

2.9 APPLICATIONS

Six Illustrative Examples complement this historical perspective of water and human interactions with the Earth.

Illustrative Example 2.1

Briefly discuss present-day atomic theory.

Solution

In ancient Greek philosophy, the word *atomos* was used to describe the smallest bit of matter that could be conceived. This "fundamental particle" was thought of as indestructible; in fact, atomos means "not divisible." Knowledge about the size and nature of the atom grew slowly throughout the centuries. As discussed, the atom consists of three subatomic particles: the *proton, neutron*, and *electron*. The charge of an electron is -1.602×10^{-19} C (coulombs), and that of a proton is $+1.602 \times 10^{-19}$ C. The quantity -1.602×10^{-19} C is defined as the *electronic charge*. Note that the charges of these subatomic particles are expressed as multiples of this charge rather than in coulombs. Thus, the charge of an electron is 1^-, and that of a proton is 1^+. Neutrons carry no charge (i.e., they are electrically neutral). Because an atom has an equal number of electrons and protons, it has zero or no net electric charge.

Both the protons and neutrons reside in the nucleus of the atom, which is extremely small. Most of the atom's volume is the space in which the electrons reside. The external electrons are attracted to the protons in the nucleus because of their opposite electrical charge.

Illustrative Example 2.2

Define the Avagadro Number.

Solution

A mole (sometimes also called a *gmol*) of a compound is an Avogadro number, or 6.023×10^{23} molecules, of that compound; a mole of an element is 6.023×10^{23} atoms of that element. Equivalently, a lbmol of an element is (454) (6.023×10^{23}) or 2.734×10^{26} atoms of the element. The number 454 is the number of grams in a pound. The atomic weights used in the Periodic Table may be given any of the following sets of units: amu/atom, g/mol, or lb/lbmol. The following conversion factors follow from these definitions:

$$1 \text{ lbmol} = 454 \text{mol}$$

$$1 \text{ g} = 6.023 \times 10^{23} \text{ amu}$$

$$1 \text{ lb} = 2.734 \times 10^{26} \text{ amu}$$

Illustrative Example 2.3

It is estimated that each human on Earth requires 4 L of water/d. Assuming there are 7 billion humans on Earth, calculate the annual volume of water consumed by the human race.

Solution

The annual volume of water consumed by the human race, AVH, can be calculated as follows:

$$\text{AVH} = (4 \text{ L/d/person})(7{,}000{,}000{,}000 \text{ people})(365 \text{ d/yr}) = 10.22 \times 10^{12} \text{ L/yr}$$

$$\text{AVH} = (10.22 \times 10^{12} \text{ L/yr})(1 \text{ gal}/3.785 \text{ L}) = 2.7 \times 10^{12} \text{ gal/yr}$$

Illustrative Example 2.4

Refer to Illustrative Example 2.3. Determine the size of a cube in units of feet that would contain the AVH quantity of water.

Solution

For a cube of side L, its volume is:

$$\text{Volume} = L^3 \tag{2.2}$$

The AVH in units of ft^3 is found to be from Illustrative Example 2.3:

$$\text{AVH} = 2.7 \times 10^{12} \text{gal/yr} \left(1 \text{ ft}^3/7.48 \text{ gal}\right) = 3.61 \times 10^{11} \text{ft}^3/\text{yr} = L^3$$

$$L = \sqrt[3]{3.61 \times 10^{11} \text{ ft}^3} = 7{,}120 \text{ ft} = 1.35 \text{ mi}$$

The cube would then be 7,120 ft or 1.35 mi on a side to contain the annual quantity of drinking water for the Earth's population.

Illustrative Example 2.5

Refer to Illustrative Example 2.3. Determine the size of a sphere in units of feet, that would contain the AVH quantity of water.

Solution

The volume of a sphere of radius, R, is given as:

$$\text{Volume} = \frac{4}{3}\pi R^3 \tag{2.3}$$

$$\text{AVH} = 3.61 \times 10^{11}\text{ft}^3/\text{yr} = \frac{4}{3}\pi R^3$$

$$R = \sqrt[3]{\frac{3}{4\pi}(3.61 \times 10^{11}\,\text{ft}^3)} = \sqrt[3]{(8.62 \times 10^{10}\,\text{ft}^3} = 4{,}417\text{ ft} = 0.84\text{ mi}$$

The sphere would then be a radius of 4,417 ft or 0.84 mi, or a diameter of 8,834 ft or 1.67 mi to contain the annual quantity of drinking water for the Earth's population.

Illustrative Example 2.6

Compare the AVH from Illustrative Example 2.3 to the volume of all water and all freshwater on Earth.

Solution

The volume of all water on earth is estimated by the USGS (2019) to be 332.5 million $\text{mi}^3 = 3.66 \times 10^{20}$ gal, whereas the volume of all freshwater on Earth is estimated to be 2.55 million $\text{mi}^3 = 2.81 \times 10^{18}$ gal. The AVH as a percentage of these volumes are as follows:

$$\text{AVH as \% of all water on Earth} = \left(2.7 \times 10^{12}\text{gal}\right)/\left(3.66 \times 10^{20}\text{gal}\right) \times 100 = 7.4 \times 10^{-7}\%$$

$$\text{AVH as \% of all fresh water on Earth} = \left(2.7 \times 10^{12}\text{gal}\right)/\left(2.81 \times 10^{18}\text{gal}\right) \times 100 = 0.0001\%$$

As can be seen in these calculations, the annual water need for human consumption is small relative even to only the total fresh water that exists on the Earth. Much of this freshwater is not readily accessible, being contained in ice caps, glaciers, and permanent snow, or in deep aquifers, but water needs just for potable use is small relative to the inventory of water on the planet. The problems arise because of local deterioration of water quality due to insufficient wastewater treatment and pollutant controls and to availability of easily accessible water for not only potable use but for all of the water demands related to irrigation, industry, and other nonpotable uses.

REFERENCES

Bartsch, A.F., and J.H. Gakstatter. 1978. Management Decisions for Lake Systems on a Survey of Trophic Status, Limiting Nutrients, and Nutrient Loadings, in *American-Soviet Symposium on Use of Mathematical Models to Optimize Water Quality Management,* 1975, EPA-600/9-78-024. Gulf Breeze, FL: U.S. Environmental Protection Agency, Office of Research and Development, Environmental Research Laboratory.

Carnegie Clean Energy. 2019. Wave. Freemantle: Australia. https://www.carnegiece.com/projects/wave/ (accessed July 1, 2019).

Defant, A. 1958. *Ebb and Flow, the Tides of Earth, Air, and Water.* Ann Arbor, MI: University of Michigan Press.

Ippen, A. 1966. *Estuary and Coastline Hydrodynamics.* New York: McGraw-Hill.

Linsley, R.R., Jr., M.A. Kohler, and J.D.H. Pualhus. 1982. *Hydrology for Engineers*, 3rd edition. New York: McGraw-Hill.

Marine Energy.biz. 2018. Japan plans ocean energy R&D demo project. https://marineenergy.biz/2018/01/12/japan-plans-ocean-energy-rd-demo-project/ (accessed July 1, 2019).

Neumann, G., and W.J. Pierson. 1966. *Principles of Physical Oceanography.* Englewood Cliffs, NJ: Prentice-Hall.

Ponting, C.A. 1991. *A Green History of the World.* New York: St. Martin's Press.

Shabecoff, P.A. 2003. *A Fierce Green Fire: The American Environmental Movement.* Revised edition Washington, DC: Island Press.

Skipka, K.J., and L. Theodore. 2014. *Energy Resources: Availability, Management, and Environmental Impacts.* Boca Raton, FL: CRC Press, Taylor & Francis Group.

Theodore, L. 2014. *Chemical Engineering: The Essential Reference.* New York: McGraw-Hill.

Theodore, L., F. Ricci, and T. VanVliet. 2007. *Thermodynamics for the Practicing Chemical Engineer.* Hoboken, NJ: John Wiley & Sons.

Theodore, M.K., and L. Theodore. 1996. *Major Environmental Issues Facing the 21st Century.* Upper Saddle River, NJ: Prentice Hall PTR.

Thomann, R.V., and J.A. Mueller. 1987. *Principles of Surface Water Quality Modeling and Control.* New York: Harper & Row.

U.S. Department of Energy. 1978. *Environmental Development Plan: Ocean Thermal Energy Conversion.* DOE/EDP-0006. Washington, DC: U.S. Department of Energy.

U.S. Geological Survey. 2019. *How Much Water Is There on Earth?* Water Science School. Reston, VA: U.S. Geological Survey. https://www.usgs.gov/special-topic/water-science-school/science/how-much-water-there-earth?qt-science_center_objects=0#qt-science_center_objects.

3 Water Properties

3.1 INTRODUCTION

There is no question that of all of the compounds on the Earth, water possesses unique properties including:

1. High dielectric constant
2. Low density
3. High melting (fusion) temperature
4. High boiling temperature
5. High heat capacity

Two other characteristics of water also need to be noted:

1. Life forms on Earth consist primarily of water
2. Water is most dense at 4°C so that it expands below this temperature and becomes less dense. Thus, water with a temperature of 4°C in a larger body of water is more dense than the surrounding water and sinks; a characteristic that has prevented the Earth's lakes from freezing from the bottom up and providing liquid habitat for lake species during the coldest season of the year.

This chapter introduces the reader to these critical physical properties of water. Section 3.4 presenting the Steam Tables is unquestionably the most often referenced data for all engineering disciplines and can justifiably be considered the key element in this chapter. An application section provides six illustrative examples related to the general subject of water's physical properties.

3.2 UNIQUE PROPERTIES OF WATER

Of those molecules that are most abundant on Earth, water is one of the lightest. In addition, being made up of only three atoms, it has a simple configuration, one oxygen atom and two hydrogen atoms at an angle of 104.5° to each other. This simple molecule has several exceptional properties that are all the result of the fact that one large oxygen atom forms a bond with two very small hydrogen atoms, the smallest atom that exists. Oxygen is the dominant partner and attracts the hydrogen electrons, which leads to the unusually strong polarity of the water molecule, which has definite positive and negative poles. The strength of the polarity of a molecule is expressed in the relative dielectric constant. Of all the natural substances, water has the largest dielectric constant. This, together with the small size of the water molecule, is the reason why water is the best known natural solvent.

This in turn leads to water's exceptionally high melting and boiling points as compared to oxygen's closest neighbors in the Periodic Table, the elements sulfur, selenium, and tellurium, which can also form similar structured molecules. Table 3.1 illustrates the fact that water, as compared to related elements in the Periodic Table, should actually have a melting point of 93°C and a boiling point of –72°C. Were this the case, then water could only exist as water vapor on Earth. Instead, due to the mutual electrical attraction among water molecules, known as hydrogen bonding, water has the well known melting point of 0°C and boiling point of 100°C. It is obvious

TABLE 3.1
Comparison of Water's Melting and Boiling Points with Other Similar Elements

Molecule	Molecular Weight, g/gmol	Melting Point, °C	Boiling Point, °C
H_2O	18	0	100
H_2S	34	–82	–61
H_2Se	80	–64	–42
H_2Te	129	–51	–4

that although water is clearly lighter than these related elements, it melts and boils at much higher temperatures. Hydrogen bonds are responsible for an additional physical property of water that is important for the Earth system: water's extremely high heat capacity and high vaporization and fusion temperatures. Large amounts of heat are needed to melt or vaporize water and are thus stored in the water molecule. This high capacity for heat storage is important not just in nature, but is also used by humans to cool machinery. To melt ice, 146 BTU/lb (340 J/g) are needed and 1,054 BTU/lb (2,450 J/g) are needed to vaporize water. This heat is stored in the water and is released by condensation or freezing, respectively.

3.3 PHASES AND THE TRIPLE POINT OF WATER

It will be helpful to review certain definitions and concepts associated with phases and the phase rule of J. Willard Gibbs before proceeding to the triple-point concept.

A homogeneous system is one for which all parts (on a macro scale) have the same physical and chemical properties. Such a system has a uniform composition and concentration throughout. Air in a cylinder is an example of such a homogeneous system. Water completely occupying a closed container is another. A heterogeneous system has nonuniform composition. Its physical and chemical properties vary from one location to another. A phase is a finite part of a system that is homogeneous throughout and is physically separated from other phases of the system by distinct boundaries called interfaces. When water stands in contact with its own vapor, the surface of the water is the interface between the liquid and vapor phases.

In stating and applying the phase rule it is essential to determine the number of so-called components. By definition, the number of components is the minimum number of independently variable chemical species needed to express the total composition of the system or any phase present in the system. The phase rule applies to all systems in which a condition of equilibrium exists (Theodore et al. 2007, 2017; Theodore 2016). According to Gibbs:

$$F = C + 2 - P \tag{3.1}$$

where P is the number of phases present, C is the number of components as previously defined, and F is the so-called degrees of freedom or variability of the system. The degrees of freedom, F, may be defined as the number of independent variables, such as temperature, pressure, or concentration, that must be specified to completely define the system. A second definition of F is the number of variables, such as temperature and pressure, that may be changed independently without causing the appearance or disappearance of a phase. The number 2 is valid only when there are two variables in addition to concentration. These two are commonly temperature and pressure. If, for example, conditions are such that pressure may be regarded as fixed throughout, then there is only one effective independent variable in addition to concentration and, in such a system

$$F = C + 1 - P \tag{3.2}$$

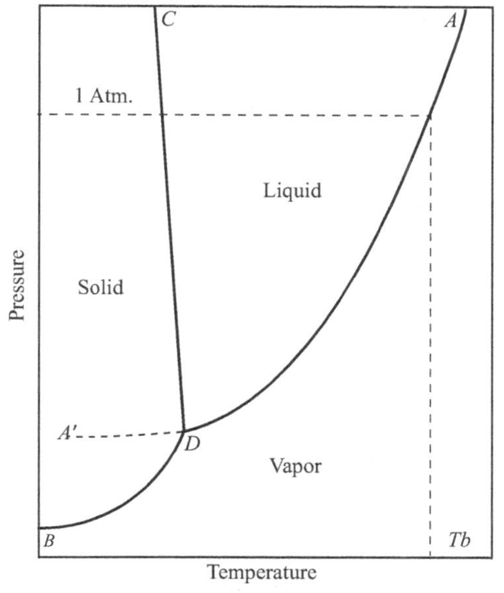

FIGURE 3.1 Solid-liquid-vapor equilibria diagram for a substance.

Regarding the triple point, the curves DA and DB in Figure 3.1 show the variation with temperature of the vapor pressure of liquid and solid forms, respectively, of a given substance. The curve DA extends only as far as the critical temperature, whereas DB continues to absolute zero, possibly with a change in direction due to a polymorphic transition. The vapor pressure curve DA of the liquid represents the conditions of equilibrium temperature and pressure, for a system of liquid and vapor, while curve DB similarly indicates the conditions under which solid and vapor phases of a substance are in equilibrium. The two curves meet at D, and hence this point is known as the triple point where the three physical states of a substance (i.e., solid, liquid, and vapor) coexist.

Further inspection of Figure 3.1 indicates that the curve DC represents the temperatures and pressures at which the solid and liquid forms can be in equilibrium (i.e., indicating the influence of pressure on the melting point of the substance). Its slope depends on whether an increase of pressure raises or lowers the melting point (i.e., on whether the specific volume of the liquid is greater or less, respectively, than that of the solid). This line in Figure 3.1 slopes to the left indicating that, as for water, the melting point is lowered by an increase of external pressure. The curve DC must obviously meet the other curves at the triple point D, where the solid, liquid and vapor phases are in equilibrium. The conditions under which two of the three phases can coexist is given by each of the curves, and all three forms are in equilibrium where the three curves meet. There is consequently only one point where this is possible, and hence there is only one triple point. The normal melting point of a solid is the temperature at which solid and liquid are in equilibrium at atmospheric pressure. At the triple point, however, the pressure is the equilibrium vapor pressure of the system, and the temperature differs slightly from the melting point. For water, the triple point is in the vicinity of 0°C, but the vapor pressure of liquid water and ice is then about 4.6 mm Hg.

3.4 VAPOR PRESSURE OF WATER

Consider the molecules of a pure liquid substance such as water in a container. The attractive forces acting among these molecules tend to keep them together. In contrast, the kinetic energy associated with their thermal motion tends to separate them from one another. As long as the average energy of attraction exceeds the average kinetic energy of translation, the molecules remain mainly in the condensed, liquid phase.

A molecule in the interior of the liquid is acted upon from all sides by the attractive forces exerted by its neighbors, but at the surface conditions are quite different. There is nothing above the liquid's surface except vapor and, perhaps, air. As a consequence, there is a net inward pull exerted on any molecule occupying the surface which accounts for the phenomenon of surface tension. It follows that for a molecule to be at the surface it must acquire energy in excess of the average energy of molecules in the system.

If the space above the liquid is sealed off and the system is held at constant temperature, equilibrium is soon reached between the liquid and its vapor, with molecules leaving and returning from each phase at the same rate from each unit of surface area. The space above the liquid now holds the greatest concentration of vapor molecules possible at the temperature of the system. In other words, it is saturated with vapor and there is exerted on the walls of the container a characteristic and constant pressure called the *equilibrium vapor pressure* or *saturation pressure* or simply the *vapor pressure* of the substance. This pressure is independent of the size of the area exposed, and the amount of liquid present, as long as some liquid is present in equilibrium with the vapor, or the shape of the container. Only on changing the temperature does this characteristic equilibrium vapor pressure change. It increases with a rise in temperature and decreases with a reduction in temperature. When the vapor pressure of a liquid in an open vessel reaches that of the atmosphere, there results the wholesale turbulent escape of molecules from the liquid and the liquid begins to boil. The temperature at which the liquid attains a vapor pressure of 760 mm Hg is its *normal boiling point*. The boiling point of a liquid may be raised by increasing, or lowered by decreasing the pressure on the system, respectively. The vapor pressure is an important property of water, and water vapor pressure data at a range of system temperatures are provided in Table 3.2.

Mixtures of gases are more often encountered than single or pure gases in environmental engineering practice. The ideal gas law is based on the *number of* molecules present in the gas volume (i.e., the type of molecule is not a significant factor), only the number. This law applies equally well to mixtures and pure gases alike. Dalton and Amagat both applied the ideal gas law to mixtures of gases.

Because pressure is caused by gas molecules colliding within the containing walls, it seems reasonable that the total pressure of a gas mixture is made up of pressure contributions from each of the component gases. These pressure contributions are called *partial pressures*. Dalton defined the partial pressure of a component as the pressure that would be exerted if the same mass of the

TABLE 3.2
Vapor Pressure of Water at 1 atm and Various System Temperatures

Temperature, °C	Vapor Pressure, mm Hg	Melting Point, °C	Boiling Point, °C
−15	1.436	35	42.17
−10	2,149	40	55.32
−5	3.163	45	71.88
0	4.579	50	92.51
1	4.926	55	118.0
2	5.294	60	149.4
3	5.685	65	187.5
4	6.102	70	233.7
5	6.543	75	289.1
10	9.209	80	355.1
15	12.79	85	433.6
20	17.54	90	525.7
25	23.76	95	633.9
30	31.82	100	760.0

component gas occupied the same total volume alone at the same temperature as the mixture. The sum of these partial pressures then equals the total pressure:

$$P = p_a + p_b + p_c + \ldots + p_n \tag{3.3}$$

where P is the total pressure, n is the number of components, and p_i is the partial pressure of component i, where $i = a, b, c \ldots n$. Equation 3.3 is known as *Dalton's Law*. Applying the ideal gas law to one component (A) only (Theodore et al. 2007, 2017; Theodore 2016), one obtains

$$p_A V = n_A RT \tag{3.4}$$

where n_A is the number of moles of Component A. Eliminating R, T, and V between Equations 3.3 and 3.4 leads to

$$p_A = y_A P \tag{3.5}$$

where y_A is the mole fraction of Component A.

Amagat's Law is similar to Dalton's Law. Instead of considering the total pressure to consist of partial pressures where each component occupies the total container volume, Amagat considered the total volume of a container to consist of the partial volumes in which each component is at (or is exerting) the total pressure. The definition of the *partial volume* is, therefore, the volume occupied by a component gas alone at the same temperature and total pressure as the mixture. For this case

$$V = V_a + V_b + V_c + \ldots + V_n \tag{3.6}$$

Applying Equation 3.4 as before, one obtains

$$V_A = y_A V \tag{3.7}$$

where V_A is the partial volume of Component A.

It is common in environmental engineering practice to describe low concentrations of components in gaseous mixtures in parts per million by volume (ppm$_v$). Because partial volumes are proportional to mole fractions, it is necessary only to multiply the mole fraction of a component by 1 million to obtain the concentration in ppm$_v$. For liquids and solids, ppm is also used to express concentration, although for these substances, concentration is based on a mass rather than volume basis.

3.5 WATER STEAM TABLES

As noted previously, a pure substance refers to a state of matter that is gas, liquid, or solid. Latent enthalpy (heat) effects are associated with phase changes (Theodore et al. 2007, 2017; Theodore 2016). These phase changes involve no change in temperature, but there is a transfer of energy to and from the substance. There are three possible latent effects, as will be detailed: (i) vapor-liquid, (ii) liquid-solid, and (iii) vapor-solid.

Vapor-liquid changes are referred to as condensation when the vapor is condensing and vaporization when liquid is vaporizing. *Liquid-solid* changes are referred to as melting when the solid melts to liquid and freezing when a liquid solidifies. *Vapor-solid* changes are referred to as sublimation. One should also note that there are enthalpy effects associated with a phase change of a solid to another solid form; however, this enthalpy effect is small compared to the other effects mentioned previously. Specific volume, enthalpy, and entropy data for saturated steam, superheated steam, and steam-ice mixtures are provided in Tables 3.3 through 3.5, respectively (Keenan and Keyes 1930; Jones and Hawkins 1960; Van Wylen and Sonntag 1965). Table 3.3

TABLE 3.3
Specific Volume, Enthalpy, and Entropy Data for Saturated Steam

		Specific Volume, ft³/lb; v			Enthalpy, Btu/lb; h			Entropy, (Btu/lb-°R)		
Temperature, °F	Absolute Pressure, lb/in²	Saturated Liquid	Evaporation Difference	Saturated Vapor	Saturated Liquid	Evaporation Difference	Saturated Vapor	Saturated Liquid	Evaporation Difference	Saturated Vapor
32	0.08854	0.01602	3306	3306	0	1075.8	1075.8	0	2.1877	2.1877
35	0.09995	0.01602	2947	2947	3.02	1074.1	1077.1	0.0061	2.1709	2.177
40	0.1217	0.01602	2444	2444	8.05	1071.3	1079.3	0.0162	2.1435	2.1597
45	0.14752	0.01602	2036.4	2036.4	13.06	1068.4	1081.5	0.0262	2.1167	2.1429
50	0.17811	0.01603	1703.2	1703.2	18.07	1065.6	1083.7	0.0361	2.0903	2.1264
60	0.2563	0.01604	1206.6	1206.7	28.06	1059.9	1088	0.0555	2.0393	2.0948
70	0.3631	0.01606	867.8	867.9	38.04	1054.3	1092.3	0.0745	1.9902	2.0647
80	0.5069	0.01608	633.1	633.1	48.02	1048.6	1096.6	0.0932	1.9428	2.036
90	0.6982	0.0161	468	468	57.99	1042.9	1100.9	0.1115	1.8972	2.0087
100	0.949 2	0.01613	350.3	350.4	67.97	1037.2	1105.2	0.1295	1.8531	0.9826
110	0.2748	0.01617	265.3	265.4	77.94	1031.6	1109.5	0.1471	0.8106	0.9577
120	0.6924	0.0162	203.25	203.27	87.92	1025.8	1113.7	0.1645	0.7694	0.9339
130	0.2225	0.01625	157.32	157.34	97.9	1020	1117.9	0.1816	0.7296	0.9112
140	0.8886	0.01629	122.99	123.01	107.89	1014.1	1122.0	0.1984	0.691	0.8894
150	0.718	0.01634	97.06	97.07	117.89	1008.2	1126.1	0.2149	0.6537	0.8685
160	0.741	0.01639	77.27	77.29	127.89	1002.3	1130.2	0.2311	0.6174	0.8485
170	0.992	0.01645	62.04	62.06	137.9	996.3	1134.2	0.2472	1.5822	0.8293
180	0.51	0.01651	50.21	50.23	147.92	990.2	1138.1	0.263	1.548	1.8109
190	0.339	0.01657	40.94	40.96	157.95	984.1	1142	0.2785	1.5147	1.7932
200	11.526	0.01663	33.62	33.64	167.99	977.9	1145.9	0.2938	1.4824	1.7762
210	14.123	0.0167	27.8	27.82	178.05	971.6	1149.7	0.309	1.4508	1.7598
212	14.696	0.01672	26.78	26.8	180.07	970.3	1150.4	0.312	1.4446	1.7566
220	17.186	0.01677	23.13	23.15	188.13	965.2	1153.4	0.3239	1.4201	1.744
230	20.78	0.01684	19.365	19.382.	198.23	958.8	1157	0.3387	1.3901	1.7288
240	24.969	0.01692	16.306	16.323	208.34	952.2	1160.5	0.3531	1.3609	1.714
250	29.825	0.017	13.804	13.82	218.48	945.5	1164	0.3675	1.3323	1.6998

(Continued)

TABLE 3.3 (*Continued*)
Specific Volume, Enthalpy, and Entropy Data for Saturated Steam

		Specific Volume, ft³/lb; v			Enthalpy, Btu/lb; h			Entropy, (Btu/lb-°R)		
Temperature, °F	Absolute Pressure, lb/in²	Saturated Liquid	Evaporation Difference	Saturated Vapor	Saturated Liquid	Evaporation Difference	Saturated Vapor	Saturated Liquid	Evaporation Difference	Saturated Vapor
260	35.429	0.01709	11.746	11.763	228.64	938.7	1167.3	0.3817	J.3043	1.686
270	41.858	0.01717	10.044	10.061	238.84	931.8	1170.6	0.3958	1.2769	1.6727
280	49.203	0.01726	8.628	8.645	249.06	924.7	1173.8	0.4096	1.2501	1.6597
290	57.556	0.01735	7.444	7.461	259.31	917.5	1176.8	0.4234	1.2238	1.6472
300	67.013	0.01745	6.449	6.466	269.59	910.1	1179.7	0.4369	1.198	1.635
310	77.68	0.01755	5.609	5.626	279.92	902.6	1182.5	0.4504	1.1727	1.6231
320	89.66	0.01765	4.896	4.914	290.28	894.9	1185.2	0.4637	1.1478	1.6115
330	103.06	0.01776	4.289	4.307	300.68	887	1187.7	0.4769	1.1233	1.6002
340	118.01	0.01787	3.77	3.788	311.13	879	1190.1	0.49	1.0992	1.5891
350	134.63	0.01799	3.324	3.342	321.63	870.7	1192.3	0.5029	1.0754	1.5783
360	153.04	0.01811	2.939	2.957	332.18	862.2	1194.4	0.5158	1.0519	0.5677
370	173.37	0.01823	2.606	2.625	342.79	853.5	11%.3	0.5286	1.0287	1.5573
380	195.77	0.01836	2.317	2.335	353.45	844.6	1198.1	0.5413	1.0059	1.5471
390	220.37	0.0185	2.0651	2.0836	364.17	835.4	1199.6	0.5539	0.9832	1.5371
400	247.31	0.01864	1.8447	1.8633	374.97	826	1201	0.5664	0.9608	1.5272
410	276.75	0.01878	1.6512	1.67	385.83	816.3	1202.1	0.5788	0.9386	1.5174
420	308.83	0.01894	1.4811	1.5	396.77	806.3	1203.1	0.5912	0.9166	1.5078
430	343.72	0.0191	1.3308	1.3499	407.79	796	1203.8	0.6035	0.8947	1.4982
440	381.59	0.01926	1.1979	1.2171	418.9	785.4	1204.3	0.6158	0.873	1.4887
450	422.6	0.0194	1.0799	1.0993	430.1	774.5	1204.6	0.628	0.8513	1.4793
460	466.9	0.0196	0.9748	0.9944	441.4	763.2	1204.6	0.6402	0.8298	1.47
470	514.7	0.0198	0.8811	0.9009	452.8	751.5	1204.3	0.6523	0.8083	1.4606
480	566.1	0.02	0.7972	0.8172	464.4	739.4	1203.7	0.6645	0.7868	1.4513
490	621.4	0.0202	0.7221	0.7423	476	726.8	1202.8	0.6766	0.7653	1.4419
520	812.4	0.0209	0.5385	0.5594	511.9	686.4	1198.2	0.713	0.7006	1.4136
540	962.5	0.0215	0.4434	0.4649	536.6	656.6	1193.2	0.7374	0.6568	1.3942

(*Continued*)

TABLE 3.3 (*Continued*)
Specific Volume, Enthalpy, and Entropy Data for Saturated Steam

		Specific Volume, ft³/lb; v			Enthalpy, Btu/lb; h			Entropy, (Btu/lb-°R)		
Temperature, °F	Absolute Pressure, lb/in²	Saturated Liquid	Evaporation Difference	Saturated Vapor	Saturated Liquid	Evaporation Difference	Saturated Vapor	Saturated Liquid	Evaporation Difference	Saturated Vapor
560	1133.1	0.0221	0.3647	0.3868	562.2	624.2	1186.4	0.7621	0.6121	1.3742
580	1325.8	0.0228	0.2989	0.3217	588.9	588.4	1177.3	0.7872	0.5659	1.3532
600	1542.9	0.0236	0.2432	0.2668	617	548.5	1165.5	0.8131	0.5176	1.3307
620	1786.6	0.0247	0.1955	0.2201	646.7	503.6	1150.3	0.8398	0.4664	1.3062
640	2059.7	0.026	0.1538	0.1798	678.6	452	1130.5	0.8679	0.411	1.2789
660	2365.4	0.0278	0.1165	0.1442	714.2	390.2	1104.4	0.8987	0.3485	1.2472
680	2708.1	0.0305	0.0810	0.1115	757.3	309.9	1067.2	0.9351	0.2719	1.2071
700	3093.7	0.0369	0.0392	0.0761	823.3	172.1	995.4	0.9905	0.1484	1.1389

TABLE 3.4
Specific Volume, Enthalpy, and Entropy Data for Superheated Steam

Absolute Pressure, lb/in² (Saturated Temperature, °F)	Phase	Temperature, °F												
		200	220	300	350	400	450	500	550	600	700	800	900	1000
	v	392.6	404.5	452.3	482.2	512.0	541.8	571.6	601.4	631.2	690.8	750.4	809.9	869.5
1	*l*	1150.4	1159.5	1195.8	1218.7	1241.7	1264.9	1288.3	1312.0	1335.7	1383.6	1432.7	1482.6	1533.4
(101.74)	*s*	2.0512	2.0647	2.1153	2.1444	2.1720	2.1983	2.2233	2.2468	2.2702	2.3137	2.3542	2.3923	2.4283
	v	78.16	80.59	90.25	96.26	102.26	108.24	114.22	120.19	126.16	138.1	150.03	161.95	173.87
5	*l*	1148.8	1158.1	1195.0	1218.1	1241.2	1264.5	1288.0	1311.7	1335.4	1383.6	1432.7	1482.6	1533.4
(162.24)	*s*	1.8718	1.8857	1.9370	1.9664	1.9942	2.0205	2.0456	2.0692	2.0927	2.1361	2.1776	2.2148	2.2509
	v	38.85	40.09	45.00	48.03	51.04	54.05	54.05	60.04	63.03	69.01	74.98	80.95	86.92
10	*l*	1146.6	1156.2	1193.9	1217.2	1240.6	1264.0	1264.0	1311.3	1335.1	1383.4	1432.5	1482.4	1533.2
(193.21)	*s*	1.7927	1.8071	1.8595	1.8892	1.9172	1.9436	1.9436	1.9924	2.0160	2.0596	2.1002	2.1383	2.1744
	v		27.15	30.53	32.62	34.68	36.73	36.73	40.82	42.86	46.94	51.00	55.07	59.13
14.696	*l*		1154.4	1192.8	1216.4	1239.9	1263.5	1263.5	1310.9	1334.8	1383.2	1432.3	1482.3	1533.1
(212.00)	*s*		1.7624	1.8160	1.8460	1.8743	1.9008	1.9008	1.9498	1.9734	2.0170	2.0576	2.0958	2.1319
	v			22.36	23.91	25.43	26.95	26.95	29.97	31.47	34.47	37.46	40.45	43.44
20	*l*			1191.6	1215.6	1239.2	1262.9	1262.9	1310.5	1334.4	1382.9	1432.1	1482.1	1533.0
(227.96)	*s*			1.7808	1.8112	1.8396	1.8664	1.8664	1.916	1.9392	1.9829	2.0235	2.0618	2.0978
	v			11.040	11.843	12.628	13.401	13.401	14.93	15.688	17.198	18.702	20.20	21.70
40	*l*			1186.8	1211.9	1236.5	1260.7	1260.7	1308.9	1333.1	1381.9	1431.3	1481.4	1532.4
(267.25)	*s*			1.6994	1.7314	1.7608	1.7881	1.7881	1.8384	1.8619	J.9058	1.9467	1.9850	2.0214
	v			7.259	7.818	8.357	8.884	8.884	9.916	10.427	I1.441	12.449	13.452	14.454
60	*l*			1181.6	1208.2	1233.6	1258.5	1258.5	1307.4	1331.8	1380.9	1430.5	1480.8	1531.9
(292.71)	*s*			1.6492	1.6830	1.7135	1.7416	1.7678	1.7926	1.8162	1.8605	1.9015	1.9400	1.9762
	v				5.803	6.220	6.624	7.020	7.410	7.797	8.562	9.322	10.077	10.830
80	*l*				1204.3	1230.7	1256.J	1281.1	1305.8	1330.5	1379.9	1429.7	1480.1	1531.3
(312.03)	*s*				1.6475	1.6791	1.7078	1.7346	1.7598	1.7836	1.8281	1.8694	1.9079	1.9442
	v				4.592	4.937	5.268	5.589	5.905	6.218	6.835	7.446	8.052	8.656
100	*l*				1200.1	1227.6	1253.7	1279.1	1304.2	1329.1	1378.9	1428.9	1479.5	1530.8
(327.81)	*s*				1.6188	1.6518	1.6813	1.7085	1.7339	1.7581	1.8029	1.8443	1.8829	1.9193

(Continued)

TABLE 3.4 (*Continued*)
Specific Volume, Enthalpy, and Entropy Data for Superheated Steam

Absolute Pressure, lb/in² (Saturated Temperature, °F)	Phase	Temperature, °F												
		200	220	300	350	400	450	500	550	600	700	800	900	1000
	v					4.081	4.363	4.636	4.902	5.165	5.683	6. 195	6.702	7.207
120	*l*					1224.4	1251.3	1277.2	1302.5	1327.7	1377.8	1428.1	1478.8	1530.2
(341.25)	*s*					1.6287	1.6591	16869	1.7127	1.7370	1.7822	1.8237	1.8625	1.8990
	v					3.468	3.715	3.954	4.186	4.413	4.861	5.301	5.738	6.172
140	*l*					1221.1	1248.7	1275.2	1300.9	1326.4	1376.8	1427.3	1478.2	1529.7
(353.02)	*s*					1.6087	1.6399	1.6683	1.6945	1.7190	1.7645	1.8063	1.8451	1.8817
	v					3.008	3.230	3.443	3.648	3.849	4.244	4.631	5.015	5.396
160	*l*					1217.6	1246.1	1273.1	1299.3	1325.0	1375.7	1426.4	1477.5	1529.1
(363.53)	*s*					1.5908	1.6230	1.6519	1.6785	1.7033	1.7491	1.7911	1.8301	1.8667
	v					2.649	2.852	3.044	3.229	3.411	3.764	4.110	4.452	4.792
180	*l*					1214.0	1243.5	1271.0	1297.6	1323.5	1374.7	1425.6	1476.8	1528.6
(373.06)	*s*					1.5745	1.6077	1.6373	1.6642	1.6894	1.7355	1.7776	1.8167	1.8534
	v					2.361	2.549	2.726	2.895	3.060	3.380	3.693	4.002	4.309
200	*l*					1210.3	1240.7	1268.9	1295.8	1322.1	1373.6	1424.8	1476.2	1528.0
(381.79)	*s*					1.5594	1.5937	1.6240	1.6513	1.6767	1.7232	1.7655	1.8048	1.8415
	v					2.125	2.301	2.465	2.621	2.772	3.066	3.352	3.634	3.913
220	*l*					1206.5	1237.9	1266.7	1294.1	1320.7	1372.6	1424.0	1475.5	1527.5
(389.86)	*s*					1.5453	1.5808	1.6117	1.6395	1.6652	1.7120	1.7545	1.7939	1.8308
	v					1.9276	2.094	2.247	2.393	2.533	2.804	3.068	3.327	3.584
240	*l*					1202.5	1234.9	1264.5	1292.4	1319.2	1371.5	1423.2	1474.8	1526.9
(397.37)	*s*					1.5319	1.5686	1.6003	1.6286	1.6546	1.7017	1.7444	1.7839	1.8209
	v						1.9183	2.063	2.199	2.330	2.582	2.827	3.067	3.305
260	*l*						1232.0	1262.3	1290.5	1317.7	1370.4	1422.3	1474.2	1526.3
(404.42)	*s*						1.5573	1.5897	1.6184	1.6447	1.6922	1.7352	1.7748	1.8118
	v						1.7674	1.9047	2.033	2.156	2.392	2.621	2.845	3.066
280	*l*						1228.9	1260.0	1288.7	1316.2	1369.4	1421.5	1473.5	1525.8
(411.05)	*s*						1.5464	1.5796	1.6087	1.6354	1.6834	1.7265	1.7662	1.8033

(*Continued*)

TABLE 3.4 (*Continued*)

Specific Volume, Enthalpy, and Entropy Data for Superheated Steam

Absolute Pressure, lb/in² (Saturated Temperature, °F)	Phase	Temperature, °F												
		200	220	300	350	400	450	500	550	600	700	800	900	1000
	v						1.6364	1.7675	1.8891	2.005	2.227	2.442	2.652	2.859
300	*l*						1225.8	1257.6	1286.8	1314.7	1368.3	1420.6	1472.8	1525.2
(417.33)	*s*						1.5360	1.5701	1.5998	1.6268	1.6751	1.7184	1.7582	1.7954
	v						1.3734	1.4923	1.6010	J.7036	J.8980	2.084	2.266	2.445
350	*l*						1217.7	1251.5	1282.1	1310.9	1365.5	1418.5	1471.1	1523.8
(431.72)	*s*						1.5119	1.5481	1.5792	1.6070	1.6563	1.7002	1.7403	1.7777
	v						1.1744	1.2851	1.3843	1.4770	1.6508	1.8161	1.9767	2.134
400	*l*						1208.8	1245.1	1277.2	1306.9	1362.7	1416.4	1469.4	1522.4
(444.59)	*s*						1.4892	J.5281	1.5607	1.5894	1.6398	1.6842	1.7247	1.7623

TABLE 3.5
Specific Volume, Enthalpy, and Entropy Data for Steam-Ice Mixtures

		Specific Volume, ft³/lb		Enthalpy, Btu/lb			Entropy, (Btu/lb-°R)		
Temperature, °F	Absolute Pressure, lb/in²	Saturated Ice, v_i	Saturated Steam, $v_g \times 10^{-3}$	Saturated Ice, h_i	Sublimation Difference, h_{sub}	Saturated Steam, h_g	Saturated Ice, s_i	Sublimation Difference, s_{sub}	Saturated Steam, s_g
32	0.0885	0.01747	3.306	−143.35	1219.1	1075.8	−0.2916	2.4793	2.1877
30	0.0808	0.01747	3.609	−144.35	1219.3	1074.9	−0.2936	2.4897	2.1961
20	0.0505	0.01745	5.658	−149.31	1219.9	1070.6	−0.3038	2.5425	2.2387
10	0.0309	0.01744	9.05	−154.17	1220.4	1066.2	−0.3141	2.5977	2.2836
0	0.0185	0.01742	14.77	−158.93	1220.7	1061.8	−0.3241	2.6546	2.3305
−10	0.0108	0.01741	24.67	−163.59	1221.0	1057.4	−0.3346	2.7143	2.3797
−20	0.0062	0.01739	42.20	−168.16	1221.2	1053.0	−0.3448	2.7764	2.4316
−30	0.0035	0.01738	74.1	−172.63	1221.2	1048.6	−0.3551	2.8411	2.4860

provides properties for saturated steam/water between 32°F and 700°F. Table 3.4 contains properties for superheated steam between 1.0 psia and 600 psia. Table 3.5 provides properties for saturated steam-ice between 32°F and −40°F. Lowercase notation is employed for specific volume (v), enthalpy (h), and entropy (s) because the values are listed on a mass basis. English units are employed throughout the three tables. Those interested in using these values with other units should refer to the Appendix for the appropriate conversion constants. Linear interpolation should be employed where necessary. Additional data are available in the three cited references. Other physical properties are addressed in the next section.

3.6 OTHER PROPERTIES OF WATER

Other important properties of water include viscosity, thermal conductivity, density, surface tension, and heat capacity. Theodore (2016) provides a detailed description of these properties. Tabulated values of these properties as a function of temperature are presented in Table 3.6 for all but heat capacity. Practicing engineers generally assume the heat capacity of water to be 1.0 Btu/lb-°F or 1 cal/g-°C and independent of temperature.

The *heat capacity* of a substance, C_P, is defined as the quantity of heat required to raise the temperature of that substance by one degree on a unit mass (or mole) basis. The term *specific heat* is frequently used in place of heat capacity. This is not strictly correct because specific heat has been defined traditionally as the ratio of the heat capacity of a substance to the heat capacity of water. However, since the heat capacity of water is approximately 1 cal/(g-°C) or 1 Btu/(lb-°F), the term specific heat has come to imply heat capacity. Heat capacities are functions of both temperature and pressure, although the effect of pressure is generally small and is neglected in almost all environmental engineering calculations. The effect of temperature on C_P can be described by

$$C_P = \alpha + \beta T + \gamma T^2 \tag{3.8}$$

or

$$C_P = a + bT + cT^2 \tag{3.9}$$

TABLE 3.6
Other Properties of Liquid Water

Temperature, °F	Viscosity, C_p	Thermal Conductivity, Btu/hr-ft-°F	Density, lb/ft³	Surface Tension, dynes/cm
32	1.794	0.320	62.42	75.6
40	1.546	0.326	62.43	74.9
50	1.310	0.333	62.42	73.5
60	1.129	0.340	62.37	73.0
70	0.982	0.346	62.30	72.7
80	0.862	0.352	62.22	72.2
90	0.764	0.358	62.11	71.0
100	0.682	0.362	62.00	69.7
120	0.559	0.371	61.71	67.9
140	0.470	0.378	61.88	66.2
160	0.401	0.384	61.00	64.4
180	0.347	0.388	60.58	62.4
200	0.305	0.392	60.13	60.6
212	0.267	0.393	59.47	58.9

Values for α, β, γ and a, b, c as well as for average heat capacity information are provided in the literature (Theodore et al. 2007; Theodore 2016).

3.7 APPLICATIONS

Six Illustrative Examples complement the material presented in this chapter.

Illustrative Example 3.1

Convert the vapor pressure of water at 20°C in mm Hg to kPa and °F.

Solution

Referring to Table 3.2, the vapor pressure of water at 20°C is 17.54 mm Hg. First the temperature can be converted to °F using the following relationship

$$°F = 1.8\,(°C) + 32 = 1.8(20) + 32 = 36 + 32 = 68°F \tag{3.10}$$

For the pressure term

$$kPa = (0.133)(mm\,Hg) = (0.133)(17.54) = 2.33\,kPa \tag{3.11}$$

Thus, the vapor pressure of water at 68°F is 2.33 kPa. The reader is left with the exercise of showing that the vapor pressure at this temperature may also be expressed as 0.00233 MPa.

Illustrative Example 3.2

Convert the result of Illustrative Example 3.1 to units of lb_f/in^2.

Solution

The conversion factor to convert pressure from mm Hg to lb_f/in^2 is 0.01934 (see Appendix A.2). Thus, the vapor pressure of water at 20°C may also be expressed as

$$lb_f/in^2 = (0.01934)(mm\,Hg) = (0.01934)(17.54) = 0.339\,lb_f/in^2 \tag{3.12}$$

Illustrative Example 3.3

Estimate the enthalpy and entropy of saturated steam (vapor) at 225°F.

Solution

Refer to Table 3.3. Linearly interpolate between temperature values of 220°F and 230°F. The enthalpy, h, is found to be

$$\begin{aligned} h\,@\,225°F &= h\,@\,230°F + 5/10(h\,@\,220°F - h\,@\,230°F) \\ &= 1157 + 0.5(1153.4 - 1157) = 1157 - 1.8 = 1155.2\,Btu/lb \end{aligned} \tag{3.13}$$

The entropy, s, is found to be

$$\begin{aligned} s\,@\,225°F &= s\,@\,230°F + 5/10\,(s\,@\,220°F - s\,@\,230°F) \\ &= 1.7288 + 0.5\,(1.744 - 1.7288) = 1.7288 - 0.0076 \\ &= 1.7364\,Btu/lb\text{-}°R \end{aligned} \tag{3.14}$$

Illustrative Example 3.4

The suspended particulate concentration, C, in a dilute water stream is 0.28 mg/L. Convert this concentration into units of μg/L, g/L, lb/ft³, and lb/gal.

Solution

Apply the appropriate conversion factors as follows:

$$C = 0.28\,\text{mg/L}\,(1{,}000\,\mu\text{g/mg}) = 280\,\mu\text{g/L}$$

$$C = 0.28\,\text{mg/L}\,(1\,\text{g}/1{,}000\,\text{mg}) = 0.00028\,\text{g/L}$$

$$C = 0.28\,\text{mg/L}\,(1\,\text{lb}/454\,\text{g})(1\,\text{g}/1{,}000\,\text{mg})(28.33\,\text{L}/1\,\text{ft}^3) = 0.175\times10^{-5}\,\text{lb/ft}^3$$

$$C = 0.28\,\text{mg/L}\,(1\,\text{lb}/454\,\text{g})\left(1\,\text{g}/1{,}000\,\text{mg}\right)(3.785\,\text{L/gal}) = 0.233\times10^{-6}\,\text{lb/gal}$$

Illustrative Example 3.5

Convert a calcium concentration of 10 ppm in water to units of mole fraction, y_{Ca}.

Solution

The water concentration of calcium is 10 g Ca/1,000,000 g water. To convert this concentration to mole fraction terms, the number of moles of Ca and of water need to be calculated as follows

$$\text{gmol Ca} = 10\,\text{g Ca}/(40\,\text{g/gmol}) = 0.25\,\text{gmol Ca}$$

$$\text{gmol water }\ 1{,}000{,}000\,\text{g water}/(18\,\text{g/gmol}) = 55{,}555\,\text{gmol water}$$

The mole fraction of this ratio of Ca to water is then determined as follows

$$\text{mole fraction}, y_{Ca} = (0.25\,\text{gmol Ca})/(55{,}555\,\text{gmol water}) = 4.5\times10^{-6}$$

Illustrative Example 3.6

Convert the calcium aqueous concentration of 10 ppm on a mass basis to units of ppm on a mole basis, ppm_m.

Solution

Refer to the solution for Illustrative Example 3.5. Because the mole fraction of Ca in this aqueous solution, y_{Ca}, is 4.5×10^{-6}, one way to express this concentration on a mole basis is:

$$\text{ppm Ca on a mole basis} = 10^6\ y_{Ca} = 10^6(4.5\times10^{-6}) = 4.5\,\text{ppm}_m$$

REFERENCES

Jones, J., and G. Hawkens. 1960. *Engineering Thermodynamics*. Hoboken, NJ: John Wiley & Sons.

Keenan, J., and F. Keyes. 1930. *Thermodynamic Properties of Steam*. Hoboken, NJ: John Wiley & Sons.

Theodore, L. 2016. *Chemical Engineering: The Essential Reference*. New York: McGraw-Hill.

Theodore, L., R. Dupont, and K. Ganesan. 2017. *Unit Operations in Environmental Engineering.* Salem, MA: Scrivener-Wiley.

Theodore, L., F. Ricci, and T. VanVliet. 2007. *Thermodynamics for the Practicing Engineer.* Hoboken, NJ: John Wiley & Sons.

Van Wylen, G., and R. Sonntag. 1965. *Fundamentals of Classical Thermodynamics.* Hoboken, NJ: John Wiley & Sons.

4 Water Chemistry

4.1 INTRODUCTION

Water chemistry deals with the fundamental chemical properties of water itself, the chemical properties of other constituents that dissolve in water, and the countless chemical reactions that take place in water. The field of natural water chemistry is concerned principally with reactions that occur in relatively dilute solution (low concentrations), although some natural waters have rather high solute concentrations (Hem 1970). During a chemical reaction, tiny subatomic particles (e.g., electrons) and atoms (e.g., hydrogen) are transferred, shared, and exchanged. When a chemical reaction occurs in water, these changes require transport through the water medium. Water is not passive in these chemical reactions. Instead, it plays an active role, constantly making and breaking chemical bonds, thereby facilitating chemical change.

This chapter introduces the reader to these water chemistry principles including the chemical properties of water, the chemical composition of natural waters, chemical reactions, and an application section that provides six Illustrative Examples related to the general subject of water chemistry.

4.2 CHEMICAL PROPERTIES OF WATER

Water is called the "universal solvent" because it is capable of dissolving many substances. This chemical properly of water arises from the dipolar nature of water molecules. Water molecules effectively surround positively charged ions (cations) and negatively charged ions (anions), which serve to prevent them from precipitating as solid. This means that wherever water goes, either through the ground or through one's body, it carries with it various solutes such as dissolved minerals, nutrients, organics, and heavy metals.

Even pure water contains some amount of hydrogen ion (H^+) and hydroxide ion (OH^-) as a result of a chemical reaction known as the autoionization of water. The concentration of these ions change as acids and bases are added to water; however, the product of their concentrations is always constant. The relative presence of H^+ and OH^- is measured by the pH, which is defined as the negative logarithm of the H^+ concentration in units of mol/L. A decrease in pH indicates an increase in the H^+ concentration and a corresponding increase in the OH^- concentration.

Pure water has a neutral pH of exactly 7.0. Values of pH less than this are considered acidic, whereas pH values greater than this are termed alkaline. The typical pH range of water is from 0 to 14, although values outside of this range can be attained under extreme conditions. The pH of water is an important chemical property that controls the distribution of chemical species among various forms. For example, the pH is buffered at precise values in animal cells to maintain functionality of specific enzymes and proteins. Likewise, in the environment, the pH controls the distribution of chemicals amongst their various forms and also controls the rate at which many chemical reactions occur. It is therefore often necessary to precisely measure the pH of water to understand its water chemistry.

4.3 CHEMICAL COMPOSITION OF NATURAL WATERS

The composition of natural waters is often described according to its physical qualities, chemical constituents, or its biological inhabitants. The focus to follow is primarily concerned with chemical constituents. Water sampling programs (see Chapter 8) are used to obtain information on the chemical characteristics of potential and existing water sources and the performance of water and wastewater treatment plants (MWH 2005). Typically, prior knowledge of the type of chemical constituent (i.e., organic versus inorganic, dissolved versus suspended) is required to design and implement effective sampling programs. Preservatives are often added to prevent degradation of certain constituents, and holding times have been recommended by the Environmental Protection Agency (EPA) and other agencies so as to maintain proper quality control (Clesceri et al. 1998).

4.3.1 Dissolved Minerals

Soil water in the ground reacts with common rock-forming minerals to release ions and form new minerals. Table 4.1 lists the most commonly occurring chemical elements present in the Earth's crust. The most abundant group of minerals in crustal rock is a family called the *feldspars*. These minerals are comprised of sodium, potassium, and calcium aluminum silicates. They react with water, thereby producing Na^+, K^+, Ca^{2+}, Al^{3+}, and H_4SiO_4. Magnesium (Mg^{2+}) and iron (Fe^{2+}) are released from other silicate minerals. Carbonate rocks, limestone ($CaCO_3$(s)), and dolomite ((Ca, Mg)CO_3 (s)) weather to release Ca^{2+}, Mg^{2+}, and HCO_3^-. Phosphorous (PO_4^{3-}) is released by the chemical weathering of apatite, a calcium phosphate mineral. The soluble constituents described, Na^+, K^+, Ca^{2+}, Mg^{2+}, HCO_3^-, and H_4SiO_4, along with Cl^- and SO_4^{2-} find their way to river and streams and eventually reach the ocean. Over time, the concentrations of these constituents have increased to the levels found in the oceans today.

Aluminum (Al) and iron (Fe), mobilized from chemical weathering processes, have low solubility, and are not transported over large distances. Under aerobic conditions, iron either stays behind as a hydroxide or oxide coating on the surface of weathered rocks, or attaches to small particles that remain suspended and are carried with the flowing water. Aluminum precipitates as $Al(OH)_3$(s) or reacts with H_4SiO_4 to form the mineral kaolinite and other clay minerals.

Hardness is a bulk chemical property that measures the presence of specific dissolved mineral ions. Calcium and magnesium dissolved in water are the two most common minerals that make water "hard." Hard water interferes with washing clothes, dishwashing, and bathing. Clothes laundered in hard water may look dingy and feel harsh or scratchy. Hard water causes a film on

TABLE 4.1
Mineral Composition of the Earth's Crust

Element	%	Element	%
O (oxygen)	46.6	Cl (chlorine)	0.1
Si (silicon)	27.7	Cr (chromium)	0.04
Al (aluminum)	8.1	C (carbon)	0.03
Fe (iron)	5.0	V (vanadium)	0.02
Ca (calcium)	3.6	Ni (nickel)	0.008
Na (sodium)	2.8	Cu (copper)	0.007
K (potassium)	2.6	Co (cobalt)	0.002
Mg (magnesium)	2.1	Pb (lead)	0.001
TI (titanium)	0.6	Sc (scandium)	0.0005
Mn (manganese)	0.1	Zn (zinc)	0.0001

Source: Lutgens, F.K. and E.J. Tarbuck, E.J., *Essentials of Geology*, 7th edition, Prentice Hall, Upper Saddle River, NJ, 2000; Nicholls, D., *Complexes and First-Row Transition Elements*, Elsevier, New York, 1975.

surfaces when it evaporates because of the presence of a thin layer of salt that remains. Water flow may be reduced by mineral deposits in pipes. Synthetic detergents are usually less effective in hard water because the active ingredient is partially inactivated by the high levels of calcium and magnesium.

Another bulk measurement of the dissolved ion content of water is total dissolved solids (TDS). TDS is a measure of all of the dissolved ions in solution and is determined by filtering out any suspended material in the water, evaporating off the water, and weighing the dry residue that remains. TDS levels indicate the potential uses for a water body because the TDS is basically an indicator of the salt content. Freshwater has a TDS of less than 1000 mg/L. Surface waters with significantly high TDS may not be usable for potable water or irrigation purposes. The TDS of sea water is approximately 35,000 mg/L. This means that for every kilogram of seawater there are approximately 35 g of dissolved salt.

Alkalinity is a measure of the buffering capacity of water. It does not refer to pH but instead refers to the ability of water to resist change in pH upon addition of acid or base. Waters with low alkalinity are very susceptible to changes in pH, whereas waters with high alkalinity are able to resist major shifts in pH. The buffering chemicals in most natural waters are bicarbonate (HCO_3^-) and carbonate (CO_3^-), although borates, silicates, phosphates, ammonium, sulfides, and organic acids can also contribute to a small degree. Water having a pH below 4.5 contains virtually no alkalinity because all of the bicarbonate and carbonate have been converted to dissolved carbon dioxide. The amount of alkalinity therefore determines the ability of a water body to neutralize acidic pollution from rainfall or wastewater. Most surface waters typically have alkalinity ranging from 10 to 200 mg/L as $CaCO_3$ (Hem 1970).

4.3.2 Dissolved Gases

The Earth's atmosphere is comprised of a layer of gases that are retained by gravity. It contains roughly 78% nitrogen ($N_2(g)$), 21% oxygen ($O_2(g)$), 0.93% argon ($Ar(g)$), 0.038% carbon dioxide ($CO_2(g)$), trace amounts of other gases, and a variable amount (about 1% on average) of water vapor. The atmosphere protects life on Earth by filtering out harmful ultraviolet solar radiation and by trapping infrared radiation (heat) from escaping which regulates surface temperatures at habitable levels.

All gases present in the Earth's atmosphere dissolve to some extent into water that is in contact with it. Thus, all surface water has small amounts of N_2, O_2, Ar, CO_2, and other gases dissolved in it. For aquatic life forms, the presence of this small amount of oxygen is essential for survival. Most species of fish, for example, require at least 5 mg/L of dissolved oxygen. CO_2 acts as a weak acid when dissolved in water thereby imparting rain water with its characteristic slightly acidic pH. N_2 and Ar, although present in all waters, do not engage in chemical reactions to any significant extent and their presence is usually ignored.

4.3.3 Heavy Metals

The term "heavy metals" is an ambiguous one and not necessarily associated with any specific set of elements and, therefore, does not imply any common set of properties (such as high toxicity, high atomic weight, etc.) (Lutgens and Tarbuck 2000). Nonetheless, the term has been used more and more in the literature. A simple, but useful, definition of a heavy metal is any metallic element on the periodic table with an atomic number larger than that of calcium (atomic number = 20). Examples of heavy metals include titanium (Ti), manganese (Mn), chromium (Cr), vanadium (V), copper (Cu), cobalt (Co), lead (Pb), scandium (Sc), and zinc (Zn).

Heavy metals are natural components of the Earth's crust (see Table 4.1). Unlike organic chemicals, they cannot be degraded or destroyed. To a small extent they enter one's body via food, drinking water, and air. Some heavy metals (e.g., copper, selenium, zinc, etc.) are micronutrients and are essential to maintain the metabolism of the human body. However, at higher concentrations they can cause adverse health effects. Some metals that are in low abundance can have a large environmental impact. Mercury (Hg), arsenic (As), selenium (Se), and silver (Ag), for example, are all widely considered to be environmental stressors. Small amounts of these compounds can be harmful to both human health and aquatic life.

Heavy metals can enter a water supply by industrial and consumer waste, from drinking water contamination (e.g., lead and copper pipes), or from acidic rain breaking down soils and leaching heavy metals into streams, lakes, rivers, and groundwater. Additional routes of exposure include inhalation of high ambient air concentrations near emission sources or ingestion of metals via the food chain (Duffus 2002).

4.3.4 Organic Constituents

Organic chemistry is the study of the properties of chemical compounds containing carbon. Nearly all chemical compounds that have carbon atoms are considered organic. The exceptions, termed inorganic carbon compounds, include carbon dioxide, bicarbonate and carbonate ions, cyanide, metal carbides, and a handful of other compounds. On the other hand, the number of organic carbon compounds is impossibly large to count, a quality that arises from the fact that carbon readily bonds with other carbon atoms. This creates countless ways in which carbon atoms can be arranged relative to one another. In many organic molecules, carbon is bound to hydrogen. In addition, carbon atoms are also able to bind with other elements such as oxygen, nitrogen, phosphorous, sulfur, and chlorine, thereby further increasing the possible number of distinct and different organic carbon constituents that may be present in water.

Organic chemicals of environmental interest are usually classified into various groups or categories based on similar chemical properties or common origins. The most basic distinction is made based on whether an organic chemical is naturally occurring or synthetic. Naturally occurring organic molecules include fossil fuels such as methane (CH_4) gas and the complex mixture of compounds present in petroleum, sugars, and starches, and biomolecules such as proteins and enzymes. The G, T, C, and A base pairs of DNA are organic, thereby making DNA an organic molecule.

When plant, animal, and microbial material in soil and water undergo decomposition, a variety of complex organic molecules are produced that are called *natural organic matter* (NOM). Although NOM is ubiquitous in the environment, the structure and properties of the molecules themselves are not well understood. NOM play an important role in aquatic toxicology because it interacts with metal ions and minerals to form complexes of a widely differing chemical and biological nature (Tipping 2002). When NOM bind with metal ions, they become less bioavailable, which lowers the potential toxicity to aquatic life. However, NOM creates problems for the water supply industry, requiring removal to minimize water color and giving rise to potentially harmful chemical by-products after chlorination. Through a process called "biofouling," NOM can also degrade the performance of membrane filtration systems used for water purification and desalination.

Synthetic organic chemistry is the science of the design, analysis, or construction of organic chemicals for practical purposes. As such, synthetic organic chemicals (SOCs) are chemicals that are produced on a large scale for use by humans. SOCs include several subclasses of chemicals such a pesticides, industrial solvents, chelating agents, and disinfection by-products. Many of these compounds are highly toxic and tested for routinely in public water supply systems.

A pesticide is any substance or mixture of substances designed to prevent, destroy, repel, or mitigating any pest. Insecticides, herbicides, and fungicides all fall under the pesticide umbrella. Organophosphate pesticides such as malathion and parathion and carbamate pesticides such a aldicarb and methomyl affect the nervous system of insects by disrupting enzymes that regulate neurotransmitters. Organochlorine insecticides such a DDT and chlordane were common in the past, but many have been banned from use in the United States and many other countries because of their health and environmental effects as well as their persistence. Pyrethroids are a class of pesticides developed as a synthetic version of the naturally occurring pesticide pyrethrin.

Organic solvents are a chemical class of compounds that are used routinely in commercial industries for dissolving other organic compounds. They have found extensive use in dry cleaning (e.g., tetrachloroethylene or PCE), as paint thinner (e.g., toluene, turpentine), as nail polish remover and glue solvents (e.g., acetone), in spot removers (e.g., hexane), in detergents (e.g., limonene), and in perfumes (e.g., ethanol). They are also particularly useful in the synthesis of other organic chemicals. Many organic solvents are recognized by the EPA as known or suspected human carcinogens. Volatile organic compounds (VOCs) are another class of organic chemical compounds that are characterized as having high enough vapor pressures to significantly vaporize and enter the atmosphere. Many of the organic solvents discussed previously are also VOCs.

Chelating agents, or chelators for short, are a class of SOCs that are used in chemical analysis as water softeners and are ingredients in many commercial products such as shampoos and food preservatives. The most commonly used synthetic chelating agents are NTA and EDTA (Grundler et al. 2004). Because of its inability to be broken down in many wastewater treatment plants, significant concentrations of EDTA have been found in the environment. Long-term accumulation of chelating agents is not a concern, however, because they are eventually broken down by bacteria (Grundler et al. 2004). Their presence in surface waters is more of a concern because of their ability to solubilize heavy metals, thereby making them mobile.

Disinfection by-products (DBPs) are a class of chemical compounds that are formed when water is disinfected. Disinfection of drinking water with chlorine has been applied since the 1900s and has prevented the spread of waterborne diseases such as cholera and typhoid. However, during the 1970s, scientists discovered that chlorination of drinking water containing moderate to high levels of NOM produced a new class of compounds, DBPs, which were later shown to be harmful to human health. Alternative disinfectants such as ozone or chlorine dioxide produce their own characteristic DBPs. Thus, switching from chlorine to ozone or chlorine dioxide is not an exhaustive remedy.

Pharmaceuticals and personal care product (PPCPs) refer to any product used by individuals for personal health or cosmetic reasons or used by agriculture to enhance growth or health of livestock. PPCPs comprise a diverse collection of thousands of individual chemicals, including prescription and over-the-counter therapeutic drugs, veterinary drugs, fragrances, cosmetics, sunscreen products, diagnostic agents, and vitamins. We all contribute PPCPs to the environment through excretion, bathing, and disposal of medication to sewers and trash. The various sources of PPCPs include human activity (e.g., bathing, shaving, swimming, etc.), illicit drug use, veterinary drug use-especially antibiotics and steroids, agriculture, and residue from pharmaceutical manufacturing and hospitals. Studies have shown that PPCPs are present in the nation's water bodies. To date, scientists have found no evidence of adverse human health effects from PPCPs in the environment. However, there is strong evidence of ecological harm. PPCPs that can affect the endocrine system in animals, which controls important function through communication of glands, hormones, and cellular receptors, are known as *endocrine-disrupting compounds* (EDCs). Many EDCs are associated with developmental, reproductive, and other health problems in fish and wildlife, both in the field and the laboratory.

4.3.5 Nutrients

Nutrients are chemical elements critical to the growth of plant and animal life. In healthy rivers and lakes, nutrients are needed for the growth of algae that form the base of a complex food web that supports the entire aquatic ecosystem. The nutrients that receive the most attention in lakes and streams are nitrogen (as nitrate and ammonia) and phosphorus (as orthophosphate or total phosphate).

If provided with an abundance of nutrients, algae and aquatic plants will continue to grow well beyond the amount needed to support the food web. The excess algae and plants will die. and consume dissolved oxygen as microorganisms break down their cellular material. As a result, other aquatic organisms may suffer from lack of oxygen. Other problems associated with excessive algal and plant growth include scum and foam formation and odor and taste problems if the water is used for drinking.

Eutrophication is the natural process of enrichment of lakes and streams with nutrients, and the associated biological and physical changes that result. Human activity has dramatically increased the rate of eutrophication in many water bodies. Lakes and ponds are particularly vulnerable to eutrophication because the nutrients carried into them continue to build up; in contrast, the nutrients present in rivers and streams can be carried away in moving water.

Phosphorus in the form of phosphates is contributed naturally from soil and dissolution of rocks, while natural sources of nitrogen include leaves and other organic debris from riparian vegetation. The primary anthropogenic sources of these nutrients are wastewater treatment plants, septic systems, stormwater runoff, suspended sediment resulting from excessive erosion (only P), acid rain (only N), animal manure, and commercial fertilizers. In the past, household detergents brought high loads of phosphorus to treatment plants, which then were discharged with the effluent. In the United States, however, laws restricting the phosphorus content of detergents have produced markedly reduced phosphate levels in many surface water bodies.

4.4 CHEMICAL REACTIONS

Chemistry is the science of making new substances out of old substances via chemical reactions. All of the chemical constituents described undergo chemical reactions that result in their being degraded, as is the case for many of the organic chemicals, or transformed to another form, as is the case for many of the inorganic chemicals. When studying a chemical reaction, chemist often poses some basic questions. What is the driving force that makes this reaction occur? What is the equilibrium state of this chemical system? These questions are answered with the help of a subject known as *chemical thermodynamics*. Why is this reaction so fast? Why is that reaction so slow? These questions are answered with the help of a subject known a *chemical kinetics*.

In the same way that water will always find its own level, chemical reactions proceed in a way that minimizes the useful energy that is available. Chemical thermodynamic calculations quantify the change in this energy (known as the Gibbs energy) as a reaction proceeds. This allows for one to determine the equilibrium state of this chemical system. The calculations are often relatively simple, and there are many commercially available computer software programs that automate the task. The results from equilibrium calculations are often a reasonable approximation for many systems and even if the system is not at equilibrium they provide information about the direction and extent to which reactions will proceed (Benjamin 2002; Theodore et al. 2007).

The subject of chemical kinetics allows one to quantify how fast chemical reactions occur and answer why certain reactions are faster than others (Stumm and Morgan 1996; Reynolds et al. 2004). Chemical kinetics is often quantified through the measurement of the rate of change in concentrations of reactant or products. The most important factors that influence rates of chemical reactions are the nature and concentration of the reactant(s).

Increasing the temperature of a system imparts more kinetic energy to molecules, thereby serving to increase rates of chemical reactions. The detailed explanation of how a reaction proceeds at a molecular level is called a reaction mechanism. Determination of reaction mechanisms requires a broad and detailed understanding of the properties of reactants and products and the changes that occur before, during, and after a chemical reaction and is often difficult if not impossible to confirm unequivocally.

4.5 WATER pH

An important chemical property of an aqueous solution is its pH. The pH measures the acidity or basicity of the solution. In a neutral solution, such as pure water, the hydrogen (H^+), and hydroxyl (OH^-) ion concentrations are equal. At ordinary temperatures, this concentration is

$$C_{H^+} = C_{OH^-} = 10^{-7} \text{ g-ion/L} \tag{4.1}$$

where C_{H^+} is the hydrogen ion concentration, and C_{OH^-} is the hydroxyl ion concentration.

The unit g-ion represents an Avogadro number of ions. In all aqueous solutions, whether neutral, basic, or acidic, a chemical equilibrium or balance is established between these two concentrations, so that

$$K_{eq} = C_{H^+} + C_{OH^-} = 10^{-14} \tag{4.2}$$

where K_{eq} is the equilibrium constant.

The numerical value for K_{eq} holds for room temperature and only when the concentrations are expressed in g-ion/L. In acid solutions, C_{H^+} is > C_{OH^-}; in basic solutions, C_{OH^-} predominates.

The pH is a direct measure of the hydrogen ion concentration and is defined as

$$pH = -\log(C_{H^+}) \tag{4.3}$$

Thus, an acidic solution is characterized by a pH below 7 (the lower the pH, the greater the acidity); a basic solution has a pH above 7; and a neutral solution, possesses a pH of 7. For example, for a pH of 1.0,

$$pH = 1.0 = -\log(C_{H^+}) \tag{4.4}$$

$$C_{H^+} = 10^{-1} = 0.1\,\text{g-ion/L} \tag{4.5}$$

$$C_{OH^-} = \frac{10^{-14}}{C_{H^+}} = 10^{-13}\,\text{g-ion/L} \tag{4.6}$$

It should be pointed out that these equations employed are not the exact definition of pH but are a close approximation of it. Strictly speaking, the *activity* of the hydrogen ion, α_{H^+}, and to the ion concentration, C_{H^+}, belongs in the equations (Theodore et al. 2007).

4.6 APPLICATIONS

Six Illustrative Examples complement the material presented in this chapter.

Illustrative Example 4.1

Describe the role of nutrients in natural waters.

Solution

Nutrients are chemical elements critical to the growth of plant and animal life. In healthy rivers and lakes, nutrients are needed for the growth of algae, which form the base of a complex food web that supports the aquatic ecosystem.

Illustrative Example 4.2

Describe the neutrality aspect of water in terms of pH.

Solution

Pure water has a neutral pH of exactly 7.0. Values of pH less than this are considered acidic, whereas pH values greater than this are termed alkaline. The typical pH range of water is from

0 to 14, although values outside of this range can be attained under extreme conditions. The pH of water is an important chemical property, which controls the distribution of chemical species among various forms.

Illustrative Example 4.3

Calculate the hydrogen ion and hydroxyl ion concentration of an aqueous solution if the pH of the solution is 1, 3, 5, 7, 8, 10, 12, and 14.

Solution

Regarding the problem statement, it was shown in Equations 4.4 through 4.6, that at pH 1.0 the $C_{H^+} = 10^{-1} = 0.1\,\text{g-ion/L}$, and that the $C_{OH^-} = \frac{10^{-14}}{C_{H^+}} = 10^{-13}\,\text{g-ion/L}$. The remaining results are calculated in a similar fashion with the results presented in Table 4.2.

Illustrative Example 4.4

An industrial wastewater has an alkalinity of 60 mg/L and a CO_2 content of 7.0 mg/L. Determine the pH of this wastewater. The first ionization constant of H_2CO_3 is

$$K_{i,1} = \frac{(H^+)(HCO_3^-)}{(H_2CO_3)} = 3.98\times10^{-7} \tag{4.7}$$

Solution

"Alkalinity" is measured as equivalents of $CaCO_3$, which has a molecular weight of 100 and an equivalent weight of 50. For example, a water solution of 17 mg/L of OH^- (equivalent weight = 17) contains 10^{-3} equivalent weights of OH^-, which is equivalent to 10^{-3} equivalents of $CaCO_3$. The alkalinity of that solution is therefore (50) (10^{-3}) g/L or 50 mg/L.

For an alkalinity of 60 mg/L, the concentration of HCO_3 is

$$(HCO_3^-) = \frac{60\text{ mg/L}}{50\text{ meq/L}} = 1.2\,\text{meq/L} = 1.2\text{ mmol/L} \tag{4.8}$$

The unit "meq" represents 10^{-3} equivalents. For the CO_2,

$$(CO_2) = \frac{7\text{ mg/L}}{44\text{ mg/mmol}} = 0.159\text{ mmol/L} \tag{4.9}$$

TABLE 4.2
Hydrogen and Hydroxyl Ion Concentration at Various pH Values

pH	C_{H^+}, g-ion/L	C_{OH^-}, g-ion/L
1	10^{-1}	10^{-13}
3	10^{-3}	10^{-11}
5	10^{-5}	10^{-9}
7	10^{-7}	10^{-7}
8	10^{-8}	10^{-6}
10	10^{-10}	10^{-4}
12	10^{-12}	10^{-2}
14	10^{-14}	$10^{-0} = 1.0$

Note that $(CO_2) = (H_2CO_3)$. Substitution in Equation (4.7) gives

$$3.98\times10^{-7}=\frac{(H^+)(1.2\times10^{-3})}{(0.159\times10^{-3})}=5.27\times10^{-8} \tag{4.10}$$

The pH is therefore given by

$$pH=-\log\left(H^+\right)=-\log(5.27\times10^{-8})=7.28 \tag{4.11}$$

Illustrative Example 4.5

An elementary school science teacher found an old bottle containing 2 L of concentrated sulfuric acid in the science room storage closet. Not having any use for the acid, the teacher considered disposing of it by pouring it down the drain with the faucet running. The principal happened to stop by and informed the teacher that federal regulations required that any material put down the drain must have a pH greater than 2.0 and less than 12.5 to protect the aquatic environment and the bacteria in the local wastewater treatment plant—not to mention the pipe in the sewerage system. The teacher considered three methods of bringing the acid to a pH of 2.1 to comply with the regulations and to dispose of the acid safely and quickly. These three methods included: dilution, neutralization to Na_2SO_4 and H_2O with NaOH, and neutralization to $CaSO_4$ and H_2O with slaked lime, $Ca(OH)_2$. Answer the following four questions.

1. Find the pH of the concentrated H_2SO_4.
2. Find the volume to which the 2 L of the 18 M H_2SO_4 would have to be diluted in order to raise the mixture pH to 2.1.
3. Calculate the quantity of NaOH needed to neutralize the 2 L of 18 M H_2SO_4.
4. Calculate the quantity of $Ca(OH)_2$ needed to neutralize the 2 L of 18 M H_2SO_4.

Solution

With the strong acid ionizing 100%, the acid dissociates completely as follows:

$$H_2SO_4 \rightarrow 2H^+ + SO_4^- \tag{4.12}$$

1. Therefore, 18 gmol/L H_2SO_4 would ionize to yield 36 gmol/L H^+ to yield a pH of:

$$pH=-\log\left[H^+\right]=-\log(36)=-1.6 \tag{4.13}$$

Note: A negative pH is possible at very high acid concentrations.

2. Next, the H^+ concentration in a sulfuric acid solution of pH = 2.1 can be calculated as:

$$\left[H^+\right]=10^{-pH}=10^{-2.1}=7.94\times10^{-3} \tag{4.14}$$

The molarity of this solution is $7.94 \times 10^{-3}/2 = 3.97 \times 10^{-3}$. Thus, to raise the pH of the 2 L of concentrated sulfuric acid from −1.6 to 2.1, the molarity of the solution must be lowered from 18 M to approximately 0.004 M. The volume of the diluted acid can be calculated from a mass balance for hydrogen ion, that is, $M_1V_1 = M_2V_2$:

$$18\,M(2\,L)=0.004\,M\left(V_2\right);\ V_2=9{,}000\,L$$

Note: Because 9,000 L equals approximately 2,250 gallons of solution this is clearly not a practical method of disposal.

3. The number of moles of NaOH required to neutralize the sulfuric acid is based on the number of moles of sulfuric acid in the 18 M solution and the stoichiometric relationship describing this neutralization reaction, or 2 L (18 gmol/L) = 36 gmol to be neutralized according to the stoichiometry:

$$H_2SO_4 + 2\,NaOH \rightarrow Na_2SO_4 + 2\,H_2O \tag{4.15}$$

This requires 2 gmol of NaOH for each gmol of sulfuric acid neutralized for a total of 72 gmol of NaOH for complete neutralization. Converting gmol to pounds of NaOH required yields:

$$\frac{(72\,\text{gmol NaOH})(40\text{ g/gmol})}{454\text{ g/lb}} = 6.3\text{ lb NaOH}$$

Note: As a laboratory scale reaction, this would be a very large reaction, generating a significant amount of heat with considerable chance of splattering and injury.

4. The number of moles of $Ca(OH)_2$ required to neutralize the sulfuric acid is again based on the number of moles of sulfuric acid in the 18 M solution and the stoichiometric relationship describing this neutralization reaction, or 2 L (18 gmol/L) = 36 gmol to be neutralized according to the stoichiometry:

$$H_2SO_4 + Ca(OH)_2 \rightarrow CaSO_4 + 2\,H_2O \tag{4.16}$$

This requires 1 gmol of $Ca(OH)_2$ for each gmol of sulfuric acid neutralized for a total of 36 gmol of $Ca(OH)_2$ for complete neutralization. Converting gmol to pounds of $Ca(OH)_2$ required yields:

$$\frac{(36\text{ gmol }Ca(OH)_2)(74\text{ g/gmol})}{454\text{ g/lb}} = 5.9\text{ lb }Ca(OH)_2$$

Note: Again, as a laboratory scale reaction, this would still be a very large reaction, generating a significant amount of heat with considerable chance of splattering and injury.

Illustrative Example 4.6

A biologically contaminated hospital liquid waste requires disinfection with chlorine prior to discharge into a nearby lake. Given a contact time of 30 minutes, a required chlorine dose for a 99.99% kill of the pathogenic organisms = 3.0 mg/L, and a 0.5 mg/L chlorine requirement to carry out oxidation-reduction reactions with compounds in the liquid waste, how many pounds of pure chlorine (in the gaseous form) would be required daily to disinfect a 100,000 gal/d waste flow?

Solution

The total chlorine concentration required to meet the oxidation-reduction reaction requirement in the water is 0.5 mg/L, while 3.0 mg/L chorine are required for the disinfection reaction. The total chlorine dose necessary then is 3.5 mg/L. The required chlorine dose (D) in terms of lb/d is found by employing appropriate conversion factors as follows:

$$Q,\text{ MGD} = (100{,}000\text{ gal/d})/(1\text{ MG}/10^6\text{ gal}) = 0.10\text{ MGD} \tag{4.17}$$

$$D = (3.5\text{ mg/L})(0.10\text{ MG})(8.34\text{ lb/MG/mg/L}) = 2.94\text{ lb/d}$$

REFERENCES

Benjamin, M.M. 2002. *Water Chemistry.* New York: McGraw-Hill.

Clesceri, L.S., A.E. Greenberg, and A.D. Eaton. 1998. *Standard Methods for the Examination of Water and Wastewater*, 20th edition. American Public Health Association Publications, Washington, DC.

Duffus, J.H. (2002) Heavy Metal-A Meaningless Term? *Pure and Applied Chemistry*, 74, 793–807. doi:10.1351/pac200274050793.

Grundler, O., J. Hans-Ulrich, and H. Witteler. 2004. Environmental impact of amniocarboxylate chelating agents. In *Handbook of Detergents, Part B: Environmental Impact (Surfactant Science)*. Ed. U. Zoller. CRC Press, Boca Raton, FL.

Hem, J.D. 1970. *Study and Interpretation of the Chemical Characteristics of Natural Water*, U.S. Geological Society, Water Supply Paper 2254. Washington, DC.

Lutgens, F.K., and E.J. Tarbuck. 2000. *Essentials of Geology*, 7th edition. Upper Saddle River, NJ: Prentice Hall.

MWH. 2005. *Water Treatment: Principles and Design*. Hoboken, NJ: John Wiley & Sons.

Nicholls, D. 1975. *Complexes and First-Row Transition Elements*. New York: Elsevier.

Reynolds, J., J. Jeris, and L. Theodore. 2004. *Handbook of Chemical and Environmental Engineering Calculations*. Hoboken, NJ: John Wiley & Sons.

Stumm, W., and J.J. Morgan. 1996. *Aquatic Chemistry: Chemical Equilibria and Rates in Natural Waters*. Hoboken, NJ: John Wiley & Sons.

Theodore, L., F. Ricci and T. Van Vliet. 2007. *Thermodynamics for the Practicing Engineer*. Hoboken, NJ: John Wiley & Sons.

Tipping, E. 2002. *Cation Binding by Humic Substances*. Cambridge, UK: Cambridge University Press.

5 Environmental Regulatory Framework

5.1 INTRODUCTION

A detailed discussion of the U.S. regulatory system and the difference between laws and regulations is first presented. The chapter is highlighted with the various laws that exist for controlling water pollution in the United States. The U.S. Environmental Protection Agency (EPA) was officially born December 1970. In many ways, the EPA is the most far-reaching regulatory agency in the federal government because its authority is so broad. The EPA is charged with the protection the nation's land, air, and water systems (Burke et al. 2000). Under a mandate of national environmental laws, the EPA strives to formulate and implement actions that lead to a compatible balance between human activities and the ability of natural systems to support and nurture life (U.S. EPA 1988a).

The EPA works with the states and local governments to develop and implement comprehensive environmental programs. Federal laws such as the Clean Air Act, the Safe Drinking Water Act, the Resource Conservation and Recovery Act, and the Comprehensive Environmental Response, Compensation, and Liability Act all mandate involvement by state and local government in the details of implementation. This chapter provides an overview of eight key environmental laws and subsequent regulations that affect the environment in the United States. The applications section provides six Illustrative Examples related to the general subject of environmental regulations.

5.2 THE REGULATORY SYSTEM

Over the past four plus decades environmental regulation has become a system in which laws, regulations, and guidelines have become interrelated. The history and development of this regulatory system has led to laws that focus principally on only one environmental medium (i.e., air, water, or land). Some environmental managers feel that more needs to be done to manage all of the media simultaneously. Hopefully, the environmental regulatory system will evolve into a truly integrated, multimedia management framework in the future.

Federal laws are the product of Congress. Regulations written to implement the law are promulgated by the executive branch of government, but until judicial decisions are made regarding the interpretations of the regulations, there may be uncertainty about what regulations mean in real situations. Until recently, environmental protection groups were most frequently the plaintiffs in cases brought to court seeking interpretation of the law. Recently, industry has become more active in this role.

Enforcement approaches for environmental regulations are environmental management oriented in that they seek to remedy environmental harm and not simply a specific infraction of a given regulation. All laws in a legal system may be used in enforcement to prevent damage or threats of damage to the environment or human health and safety. Tax laws (e.g., tax incentives) and business regulatory laws (e.g., product claims, liability disclosure) are examples of laws not directly focused on environmental protection but that may also be used to encourage compliance and discourage non-compliance with environmental regulations.

Common law also plays an important role in environmental management. Common law is the set of rules and principles relating to the government and security of persons and property. Common law authority is derived from the usages and customs that are recognized and enforced by the courts. In general, no infraction of the law is necessary when establishing a common law court action. A common law "civil wrong" (e.g., environmental pollution) that is brought to court is called a *tort*. Environmental torts may arise because of nuisance, trespass, or negligence.

Laws tend to be general and contain uncertainties relative to the implementation of principles and concepts they contain. Regulations derived from laws may be more specific but are also frequently too broad to allow clear translation into environmental technology practice. Permits may be used to bridge this gap and prescribe specific technical requirements concerning the discharge of pollutants or other activities carried out by a facility that may impact the environment.

Most major federal environmental laws provide for citizen law suits. This empowers individuals to seek compliance or monetary penalties when these laws are violated and regulatory agencies do not take enforcement action against the violator.

5.3 LAWS AND REGULATIONS: THE DIFFERENCES

The following is a listing of some of the major differences between a federal law and a federal regulation, as briefly described in the previous section:

1. A law (or act) is passed by both houses of Congress and signed by the president. A regulation is issued by a government agency such as the EPA or the Occupational Safety and Health Administration (OSHA).
2. Congress can pass a law on any subject it chooses. It is only limited by the restrictions in the Constitution. A law can be challenged in court only if it violates the Constitution. It may not be challenged if it is merely unwise, unreasonable or even silly. If, for example, a law was passed that placed a tax on sneezes, it could not be challenged in court just because it was unenforceable. A regulation can be issued by an agency only if the agency is authorized to do so by the law passed by Congress. When Congress passes a law, it usually assigns an administrative agency to implement that law. A law regarding radio stations, for example, may be assigned to the Federal Communications Commission (FCC). Sometimes a new agency is established to implement a law. This was the case with the Consumer Product Safety Commission (CPSC). OSHA is authorized by the Occupational Safety and Health Act to issue regulations that protect workers from exposure to the hazardous chemicals they use in manufacturing processes.
3. Laws include a congressional mandate directing EPA to develop a comprehensive set of regulations. Regulations, or rulemakings, are issued by an agency, such as EPA, that translate the general mandate of a statute into a set of requirements for the agency and the regulated community.
4. Regulations are developed by EPA in an open and public manner according to an established process. When a regulation is formally proposed, it is published in an official government document called the *Federal Register* (Figure 5.1) to notify the public of EPA's intent to create new regulations or modify existing ones. EPA provides the public, which includes the potentially regulated community, with an opportunity to submit comments. Following an established comment period, EPA may revise the proposed rule based on both an internal review process and public comments.
5. The final regulation is published, or promulgated, in the *Federal Register*. Included with the regulation is a discussion of the agency's rationale for the regulatory approach, known as preamble language. Final regulations are compiled annually and incorporated in the Code of Federal Regulations (CFR) according to a highly structured format based on the topic of the regulation. This latter process is called *codification*, and each CFR title corresponds to a different regulatory authority. For example, EPA's regulations are in Title 40 of the CFR. The codified RCRA regulations can be found in Title 40 of the CFR, Parts 240 through 282. These regulations are often cited as 40 CFR, with the part listed afterward (e.g., 40 CFR Part 264), or the part and section (e.g., 40 CFR §264.10).

FIGURE 5.1 *Federal register* cover, November 29, 1996.

6. A regulation may be challenged in court on the basis that the issuing agency exceeded the mandate given to it by Congress. If the law requires the agency to consider costs versus benefits of their regulation, the regulation could be challenged in court on the basis that the cost-benefit analysis was not correctly or adequately done. If OSHA issues a regulation limiting a worker's exposure to a hazardous chemical to 1 part per million (ppm), OSHA could be called on to prove in court that such a low limit was needed to prevent a worker from being harmed. Failure to prove this would mean that OSHA exceeded its mandate under the law, as OSHA is charged to develop standards only as stringent as those required to protect worker health and provide worker safety.
7. Laws are usually brief and general. Regulations are usually lengthy and detailed. The Hazardous Materials Transportation Act, for example, is approximately 20 pages long. It speaks in general terms about the need to protect the public from the dangers associated with transporting hazardous chemicals and identifies the Department of Transportation (DOT) as the agency responsible for issuing regulations implementing the law. The regulations issued by the DOT are several thousand pages long and are very detailed, down to

the exact size, shape, design, and color of the warning placards that must be used on trucks carrying any of the thousands of regulated chemicals.

8. Generally, laws are passes infrequently. Often years pass between amendments to an existing law. A completely new law on a given subject already addressed by an existing law is unusual. Laws are published as a "Public Law#_-_" and are eventually codified into the United States Code of Federal Regulations.
9. Regulations are issued and amended frequently. Proposed and final new regulations and amendments to existing regulations are published daily in the *Federal Register.* Final regulations have the force of law when published.

5.4 THE ROLE OF THE STATES

The Resource Conservation and Recovery Act (RCRA), for example, like most federal environmental legislation, encourages states to develop and run their own hazardous waste programs as an alternative to EPA management. Thus, in a given state, the hazardous waste regulatory program may be run by the EPA or by a state agency. For a state to have jurisdiction over its environmental management programs (i.e., state primacy), it must receive approval from the EPA by showing that its program is at least as stringent as the EPA program would be.

States that are authorized to operate the RCRA program for example, oversee the hazardous waste tracking system in their state, operate the permitting system for hazardous waste facilities, and act as the enforcement arm in cases where an individual or a company practices illegal hazardous waste management. If needed, the EPA steps in to assist the states in enforcing the law. The EPA can also act directly to enforce RCRA or other laws in states that do not yet have authorized programs. The EPA and the states currently act jointly to implement and enforce environmental regulations (U.S. EPA 1986).

5.5 THE RESOURCE CONSERVATION AND RECOVERY ACT (RCRA)

Defining what constitutes a "hazardous waste" requires consideration of both legal and scientific factors. The basic definitions used here are derived from: The Resource Conservation and Recovery Act (RCRA) of 1976, as amended in 1978, 1980, and 1986; the Hazardous and Solid Waste Amendments (HSWA) of 1984; and, the Comprehensive Environmental Response, Compensation and Liability Act (CERCLA) of 1980, as amended by the Superfund Amendments and Reauthorization Act (SARA) of 1986. Within these statutory authorities a distinction exists between a hazardous waste and a hazardous substance. The former is regulated under RCRA, whereas the latter is regulated under the Superfund program (Section 5.7).

Hazardous waste refers to "a solid waste, or combination of solid wastes, which because of its quantity, concentration, or physical, chemical or infectious characteristics may [pose a] substantial present or potential hazard to human health or the environment when improperly ... managed" [RCRA, Section 1004(5)]. Under RCRA regulations, a waste is considered hazardous if it is reactive, ignitable, corrosive, or toxic or if the waste is listed as a hazardous waste in Title 40 Parts 261.31-33 of the Code of Federal Regulations.

In addition to hazardous wastes defined under RCRA, there are "hazardous substances" defined by Superfund. Superfund's definition of a hazardous substance is broad and grows out of the lists of hazardous wastes or substances regulated under the Clean Water Act (CWA), the Clean Air Act (CAA), the Toxic Substances Control Act (TSCA), and RCRA. Essentially, Superfund considers a hazardous substance to be any hazardous substance or toxic pollutant identified under the CWA and applicable regulations, any hazardous air pollutant listed under the CAA and applicable regulations, any imminently hazardous chemical for which a civil action has been brought under TSCA, and any hazardous waste identified or listed under RCRA and applicable regulations.

The RCRA of 1976 completely replaced the previous language of the Solid Waste Disposal Act of 1965 to address the enormous growth in the production of waste. The objectives of this act were to promote the protection of health and the environment and to conserve valuable materials and energy resources by (Bureau of National Affairs 1976; Cheremisinoff and Ellerbusch 1979):

1. Providing technical and financial assistance to state and local governments and interstate agencies for the development of solid waste management plans (including resource recovery and resource conservation systems) that promote improved solid waste management techniques (including more effective organizational arrangements), new and improved methods of collection, separation, and recovery of solid waste, and the environmentally safe disposal of nonrecoverable residues.
2. Providing training grants in occupations involving the design, operation, and maintenance of solid waste disposal systems.
3. Prohibiting future open dumping on the land and requiring the conversion of existing open dumps to facilities that do not pose danger to the environment or to health.
4. Regulating the treatment, storage, transportation, and disposal of hazardous waste that have adverse effects on health and the environment.
5. Providing for the promulgation of guidelines for solid waste collection, transport, separation, recovery, and disposal practices and systems.
6. Promoting a national research and development program for improved solid waste management and resource conservation techniques; more effective organization arrangements; and, new and improved methods of collection, separation, recovery, and recycling of solid wastes and environmentally safe disposal of nonrecoverable residues.
7. Promoting the demonstration, construction, and application of solid waste management, resource recovery, and resource conservation systems that preserve and enhance the quality of air, water, and land resources.
8. Establishing a cooperative effort among federal, state, and local governments and private enterprises to recover valuable materials and energy from solid waste.

Structurewise, RCRA is divided into eight subtitles. These subtitles are (A) General Provisions; (B) Office of Solid Waste; Authorities of the Administrator; (C) Hazardous Waste Management; (D) State or Regional Solid Waste Plans; (E) Duties of Secretary of Commerce in Resource and Recovery; (F) Federal Responsibilities; (G) Miscellaneous Provisions; and, (H) Research, Development, Demonstration, and Information. Subtitles C and D generate the framework for regulatory control programs for the management of hazardous and solid nonhazardous wastes, respectively. The hazardous waste program outlined under Subtitle C is the one most people associate with the RCRA (Bureau of National Affairs 1976).

5.6 MAJOR TOXIC CHEMICAL LAWS ADMINISTERED BY THE U.S. EPA

People have long recognized that sulfuric acid, arsenic compounds, and other chemical substances can cause fires, explosions, or poisoning. More recently, researchers have determined that many chemical substances such as benzene and a number of chlorinated hydrocarbons may cause cancer, birth defects, and other long-term health effects. Today, the hazards of new substances, including genetically engineered microorganisms and nanoparticles are being evaluated. The EPA has a number of legislative tools, summarized in Table 5.1, to use in controlling the risks from toxic substance.

The Federal Insecticide, Fungicide, and Rodenticide Act of 1972 (FIFRA) encompasses all pesticides used in the United States. When first enacted in 1947, FIFRA was administered by the

TABLE 5.1
Major Toxic Chemical Laws Administered by the EPA

Statue	Provisions
Toxic Substances Control Act	Requires that the EPA be notified of any new chemical prior to its manufacture and authorizes EPA to regulate production, use, or disposal of a chemical.
Federal Insecticide, Fungicide, and Rodenticide Act	Authorizes the EPA to register all pesticides and specify the terms and conditions of their use, and remove unreasonably hazardous pesticides from the marketplace
Federal Food, Drug, and Cosmetic Act	Authorizes the EPA in cooperation with the FDA to establish tolerance levels for pesticide residues on food and food producers.
Resource Conservation and Recovery Act	Authorizes the EPA to identify hazardous wastes and regulate their generation, transportation, treatment, storage, and disposal.
Comprehensive Environmental Response, Compensation, and Liability Act (CERCLA)	Requires the EPA to designate hazardous substances that can present substantial danger and authorizes the cleanup of sites contaminated with such substances.
Clean Air Act	Authorizes the EPA to set emission standards to limit the release of hazardous air pollutants.
Clean Water Act	Requires the EPA to establish a list of toxic water pollutants and set standards.
Safe Drinking Water Act	Requires the EPA to set drinking water standards to protect public health from hazardous substances.
Marine Protection, Research, and Sanctuaries Act	Regulates ocean dumping of toxic contaminants.
Asbestos School Hazard Act	Authorizes the EPA to provide loans and grants to schools with financial need for abatement of severe asbestos hazards.
Asbestos Hazard Emergency Response Act	Requires the EPA to establish a comprehensive regulatory framework for controlling asbestos hazards in schools.
Emergency Planning and Community Right-to-Know Act	Requires states to develop programs for responding to hazardous chemical releases and requires industries to report on the presence and release of certain hazardous substances.

Abbreviations: EPA, Environmental Protection Agency; FDA, Food and Drug Administration.

U.S. Department of Agriculture and was intended to protect consumers against fraudulent pesticide products. When many pesticides were registered, their potential for causing health and environmental problems was unknown. In 1970, the EPA assumed responsibility for FIFRA, which was amended in 1972 to shift emphasis to health and environmental protections. Allowable levels of pesticides in food are specified under the authority of the federal Food, Drug, and Cosmetic Act of 1954. Today, FIFRA contains registration and labeling requirements for pesticide products. The EPA must approve any use of a pesticide, and manufacturers must clearly state the conditions of that use on the pesticide label. Some pesticides are listed as hazardous wastes and are subject to RCRA rules when discarded.

The Toxic Substances Control Act (TSCA) authorizes EPA to control the risks that may be posed by the thousands of commercial chemical substances and mixtures (chemicals) that are not regulated as either drugs, food additives, cosmetics, or pesticides. Under TSCA, the EPA can, among other things, regulate the manufacture and use of a chemical substance and require testing for cancer and other effects. TSCA regulates the production and distribution of new chemicals and governs the manufacture, processing, distribution, and use of existing chemicals. Among the chemicals controlled by TSCA regulations are PCBs, chloroflurocarbons, and asbestos. In specific cases, there is an interface with RCRA regulations. For example, PCB disposal is generally regulated by TSCA. However, hazardous wastes mixed with PCBs are regulated under RCRA.

Under both TSCA and FIFRA, the EPA is responsible for regulating certain biotechnology products, such as genetically engineered microorganisms designed to control pests or assist in industrial processes.

The Clean Air Act (CAA), in Section 112, lists 189 hazardous air pollutants. The CAA also requires emission standards for many types of air emission sources, including RCRA regulated incinerators and industrial boilers or furnaces.

The Clean Water Act (CWA) lists substances to be regulated by effluent limitations in 21 primary industries. The CWA substances are incorporated into both RCRA and CERCLA. In addition, the CWA regulates discharges from publicly owned treatment works (POTWs) to surface waters, and indirect discharges to municipal wastewater treatment systems (through a pretreatment program). Some hazardous wastewaters which would generally be considered RCRA-regulated wastes are covered under the CWA because of the use of treatment tanks and a National Pollutant Discharge Elimination System (NPDES) permit to legally dispose of the wastewaters. Sludges from these tanks, however, are subject to RCRA regulations when they are removed.

The Safe Drinking Water Act (SDWA) regulates underground injection systems, including deep-well injection systems. Prior to underground injection, a permit must be obtained which imposes conditions that must be met to prevent the endangerment of underground sources of drinking water.

The Marine Protection, Research, and Sanctuaries Act of 1972 has regulated the transportation of any material for ocean disposal and prevents the disposal of any material in oceans that could affect the marine environment. Amendments enacted in 1988 were designed to end ocean disposal of sewage sludge, industrial waste, and medical wastes.

5.6.1 The Superfund Amendments and Reauthorization Act (SARA) of 1986

The 1986 amendments to the Comprehensive Environmental Response, Compensation, and Liability Act (CERCLA), known as the Superfund Amendments and Reauthorization Act (SARA), authorized \$8.5 billion for both the emergency response and longer term (or remedial) cleanup programs. The Superfund amendments focused on:

1. Permanent Remedies. The EPA must implement permanent remedies to the maximum extent practicable. A range of treatment options are considered whenever practicable.
2. Complying with Other Regulations. Applicable or relevant and appropriate standards from other federal, state, or tribal environmental laws must be met at Superfund sites where remedial actions are taken. In addition, state standards that are more stringent than federal standards must be met in cleaning up sites.
3. Alternative Treatment Technologies. Cost-effective treatment and recycling must be considered as an alternative to the land disposal of wastes. Under RCRA, Congress banned land disposal of some wastes. Many Superfund site wastes, therefore, are banned from disposal on the land; alternative treatments are under development and will be used where possible.
4. Public Involvement. Citizens living near Superfund sites are involved in the site decision-making process. They are also able to apply for technical assistance grants that further enhance their understanding of site conditions and activities.
5. State Involvement. States and tribes are encouraged to participate actively as partners with EPA in addressing Superfund sites. They assist in making the decisions at sites, can take responsibility in managing cleanups, and can play an important role in oversight of responsible parties.
6. Enforcement Authorities. Settlement policies were strengthened through congressional approval and inclusion in SARA. Different settlement tools, such as *de minimis* settlements (settlements with minor contributors) are part of the act.

7. Federal Facility Compliance. Congress emphasized that federal facilities "are subject to, and must comply with, this Act in the same manner and to the same extent … as any non-government entity." Mandatory schedules have been established for federal facilities to assess their sites, and if listed on the National Priority List (NPL), to clean up such sites. EPA assists and over-seeing federal agencies with these requirements.

These amendments also expanded research and development, especially in the area of alternative technologies. They also provided for more training for state and federal personnel in emergency preparedness, disaster response, and hazard mitigation.

A significant development in the control of hazardous materials in communities was the establishment of Title III of SARA (also known as Emergency Planning and Community Right-to-Know Act or EPCRA). EPCRA requires states to develop programs for responding to hazardous chemical releases and requires industries to report on the presence and release of certain hazardous substances. Major provisions of Title Ill of EPCRA include:

1. Emergency Planning. EPCRA establishes a broad-based framework at the state and local levels to compile chemical information and use that information in communities for chemical emergency planning.
2. Emergency Release Notification. EPCRA requires facilities to report certain releases of extremely hazardous chemicals and hazardous substances to their state and local emergency planning and response officials.
3. Hazardous Chemical Inventory Reporting. EPCRA requires facilities to maintain a safety data sheet (SDS) for any hazardous chemicals stored or used in the work place and to submit those sheets to state and local authorities. It also requires them to submit an annual inventory report for those same chemicals to local emergency planning and fire protection officials, as well as state officials.
4. Toxic Release Inventory Reporting. EPCRA requires facilities to annually report on routine emissions of certain toxic chemicals to the air, land, or water. Facilities must report if they are in Standard Industrial Classification (SIC) Codes 20 through 39 (i.e., manufacturing facilities) with 10 or more employees and manufacture or process any of 650 listed chemical compounds in amount greater than specified threshold quantities. If the chemical compounds are considered persistent, bioaccumulative, or toxic, the thresholds are much lower. EPA is required to use these data to establish a national chemical release inventory database, making the information electronically available to the public.

5.6.2 The Clean Air Act (CAA)

The Clean Air Act defines the national policy for air pollution abatement and control in the United States. It establishes goals for protecting health and natural resources and delineates what is expected of federal, state, and local governments to achieve those goals. The Clean Air Act, which was initially enacted as the Air Pollution Control Act of 1955, has undergone several revisions over the years to meet the ever-changing needs and conditions of the nation's air quality. On November 15, 1990, the president signed the most recent amendments to the Clean Air Act, referred to as the 1990 Clean Air Act Amendments. Embodied in these amendments were several progressive and creative new themes deemed appropriate for effectively achieving the air quality goals and for reforming the air quality control regulatory process. Specifically, the amendments:

1. Encouraged the use of market-based principles and other innovative approaches similar to performance-based standards and emission banking and trading.
2. Promoted the use of clean low-sulfur coal and natural gas, as well as innovative technologies to clean high-sulfur coal through the acid rain program.

3. Reduced energy waste and created enough of a market for clean fuels derived from grain and natural gas to cut dependency on oil imports by one million barrels per day.
4. Promoted energy conservation through an acid rain program that gave utilities flexibility to obtain needed emission reductions through programs that encouraged customers to conserve energy.

Several of the key provisions of the act are reviewed next.

5.6.2.1 Provisions for Attainment and Maintenance of National Ambient Air Quality Standards (NAAQS)

Although the Clean Air Act brought about significant improvements in the nation's air quality, the urban air pollution problems of ozone (smog), carbon monoxide (CO), and particulate matter persist. In 1995, approximately 70 million U.S. residents were living in counties with ozone levels exceeding the EPA's current ozone standard.

The Clean Air Act, as amended in 1990, established a more balanced strategy for the nation to address the problem of urban smog. Overall, the amendments revealed the Congress's high expectations of the states and the federal government. Although it gave states more time to meet the air quality standard (up to 20 years for ozone in Los Angeles), it also required states to make constant progress in reducing emissions. It required the federal government to reduce emissions from cars, trucks, and buses; from consumer products such as hair spray and window-washing compounds; and from ships and barges during loading and unloading of petroleum products. The federal government also developed the technical guidance that states need to control stationary sources.

Specifically, the Clean Air Act clarifies how areas are designated and re-designated "attainment." It also allows the EPA to define the boundaries of "nonattainment" areas (i.e., geographical areas whose air quality does not meet federal ambient air quality standards designed to protect public health). The law also establishes provisions defining when and how the federal government can impose sanctions on areas of the country that have not met certain conditions. For the pollutant ozone, the Clean Air Act established nonattainment area classifications ranked according to the severity of the area's air pollution problem. These classifications are marginal, moderate, serious, severe, and extreme. The EPA assigns each nonattainment areas one of these categories, thus triggering varying requirements the area must comply with to meet the ozone standard.

As mentioned, nonattainment areas have to implement different control measures, depending on their classification. Marginal areas, for example, are the closest to meeting the standard. They are required to conduct an inventory of their ozone-causing emissions and institute a permit program. Nonattainment areas with more serious air quality problems must implement various control measures. The worse the air quality, the more controls these areas will have to implement.

The Clean Air Act also established similar programs for areas that do not meet the federal health standards for carbon monoxide and particulate matter. Areas exceeding the standards for these pollutants are divided into "moderate" and "serious" classifications. Depending upon the degree to which they exceed the carbon monoxide standard, areas are then required to implement programs such as introducing oxygenated fuels or enhanced emission inspection programs, among other measures. Depending on their classification, areas exceeding the particulate matter standard have to implement reasonably available control measures (RACM) or best available control measures (BACM), among other requirements.

5.6.2.2 Provisions Relating to Mobile Sources

Although motor vehicles built today emit fewer pollutants (60%–80% less, depending on the pollutant) than those built in the 1960s, cars and trucks still account for almost half the emissions of the ozone precursors (volatile organic carbons [VOCs] and nitrogen oxides [NOx]), and up to 90% of the CO emissions in urban areas. The principal reason for this problem is the rapid growth in the number of vehicles on the roadways and the total miles driven. This growth has offset a large portion of the emission reductions gained from motor vehicle controls.

In view of the continuing growth in automobile emissions in urban areas, combined with the serious air pollution problems in many urban areas, Congress made significant changes to the motor vehicle provisions of the Clean Air Act and established tighter pollution standards for emissions from automobiles and trucks. These standards were set so as to reduce tailpipe emissions of hydrocarbons, carbon monoxide, and nitrogen oxides on a phased-in basis beginning in model year 1994. Automobile manufacturers also were required to reduce vehicle emissions resulting from the evaporation of gasoline during refueling.

Fuel quality was also controlled. Scheduled reductions in gasoline volatility and sulfur content of diesel fuel, for example, were required. Programs requiring cleaner (so-called "reformulated") gasoline were initiated in 1995 for the nine cities with the worst ozone problems. Higher levels (2.7%) of alcohol-based oxygenated fuels were produced and sold in those areas that exceed the federal standard for carbon monoxide during the winter months.

The 1990 amendments to the Clean Air Act also established a clean fuel car pilot program in California, requiring the phase-in of tighter emission limits for 150,000 vehicles in model year 1996 and 300,000 by the model year 1999. These standards were to be met with any combination of vehicle technology and cleaner fuels. The standards became even more strict in 2001. Other states were able to "opt in" to this program, through incentives and not sales or production mandates.

5.6.2.3 Air Toxics

Toxic air pollutants are those pollutants which are hazardous to human health or the environment. These pollutants are typically carcinogens, mutagens, and reproductive toxins. The toxic air pollution problem is widespread. Information generated in 1987 from the Superfund "Right to Know" rule (SARA Section 313) discussed earlier indicated that more than 2.7 billion pounds of toxic air pollutants were emitted annually in the United States in that year. The EPA studies indicated that exposure to such quantities of toxic air pollutants may result in 1000–3000 cancer deaths each year.

Section 112 of the Clean Air Act includes a list of 189 substances that are identified as hazardous air pollutants (HAPs). A list of categories of sources that emit these pollutants was prepared [The list of source categories included (i) major sources, or sources emitting 10 tons per year of any single HAPs or 25 tons per year of all HAPs; and (ii) area sources (smaller sources, such as dry cleaners and auto body refinishing)]. In turn, EPA promulgated emission standards, referred to as maximum achievable control technology (MACT) standards, for each listed source category. These standards were based on the best demonstrated control technology or practices used by sources that make up each source category. Within 8 years of promulgation of a MACT standard, EPA must evaluate the level of risk that remains (residual risk) because of exposure to emissions from a source category and determine if the residual risk is acceptable. If the residual risks are determined to be unacceptable, additional standards are required.

5.6.2.4 Acid Deposition Control

Acid rain occurs when sulfur dioxide (SO_2) and nitrogen oxide emissions are transformed in the atmosphere and return to the earth in rain, fog, or snow. Approximately 20 million tons of sulfur dioxide are emitted annually in the United States, mostly from the burning of fossil fuels by electric utilities. Acid rain damages lakes, harms forests and buildings, contributes to reduced visibility, and is suspected of damaging health.

It was hoped that the Clean Air Act would bring about a permanent 10 million-ton reduction in SO_2 emissions from 1980 levels. To achieve this, the EPA allocated allowances in two phases, permitting utilities to emit 1 ton of sulfur dioxide. The first phase, which became effective January 1,

1995, required 110 power plants to reduce their emissions to a level equivalent to the product of an emissions rate of 2.5 lb of SO_2/MM Btu × an average of their 1985–1987 fuel use. Emissions data indicate that 1995 SO_2 emissions at these units nationwide were reduced by almost 40% below the required level.

The second phase, which became effective January 1, 2000, required approximately 2,000 utilities to reduce their emissions to a level equivalent to the product of an emissions rate of 1.2 lb of SO_2/MM Btu × the average of their 1985–1987 fuel use. In both phases, affected sources were required to install systems that continuously monitor emissions to track progress and assure compliance.

The Clean Air Act allowed utilities to trade allowances within their systems or buy or sell allowances to and from other affected sources. Each source must have had sufficient allowances to cover its annual emissions. If not, the source was subject to a $2,000/ton excess emissions fee and a requirement to offset the excess emissions in the following year.

The Clean Air Act also included specific requirements for reducing emissions of nitrogen oxides.

5.6.2.5 Operating Permits

The act requires the implementation of an operating permits program modeled after the National Pollutant Discharge Elimination System (NPDES) of the Clean Water Act (to be discussed in the next chapter). The purpose of the operating permits program is to ensure compliance with all applicable requirements of the Clean Air Act. Air pollution sources, subject to the program, must obtain an operating permit; states must develop and implement an operating permit program consistent with the act's requirements; and the EPA must issue permit program regulations, review each state's proposed program, and oversee the state's effort to implement any approved program. The EPA must also develop and implement a federal permit program when a state fails to adopt and implement its own program.

In many ways this program is the most important procedural reform contained in the 1990 Amendments to the Clean Air Act. It enhanced air quality control in a variety of ways and updated the Clean Air Act, making it more consistent with other environmental statutes. The Clean Water Act, the RCRA, and the FIFRA all require operating permits.

5.6.2.6 Stratospheric Ozone Protection

The Clean Air Act requires the phase out of substances that deplete the ozone layer. The law required a complete phase-out of chlorofluorocarbons (CFCs) and halons, with stringent interim reductions on a schedule similar to that specified in the Montreal Protocol, including CFCs, halons, and carbon tetrachloride by 2000 and methyl chloroform by 2002. All Class II chemicals (hydrochlorofluorocarbons [HCFCs]) will be phased out by 2030.

The law required nonessential products releasing Class I chemicals to the banned. This ban went into effect for aerosols and non-insulating foams using Class II chemicals in 1994. Exemptions were included for chemicals controlling flammability in products and for safety concerns.

5.6.2.7 Provisions Relating to Enforcement

The Clean Air Act contains provisions for a broad array of authorities to make the law readily enforceable. EPA has authorities to:

1. Issue administrative penalty orders up to $200,000 and field citations up to $5000.
2. Obtain civil judicial penalties.
3. Secure criminal penalties for knowing violations and for knowing and negligent endangerment.
4. Require sources to certify compliance.
5. Issue administrative subpoenas for compliance data.
6. Issue compliance orders with compliance schedules of up to 1 year.

Citizen suit provisions are also included to allow citizens to seek penalties against violators, with penalties going to a U.S. Treasury fund for use by the EPA for compliance and enforcement activities.

The following EPA actions represent examples of recent regulations promulgated to implement the requirements of the Clean Air Act:

1. Clean Air Interstate Rule published on May 12, 2005 (70 FR 25161) amends requirements for State Implementation Plans (SIPs) and for the provisions for the Acid Rain Program.
2. Mercury Rules published on May 18, 2005 (70 FR 28605) amends New Source Performance Standards for electric utility steam generating units and some provisions of the Acid Rain Program.
3. Non-road Diesel Rule published on May 11, 2004 (69 FR 38957) amends provisions for mobile sources and for highway vehicles and engines.
4. Ozone Rules promulgated on October 1, 2015 (83 FR 62998), identified those areas that are designated as not attaining the ambient air quality standards for ozone.
5. Fine Particle Rules published on July 6, 2011 (76 FR 48208) took the form of the Cross-State Air Pollution Rule (CSAPR) to address air pollution from upwind states that crosses state lines and affects the formation of fine particle pollution and Regional Haze.

5.6.3 The Occupational Safety and Health Act (OSHA)

The Occupational Safety and Health Act (OSHAct) was enacted by Congress in 1970 and established the Occupational Safety and Health Administration (OSHA), which addressed safety in the workplace. At the same time, the EPA was established. Both EPA and OSHA are mandated to reduce the exposure of hazardous substances to land, water, and air. The OSHAct is limited to conditions that exist in the workplace, where its jurisdiction covers both safety and health. Frequently, both OSHA and EPA regulate the same substances but in a different manner as they are overlapping environmental organizations.

Congress intended that OSHA be enforced through specific standards in an effort to achieve a safe and healthful working environment. A "general duty clause" was added to the OSHAct to attempt to cover those obvious situations that were admitted by all concerned but for which no specific standard existed. The OSHA standards are an extensive compilation of regulations, some that apply to all employers (such as eye and face protection) and some that apply to workers who are engaged in a specific type of work (such as welding or crane operation). Employers are obligated to familiarize themselves with the standards and comply with them at all times.

Health issues, most importantly, contaminants in the workplace, have become OSHA's primary concern. Health hazards are complex and difficult to define. Because of this, OSHA has been slow to implement health standards. To be complete, each standard requires medical surveillance, record keeping, monitoring, and physical reviews. On the other side of the ledger, safety hazards are aspects of the work environment that are expected to cause death or serious physical harm immediately or before the imminence of such danger can be eliminated.

Probably one of the most important safety and health standards ever adopted is the OSHA hazard communication standard, more properly known as the "right to know" laws. The hazard communication standard requires employers to communicate information to the employees on hazardous chemicals that exist within the workplace. The program requires employers to craft a written hazard communication program, keep *safety data sheets (SDSs)* for all hazardous chemicals at the workplace and provide employees with training on those hazardous chemicals, and assure that proper warning labels are in place (Theodore and Theodore 1995).

5.6.4 USEPA's Risk Management Program (RMP)

Developed under the Clean Air Act's Section 112(r), the *Risk Management Program (RMP)* rule (40 CFR Part 68) is designed to reduce the risk of accidental releases of acutely toxic, flammable, and explosive substances. A list of the regulated substances (138 chemicals) along with their threshold quantities is provided in the Code of Federal Regulations at 40 CFR 68.130.

In the RMP rule, EPA requires a *Risk Management Plan* that summarizes how a facility is to comply with EPA's RMP requirements. It details methods and results of hazard assessment, accident prevention, and emergency response programs instituted at the facility. The hazard assessment shows the area surrounding the facility and the population potentially affected by accidental releases. EPA requirements include a three-tiered approach for affected facilities. A facility is affected if a process unit manufactures, processes, uses, stores, or otherwise handles any of the listed chemicals at or above the threshold quantities.

The EPA approach is summarized in Table 5.2. For example, EPA defined Program 1 facilities as those processes that have not had an accidental release with offsite consequences in the 5 years prior to the submission date of the RMP and have no public receptors within the distance to a specified toxic or flammable endpoint associated with a worst-case release scenario. Program 1 facilities have to develop and submit a RMP and complete a registration that includes all processes that have a regulated substance present in more than a threshold quantity. They also have to analyze the worst-case release scenario for the process or processes; document that the nearest public receptor is beyond the distance to a toxic or flammable endpoint; complete a 5-year accident history for the process or processes; ensure that response actions are coordinated with local emergency planning and response agencies; and certify that the source's worst-case release would not reach the nearest public receptors.

Program 2 applies to facilities that are not Program 1 or Program 3 facilities. Program 2 facilities have to develop and submit the RMP as required for Program 1 facilities plus develop and implement a management system; conduct a hazard assessment; implement certain prevention steps; develop and implement an emergency response program; and submit data on prevention program elements for Program 2 processes.

Program 3 applies to processes in Standard Industrial Classification (SIC) codes 2611 (pulp mills), 2812 (chloralkali), 2819 (industrial inorganics), 2821 (plastics and resins), 2865 (cyclic crudes), 2869 (industrial organics), 2873 (nitrogen fertilizers), 2879 (agricultural chemicals), and 2911 (petroleum refineries). These facilities belong to industrial categories identified by EPA as historically accounting for most industrial accidents resulting in off-site risk. Program 3 also applies to all processes subject to the OSHA Process Safety Management (PSM) standard (29 CFR 1910.119).

TABLE 5.2
EPA's RMP Approach

Program	Description
1	Facilities submit RMP, complete registration of processes, analyze worst-case release scenario, complete 5-year accident history, coordinate with local emergency planning and response agencies; and certify that the source's worst-case release would not reach the nearest public receptors.
2	Facilities submit RMP, complete registration of processes, develop and implement a management system; conduct a hazard assessment; implement certain prevention steps; develop and implement an emergency response program; and submit data on prevention program elements.
3	Facilities submit RMP, complete registration of processes, develop and implement a management system; conduct a hazard assessment; implement prevention requirements; develop and implement an emergency response program; and provide data on prevention program elements.

Abbreviations: EPA, Environmental Protection Agency; RMP, Risk Management Program.

Program 3 facilities have to develop and submit the RMP as required for Program 1 facilities plus: develop and implement a management system; conduct a hazard assessment; implement prevention requirements; develop and implement an emergency response program; and provide data on prevention program elements for the Program 3 processes.

5.6.5 The Pollution Prevention Act (PPA) of 1990

The Pollution Prevention Act (PPA) represents a clear breakthrough in this nation's understanding of environmental problems. The PPA calls pollution prevention a "national objective" and establishes a hierarchy of environmental protection priorities as national policy.

Under the PPA, it is the national policy of the United States that pollution should be prevented or reduced at the source whenever feasible; where pollution cannot be prevented, it should be recycled in an environmentally safe manner. In the absence of feasible prevention and recycling opportunities, pollution should be treated; and disposal should be used only as a last resort.

Among other provisions, the Act directed the EPA to facilitate the adoption of source reduction techniques by businesses and federal agencies, to establish standard methods of measurement for source reduction, to review regulations to determine their effect on source reduction, and to investigate opportunities to use federal procurement to encourage source reduction. The act initially authorized an $8 million state grant program to promote source reduction, with a 50% state match requirement (U.S. EPA 1991).

The EPA's pollution prevention initiatives are characterized by its use of a wide range of tools, including market incentives, public education and information, small business grants, technical assistance, research and technology applications, as well as the more traditional regulations and enforcement to encourage and support pollution prevention initiatives across all sectors of the U.S. economy. In addition, there are other significant behind-the-scenes achievements: identifying and dismantling barriers to pollution prevention; laying the groundwork for a systematic prevention focus; and creating advocates for pollution prevention that serve as catalysts in a wide variety of settings. Additional details of the PPA are provided in Chapter 20.

5.7 LEGISLATIVE TOOLS FOR CONTROLLING WATER POLLUTION

Congress has provided the EPA and the states with three primary statutes to control and reduce water pollution: the Clean Water Act; the Safe Drinking Water Act; and the Marine Protection, Research, and Sanctuaries Act. Each statute provides a variety of tools that can be used to meet the challenges and complexities of reducing water pollution in the nation (U.S. EPA 1988b). Details of the Clean Water Act and the Safe Drinking Water Act are provided in the next two chapters.

5.8 APPLICATIONS

Six Illustrative Examples complement the material presented in this chapter.

Illustrative Example 5.1

List and discuss various "concentration" terms employed by OSHA.

Solution

OSHA uses the following concentration terms in its regulatory programs:

1. LC_{50}. Lethal Concentration 50%. This is similar to LD_{50} except that the route of entry is inhalation. The concentrations of the inhaled chemicals (usually gases) are expressed as parts per million (ppm) or milligrams per cubic meter (mg/m^3).
2. *LDLo*. Lethal Dose Low. The lowest dose that killed any animals in a chemical dose study when administered by a route of entry other than inhalation.

3. *LCLo*. Lethal Concentration Low. Same as *LDLo* except that the route of entry is inhalation.
4. *TDLo*. Toxic Dose Low. The lowest dose used in a chemical dose study that caused any toxic effect (not just death) when administered by a route of entry other than inhalation.
5. *TCLo*. Toxic Concentration Low. Same as *TDLo* except that the route of entry is inhalation.
6. EC_{50}. This is the median effective concentration calculated to affect 50% of a test population during continuous exposure over a specified period of time.

Illustrative Example 5.2

Some wastewater and water standards and regulations are based on a term defined as *parts per million, ppm, or parts per billion, ppb*. Define the two major classes of these terms and describe the interrelationship from a calculational point of view. Also convert 5.0 parts calcium per million parts of water on a mass basis to parts per million on a mole-basis.

Solution

Water streams seldom consist of a single component. It may also contain two or more phases (a dissolved gas or suspended solids), or a mixture of one or more solutes. For mixtures of substances, it is convenient to express compositions in mass fractions or mole fraction terms. The following definitions are often used to represent the composition of component. A in a mixture of components:

$$w_A = \frac{\text{Mass of A}}{\text{Total Mass of Water Stream}} = \text{Mass Fraction of A} \tag{5.1}$$

$$y_A = \frac{\text{Mass of A}}{\text{Total Mass of Water Stream}} = \text{Mass Fraction of A} \tag{5.2}$$

Trace quantities of substances in water streams are often expressed in parts per million by weight (ppm_w) or as parts per billion (ppb_w) on a mass basis. These concentrations can also be provided on a mass per volume basis for liquids and on a mass per mass basis for solids. Gas concentrations are usually represented on a mole or volume basis (e.g., ppm_m or ppm_v, respectively). The following equations apply:

$$ppm_w = 10^6 w_A = 10^3 ppb_w \tag{5.3}$$

$$ppm_m = 10^6 y_A = 10^3 ppb_m \tag{5.4}$$

The two terms ppm_w and ppm_m are related through the molecular weight. To convert pm_w of Ca to ppm_m, select a basis of 10^6 g of solution. The mass fraction of Ca is first obtained by the following equation:

$$\text{Mass of Ca} = 5\text{ g; Moles of Ca} = 5\text{ g}/\left(40\text{ g/gmol}\right) = 0.125\text{ gmol}$$

$$\text{Moles of Water} = 10^6\text{g}/\left(18\text{ g/gmol}\right) = 55{,}555\text{ gmol}$$

$$y_{Ca} = \frac{\text{Moles of Ca}}{\text{Total Moles of Water Stream}} = \frac{0.125\text{ gmol}}{55{,}555\text{ gmol}} = 2.25 \times 10^{-6} \tag{5.5}$$

$$ppm_m \text{ of Ca} = 10^6 y_{Ca} = 10^6\left(2.25\times10^{-6}\right) = 2.25\text{ ppm}_m \tag{5.6}$$

Illustrative Example 5.3

Express the regulatory concentration values for the solutions given in terms of percentage by weight, ppm_w, and molarity.

36 g of HCl in 64 cm^3 of water
0.003 g of ethanol in 1 kg of water
34 g of ammonia in 2000 g of water

Note that molarity is defined as the moles of solute per volume of solutions.

Solution

These calculations are left as an exercise for the reader. The concentration conversions for hydrochloric acid, HCl, ethanol, and ammonia are given in Table 5.3.

Illustrative Example 5.4

Discuss the impact of the Clean Air Act on wastewater treatment plants.

Solution

A wide range of residential, commercial, and industrial dischargers contribute VOCs and toxic pollutants to publicly owned treatment works (POTWs). An even wider range of pollutants is potentially discharged to industrial wastewater treatment plants, depending on the specific type of industrial activity generating the wastewater. Limited information on air emissions of VOCs and air toxics from industrial wastewater treatment plants and POTWs is currently available. However, more extensive information on treatment plant wastewater influent quality is more readily available in the literature, particularly for POTWs, because of the monitoring requirements for all wastewater treatment plants under the National Pollutant Discharge Elimination System (NPDES) permit program. In addition to those pollutants commonly present in the influent of POTWs, by-products of various wastewater treatment processes considered to be VOCs or toxic air pollutants can be potentially emitted from POTWs. For example, chloroform can be formed as a by-product of wastewater chlorination.

The provisions of the Clean Air Act dealing with VOCs for reducing urban smog and the control of air toxics significantly affect the water pollution control field in the areas of water quality and wastewater treatment. Although the Clean Air Act does not specifically require that industrial or municipal wastewater treatment plants control VOCs and air toxic emissions, federal, state, and local air quality laws and regulations developed as a result of the Clean Air Act requirements, focus on stationary sources of ozone precursors. State air pollution control agencies in ozone nonattainment areas, particularly those classified as extreme and severe, may require that large wastewater treatment plants install air pollution control devices to limit emissions of VOCs from their treatment units.

In addition, it is possible that some of the large POTWs may fall into the category of a major source of hazardous air pollutants (HAPs) (annual potential to emit >10 T of a single HAP, or >25 T of all HAPs) if all 189 HAPs are included in the calculations. In Los Angeles, one of two extreme

TABLE 5.3
Concentration Conversions for Illustrative Example 5.3

	% Weight	ppm_v	Molarity
HCl	36	562,500	15.4
Ethanol	3×10^{-4}	3	6.5×10^{-5}
Ammonia	1.67	17,000	1.0

ozone nonattainment areas, the local air quality agency adopted Rule 1401, which regulates emissions of known or suspected carcinogenic air toxics based on a human cancer risk assessment. Local POTWs are covered under this rule.

Illustrative Example 5.5

The following total suspended solids (TSS) data were collected from a clarifier at a local municipal wastewater treatment plant over a 7-day period (see Table 5.4). The NPDES permit limitations for TSS effluent concentrations from this wastewater treatment plant is 45 mg/L on a 7-day average. Based on this information, is the treatment plant within its NPDES permit limits? This information is being requested since concern has arisen regarding the potential of the wastewater discharge to affect a local drinking water supply.

The 7-day average concentration for TSS is:

$$(\text{TSS})_7 = (20 + 100 + 50 + 42 + 33 + 25 + 25)/7 = 40.7\ \text{mg/L}$$

The wastewater treatment plant is still within its NPDES permit limit (but only marginally) for an average 7-day maximum concentration of 45 mg/L for TSS.

Illustrative Example 5.6

A regulatory agency stipulates that the maximum concentration of benzo(a)pyrene in drinking water should not exceed 200 ng/L. Express this concentration in lb/10,000 US gal.

Solution

This unit conversion may be carried out as follows:

$$(200\ \text{ng/L})(1\ \text{g}/10^9\ \text{ng})(1\ \text{lb}/454\ \text{g})(3.785\ \text{L}/1\ \text{gal}) = 1.67 \times 10^{-9}\ \text{lb/gal}$$

$$= (1.67 \times 10^{-9}\ \text{lb/gal})(10^4\ \text{gal}/10{,}000\ \text{gal})$$

$$= 1.67 \times 10^{-5}\ \text{lb}/10{,}000\ \text{gal}$$

TABLE 5.4
Daily Effluent TSS Concentration Data Collected over a 7-Day Period at a Municipal Wastewater Treatment Plant for Illustrative Example 5.3

Day	TSS (mg/L)
1	20
2	100
3	50
4	42
5	33
6	25
7	15

Abbreviation: TSS, Total Suspended Solids.

Alternatively, knowing 1 mg/L = 1 ppm = 8.34×10^{-6} lb/gal, the solution can be calculated as follows:

$$200 \text{ ng/L} \left(1 \text{ mg}/10^6 \text{ng}\right)\left(\left(8.34 \times 10^{-6} \text{lb/gal}\right)/\left(1 \text{ mg/L}\right)\right)\left(10^4 \text{gal}/10{,}000 \text{ gal}\right)$$
$$= 1.67 \times 10^{-5} \text{lb}/10{,}000 \text{ gal}$$

REFERENCES

Bureau of National Affairs. 1976. Resource Conservation and Recovery Act of 1976. *International Environmental Reporter.* Washington, DC.

Burke, G., B. Singh, and L. Theodore. 2000. *Handbook of Environmental Management and Technology*, 2nd edition. John Wiley & Sons, Hoboken, NJ.

Cheremisinoff, P. N., and F. Ellerbusch. 1979. *Solid Waste Legislation, Resource Conservation & Recovery Act, A Special Report.* CRC Press, Washington, DC.

Theodore, M.K., and L. Theodore. 1995. *Major Environmental Issues Facing the 21st Century.* Theodore Tutorials (originally published by Simon & Schuster), East Williston, NY.

U.S. Environmental Protection Agency. 1986. *Solving the Hazardous Waste Problem.* EPA/530-SW-86-037, Office of Solid Waste, Washington, DC.

U.S. Environmental Protection Agency. 1988a. EPA: A Regulatory Agency. *EPA Journal*, 14(2):1.

U.S. Environmental Protection Agency. 1988b. *Environmental Progress and Challenges: EPA's Update.* EPA-230-07-88- 033, Office of Policy, Planning and Evaluation (PM-219), Washington, DC.

U.S. Environmental Protection Agency. 1991. *Pollution Prevention News.* Office of Pollution Prevention, Washington, DC.

6 The Clean Water Act

6.1 INTRODUCTION

Congress has provided the U.S. Environmental Protection Agency (EPA) and the states with three primary statutes to control and reduce water pollution: the Clean Water (CWA), the Safe Drinking Water Act (SDA), and the Marine Protection, Research, and Sanctuaries Act (MPRSA). Each statute provides a variety of tools that can be used to meet the challenges and complexities of reducing water pollution in the United States. This chapter focuses on the CWA.

Under the CWA, the states adopt water quality standards (WQSs) for every water body within their borders. These standards include a designated use such as fishing, swimming, boating, wildlife habitat, agriculture, and drinking water supply and prescribe numeric criteria to protect that beneficial use. The numeric criteria are pollutant specific and represent the permissible levels of substances in the water that enable the designated beneficial use to be achieved. WQSs are the basis for nearly all water quality management decisions. Depending on the standard adopted for a particular water body, controls may be needed to reduce the pollutant levels discharged to it.

The law focuses primarily on surface water. Groundwater is included indirectly through regulations that cover the interaction between surface water and groundwater, and through requirements in the act for groundwater protection strategies. The act requires the EPA and those states with delegated authority to regulate industrial and other discharges of wastewater to rivers, streams, and the ocean; determine allowable levels of contaminants in discharges; and develop surface water quality criteria and standards for designated beneficial uses of their water bodies.

This chapter provides an overview of the CWA through a discussion of its early history, details of the act, WQSs, water quality criteria, total maximum daily load (TMDL) assessments, the National Pollutant Discharge Elimination System (NPDES) permitting system, and the CWA grants program. The applications section provides three Illustrative Examples related to the general subject of the CWA.

6.2 EARLY HISTORY OF WATER POLLUTION CONTROL

The Cuyahoga River was one of the most polluted rivers in the United States, being devoid of fish throughout the 1950s and 1960s in the reach from Akron to Cleveland. There have been at least 13 major fires on the Cuyahoga River, the first occurring in 1868. The largest river fire, in 1952, caused more than $1 million in damage to boats and a riverfront office building. Fires erupted on the river several more times before June 22, 1969, when on that date a river fire captured the attention of *Time* magazine, which described the Cuyahoga as the river that "oozes rather than flows" and in which a person "does not drown but decays."

The 1969 Cuyahoga River fire mobilized public concern across the nation and helped spur an proactive water pollution control efforts resulting in the CWA, Great Lakes Water Quality Agreement, and the creation of the federal EPA and the Ohio Environmental Protection Agency (OEPA).

Although the Cuyahoga River fires dramatically focused public opinion and motivated action, several federal laws had been in place to regulate activities in surface waters. The Rivers and Harbors Act of 1899 addressed activities that could potentially impede navigation such as placing dredged or fill material in waterways, altering channels, and constructing dams, bulkheads, jetties, and other structures. In 1948, Congress enacted the Water Pollution Control Act to "enhance the quality and value of our water resources and to establish a national policy for the prevention, control and abatement of water pollution." This was an important step in establishing the basic legal authority for federal regulation of water quality. The act was amended in 1956 to strengthen enforcement provisions and again in 1965 to establish WQSs for surface waters enforceable by state and federal authorities. Incremental adjustments through 1970 beefed-up reporting requirements, enforcement provisions, and added an antidegradation component to national water quality protection laws (U.S. EPA 2019).

6.3 THE CLEAN WATER ACT

Growing public awareness and concern for controlling water pollution led to enactment of the Federal Water Pollution Control Act amendments of 1972 (U.S. EPA 2019). As amended in 1977, this law became commonly known as the Clean Water Act. The act established the basic structure for regulating discharges of pollutants into the waters of the United States. It gave EPA the authority to implement pollution control programs such as setting wastewater standards for industry. The CWA also continued to set WQSs for all contaminants in surface waters. The act made it unlawful for any person to discharge any pollutant from a point source into navigable waters unless a permit was obtained under its provisions. It also funded the construction of sewage treatment plants under the construction grants program and recognized the need for planning to address the critical problems posed by nonpoint source pollution.

Subsequent enactments modified some of the previous CWA provisions. Revisions in 1981 streamlined the municipal construction grants process, improving the capabilities of treatment plants built under the program. Changes in 1987 phased out the construction grants program, replacing it with the State Water Pollution Control Revolving Fund, more commonly known as the Clean Water State Revolving Fund. This new funding strategy addressed water quality needs by building on existing EPA-state partnerships.

Over the years, many other laws have changed parts of the CWA. Title I of the Great Lakes Critical Programs Act of 1990 put into place parts of the Great Lakes Water Quality Agreement of 1978, signed by the United States and Canada, where the two nations agreed to reduce certain toxic pollutants in the Great Lakes. That law required the EPA to establish water quality criteria for the Great Lakes, addressing 29 toxic pollutants with maximum levels that are safe for humans, wildlife, and aquatic life. It also required EPA to help the states implement the criteria on a specific schedule.

The CWA is the cornerstone of surface water quality protection in the United States. The statute employs a variety of regulatory tools to sharply reduce direct pollutant discharges into waterways, finance municipal wastewater treatment facilities, and manage polluted runoff. Those tools are employed to achieve the broader goal of restoring and maintaining the chemical, physical, and biological integrity of the nation's waters so that they can support "the protection and propagation of fish, shellfish, and wildlife and recreation in and on the water."

For many years following the passage of CWA in 1972, the EPA, states, and Native American tribes focused mainly on the chemical aspects of the "integrity" goal. During the last decade, however, more attention has been given to physical and biological integrity. Also, in the early decades of the act's implementation, efforts focused on regulating discharges from traditional "point source" facilities, such as municipal sewage plants and industrial facilities, with little attention paid to nonpoint discharges from streets, construction sites, farms, and other "wet-weather" sources.

Starting in the late 1980s, efforts to address polluted runoff from nonpoint sources have increased significantly. For nonpoint runoff from agricultural sources, voluntary programs, including cost sharing with landowners, are the key tool. For wet weather sources from urban stormwater runoff, increasingly stringent treatment requirements for Municipal Separate Storm Sewer Systems (MS4) have come into existence through Phase I regulations on medium and large cities in 1990, and Phase 2 regulations for small urban and rural MS4s in 1999 to obtain permits for stormwater discharge that require stormwater management programs to be developed to limit the impact of stormwater discharges on receiving water quality. A significant emphasis on the EPA's current stormwater guidance is on the use of Green Infrastructure to collect, treat, and infiltrate stormwater throughout a watershed, rather than using the conventional approach of collection and conveyance of stormwater for discharge to surface water bodies. More information regarding the EPA's Green Infrastructure program can be found at https://www.epa.gov/green-infrastructure.

Evolution of CWA programs over the last decades has also included something of a shift from a program-by-program, source-by-source, pollutant-by-pollutant approach to more holistic watershed-based strategies. Under the watershed approach, equal emphasis is placed on protecting healthy waters and restoring impaired ones. A full array of issues are addressed and not just those subject to CWA regulation (U.S. EPA 2019).

6.4 WATER QUALITY STANDARDS

WQS are aimed at translating the broad goals of the CWA into waterbody-specific objectives. Ideally, WQSs should be expressed in terms that allow quantifiable measurement (U.S. EPA 2019). WQSs, like the CWA overall, apply only to the waters of the United States. As defined in the CWA, "waters of the United States" apply only to surface waters-rivers, lakes, estuaries, coastal waters, and wetlands. Not all surface waters are legally "waters of the United States." Generally, however, those waters include the following:

- All interstate waters
- Intrastate waters used in interstate or foreign commerce
- Tributaries of the above
- Territorial seas at the cyclical high tide mark
- Wetlands adjacent to all the above

The exact dividing line between "waters of the United States" according to the CWA and other waters can be hard to determine, especially with regard to smaller streams, ephemeral water bodies, and wetlands not adjacent to other "waters of the United States." In fact, the delineation changes from time to time, as new court rulings are handed down, new regulations are issued, or the act itself is modified.

Designated uses, water quality criteria, and an antidegradation policy constitute the three major components of a water quality standards program. The designated uses (DUs) of a waterbody are those uses that society, through various units of government, determines should be attained in the waterbody. The DUs are the goals set for the waterbody. In some cases, these uses have already been attained, but sometimes conditions in a waterbody do not support all the DUs. These waters not supporting the DUs are classified as Section 303(d) Threatened and Impaired Waters.

Water quality criteria (WQC) are descriptions of the conditions in a water body necessary to support the DUs. These can be expressed as concentrations of pollutants, temperature, pH, turbidity units, toxicity units, or other quantitative measures. WQC can also be narrative statements such as "no toxic chemicals in toxic amounts."

Antidegradation policies are a component of state/tribal WQS that establish a set of rules that should be followed when addressing proposed activities that could lower the quality of high quality waters (i.e., those with conditions that exceed those necessary to meet the DUs). The antidegradation regulations help to ensure that: (i) all waters continue to support their designated uses (Tier 1 waters); (ii) waters with higher quality than the minimum are protected, unless there are important benefits associated with carefully considered actions that could cause additional degradation (Tier 2 waters); and (iii) highly valued, high-quality waters are not degraded at all (Tier III Outstanding National Resource Waters [ONRW]). Water quality cannot be degraded for "Tier 3" ONRWs, nor when the water is barely meeting the applicable criteria for "Tier 1" beneficial use support. Activities that would degrade high quality "Tier 2" waters must be justified, through alternatives analyses and a demonstration of "important" economic or social benefits in the area where the water is located (U.S. EPA 2019).

6.5 WATER QUALITY CRITERIA

WQC are levels of individual pollutants or water quality characteristics, or descriptions of conditions of a water body that, if met, will protect the designated beneficial use of the water (U.S. EPA 2019). For a given DU, there are likely to be a number of criteria dealing with different types of conditions, as well as levels of specific chemicals. Because most water bodies have multiple DUs, the number of WQC applicable to a given waterbody can be numerous.

WQC must be scientifically consistent with the attainment of DUs. This means that only scientific considerations should be taken into account when determining what water quality conditions are consistently meeting a given DU. Economic and social impacts are not considered when developing WQC.

WQC can be divided up for descriptive purposes in many ways. For instance, numeric criteria (weekly average of 5 mg/L dissolved oxygen) can be contrasted with narrative criteria (no putrescent bottom deposits). Criteria can also be categorized according to what portion of the aquatic system they can be applied to: the water itself (water column), the bottom sediments, or the bodies of aquatic organisms (fish tissue). The duration of time to which they apply is another way of dividing WQC, with those dealing with short-term exposures (acute) being distinguished from those addressing long-term exposure (chronic), or those addressing young spawning organisms distinguished from those addressing adult organisms.

Criteria can also be distinguished according to the types of organisms they are designed to protect. Aquatic life criteria are aimed at protecting entire communities of aquatic organisms, including a wide array of animals and various plants and microorganisms. These can be expressed as parameter-specific (daily average of 30 µg/L of copper) or in terms of various "metrics" that directly measure numbers, weight, and diversity of plants and animals in a water body (community indices). Human health criteria can apply to two exposure routes: (i) ingesting water and (ii) consuming aquatic foodstuffs.

Wildlife criteria, like human health and fish consumption criteria, deal with the effects of pollutants with high bioaccumulation factors. To date, the EPA has issued or adopted fewer wildlife criteria than aquatic life or human health criteria. Such criteria are designed to protect terrestrial animals that feed on aquatic species. Examples are ospreys, herons, and other wading birds, mink, and otters.

Numeric criteria are usually parameter specific (i.e., they express conditions for specific measures), such as dissolved oxygen, temperature, turbidity, nitrogen, phosphorus, heavy metals such as mercury and cadmium, and synthetic organic chemicals like dioxin and polychlorinated biphenyls (PCBs). They do not consist merely of stated levels or concentrations, such as 15 µg/L or a pH above 5.0. They should also specify the span of time over which conditions must be met. This is the "duration" component of a WQC. Combining the concentration/magnitude and duration components of a WQC results in wording such as "the average 4-day concentration of pollutant X shall not exceed 50 µg/L."

A numeric WQC should also indicate how often it would be acceptable to exceed specified concentration or duration combinations. This is often called the *frequency* or the *recurrence interval component* of the WQC. For instance, for protection of aquatic life, as a general rule, the EPA recommends a recurrence interval of once in 3 years. The purpose of the recurrence interval is to recognize that aquatic systems can recover from impacts of exposure to harmful conditions, but such conditions must be sufficiently rare to keep the community from being in a constant state of recovery.

Simply because one sample has exceeded the concentration component of a WQC does not necessarily mean the WQC has been violated and a DU affected (U.S. EPA 2019). This is true only in the case of "instantaneous criteria" levels that are never to be exceeded. If there is a criterion of 50 mg/L of "X," for a 7-day average, then having one sample at a concentration above 50 mg/L would not "prove" that this criterion had actually been exceeded. Likewise, having just one or two samples below 50 mg/L is not a good basis for concluding a waterbody is indeed meeting WQS.

6.6 TOTAL MAXIMUM DAILY LOADS (TMDLs)

If monitoring and assessment indicate that a water body or segment is impaired by one or more pollutants, and it is therefore placed on the Section 303(d) Threatened and Impaired Waters list, then the relevant entity (state, territory, or authorized tribe) is required to develop a strategy that would lead to attainment of the WQS (U.S. EPA 2019).

The CWA requires that a TMDL study be conducted for waters affected by pollutants where implementation of the technology-based controls imposed upon point sources by tbc CWA and EPA regulations do not result in achievement of WQS. At this point in the history of the CWA, most point sources have been issued NPDES permits with technology-based discharge limits. In addition, a substantial fraction of point sources also have more stringent water quality–based permit limits. But because nonpoint sources are major contributors of pollutant loads to many water bodies, even these more stringent limits on point sources have not resulted in attainment of WQS.

Strategies to address impaired waters must consist of a TMDL or another comprehensive strategy that includes a functional equivalent of a TMDL. In essence, TMDLs are "pollutant budgets" for a specific waterbody or segment, that if not exceeded, would result in attainment of WQS.

TMDLs are required for "pollutants" but not for forms of "pollution." Pollutants include clean sediments, nutrients (nitrogen and phosphorus), pathogens, acids/bases, heat, metals, cyanide, and synthetic organic chemicals. Pollution includes not only all pollutants but also flow alterations and physical habitat modifications.

At least one TMDL must be performed for every water body or segment impaired by one or more pollutants. TMDLs are performed pollutant by pollutant, although if a waterbody or segment were impaired by two or more pollutants, the TMDLs for each pollutant could be performed simultaneously (U.S. EPA 2019).

The EPA is encouraging states, tribes, and territories to do TMDLs on a "watershed basis" (e.g., to "bundle" TMDLs together) to realize program efficiencies and foster more holistic analysis. Ideally, TMDLs would be incorporated into comprehensive watershed strategies. Such strategies would address the protection of high-quality waters (antidegradation) as well as restoration of impaired segments. They would also address the full array of activities affecting the waterbody. Finally, such strategies would be the product of collaborative efforts between a wide variety of stakeholders.

TMDLs must be submitted to the EPA for review and approval/disapproval. If the EPA ultimately decides that it cannot approve a TMDL that has been submitted, the agency would need to develop and promulgate what it considers to be an acceptable TMDL on its own. Doing so requires going through the formal federal rulemaking process.

The first element of a TMDL is "the allowable load," also referred to as pollutant "cap" (U.S. EPA 2019). It is basically a budget for a particular pollutant in a particular body of water, or an expression of the "carrying capacity." This is the loading rate that would be consistent with meeting the WQC for the pollutant in question. The cap is usually derived through use of mathematical models, commonly QUAL2Kw (Pelletier et al. 2006) for stream modeling and HSPF (Bicknell et al. 1997) for watershed modeling.

The CWA requires that all TMDLs include a safety factor as an extra measure of environmental protection, taking into account uncertainties associated with estimating the acceptable cap or load. This is referred to as the margin of safety (MOS) (U.S. EPA 2019). Once the cap has been set (with the MOS factored in), the next step is to allocate that total pollutant load among various sources. This is in essence the "slicing of the pie."

TMDLs set loading caps for individual pollutants such as clean sediments, nitrogen, phosphorus, coliform bacteria, temperature, biochemical oxygen demand (BOD), copper, mercury, PCBs, etc. (Again, TMDLs are not required for nonpollutant forms of pollution, such as stream-flow patterns and stream-channel modification.). States, territories, and authorized tribes are free to develop TMDLs for such pollutants as they see fit. The CWA and EPA regulations put no limits on these other government entities going beyond what the act requires (U.S. EPA 2019).

Though the CWA itself uses the term *total maximum daily loads*, the EPA has determined that loadings rates (caps) can be expressed as weekly, monthly, or even yearly loads. Which time period to use depends on the type of pollutant for which the TMDL is being performed. Toxic chemicals that exhibit acute effects would probably call for daily or weekly loads, whereas nutrients and sediments could be expressed as monthly or yearly loading rates.

The CWA allows for seasonal TMDLs (i.e., it allows different rates of loading at different times of the year). For example, colder waters can absorb more oxygen-demanding substances than can warm water, so allowable loading could be higher in the winter than in the summer.

EPA regulations use the terms *wasteload allocations* (WLA) and *load locations* (LA) to describe loading assigned to point and nonpoint sources, respectively (U.S. EPA 2019).

Generally, point sources that are required to have individual NPDES permits are required to be assigned individual WLAs. On the other hand, a group of sources covered under a "general" NPDES permit would be assigned one collective WLA.

Although load allocations should ideally be assigned to individual nonpoint sources, this is often not practical or even scientifically feasible; hence, loads can be assigned to categories of nonpoint sources (all soybean fields in the watershed, for example), or to geographic groupings of nonpoint sources (all in a particular subwatershed) (U.S. EPA 2019).

Even though the CWA provides no federal authority for requiring nonpoint sources to reduce their loadings of pollutants to the nation's waters, the act does require states (and authorized territories and tribes) to develop TMDLs for waters where nonpoint sources are significant sources of pollutants. TMDLs do not create any new federal regulatory authority over any type of sources. Rather, with regard to nonpoint sources, TMDLs are simply a source of information that, for a given waterbody, should answer such questions as the following (U.S. EPA 2019):

- Are nonpoint sources a significant contributor of pollutants to this impaired water body?
- What are the approximate total current loads of impairment-causing pollutants from all nonpoint sources in the watershed?
- What fraction of total loads of the pollutant(s) of concern come from nonpoint sources versus point sources?
- What are the approximate loadings from the major categories of nonpoint sources in the watershed?
- How much do loads from nonpoint sources need to be reduced to achieve the water quality standards for the waterbody?
- What kinds of management measures and practices would need to be applied to various types of nonpoint sources to achieve the needed load reductions?

A common misconception about TMDLs is that EPA has issued regulations specifying how pollutant caps in a TMDL should be allocated among sources: equal reductions for all or equal loadings from each, for example. The EPA has no such regulations (U.S. EPA 2019). States, territories, and tribes are free to allocate to sources in any way they see fit, so long as the sum of all the allocations is no greater than the overall loading cap. However, when thinking about changing the share of allowed loads among sources, it is important to realize that in all but very small water body segments, load location matters. In many cases, the farther away from the zone of impact that a loading enters into the waterbody system, the less of an effect that load will have on the impaired zone. For example, studies of large watersheds, such as Long Island Sound (local to one of the authors), have indicated that 1 lb of pollutant (nitrogen in the case of the Sound) discharged close to the impaired zone has the same impact on that zone as 10 lb discharged substantially farther away. Furthermore, even after accounting for location-related relative impacts on a particular segment or zone, care must be taken to ensure that localized exceedances of WQS do not result from moving loads from one tributary or segment to another.

6.7 NATIONAL POLLUTANT DISCHARGE ELIMINATION SYSTEM (NPDES)

Under the CWA, the discharge of pollutants into the waters of the United States is prohibited unless a permit is issued by the EPA or a state under the NPDES. These permits must be renewed at least once every 5 years. There are more than 50,000 industrial and 16,000 municipal facilities that currently have NPDES permits.

An NPDES permit contains effluent limitations and monitoring and reporting requirements. Effluent limitations are restrictions on the amount of specific pollutants that a facility can discharge into a stream, river, or harbor. Monitoring and reporting requirements are specific instructions on how sampling of the effluent should be done to check whether the effluent limitations are being met. Instructions may include required sampling frequency (i.e., daily, weekly, or monthly) and the type of monitoring required. The permittee may be required to monitor the effluent on a daily, weekly, or monthly basis. The monitoring results are then regularly reported to the EPA and state authorities. When a discharger fails to comply with the effluent limitations or monitoring and reporting requirements, the EPA or the state may take enforcement action.

Congress recognized that it would be an overwhelming task for the EPA to establish effluent limitations for each individual industrial and municipal discharger. Therefore, Congress authorized the EPA to develop uniform effluent limitations for each category of point sources such as steel mills, paper mills, and pesticide manufacturers. The EPA develops these effluent limitations on the basis of many factors, most notably efficient treatment technologies. Once the EPA proposes an effluent limit and public comments are received, the EPA or the states issue all point sources within that industry category NPDES permits using the technology-based limits.

Wastewater treatment plants also are provided with effluent limitations based on technology performance. These federal effluent limitations are minimum performance standards for a given source of wastewater discharge regardless of the location of the facility.

Limitations that are more stringent than those based on technology are often necessary to ensure that state-developed water quality standards are met. As discussed in Section 6.6 if these technology-based effluent limitations are not adequate to protect the designated beneficial use of a surface water body, NPDES permits reflect more stringent water quality–based limitations as determined through the TMDL process.

The EPA and the U.S. Army Corps of Engineers implement jointly a permit program regulating the discharge of dredged or fill material into waters of the United States, including wetlands. As part of this program, the EPA's principal responsibility as set forth in the CWA is to develop the substantive environmental criteria by which permit applications are evaluated. The EPA also reviews the permit applications and, if necessary, can veto permits that would result in significant environmental damage.

The National Estuary Program is also regulated under the CWA. States nominate and the EPA selects estuaries of national significance that are threatened by pollution, development, or overuse. The EPA and the involved state(s) form a management committee consisting of numerous workgroups to assess the problems, identify management solutions, and develop and oversee implementation of plans for addressing the problems.

6.8 GRANTS

The CWA authorized the EPA to provide financial assistance to states (the Construction Grants Program) to support programs such as (a) the construction of municipal sewage treatment plants; (b) water quality monitoring, permitting, and enforcement; and (c) implementation of nonpoint-source controls. These funds could also support development and implementation of state groundwater protection strategies.

In 1987, Congress voted to phase out the old Construction Grants Program, replacing it with the Clean Water State Revolving Fund (CWSRF). Under the CWSRF, the EPA provides annual capitalization grants to states, who in turn provide low interest loans for a wide variety of water quality projects that are prioritized within each state. States must match the federal funds with $1 for every $5. The program was amended in 2014 by the Water Resources Reform and Development Act.

Some funds also are provided to territories and tribes to be used as grants for municipal wastewater treatment projects. Territories must match the federal funds with a 20% match, but the tribes are not required to provide a match. Loans are usually made at low, sometimes no, interest. Although most loans have gone to local governments, they also can go to businesses or nonprofit organizations. Payback periods for loans extend to 20 years.

Most of the CWSRF dollars loaned to date have gone for construction expansion, repair, or upgrading of municipal sewage collection and treatment systems. But CWSRF loans can be made for the following: (i) National Park Service (NPS) control projects consistent with a state, territorial, or tribal Section 319 program; or (ii) implementing a management plan developed under the National Estuary Program.

As a result of federal capitalization grants, state match, loan repayments, and leverage bonds, the total amount of assets in all the CWSRFs has exceeded $43 billion. Building on a federal investment of $43.3 billion, the state CWSRFs have provided $3 and $4 billion annually, with a total of $133 billion loaned to communities for water quality improvement projects through 2018. States have provided 39,948 low-interest loans to protect public health, protect valuable aquatic resources, and meet environmental standards benefiting hundreds of millions of people.

The CWA also provides additional grant funding through Section 319 dealing with nonpoint source pollution management. States, territories, and delegated tribes are required to develop nonpoint source pollution management programs (if they wish to receive 319 funds). Once it has approved a state's nonpoint source program, the EPA provides grants to these entities to implement NPS management programs under Section 319(h). Section 319 is a significant source of funding for implementing NPS management programs, but there are other federal (e.g., Farm Bill), state, local, and private programs.

Congressional appropriations for the CWA Section 319 program peaked at $238.5 million in 2003, but it averaged approximately $175 million between 2008 and 2017. Recipients of CWA Section 319 grant funds must provide a 40% match, either in dollars or in-kind services. States and territories "pass on" a substantial fraction of the 319 funds they receive from the EPA to support local nonpoint source pollution management efforts. Depending on the state or territory, a "local match" may be required.

Though there is no CWA federal regulatory authority over nonpoint sources of pollution and the act does not require states to develop their own regulatory programs to obtain 319 grants, states, territories, and tribes may, at their discretion, use 319 funds to develop their own NPS regulatory programs. To date, however, few have done so.

Section 319 funds can be used to conduct activities to ensure the use of Best Management Practices (BMPs), develop strategies for collaborating with other agencies, and draft monitoring and evaluation plans. Section 319 funds also can be used for developing and implementing TMDLs in watersheds where nonpoint sources are a substantial contributor of loadings of the pollutants causing impairment. A state, tribe, or territory receiving Section 319 funds must complete and update an NPS management plan every 5 years.

6.9 APPLICATIONS

Three Illustrative Examples complement the material presented in this chapter on the CWA.

Illustrative Example 6.1

List and describe some key wastewater constituents that appear in the CWA regulations.

Solution

Key wastewater constituents regulated by the CWA include:

Suspended Solid (SS): A measure of solids that are suspended (not dissolved) in the wastewater. SSs can lower the amount of light and oxygen in a body of water. Over time SSs will settle from the water and can form a layer of solids on the bottom of a stream or lake. These solids may be biodegradable and may have an oxygen demand associated with them that could deplete the oxygen level in a body of water below levels necessary to support diverse aquatic life.

Biochemical Oxygen Demand (BOD): A measure of the amount of oxygen required to stabilize the biodegradable organic material in a water sample. The three major classes of biodegradable organics in wastewater are composed principally of proteins, carbohydrates, and fats. If discharged untreated to the environment, the stabilization of high BOD wastewater can lead to the depletion of oxygen and to the development of septic conditions in rivers and other natural bodies of water that would not be supportive of diverse aquatic life.

Pathogens: Pathogenic organisms that can transmit communicable diseases via wastewater. Typical infectious diseases reported are cholera, typhoid, paratyphoid fever, salmonellosis, and shigellosis. *Escherichia coli* and fecal coliform are indicators of potential presence of pathogens, and effluent limitations are set for these indicator organisms to reduce the potential of disease transmission to acceptable levels.

Other key constituents include:

Nutrients: Nitrogen and phosphorus are the major nutrient sources in wastewater. When discharged to the receiving water, these nutrients can lead to the excessive growth of undesirable aquatic life, primarily algae, with resulting excessive oxygen consumption and deteriorated water quality. This excessive algae growth and water quality deterioration is labeled *eutrophication*. When discharged in excessive amounts on land, they can also lead to the pollution of groundwater.

Heavy Metals: Heavy metals are usually added to wastewater from municipal commercial and industrial activities and may have to be removed if the wastewater is to be reused or discharged into a water body because of their potential toxicity to both humans and aquatic life.

Priority Pollutants: Organic and inorganic compounds designated on the basis of their known or suspected carcinogenicity, mutagenicity, or high acute toxicity to either humans or aquatic life.

Illustrative Example 6.2

Discuss the difference between BOD and COD.

Solution

BOD and COD are two important water quality parameters in wastewater engineering. The accurate measurement of these parameters is essential in the proper design of wastewater treatment systems and in the study of the transport and fate of contaminants in the aquatic environment.

Biochemical oxygen demand, or BOD, is the quantity of dissolved oxygen required to stabilize biodegradable organic material in water. BOD of the incoming wastewater is a measure of the oxygen needed in the aerobic biological treatment of the wastewater. BOD of the effluent from the wastewater treatment plant is an important measure of the impact of the wastewater on the receiving water quality and is a critical factor in the viability of the aquatic system's ecology. The EPA and state environmental regulatory agencies routinely require BOD monitoring of all municipal and industrial discharges.

BOD is generally measured in a three-step process. First, a sample is quantitatively diluted so that there is an appropriate concentration of oxidizable substrate and dissolved oxygen in the sample volume to be measured. For well-treated or "polished" wastewaters (5–30 mg/L BOD), dilution is small. For raw, untreated wastewaters (100–400 mg/L BOD), a high level of dilution is necessary. Second, the sample is inoculated with microorganisms (seed), sealed, and incubated at 20°C for a set period of time (the regulatory standard is 5 days). During incubation the dissolved oxygen (DO) and substrate are consumed through microbial activity. Third, DO is measured at the start of incubation and again at the end. The BOD is equal to the difference in DO concentration after it has been adjusted for the dilution factor. There is also an adjustment factor necessary to compensate for extra BOD introduced to the sample with the inoculating culture material (seed).

BOD is most commonly expressed as "five-day BOD" (BOD_5) and ultimate BOD (BOD_U). BOD_5 is the oxygen demand that is exerted after a standard period of 5 days of incubation and is the value required by regulatory agencies. BOD_U represents the oxygen demand that would be exerted if incubation was allowed to occur long enough for virtually all the biologically oxidizable substrate to be consumed. Often, BOD_U is approximated by allowing incubation to occur for a standard period of 20 days, or longer.

Typically, BOD_5 is equal to approximately two-thirds of the BOD_U. Although this ratio is a good approximation for typical domestic wastewaters, it can also vary significantly depending on the nature of the wastewater source. Lower ratios are typical of industrial wastewaters with a lower relative composition of readily oxidizable substrate. For example, such a wastewater typically might contain higher colloidal or SSs substrates (such as petroleum refinery or paper and pulp mill effluents) or higher nitrogenous substrates (such as meat processing effluents). Higher ratios are indicative of industrial wastewaters with a higher relative composition of readily oxidizable soluble substrate (such as brewery or carbohydrate food processing wastewaters).

BOD can also be distinguished by the type of oxidizable substrate that is consumed during the incubation period. Carbonaceous BOD (CBOD) represents the oxygen demand exerted by the carbon-based components of the substrate. Nitrogenous BOD (NBOD) represents oxygen demand exerted by the process of nitrification, or the oxidation of ammonia to nitrite and nitrate. To measure CBOD only, an inhibitory chemical is added to the sample to stop the nitrification portion of oxidation from occurring during the incubation period. To measure NBOD, it is necessary to measure both total BOD (TBOD) and CBOD in samples that have incubated 5 or more days. NBOD is calculated as the difference between total BOD and CBOD. Thus,

$$\text{Total BOD} = \text{CBOD} + \text{NBOD} \tag{6.1}$$

$$\text{TBOD (without inhibitor)} - \text{CBOD (with inhibitor)} = \text{NBOD} \tag{6.2}$$

Because CBOD is the primary substrate consumed during the first 5 days of incubation, BOD_5 is approximately equal to $CBOD_5$, except in highly treated waters. The measurement of NBOD with unacclimated organisms typically requires longer incubation periods because nitrifying organisms are slower growing and do not start consuming oxygen to a measurable degree until well after carbonaceous oxidation has begun. This delay in nitrification is generally considered to be approximately 8 to 10 days in raw municipal wastewater.

Chemical oxygen demand (COD) is equal to the equivalent oxygen concentration required to chemically oxidize organic materials in water. Because the test can be completed in a little more than 2 hours, it quickly provides the concentration of chemically oxidizable organic material in a wastewater. When used in conjunction with the BOD test it also is an indicator of nonbiodegradable organics in a sample. To chemically oxidize all the substrate in wastewater, a strong chemical oxidant (such as dichromate in sulfuric acid) to oxidize organic material at high temperature. The COD is reported as the DO concentration equivalent to the decrease in the acidic dichromate concentration. The level of oxidation provided by this reagent is generally sufficient to oxidize almost all of the oxidizable organics in water. Because there are additional (albeit less commonly used) chemical oxidants other than acidic dichromate, which can serve as COD standard oxidants, COD measured using dichromate is often referred to as "dichromate COD."

COD is an important water quality parameter because it often serves as a reliable measure of the CBOD once the wastewater being examined has been characterized. It is a faster, less expensive analysis than CBOD, and is therefore used where frequent, routine process control analyses are required. A wastewater that is well suited to biological treatment generally has a $CBOD_U$ concentration approximately equal to the COD concentration. In effect, the bulk of the substrate that can be oxidized ultimately by process microorganisms is also oxidized by a strong chemical oxidant. $CBOD_U$ will be less than but approximately equal to COD for typical, primary-treated domestic wastewater. Industrial wastewater, or domestic sewage with a significant industrial component, will often have $CBOD_U$/COD ratios much less than 1 if they contain substrates that are chemically reactive, but nonbiodegradable. Chemical oxygen demand is of direct significance in industrial treatment when there are relatively few biologically oxidizable components.

Regulatory agencies will not accept COD concentrations for permit limits, but COD/BOD relationships can be useful in predicting effluent BOD values. COD measurements provide near real-time knowledge of process efficiency, whereas BOD requires 5 days to generate results. BOD and COD data used together can also indicate the relative biodegradability or recalcitrance of a waste stream and can indicate potential toxicity from industrial waste sources.

Illustrative Example 6.3

Describe the interrelationship between the CWA and the Priority Pollutants list.

Solution

The CWA addresses a large number of issues related to water pollution control, including the management of industrial wastewaters. Any municipality or industry that discharges wastewater in the United States must obtain a discharge permit under the regulations set forth by the NPDES. Under this system, there are three classes of pollutants (conventional pollutants, priority pollutants, and nonconventional/nonpriority pollutants). Conventional pollutants are substances such as BOD, SSs, pH, oil and grease, and coliform bacteria. A Toxic Pollutants List was originally set forth in a consent decree between the EPA and several environmental organizations and contained 65 broad categories of compounds considered toxic and hazardous to human health and the environment. This list was originally incorporated into the 1977 amendments to the Clean Water Act (Section 307(a)(1)) as the Toxic Pollutant List and has since been redesignated as the Priority Pollutant List, which now contains 126 specific individual toxic substances. Most of the substances on this list are organics, but it does include most of the heavy metals. These substances are generally considered to be toxic. However, the toxicity is not absolute; it primarily depends on the concentration. In recent years, pollution prevention programs have been implemented to reduce their use in industrial processes. The third class of pollutants could include any pollutant not in the first two categories. Examples of substances that are presently regulated in the third category are nitrogen, phosphorus, and sodium.

REFERENCES

Bicknell, B.R., J.C. Imhoff, J.L. Kittle, A.S. Donigian, and R.C. Johanson. 1997. *Hydrological Simulation Program: Fortran, User's Manual for Version 11.* EPA/600/R-97/080, U.S. Environmental Protection Agency, National Exposure Research Laboratory, Athens, GA. 755 p.

Pelletier, G.J., S.C. Chapra, and H. Tao. 2006. QUAL2Kw–A framework for modeling water quality in streams and rivers using a genetic algorithm for calibration. *Environmental Modeling & Software* 21:419–425.

U.S. Environmental Protection Agency. 2019. *Introduction to the Clean Water Act.* Watershed Academy, Online Training in Watershed Management, Water Law Modules, Washington, DC. http://cfpub.epa.gov/watertrain/pdf/modules/IntrotoCWA.pdf.

7 The Safe Drinking Water Act

7.1 INTRODUCTION

The first legislation enacted in the United States to protect the quality of drinking water was the Public Health Service (PHS) Act of 1912. The PHS Act brought together the various federal health authorities and programs, such as the Public Health Service and the Marine Hospital Service, under one statute. The PHS Act authorized scientific studies on the impact of water pollution and human health and introduced the concept of water quality standards. True national drinking water standards were not established, however, until 60 years later with the Safe Drinking Water Act (SDWA).

The SDWA, Title XIV of the Public Health Service Act, is the key federal law for protecting public water supplies from harmful contaminants. The SDWA was originally passed by Congress in 1974 to protect public health by regulating the nation's public drinking water supply. Since its enactment, there have been more than 10 major revisions and additions to it, with substantial changes occurring in the amendments in 1986, 1996, and 2016. The SDWA requires many actions to protect drinking water and its sources: rivers, lakes, reservoirs, springs, and groundwater wells. The SDWA applies to every public water system (PWS) in the United States, which amounts to approximately 87% of all water used in the United States. The SDWA does not regulate private wells, which serve fewer than 25 individuals, but does include both public and private municipal water companies, homeowner associations, schools, businesses, campgrounds, and shopping malls. The SDWA authorizes the EPA to set national health-based standards for drinking water to protect against both naturally occurring and man-made contaminants that may be found in drinking water. The EPA, states, and PWSs then work together to make sure that these standards are met.

Originally, the SDWA focused primarily on treatment as the means of providing safe drinking water at the tap. The 1996 amendments significantly enhanced existing law by recognizing source water protection, operator training, funding for water system improvements, public information, and a risk-based approach for selecting contaminants for regulation as important components of a safe drinking water program. This approach ensures the quality of drinking water by protecting it from source to tap.

The 2016 revisions to the SDWA authorized new grant programs to: (i) help public water systems serving small or disadvantaged communities meet SDWA requirements; (ii) support lead reduction projects, including lead service line replacement; and (iii) establish a voluntary program for testing for lead in drinking water at schools and child care programs (Tiemann 2017).

This chapter provides an overview of the SDWA through a discussion of regulated public water systems, details of the act, drinking water standards, primary and secondary drinking water regulations, and unregulated contaminants. The applications section provides four Illustrative Examples related to the general subject of the SDWA.

7.2 REGULATED PUBLIC WATER SYSTEMS

There are more than 151,000 public water systems (PWS) in the United States (U.S. EPA 2019a) that provide drinking water to most Americans. The EPA classifies these water systems according to the number of people they serve, the source of their water, and whether they serve the same customers year-round or on an occasional basis. A PWS may be publicly or privately owned and provides water for human consumption through pipes or other constructed conveyances to at least 15 service connections or serves an average of at least 25 people for at least 60 days a year.

The EPA has defined three types of PWSs: *Community Water Systems* (CWS), which are PWSs that supply water to the same population year-round; *Non-Transient Non-Community Water System* (NTNCWS), which are PWSs that regularly supply water to at least 25 of the same people at least 6 months per year. Some examples of NTNCWSs are schools, factories, office buildings, and hospitals, which have their own water systems; and *Transient Non-Community Water System* (TNCWS), which are public water systems that provide water in a place such as a gas station or campground where people do not remain for long periods of time.

Some 51,350 of the regulated PWSs are CWSs (Tiemann 2017). These water systems provide water to more than 299 million people (Tiemann 2017). All federal regulations apply to these systems. Most CWSs (82%) are relatively small, serving ≤3,300 individuals and provide water to just 9% of the total population served by CWSs. A total of 92% of CWSs serve populations of ≤10,000, and 55% serve populations ≤500. The few large CWSs serving populations of >10,000 provide water to 82% of the total population served by CWSs. Among the CWSs, 71% rely on groundwater, and 29% rely on surface water.

A total of 18,178 PWSs were classified as NTNCWSs (Tiemann 2017) such as schools or factories, which have their own water supplies and generally serve the same individuals for more than 6 months but not year-round. Most drinking water regulations apply to these systems, most, 99%, of which serve ≤3,300 and provide water to 83% of the population served by NTNCWSs.

More than 83,000 other PWSs are TNCWSs, which provide their own water to transitory customers. Only regulations for contaminants that pose immediate health risks apply to these systems (Tiemann 2017).

Drinking water supply in the United States is characterized, then, by a few very large systems (0.3%) providing drinking water to a large portion of the population (45%) and a very large number of very small systems (81%) providing drinking water to a small number of Americans (4.5%) (Tiemann 2017). Ensuring clean, safe, and aesthetically pleasing drinking water supplies under these demographics is a challenge that the EPA has been addressing for many years.

7.3 DETAILS OF THE SAFE DRINKING WATER ACT

The Clean Water Act and the Safe Drinking Water Act place great reliance on state and local initiatives in addressing water problems (U.S. EPA 2004, Tiemann 2017). With the enactment of the 1986 Safe Drinking Water Act amendments and the 1987 Water Quality Act, significant additional responsibilities were assigned to the EPA and the states. To ensure that drinking water is safe, the SDWA sets up multiple barriers against pollution. These barriers include source water protection, treatment, distribution system integrity, and public information. PWSs are responsible for ensuring that contaminants in tap water do not exceed the standards. Water systems treat the water and must test their water frequently for specified contaminants and report the results to state drinking water regulatory agencies. If a water system is not meeting these standards, it is the water supplier's responsibility to notify its customers. Many water suppliers now are also required to prepare annual reports for their customers. The public is responsible for helping local water suppliers to set priorities, make decisions on funding and system improvements, and establish programs to protect drinking water sources.

Essential components of safe drinking water include protection and prevention. States and water suppliers must conduct assessments of water sources to see where they may be vulnerable to contamination. Water systems may also voluntarily adopt programs to protect their watershed or wellhead, and states can use legal authorities from other laws to prevent pollution. The SDWA mandates that states have programs to certify water system operators and make sure that new water systems have the technical, financial, and managerial capacity to provide safe drinking water. The SDWA also sets a framework for the Underground Injection Control (UIC) program to control the injection of wastes into groundwater. The EPA and states implement the UIC program, which sets standards for safe waste injection practices and bans certain types of injection altogether. All of these programs help prevent the contamination of drinking water.

The most direct oversight of water systems is conducted by state drinking water programs. States are given "primacy," the authority to implement the SDWA within their jurisdictions, from the EPA if they can show that they will adopt standards at least as stringent as EPA's and make sure water systems meet these standards. All states and territories, except Wyoming and the District of Columbia, have received primacy. Although no Indian tribe has yet applied for and received primacy, four tribes currently receive "treatment as a state" status and are eligible for primacy. States, or the EPA acting as a primacy agent, make sure water systems test for contaminants, review plans for water system improvements, conduct on-site inspections and sanitary surveys, provide training and technical assistance, and take action against water systems not meeting standards. The SDWA also authorizes the EPA to award grants to states (Drinking Water State Revolving Fund) for developing and implementing programs to protect drinking water at the tap and groundwater resources. These grant programs may be used for supporting state public water supply, wellhead protection, and underground injection programs, including compliance and enforcement.

The SDWA recognizes that since everyone drinks water, everyone has the right to know what is in their drinking water and where it comes from. All water suppliers must notify consumers quickly when there is a serious problem with water quality. Water systems serving the same people year-round must provide annual consumer confidence reports on the source and quality of their tap water. States and the EPA must prepare annual summary reports of water system compliance with drinking water safety standards and make these reports available to the public. The public must also have the chance to be involved in developing source water assessment programs, state plans to use drinking water state revolving loan funds, state capacity development plans, and state operator certification programs.

There are a large number of specific rules within the SDWA that PWSs must comply with based on their size and the population they serve. Table 7.1 summarizes these rules, along with PWSs they apply to, and a reference to an EPA Quick Reference Guide for more information and details regarding the rule.

The EPA establishes primary standards for drinking water quality that represent the maximum contaminant levels (MCLs) allowable to protect the health and safety of consumers. These primary standards consist of numerical criteria for specified contaminants. Local water supply systems are required to monitor their drinking water periodically for contaminants with MCLs and for a broad range of other contaminants as specified by the EPA. The next two sections discuss these drinking water standards and the drinking water regulations that have been developed for human health protection and for minimum aesthetic quality for finished water in the United States.

7.4 DRINKING WATER STANDARDS

Drinking water standards are regulations that EPA has established to control the concentration of contaminants in the U.S. drinking water supply for all PWSs regardless of the size of the population they serve. In most cases, the EPA delegates responsibility for implementing drinking water standards to states and tribes.

The SDWA requires EPA to identify potential drinking water problems, establish a prioritized list of chemicals of concern, and set standards where appropriate. Peer-reviewed science and data support an intensive technological evaluation, which includes many factors such as the occurrence of the chemicals in the environment; human exposure and risks of adverse health effects in the general population and sensitive subpopulations; analytical methods of detection; technical feasibility; and impacts of regulation on water systems, the economy, and public health.

After reviewing health effect studies, the EPA sets a maximum contaminant level goal (MCLG). The MCLG is the maximum level of a contaminant in drinking water at which no known or anticipated adverse effect on the health of persons would occur and which allows an adequate margin

TABLE 7.1
Drinking Water Rules from the SDWA

Rule	Applicable PWS	Quick Reference Guide Number
Arsenic Rule—2001	CWSs, NTNCWSs	EPA 816-F-01-004
Consumer Confidence Report Rule—2009	All CWS all sizes	EPA 816-F-09-009
Filter Backwash Recycle Rule—2001	All PWSs using SW or GW under influence of SW and recycle backwash from filtration	EPA 816-F-01-019
Groundwater Rule—2008	All PWSs that use GW. Does not apply to PWSs that combine all of their GW with SW or with GW under the direct influence of SW	EPA 816-F-08-029
Interim Enhanced Surface Water Rule—2001	All PWSs using SW or GW under the direct influence of SW—Sanitary survey All PWSs using SW or GW under the direct influence of SW serving ≥10,000—All remaining requirements	EPA 816-F-01-011
Lead and Copper Rule—2008	All CWSs and NTNCWSs	EPA 816-F-08-018
Long Term 1 Enhanced Surface Water Treatment Rule—2002	All PWSs using SW or GW under the direct influence of SW serving <10,000	EPA 816-F-02-001
Long Term 2 Enhanced Surface Water Treatment Rule—2006	PWSs using SW or GW under the direct influence of SW Schedule 1—PWS serving >100,000 Schedule 2—PWS serving 50,000 to 99,999 Schedule 3—PWS serving 10,000 to 49,999 Schedule 4—PWS serving <10,000	EPA 816-F-06-005 EPA 816-F-06-006 EPA 816-F-06-007 EPA 816-F-06-008
Public Notification Rule—2009	All PWSs	EPA 816-F-09-101
Stage 2 Disinfectants and Disinfection Byproducts Rule—2006	All CWSs and NTNCWSs adding disinfectant Stage 1—CWSs and NTNCWSs serving >100,000 Stage 2—CWSs and NTNCWSs serving 50,000 to 99,999 Stage 3—CWSs and NTNCWSs serving 10,000 to 49,999 Stage 4—CWSs and NTNCWSs serving <10,000	EPA 816-F-06-001 EPA 816-F-06-002 EPA 816-F-06-003 EPA 816-F-06-004
Surface Water Treatment Rule—2010	PWSs that use SW Conventional or Direct Filtration Slow Sand or Diatomaceous Earth Filtration No Filtration	EPA 816-F-10-074 EPA 816-F-10-076 EPA 816-B-10-001
Revised Total Coliform Rule—2013	All PWSs	EPA 816-F-01-035
Radionuclides Rule—2001	All CWSs	EPA 816-F-01-003
Record Keeping Rules—2006	Summary of Record Keeping Rules for all PWSs	EPA 816-F-06-033
Aircraft Drinking Water Rule—2009	Aircraft PWSs	EPA-816-F-10-077
Unregulated Contaminant Monitoring Rule—1988 to 2021	All CWSs	https://www.epa.gov/dwucmr

Source: U.S. Environmental Protection Agency, *Drinking Water Rule Quick Reference Guides*, Drinking Water Requirements for States and Public Water Systems, Office of Ground Water and Drinking Water, Washington, DC, https://www.epa.gov/dwreginfo/drinking-water-rule-quick-reference-guides, 2019b.

Abbreviations: CWSs, Community Water Systems; EPA, Environmental Protection Agency; GW, Groundwater; NTNCWS, Nontransient, Noncommunity Water System; PWSs, Public Water Systems; SDWA, Safe Drinking Water Act; SW, Surface Water.

of safety. MCLGs are not enforced but, instead, are public health goals. Because MCLGs consider only public health and not the limits of detection and treatment technology, they are sometime set at a level at which water systems cannot meet. When determining an MCLG, the EPA considers the risk to sensitive subpopulations (infants, children, the elderly, and those with compromised immune systems) of experiencing a variety of adverse health effects.

For chemicals that can cause noncancer adverse health effects (noncarcinogens), the MCLG is based on the reference dose. A reference dose (RFD) is an estimate of the amount of a chemical that a person can be exposed to on a daily basis that is not anticipated to cause adverse health effects over a person's lifetime. In RFD calculations, an uncertainty factor is used to account for sensitive subgroups of the population. The RFD (mg/kg-d) is multiplied by typical adult body weight (70 kg) and divided by average daily water consumption (2 L/day) to provide a drinking water equivalent level (DWEL) in units of mg/L. The DWEL is multiplied by a percentage of the total daily exposure contributed by drinking water (usually 20%) to determine the numeric value of the MCLG (in mg/L).

If there is evidence that a chemical may cause cancer (carcinogens), it is usually assumed that there is no dose below which the chemical is considered safe (i.e., no threshold), and the MCLG is set to zero. For microbial contaminants that may present public health risk, the MCLG is also set at zero because ingestion of one protozoa, virus, or bacterium may cause an adverse health effect.

Once the MCLG is determined, EPA starts the process of setting an enforceable standard. In most cases, the standard is the MCL, the maximum permissible level of a contaminant in water that is delivered to any user of a PWS. The MCL is set as close to the MCLG as feasible, which the SDWA defines as the level that may be achieved with the use of the best available technology, treatment techniques, and other means which EPA finds are available, while taking cost into consideration.

When there is no economically and technically feasible method to measure a contaminant at low concentrations, a treatment technique (TT) is set rather than an MCL. A TT is an enforceable procedure or level of technological performance that a PWS must follow to ensure control of a contaminant. Examples of TT rules are the Surface Water Treatment Rule and the Lead and Copper Rule listed in Table 7.1. The Surface Water Treatment Rule requires systems using surface water or groundwater under the direct influence of surface water to disinfect their water, filter their water, or meet criteria for avoiding filtration to control levels of contamination by bacteria, viruses, and protozoa. The Lead and Copper Rule states that if lead or copper concentrations exceed an action level in more than 10% of customer taps sampled, the system must undertake a number of additional actions to control corrosion.

EPA gathers input from many external groups during the process of establishing standards. For example, the National Drinking Water Advisory Council (NDWAC) is a committee created by the SDWA that advises the EPA Administrator on everything that the EPA does related to drinking water. It is comprised of five members of the general public, five representatives of state and local agencies, and five representations of private organizations and groups that are active in the field of public health and public water supply. Representatives from water utilities, community environmental groups, states, and the general public are encouraged to participate in public meetings.

7.5 PRIMARY AND SECONDARY DRINKING WATER REGULATIONS

National Primary Drinking Water Regulations (NPDWRs or primary standards) are the legally enforceable MCLs and TTs that apply to PWSs. The EPA sets primary drinking water standards through a three-step process. First, the EPA identifies contaminants that may adversely affect public health and occur in drinking water with a frequency and at levels that pose a threat to public health. The EPA identifies these contaminants for further study and determines contaminants to potentially regulate. Second, the EPA determines an MCLG for contaminants it decides to regulate. Third, the EPA specifies an MCL in drinking water or the required TT for water which is delivered to any user of a PWS.

These national standards for drinking water are based on sound science to protect against health risks, considering available technology, and costs. Each standard also includes requirements for water systems to test for contaminants in the water to make sure standards are achieved. In addition to setting these standards, EPA provides guidance, assistance, and public information about drinking water, collects drinking water data, and oversees state drinking water programs.

The contaminants for which primary drinking water standards have been developed are broken up into the following groupings according to the type of contaminants being regulated: inorganic chemicals, organic chemicals, microorganisms, disinfectants, disinfection by-products, and radionuclides. A list of these contaminants and their respective standards is shown in Table 7.2. Additional information regarding potential health effects, common sources of these contaminants in drinking water, and the public health or MCLGs is provided by the EPA (2009). Primary standards go into effect 3 years after they are finalized. If capital improvements of PWSs are required to meet NPDWRs, the EPA administrator or state agency may allow this period to be extended up to 2 additional years before requiring full compliance with the standards.

TABLE 7.2
National Primary Drinking Water Regulations

		Inorganic Contaminants			
Antimony	0.006	Chromium (Total)	0.1	Mercury (inorganic)	0.002
Arsenic	0.010	Copper	TT	Nitrate as N	10
Asbestos	7×10^6 fibers/L	Cyanide (free)	0.2	Nitrite as N	1
Barium	2	Fluoride	4.0	Selenium	0.05
Beryllium	0.004	Lead	TT	Thallium	0.002
Cadmium	0.005				
		Organic Contaminants			
Acrylamide	TT	Dichloromethane	0.005	Methoxychlor	0.04
Alachlor	0.002	1,2-Dichloropropane	0.005	Oxamyl (Vydate)	0.04
Atrazine	0.003	Di(2-ethylhexyl) adipate	0.4	Pentachlorophenol	0.001
Benzene	0.005	Di(2-ethylhexyl) phthalate	0.006	Picloram	0.5
Benzo(a)pyrene	0.0002	Dinoseb	0.007	Polychlorinated byphenyls (PCBs)	0.0005
Carbofuran	0.04	Dioxin (2,3,7,8-TCDD)	3×10^{-8}	Simazine	0.004
Carbon tetrachloride	0.005	Diquat	0.02	Styrene	0.1
Chlordane	0.002	Endothall	0.1	Tetrachloroethylene	0.005
Chlorobenzene	0.1	Endrin	0.002	Toluene	1
2,4-D	0.07	Epichlorohydrin	TT	Toxaphene	0.003
Dalapon	0.2	Ethylbenzene	0.7	2,4,5-TP (Sivex)	0.05
1,2-Dibromo-3-chloropropane (DBCP)	0.0002	Ethylene dibromide	3×10^{-5}	1,2,4 Trichlorobenzene	0.07
o-Dichlorobenzene	0.6	Glyphosate	0.07	1,1,1-Trichloroethane	0.2
p-Dichlorobenzene	0.075	Heptachlor	0.0004	1,1,2-Trichloroethane	0.005
1,2-Dichloroethane	0.005	Heptachlor epoxide	0.0002	Trichloroethylene	0.005
1,1-Dichloroethylene	0.007	Hexachlorobezene	0.001	Vinyl chloride	0.002
cis-1,2-Dichloroethylene	0.07	Hexachlorocyclo-pentadiene	0.05	Xylenes (total)	10
trans-1,2-Dichloroethylene	0.1	Lindane	0.0002		

(Continued)

TABLE 7.2 (*Continued*)
National Primary Drinking Water Regulations

		Radionuclides			
Radium 226 + 228	5 pCi/L	Alpha emitters	15 pCi/L	Beta particles and photon emitters	4 mrem/yr
Uranium	0.030				
		Microbiological Contaminants			
Cryptosporidium	TT	Heterotrophic plate count	TT	Total coliform	<5% +
Enteric viruses	TT	*Legionella*	TT	Turbidity	TT
Giardia lamblia	TT				
		Disinfectants/Disinfectant By-Products			
Bromate	0.010	Chlorine dioxide as Cl_2	0.8	Haloacetic acid (HAAs)	0.060
Chloramines as Cl_2	4.0	Chlorite	1.0	Total trihalomethanes (TTHMs)	0.080
Chlorine as Cl_2	4.0				

Units are in mg/L unless otherwise noted.

National Secondary Drinking Water Regulations (NSDWRs or secondary standards) are nonenforceable guidelines regulating contaminants that may cause cosmetic effects (such as skin or tooth discoloration) or aesthetic effects (such as taste, odor, or color) in drinking water. The EPA recommends secondary standards to water systems but does not require systems to comply. State and local agencies may choose to adopt these as enforceable standards. A list of these contaminants, their associated secondary MCLs, and contaminant effects are shown in Table 7.3.

TABLE 7.3
National Secondary Drinking Water Regulations

Contaminant	Secondary MCL	Contaminant Effect
Aluminum	0.05 to 0.02	Colored water
Chloride	250	Salty taste
Color	15 color units	Visible tint
Copper	1.0	Metallic taste, blue-green staining
Corrosivity	Noncorrosive	Metallic taste, pipe and fixture corrosion
Fluoride	2.0	Tooth discoloration
Foaming Agents	0.5	Frothy, cloudy, bitter taste, odor
Iron	0.3	Rusty color and staining, metallic taste
Manganese	0.05	Black or brown staining, bitter metallic taste
Odor	3 threshold odor number	"Rotten egg," musty, chemical smell
pH	6.5 to 8.5	Bitter metallic taste, slippery feel
Silver	0.10	Skin discoloration, graying of white of eye
Sulfate	250	Salty taste
Total dissolved solids	500	Deposits, colored water, staining, salty taste
Zinc	5	Metallic taste

Units are in mg/L unless otherwise noted.

7.6 UNREGULATED CONTAMINANTS

The SDWA includes a process where new contaminants are identified that may require regulation in the future with a primary standard. The EPA is required to periodically release a Contaminant Candidate List (CCL), which is used to prioritize research and data collection efforts to help determine whether a specific contaminant should be regulated (U.S. EPA 2019c). On March 2, 1998, the first Drinking Water CCL as part of the Unregulated Contaminant Monitoring Rule (UMCR 1) was released which contained 60 contaminants (10 microbiological contaminants and 50 chemical contaminants). After the list was released, research was undertaken to develop analytical methods for detecting the contaminants, to determine whether they occur in drinking water, to evaluate treatment technologies to remove them from drinking water supplies, and to assess potential health effects resulting from exposure to the contaminants. In July 2003, the EPA announced its final determination for nine contaminants from the CCL 1, which was not to regulate the following contaminants: *Acanthamoeba*, aldrin, dieldrin, hexachlorobutadiene, manganese, metribuzin, naphthalene, sodium, and sulfate.

EPA announced the second Drinking Water Contaminant Candidate List (CCL 2) on February 23, 2005. The CCL 2 list included the remaining 51 contaminants from the CCL 1. In July 2008, after consideration of public comments, the EPA published its final determination that no regulatory action was appropriate or necessary for the following 11 contaminants: boron; dacthal mono-acid (MTP) degradate; dacthal di-acid (TPA) degradate; 1,1-dichloro-2,2-bis(*p*-chlorophenyl) ethylene (DDE); 1,3-Dichloropropene (Telone); 2,4-Dinitrotoluene; 2,6-Dinitrotoluene; s-Ethyl propylthiocarbamate (EPTC); Fonofos; Terbacil; and 1,1,2,2-Tetrachloroethane.

In February 2008, the EPA announced the draft CCL 3. Approximately 7,500 potential chemical and microbial contaminants were considered for inclusion in the CCL 3. Of these, 560 were further evaluated based on the potential for a contaminant to occur in a PWS and the potential for public health concern. Of these, 104 chemical contaminants and 12 microbial contaminants were included in the CCL 3 based on more detailed evaluation of occurrence and potential health effects. On January 4, 2016, the EPA announced the final determinations not to regulate four of the 116 CCL 3 contaminants—dimethoate; 1,3-dinitrobenzene; terbufos; and terbufos sulfone. The agency delayed the final regulatory determination on strontium to consider additional data and decide whether there is a meaningful opportunity for health risk reduction by regulating strontium in drinking water.

The EPA announced the Final CCL 4 on November 17, 2016, which included 97 chemicals or chemical groups and 12 microbial contaminants. The list includes, among others, chemicals used in commerce, pesticides, biological toxins, disinfection by-products, pharmaceuticals, and waterborne pathogens. Evaluation of the CCL 4 group is complete, and the EPA final regulatory determination for this CCL 4 includes 97 chemicals or chemical groups and 12 microbial contaminants (U.S. EPA 2019d). Nominations for candidate chemicals and microbial contaminants for CCL 5 ended in December 2018, and EPA is evaluating the nominations and other contaminant data and information and plans to publish a draft CCL 5 for public review and comment in the near future.

7.7 APPLICATIONS

Four Illustrative Examples complement the material presented in this chapter on the SDWA.

Illustrative Example 7.1

A long list of rules has been promulgated through the SDWA as summarized in Table 7.1. Select one of these rules and provide a summary of when the rule was enacted, what PWSs the rule applies to, what the basic requirements are, and what actions must be taken in response to the rule.

Solution

This is an open-ended question the solution of which will depend on the specific rule being chosen to describe. Two example responses are provided.

The Arsenic Rule: This rule was promulgated as a final rule on January 22, 2001 (66 FR 6976-7066). This rule revised the existing arsenic standard in finished drinking water set in 1975 from 50 parts per billion (ppb) to 10 ppb. In setting the new standard, EPA used its discretionary authority under the 1996 SDWA amendments to set the standard at a level that "maximizes health risk reduction benefits at a cost that is justified by the benefits." In other words, although technology will allow lower levels of arsenic to be reached, the EPA determined that the potential health benefits did not justify the added cost. On May 22, 2001, the EPA announced that it would delay the effective date of the arsenic rule until February 22, 2002, allowing time to complete a reassessment and to afford the public a full opportunity to provide further input. On July 19, 2001, the EPA requested comment on whether the data and technical analyses associated with the January 2001 arsenic rule supported setting the standard at 3, 5, 10, or 20 ppb. The EPA and others reviewed the technical analyses supporting the standard, and in October 2001, the EPA determined the promulgated arsenic standard of 10 would be the final standard. All finished water delivered in the United States by CWSs and NTNCWSs must meet this 10 ppb arsenic standard through implementation of appropriate arsenic removal technology or be found to be non-compliant with the SDWA requirements.

The Filter Backwash Recycle Rule: This rule was promulgated on June 8, 2001 (66 FR 31085-31105). It requires spent filter backwash, sludge thickener supernatant, and liquids from dewatering processes to pass through all of the unit processes of a system's existing conventional or direct filtration system, or through an alternative location approved by the state to avoid possible disruption of the treatment process by hydraulic surges through the facility, creation of a coagulation chemistry imbalance, or return of concentrated amounts of disinfection–resistant pathogens (such as Cryptosporidium) through the plant. This rule applies to all PWSs using surface water or groundwater under the direct influence of surface and that recycle backwash from conventional or direct filtration, sludge thickener supernatant, or liquids from dewatering processes. Systems were required to begin compliance with notification requirements in December 2003 and actually meet recycle return location requirements by June 8, 2006.

Illustrative Example 7.2

The SDWA specifically calls out "Groundwater under the direct influence of surface water" (GWUDI) to be differentiated from simply groundwater. Describe what the term GWUDI refers to and why the EPA is concerned about this category of a drinking water supply.

Solution

Groundwater is a water supply that is pumped from underground aquifers through drilled wells or reaches the surface from springs where the water table intersects the ground surface. Groundwater from shallow aquifers, from aquifers near surface water sources, or from sources not well protected through the natural geology may be subject to influence from surface water sources. This groundwater may have characteristics commonly associated with surface water (e.g., presence of large microbiological contaminants such as cysts), and as such, this groundwater is defined as groundwater under the influence of surface water and is regulated like surface water.

The EPA explicitly defines GWUDI of surface water as water beneath the surface of the ground with: (i) significant occurrence of insects or other macroorganisms, algae, or large-diameter pathogens such as *Giardia lamblia*, or (ii) significant and relatively rapid shifts in water characteristics such as turbidity, temperature, conductivity, or pH that closely correlate to climatological or surface water conditions. GWUDI faces the same risks as surface water and the same precautions should be taken before using GWUDI as a drinking water source.

Surface water is vulnerable to both chemical and microbiological contamination and in most cases requires filtration and disinfection before it is safe to drink. Groundwater, which is protected by layers of soils and other subsurface materials, sometimes does not require treatment other than being disinfected before it is used as drinking water to provide disinfectant residual in the distribution system to reduce the risk of microbiological contamination. Groundwater can become contaminated, however, through infiltration from the surface, injection of contaminants, or by naturally occurring substances in the soil or rock through which it flows. Groundwater generally is much less susceptible to contamination than surface water and generally requires much less treatment than surface water or GWUDI.

Illustrative Example 7.3

The SDWA defines a number of different types of water systems that provide drinking water to the public. These different types of systems must comply with some Drinking Water Rules but not others, so it is important to understand the distinction among these different types of systems. Describe the various water systems defined by the SDWA, give examples of each, and state why they are regulated.

Solution

The first delineation of water systems by the SDWA is between PWSs and non-PWSs. PWSs are defined as "a system for the provision to the public of water for human consumption through pipes or other constructed conveyances, if such system has at least 15 service connections, or regularly serves at least 25 individuals." [Section 1401(4)(a)]. Thus, individuals on wells and systems that serve fewer people are not covered under federal regulations, though some states regulate smaller systems. Federally regulated systems are called "public water systems" because they serve water to the public and not because they are publicly owned. It is important to note that a PWS *may* be publicly owned (e.g., owned by a municipality) but can also be privately owned (e.g., owned by an investor-owned utility or by the owner of a mobile home court).

Within the realm of PWSs, there are two distinct types of systems. *Community water systems* (CWSs) and *non-community water systems* (NCWSs). CWSs include any PWS that serves 25 people or 15 connections year-round. Examples of CWSs include municipal water systems or water systems that serve a mobile home park or other groups of residents. NCWSs are PWSs that do not serve a permanent resident population. NCWSs are further divided into two water system types, *non-transient, non-community water systems (NTNCWSs)* and *transient, non-community water systems (TNCWSs)*. NTNCWSs include systems serving at least the same 25 people at least 6 months of the year, such as some churches, schools, and factories. TNCWSs, includes facilities such as roadside stops, commercial campgrounds, hotels, and restaurants that have their own water supplies and serve a transient population at least 60 days per year.

The decision to regulate systems serving 15 service connections or 25 people was somewhat arbitrarily decided during the debate in Congress for the 1974 SDWA before the actual number of water systems was known. PWSs are divided into the categories because the risks to the populations these systems serve vary. The majority of PWSs are TNCWSs. Although these systems are numerous, they do not serve the majority of the population because each system only serves a small number of people. However, almost everyone is served by TNCWSs at some point in their life. (Remember that TNCWSs include roadside stops, commercial campgrounds, hotels, restaurants, and other facilities that have their own water supplies and serve a transient population at least 60 days per year). For example, water that you drink at a campground or a restaurant may be from a TNCWS. Therefore, it is important to regulate these systems even though a given TNCWS will generally serve only a small population.

Illustrative Example 7.4

The SDWA uses a comprehensive, preventive approach to drinking water source protection using a multiple-barrier approach combined with standards, monitoring, and enforcement. Discuss this preventative source water protection approach that includes watershed and wellhead protection programs, along with Sole Source Aquifer designation.

Solution

The 1996 Amendments to the SDWA established a Source Water Protection Program, which includes measures to identify and protect all sources (both surface water and groundwater) of drinking water in the United States. A Source Water Protection Area (SWPA) is defined as the watershed or groundwater area that may contribute pollution to a water supply. The entire SWPA needs to be protected to minimize pollution of the source water.

A watershed protection area may include wellhead protection areas because protection of surface water sources may encompass areas that recharge a ground water well. A wellhead protection area is the area surrounding a drinking water well or well field, including the "recharge zone," the land area that replenishes the aquifer, that is protected to prevent contamination of the wells.

Whether a PWS relies on surface water, groundwater, or a combination of the two, protection of a water system's source is important. Prevention of contamination is one of the most cost-effective methods of ensuring safe drinking water supplies. If source water becomes contaminated, expensive treatment or replacement of the water source may be required before safe drinking water can be delivered again to users.

The Sole Source Aquifer (SSA) program is a special case of watershed protection to ensure the sustainability of high-quality and reliable source water. An SSA is one that supplies at least 50% of the drinking water consumed in the area overlying the aquifer. These areas can have no alternative drinking water source that could physically, legally, and economically supply all those who depend on the aquifer for drinking water. As of 2017 there were 89 SSAs identified throughout the United States.

No federal financial assistance may be provided for any project that may contaminate an area designated as an SSA, and any proposed projects with federal financial assistance that have the potential to contaminate SSAs are subject to EPA review. This review is coordinated with National Environmental Policy Act reviews and with relevant federal, state and local agencies. Examples of projects that might be subject to review include highways, wastewater treatment facilities, construction projects that involve storm water disposal, public water supply wells and transmission lines, agricultural projects that involve the management of animal waste, and projects funded through Community Development Block Grants.

Careful source selection and source water protection through watershed and wellhead protection and SSA designation are a component of this multiple barrier approach that also includes water treatment and distribution system monitoring and protection to ensure safe water supplies for the public.

REFERENCES

Tiemann, M. 2017. *Safe Drinking Water Act (SDWA): A Summary of the Act and Its Major Requirements.* RL31243. Washington, DC: Congressional Research Service. https://fas.org/sgp/crs/misc/RL31243.pdf.

U.S. Environmental Protection Agency. 2004. *Understanding the Safe Drinking Water Act.* EPA 816-F-04-030. Washington, DC: Office of Water. https://www.epa.gov/sites/production/files/2015-04/documents/epa816f04030.pdf.

U.S. Environmental Protection Agency. 2009. *National Primary Drinking Water Regulations.* EPA 816-F-09-004. Washington, DC: Office of Ground Water and Drinking Water. https://www.epa.gov/sites/production/files/2016-06/documents/npwdr_complete_table.pdf.

U.S. Environmental Protection Agency. 2019a. *Information about Public Water Systems.* Drinking Water Requirements for States and Public Water Systems. Washington, DC: Office of Ground Water and Drinking Water. https://www.epa.gov/dwreginfo/information-about-public-water-systems.

U.S. Environmental Protection Agency. 2019b. *Drinking Water Rule Quick Reference Guides.* Drinking Water Requirements for States and Public Water Systems. Washington, DC: Office of Ground Water and Drinking Water. https://www.epa.gov/dwreginfo/drinking-water-rule-quick-reference-guides.

U.S. Environmental Protection Agency. 2019c. *Monitoring the Occurrence of Unregulated Drinking Water Contaminants.* Washington, DC: Office of Ground Water and Drinking Water. https://www.epa.gov/dwucmr.

U.S. Environmental Protection Agency. 2019d. Contaminant Candidate List (CCL) and Regulatory Determination, Contaminant Candidate List 4 - CCL4. Washington, DC: Office of Ground Water and Drinking Water. https://www.epa.gov/ccl/contaminant-candidate-list-4-ccl-4-0.

8 Water Monitoring and Analysis

8.1 INTRODUCTION

Human and natural activities can change the physical, chemical, and biological characteristics of water and have specific ramifications for human and ecosystem health. Water quality is affected by a host of factors, such as changes in nutrients, sedimentation, temperature, and pH. These changes can pose as a threat to the quality of drinking water. A host of water properties that should be considered in any water analysis include pH, acidity, alkalinity, hardness, turbidity, and dissolved oxygen (DO). For wastewater samples, standard water analysis includes pH, DO, total suspended solids (TSS), volatile suspended solids (VSS), nutrients (nitrogen [N] and phosphorus [P]), chemical oxygen demand (COD), and biochemical oxygen demand (BOD) as well as general water quality indicators such as oil and grease. Elements that should also be included in the analysis are iron, manganese, and a range of potentially toxic metals (lead, arsenic, copper, etc.) that may have ecological impacts as well as human health standards. The monitoring and measurement of these pollutants in water and wastewater is a necessary component of effective water management.

Compounding monitoring and analysis complexities is that some pollutants, when combined, can cause worse, or different impacts than the cumulative effects of a single pollutant. In addition, concerns over potential terrorist acts against drinking water infrastructure (see Chapter 23) has elevated the need for robust water system monitoring and analysis programs for both routine and emergency monitoring situations.

One of the authors, resides in the Incorporated Village of East Williston, New York. Interestingly, the Village routinely monitors, in accordance with state and federal regulations, numerous parameters well beyond the basic analyte list. During 2017, the Village conducted water quality tests for more than 130 contaminants in their drinking water. A number of those contaminants are listed in Table 8.1 (City of East Williston 2018).

This chapter highlights sampling method selection for surface and groundwater; standard practice for sample collection, documentation, and handling; sample containers and preservation methods; analytical methods; and appropriate statistical analyses of results. An applications section includes six Illustrative Examples related to water monitoring and analysis concepts.

8.2 SELECTING A SAMPLING METHOD

The method used to sample a given media will directly affect the accuracy and representativeness of the analysis. It is imperative that an appropriate methodology be selected to obtain the most reliable results possible; in effect, to be representative of the population from which the sample is drawn (see Section 8.8 for reference). Several factors should be considered when selecting a sampling method, including:

1. The sampling program objectives (documenting exposures, determining regulatory compliance, locating a source)
2. The media to be sampled (soil, vegetation, air, water, sludge, etc.)
3. The physical and chemical properties of the contaminant(s) of interest
4. Potential matrix effects as a result of characteristics of other contaminants in the sample

---

Contributing Author: Jake McDermott

TABLE 8.1
Water Analytes in Drinking Water of East Williston, New York

1,1,1,2-Tetrachloroethane	4-Isopropyltoluene (P-Cumene)	Chlordane	Ethylbenzene	Oxamyl
1,1,2,2-Tetrachloroethane	Alachlor	Chloroacetic Acid	Fluoride	o-Xylene
1,1,2-Trichloroethane	Aldicarb	Chlorobenzene	Free Cyanide	Pentachlorophenol
1,1-Dichloroethane	Aldicarb Sulfone	Chlorodifluoromethane	Glyphosate	Perchlorate
1,1-Dichloropropene	Aldicarb Sulfoxide	Chloroethane	Heptachlor	Picloram
1,2,3-Trichlorobenzene	Aldrin	Chloromethane	Heptachloro Epoxide	Propachlor
1,2,3-Trichloropropane	Ammonia	Chromium	Hexachlorobenzene	sec-Butylbenzene
1,2,4-Trimethylbenzene	Antimony	cis-1,2-Dichloroethene	Hexachlorobutadiene	Silver
1,24-Trichlorobenzene	Arsenic	cis-1,3-Dichloropropene	Hexachlorocycleopentadiene	Simazine
1,2-Dibromo-3-Chloropropane	Atrazine	Color	Isopropylbenzene (Cumene)	Styrene
1,2-Dibromoethane (EDB)	Benzene	Dalapon	Lindane	Tert-Butylbenzene
1,2-Dichlorobenzene	Benzol(A)Pyrene	Detergents (MBAS)	M, P-Xylene	Tetrachloroethene
1,2-Dichloroethane	Beryllium	Dibromoacetic Acid	Manganese	Thallium
1,2-Dichloropropane	bis(2-Ethylhexyl)adipate	Dibromomethane	Mercury	Toluene
1,3,5-Trimethylbenzene	bis(2-Ethylhexyl)phthalate	Dicamba	Methomyl	Total Aldicarbs
1,3-Dichlorobenzene	Bromoacetic Acid	Dichloroacetic Acid	Methoxychlor	Total Coliform Bacteria
1,3-Dichloropropane	Bromobenzene	Dichlorodifluoromethane	Methyl Tert. Butyl Ether (MTBE)	Total Halocecetic Acid
1,4-Dichlorobenzene	Bromoform	Dieldrin	Methylene Chloride	Total PCBs
2,2-Dichloropropane	Bromomethane	Dinoseb	Metolachlor	Toxaphene
2,4,5-TP (Silvex)	Butachlor	Dioxin	Metribuzin	Trans-1,2-Dichloroethane
2,4-D	Cadmium	Diquat	N-Butylbenzene	Trans-1,3-Dichloropropene
2-Chlorotoluene	Carbaryl	*E. coli* Bacteria	Nickel	Trichloroacetic Acid
3-Hydroxycarbofuran	Carbofuran	Endothall	N-Propylbenzene	Trichlorofluoromethane
4-Chlorotoluene	Carbon Tetrachloride	Endrin	Odor	Turbidity
				Vinyl Chloride

Source: City of East Williston, *East Williston Annual Water Report*, East Williston, NY, 2019.

5. Regulatory requirements
6. Safety requirements
7. Difficulty of using a specific method
8. Cost
9. Scale of sample area (small-scale site related to individual persons vs a large-scale site)
10. Accessibility to the sample site
11. Short- versus long-term sampling requirements

In addition to these, basic professional judgment also plays a role in selecting a sampling method. The New England Waterworks Association (2004) has produced a comprehensive *Pocket Sampling Guide for Operators of Small Water Systems* that provides both general potable water system sampling guidelines as well as guidelines addressing requirements for specific contaminants regulated by the Safe Drinking Water Act. The U.S. EPA (2016) also provides guidelines for routine drinking water sample collection. Comprehensive guidelines for sampling in response to potable water system threats and terrorist incidents can be found at U.S. EPA (2004, 2017a).

8.3 STANDARD PRACTICES FOR SAMPLING OF WATER

The goal of sampling is to obtain an aliquot for analysis that is truly representative of the larger volume of media being sampled. The most critical factors necessary to achieve this goal are points of sampling, time of sampling, frequency of sampling, and maintenance of the integrity of the sample after sampling and prior to analysis. Homogeneity in the media being sampled is frequently lacking, necessitating multiple-point composite sampling. If it is impractical to use a single representative sampling point, it may be practical to determine and understand interrelationships among various sampling points so that results obtained with a minimum number of sampling locations may be extrapolated to the larger media volume and properly interpreted. With adequate interpretation, a nonrepresentative sample can yield valuable information about trends and can indicate locations where more representative data could be collected. Most samples collected from a single point in a system must be recognized as being nonrepresentative to some degree. Therefore, it becomes important to recognize the degree of a representation in the sample and to make it a part of the permanent record. Otherwise, an artificial degree of precision is assigned to the data when it is recorded.

The following general rules are applicable to all sampling procedures.

1. The samples must represent the conditions existing at the collection point.
2. The samples must be of sufficient volume and must be taken frequently enough to permit reproducibility for the desired objective, as dictated by the method of analysis to be employed.
3. The samples must be collected, preserved, labeled, packaged, shipped, stored, and handled before analysis in a manner that safeguards against change in the particular constituents or properties to be quantified.

Underlying all sampling and analysis programs should be a well thought out and executed sampling and analysis plan (SAP), which describes the sampling locations, numbers, and types of samples to be collected, the quality control requirements of the project, sample labeling conventions, and proper preservation and shipping requirements to ensure sample integrity. The EPA provides examples of SAPs and sample collection documentation in EPA (2004, 2017a).

8.4 SAMPLING OPTIONS FOR WATER

There are generally three procedures for water sample collection. The first is the collection of a grab sample at a specific location representing conditions at the time of sampling. It is the only procedure suitable for bacteriological analyses, VOC compound analyses, and some radiological test procedures. The second is collection of a composite sample at a specific site, portions of which are collected at varied time intervals on a time or volume-weighted basis. Alternatively, the composite may consist of subsamples collected at various locations within a site, a procedure particular appropriate for sludge and contaminated soil samples. The third procedure collects data on a continuous basis, from one or more sampling locations, generated from on-stream analyzers (ASTM 1990) in real-time in situ data-collection networks in many watersheds throughout the country (Michael Baker International et al. 2015).

A number of state and federal guidelines for sampling for compliance with the Safe Drinking Act, the Water Quality Act, and so on, have been provided to the regulated community over the years, and Table 8.2 provides a list of a number of these relevant sampling documents. All field activities should be performed in accordance with standard operating procedures (SOPs) prepared for the entire sampling process. Details, including information on groundwater monitoring well sampling, follow.

8.4.1 Grab Sampling

The procedure for collecting grab samples is applicable to sampling water from potable water and wastewater treatment processes as well as sources such as wells, rivers, streams, lakes, oceans, springs, and reservoirs for chemical, physical, bacteriological, or radiological analysis.

TABLE 8.2
Federal and State Sampling Guidelines Provided to Support Regulatory Compliance

Author	Title	Report Number, Date
U.S. EPA, Robert S. Kerr Environmental Research Laboratory	Manual of Ground-Water Quality Sampling Procedures	EPA-600/2-81-160, 1981
U.S. EPA, Environmental Monitoring and Support Laboratory	Handbook for Sampling and Sample Preservation of Water and Wastewater	EPA-600/4-82-029, 1982
U.S. EPA, Office of Water	NPDES Storm Water Sampling Guidance Document	EPA 833-8-92-001, 1992
U.S. EPA, Office of Research and Development	Subsurface Characterization and Monitoring Techniques, A Desk Reference Guide, Volume I: Solids and Ground Water Appendices A and B	EP A/625/R-93/003a, 1993
New England Water Works Association	Pocket Sampling Guide for Operators of Small Water Systems	2004
U.S. EPA, Office of Ground Water and Drinking Water, Water Security Division	Response Protocol Toolbox: Planning for and Responding to Drinking Water Contamination Threats and Incidents, Interim Final—August 2004. Response Guidelines	EPA 817-D-04-001, 2004
U.S. EPA, Region 8 Laboratory	Quick Guide to Drinking Water Sample Collection	2016
U.S. EPA, Office of Water	Sampling Guidance for Unknown Contaminants in Drinking Water	EPA-817-R-08-003, 2017
U.S. EPA, Office of Research and Development	Sample Collection Information Document for Pathogens	EPA/600/R-17/374, 2017
U.S. EPA, Office of Research and Development	Sample Collection Information Document for Chemicals, Radiochemicals and Biotoxins	EPA/600/R-17/389, 2017

A reasonably accurate estimate of the composition of a raw water in a large lake or reservoir should be far enough from the shoreline to avoid influence from inflowing tributaries and other discharges and may be made by taking individual samples at infrequent intervals (such as biweekly or monthly) sufficient to cover seasonal changes. Samples should be taken on a more frequent basis (such as daily) to provide specific data on the variations in composition where this information may be important based on the planned use of the water.

The laboratory that will be used for sample analyses will normally prescribe the volume and bottle type for the constituent(s) of interest based on EPA (2004, 2016) guidelines and requirements. The minimum sample volume should be three to four times the minimum sample volume required. This is to ensure that there will be enough volume remaining for analysis in the case of spillage, additional laboratory dilution requirements, etc.

When sampling large, open bodies of water, it is best practice to attempt to sample from the same location over time. This can be accomplished by installing a metal post (such as a T-Post) on the bank far above the high-water mark and collecting the sample directly below the post. Multiple stakes may be installed around the body of water, if collecting multiple samples. A survey of each stake should be taken in case of vandalism or specific GPS locations can be used to ensure repetitive sampling from the same location over time. One should also avoid collecting surface scum, and if a boat is used, to be sure that the propellers or oars do not stir up sediment to contaminate the representative sample.

If sampling large rivers or streams, multiple samples are usually taken at specific depths. This is as a result of the water not being uniformly mixed vertically, so that a better representation of the main body of water is generated when all samples are analyzed.

When sampling springs (U.S. EPA 1982), a hole deep enough (typically 6 to 10 inches) to sufficiently fill the bottle sets should be dug at the point source of the spring, days or weeks prior to the sampling date. This is to ensure that no sediment is stirred up when the collection of sampling occurs. Surface scum, insects, and vegetation should be avoided when collecting the sample. It is typically best to use a 250-mL glass beaker that has been rinsed with deionized water to collect the sample, before transferring the water into the bottle sets. Upon collection of a spring sample, a measurement of flow should also be taken. This can be done with a flow meter, provided the water is deep enough. If the water level is not deep enough, or the current not fast enough, a portable weir can be used. It is best practice to take the flow measurement at the same location every time the site is visited by installing a T-Post with a GPS survey of the location.

For potable water system sampling (U.S. EPA 2016), select a cold water faucet for sampling that is free of contaminating devices such as screens, aeration devices, hoses, purification devices, or swiveled faucets. Check the faucet to be sure it is clean. If the faucet is in a state of disrepair, select another sampling location. Collect samples from faucets that are high enough to put a bottle underneath without contacting the mouth of the container with the faucet. Open the faucet and thoroughly flush, generally 2 to 3 minutes. However longer times may be needed, especially in the case of lead distribution lines. Typically, water temperature will stabilize, which indicates flushing is completed. Once the lines are flushed, adjust the flow so it does not splash against the walls of the bathtub, sink, or other surfaces. Specific collection instructions should be followed as provided by the U.S. EPA (2016) for the analytes of interest.

Specific sampling guidance is provided by the EPA (2016) to evaluate compliance with the *Lead and Copper Rule*. For this sample collection, a cold water faucet that is free from devices that are designed to change the water composition, such as water softeners or point-of-use filters, should be selected, but any screens or aeration devices should not be removed prior to sampling. If a first-flush sample is being collected for lead/copper, allow the water to sit in the pipe network, undisturbed for at least 6 hours, and do not intentionally flush the water line before the start of the 6-hour period. A wide-mouth 1-L bottle is placed under the faucet and used to collect the first water out of the tap.

A detailed discussion of the complexities of wastewater treatment plant sampling, in terms of sampling location within a process as well as significant variability expected daily as well as seasonally, is presented in EPA's *Handbook for Sampling and Sample Preservation of Water and Wastewater* (EPA 1982), and the reader is referred to this reference to explore the intricacies of this sampling challenge.

8.4.2 Composite Sampling

This type of sampling is generally applicable for subsequent chemical and physical analyses but may not be suitable for radiological examination. Also, composite samples are not suited for bacteriological examination, or constituents that change when exposed to air, such as purgeable organics.

Automatic samplers, such as marketed by ISCO or YSI for example, are commonly employed in the flow or time-weighted composite sampling of wastewater and stormwater samples to augment grab samples for generating representative, composite samples of a flow stream of interest. One should consult individual test methods for the effect of holding time and temperature on analysis results (U.S. EPA 1982, 2016, 2017a; New England Waterworks Association 2004). Changes of unstable constituents that become part of a composite sample may be determined by the analysis on individual samples separately collected over time.

In sampling process waters, collected composite samples should be over one 24-hour period. Samples should be collected in increments of 15 minutes to 1 hour and in proportion to the rate of flow. A volume of 4 L is suitable for a composite sample. The point of sampling should be the same as described in the section of grab samples.

8.4.3 Continuous Sampling

Using this method, a sample or a parameter measurement is collected on a continual basis for chemical, physical, or radiological analytes. Analytes lending themselves to continuous data collection are those for which probes or rapid sample analysis are available. The most common are pH, temperature, dissolved oxygen, turbidity, flow rate, electrical conductivity, and residual chlorine in potable water. Additional sensors for specific analytes of interest in surface water quality (nitrate specific electrodes, fluorescent dissolved organic matter [fDOM] probes, chlorophyll sensor) and wastewater (oxygen respiration sensor) are at various stages of development and use. Researchers (Horsburgh et al. 2010; Schilling et al. 2017) are also investigating the use of these readily available water quality sensors as surrogates for other, more difficult to measure water quality parameters (i.e., turbidity as a surrogate measure for phosphorous, etc.). These continuous monitoring data have proven to provide insights into a wide range of environmental systems not possible through grab or limited composite sampling (Dupont et al. 2018) and will continue to expand in application in the future.

8.4.4 Groundwater Monitoring Wells

Groundwater quality is important and is of national and international concern. The quality of groundwater, delineating contaminant plumes, and establishing the integrity of hazardous material management facilities are some of the uses of groundwater monitoring well data. As with surface water sampling, the goal of sampling groundwater monitoring wells is to obtain samples that are truly representative of the aquifer or groundwater in question. Because of the nature of groundwater flow, what water that may be contained within monitoring wells can become stagnate and unrepresentative of the true aquifer conditions because of chemical and biochemical changes within the well that can cause water quality alterations. Sample quality can also be greatly changed in the process of sample collection, particularly from deep wells that might be impacted by microbial activity. Anaerobic water can be oxygenated through improper sampling and handling techniques for example, giving rise to redox reactions at the surface that are not representative of the chemistry and water quality existing in the aquifer at depth. Because of the sensitivity of groundwater quality to sampling techniques, the U.S. EPA (1993) and the U.S. Geological Survey (USGS 2006) have published guidance documents carefully describing appropriate well purging and sampling techniques for the preservation of groundwater quality during sample collection.

General guidelines for monitoring well sampling to maintain the integrity of the collected sample and ensure its representativeness and reliability are summarized by USGS (2006) and include:

1. Plan sampling at sites in a sequence that avoids contamination. Start with pristine sites or those least likely to be contaminated or with lowest concentrations of dissolved solids or target analytes. End at the site with the highest expected concentrations of target analytes and revise the sampling plan accordingly based on findings.
2. Clean equipment. Sample only with decontaminated equipment and quality assure the efficacy of the cleaning procedures using equipment blanks.
3. Purge the well of standing water. Purge the well to reduce artifacts from well installation or sampler deployment. If possible, pump at a rate that does not overly stress the aquifer, creating drawdown and mobilizing particulates. Protocols for purging and pumping rate can depend on well type and study objectives.
4. Isolate the sample. For example, use packers downhole and processing and preservation chambers at land surface.
5. Avoid temperature changes. Keep sample tubing as short as possible and shaded from direct sunlight.
6. Avoid sample aeration. Filter in-line; use thick, nonpermeable sample tubing; completely fill filtration assemblies and sample tubing with sample; fill sample bottles from bottom up to overflowing whenever appropriate; handle anoxic water under an inert gas atmosphere, if necessary.
7. Collect quality-control samples. Review the analytical results and adjust field procedures, if necessary, before the next sampling round.

8.5 SAMPLE DOCUMENTATION AND HANDLING

Each sample must be properly documented to ensure timely, correct, and complete analysis for all parameters requested, but more importantly, to support the use of sample data in decision making and regulatory compliance reporting (U.S. EPA 2017). The documentation system provides the means to individually identify, track, and monitor each sample from the point of collection through final data reporting. Used herein, a sample is defined as a representative specimen that is collected at a specific location of a site at a particular point in time for a specific analysis and that may refer to field samples, duplicates, replicates, splits, spikes, or blanks that are shipped from the field to the laboratory.

8.5.1 Sample Identification Number

A sample identification number should be created by the sample collector, the receiving laboratory, or a program or project manager that is unique for each sample. Sample identification numbers often consist of elements describing the sample type, matrix, location, and date and time of collection. This number is unique to each sample. It is generally included in the sample documentation, and used to identify the sample in field reports and log books, chain-of-custody (COC) forms, and sample containers and labels. The number can also be used on corresponding analytical data reports or evaluations.

8.5.2 Sample Container Labels

Each sample container should have a label that clearly provides information identifying and describing the sample. Ideally, sample container labels should provide the following information:

1. Site name
2. Sample identification number
3. Date and time the sample was collected (including time zone)
4. Sampling location (e.g., site name or address)
5. Container size
6. Container type
7. Type of sample (grab or composite)
8. Analysis required
9. Preservatives added, if applicable
10. Dechlorination method, if applicable
11. Name or initials of sample collector(s)

All of the information on the sample label should be identical to the information on the COC form. The sample collector should be able to recollect where and when the samples were taken in case additional sampling or analysis is necessary. To facilitate sample collection activities and ensure proper labeling, sample containers should be prelabeled as much as is practical before sample collection. Sample labels should be completed with a waterproof pen and securely affixed to each sample container to identify each sample clearly. If a waterproof pen is not used, it is recommended that the sampler cover the label(s) with clear packaging tape after writing the sampling information onto the label. An example sample label is provided in Figure 8.1. Samples are then shipped under COC procedures described next.

8.5.3 Chain-of-Custody Record

A COC form creates an accurate written record that can be used to trace the creation, possession, and handling of a sample from the moment it is collected through analysis. A COC form is used and required, without exception, for the tracking and recording of on-site or off-site sample collection, transport, and analysis. A COC form accompanies each sample and group of samples as custody of the sample(s) is transferred from one custodian to another. One copy of the form is retained by the original sample collector, and the original is obtained by the receiving laboratory. If multiple laboratories are receiving a sample, individual COC forms should be submitted to each laboratory describing the samples each are to receive and analyze. Each COC form represents the contents of the sample shipment. Each laboratory

Site Name: ____________________
Sample ID Number: ____________
Date: ________ Time: ________
Location: ____________________
Container Size: ______________
Container Type: ______________
Sample Type (e.g., grab, composite):
____
Analysis: ____________________
Preservative: ________________
Dechlorination: ______________
Collected by (initials): ________

FIGURE 8.1 Example sample label.

or facility representative who accepts an incoming sample shipment signs and dates the COC form. It is typically the laboratory or facility's responsibility to maintain internal logbooks and custody records throughout sample preparation and analysis. Sample custodians are typically responsible for initiating, maintaining, and completing COC tracking. A sample custodian is the person responsible for the custody of a sample or samples at a particular time, until custody is transferred to another person (and so documented), who then becomes the new custodian. A sample is under a person's custody if (U.S. EPA 2017):

- It is in that person's possession
- It is in that person's view, after being in that person's physical possession
- It was in that person's physical possession and then he or she locked it up to prevent tampering, or
- It was in that person's physical possession and then he or she placed it in a designated and identified secure area.

Under this definition the sampler is personally responsible for the care and custody of the samples collected until they are transferred or dispatched properly. Handling of COC forms during sample transportation depends on the method of transport. If the laboratory is within driving distance, the sample containers can be couriered to the laboratory. It is important to note that common commercial carriers will not usually accept responsibility for handling and completing COC forms. This often necessitates packing the COC form in the shipping container (enclosed with other documentation in a plastic zipper-type bag). As long as COC forms are sealed inside the shipping or transport container and the container's custody seals are intact, commercial carriers are not required to sign the COC form.

Although COC forms vary in style, format, and detail, the forms should contain the same minimal information required to identify the sample. Procedures for filling out other styles of COC forms should be similar. It is best for the samplers to fill out the COC form provided by the party receiving the samples. An example COC form is shown in Figure 8.2.

The following information should be provided and the following steps should be followed to complete COC forms:

- General incident information (sample owners, contact information, site name)
- Sample specific information for each sample that will be traveling in the same cooler/transport container (i.e., sample identification number, sample type [matrix], grab or composite, number and type of sample containers, and date and time sample was collected)

Chain-of-Custody Form

Site Name:	Sample Owner/Collector:
Contact Information:	Signature:

Sample ID	Collection Date/Time (24 h)	Sample Location	Sample Type (Matrix)	Grab/ Composite	Preservative(s)	No. of Bottles	Bottle Type	Analysis Requested

Matrix: DW = Drinking Water, RW = Reservoir Water, UW=Untreated Water, GW = Groundwater, SW = Surface Water

Relinquished By:	Received by:	Date/Time:
Relinquished By:	Received by:	Date/Time:
Relinquished By:	Received by:	Date/Time:
Relinquished By:	Received by:	Date/Time:
Relinquished By:	Received by:	Date/Time:

Dispatched by:	Date/Time:	Received by:	Date/Time:

Analytical Service Requestor (ASR):	Contact Information:
Method of Sample Transport:	

Shipper:	Phone No.:	Tracking No.:

FIGURE 8.2 Example chain of custody form.

- Contact information and affiliation of the sampler or responsible party
- Sign, date, and enter the time under "Relinquished by" entry. Have the person receiving the sample sign the "Received by" entry. If shipping samples by a common carrier, print the carrier to be used in this space (e.g., FedEx, UPS).
- If a common carrier is used, a copy of the airbill is to be kept for recording purposes by both the sender and recipient.
- Place the original signed copy of the COC form in a plastic zipper-type bag or other appropriate waterproof sample shipping package. Retain a copy with the field records.
- Complete carrier-required shipping papers.
- If possible, fax or scan and e-mail a copy of the COC form and field report to the party receiving the samples.

8.5.4 Sample Packaging and Shipping

One should always consult the laboratory regarding packaging requirements before the initiation of a sampling program. Packing lists, COC records, and any other shipping/sample documentation accompanying the shipment must be enclosed in a waterproof plastic bag and taped to the underside of the shipping cooler lid. Waterproof, metal, or hard plastic ice chests or coolers are the only acceptable type of sample shipping container. Cardboard and styrofoam coolers should not be used. Coolers must be sealed with custody seals in such a manner that the custody seal would be broken if the cooler were opened. Shipping coolers must clearly have a visible return address on the outside.

Shipping coolers are then secured, and custody seals placed across the cooler's openings. As long as custody forms are sealed inside of the sample cooler, and custody seals remain intact, commercial carriers are not required to sign off on the custody form. Inside the cooler, sample containers should be enclosed in clear plastic bags in which the sample tags and labels are visible on the sample bottle sets. The next section describes the containers that are used for collection of various samples.

Water samples for low or medium-level organic analysis and low-level inorganics analysis must be shipped cooled to 4°C with ice. No ice is required when shipping inorganic low-level soil samples or medium- to high-level water samples. No ice is required when shipping soil samples but may be used if the sampler chooses. All cyanide samples must also be shipped cooled to 4°C with ice. Low- and medium-level water samples for inorganic analysis require chemical preservation.

Sufficient packing material (vermiculite, cardboard, packing foam, etc.) should be used in a way so that sample containers will not make contact during shipment. Earth or ice should never be used to pack samples. Ice should be sealed in plastic bags or reusable cold packs should be used to prevent melted ice from soaking the packing material because this makes it difficult to handle when in the lab.

Samples for organic analysis must be shipped "Priority One/Overnight." If shipment requires more than a 24-hour period, sample holding times may be exceeded, and consequently the integrity of the samples could be compromised. Samples for inorganics analysis should be held until sampling for the entire area is complete and shipped for delivery within 2 days. Two days is the recommended period for collection of the samples. All samples should be shipped through a reliable carrier or hand delivered to an appropriate lab for analysis.

8.6 SAMPLE CONTAINERS AND PRESERVATION

Guidance is available from a variety of sources (see Table 8.2) regarding appropriate sample container types, sample volumes, recommended holding times and preservation methods, and appropriate analytical methods for specific analyte determinations in a given medium. It should also be noted that sample handling, preservation, and analytical methods are updated on occasion, and current and relevant methods for a particular application should be verified by consulting the appropriate EPA analytical reference guidelines at https://www.epa.gov/measurements-modeling/index-epa-test-methods.

TABLE 8.3
Pathogen Sampling and Preservation Guidelines

Contaminant	Container Volume and Type	Disinfection Reducing Agent	Preservative	Holding Time	Analytical Technique
Fecal coliforms, *Escherichia coli*	125–250 mL, Plastic	Sodium thiosulfate (0.05% final)	Transport on ice; Store <10°C, do not freeze	24–30 hours	Culture
Bacteria	500 mL–2 L, Plastic	Sodium thiosulfate (0.05% final)	Transport on ice; Store <10°C, do not freeze	Minimize Storage & Transport Time	Culture, PCR & Immunoassay
Viruses	300 mL–1.5 L, Plastic	Sodium thiosulfate (method specific)	Transport on ice; Store filters/cartridges <10°C, do not freeze	24–72 hours	Culture & PCR
Protozoa	≥10 L, Plastic	Sodium thiosulfate (0.05% final)	Transport on ice; Store at <10°C, do not freeze	Elution must begin w/in 96 hr from collection/ filtration	EPA Methods 1623.1, 1623, 1622

Source: U.S. EPA (2017b).
Abbreviation: PCR, Polymerase Chain Reaction.

Tables 8.3 through 8.5 present a useful summary of sample container, preservative, holding time, and analytical methods recommended for a wide range of relevant analytes of interest (pathogens, chemicals, and radiochemicals, respectively) describing both water and wastewater quality conditions. Additional information regarding standard analytical methods for these analytes is provided in the next section.

8.7 ANALYTICAL METHODS

The laboratory being used for sample analysis will provide the approved analytical portion of the sample's processing. The U.S. EPA has been developing and refining methods for parameter analysis in environmental samples since its inception in 1970. Current EPA methods of analysis can be located through the EPA's Environmental Measurements and Modeling portal at https://www.epa.gov/measurements-modeling/collection-methods. Of particular interest to this topic of water are the standard methods developed for compliance with drinking water regulations (https://www.epa.gov/dwanalyticalmethods) and those analytical methods specified for the Clean Water Act (https://www.epa.gov/cwa-methods).

To respond to frequent requests for details and clarification regarding the EPA test methods, Region 1 EPA Library staff developed a methods index (U.S. EPA 2003) as a tool to help locate copies of requested test methods. Confirming that there was no one volume containing all agency methods and no comprehensive list of them, the project commenced and in 1988 the first printed Index to EPA Test Methods was published. It has been updated periodically to reflect new procedures and revoked methods and the most recent 2003 edition includes nearly 1600 method references. The index includes only EPA methods and its primary goal remains as a reference tool to identify a source from which the actual method can be obtained, either free or for a fee. These nearly 1600 reference methods cover approved methods for the analysis of chemical, microbiological, and radiochemical compounds of human health and environmental interest, as well as industry-specific and regulatory program specific methods including Whole Effluent Toxicity (WET) screening to support the Clean Water Act and specific methods for chemical and microbiological contaminants of importance to the Safe Drinking Water Act. Specific details about any one of these methods is beyond the scope of this text, but the reader is encouraged to access the various links and reference documents listed previously to identify the specific methods and method details that might be relevant for a particular water-related application.

TABLE 8.4
Chemical Sampling and Preservation Guidelines

Contaminant	Container Volume and Type	Disinfection Reducing Agent	Preservative	Holding Time	Analytical Technique
Volatiles	40 mL, Glass w/PTFE-lined septa	Ascorbic acid (0.25–0.5 g)	1:1 HCl to pH <2; Store at <4°C	14 days	Purge & Trap, (Methods 502.2, 8021B, 524.3, 8260B)
Carbamate Pesticides (Methods 531.1, 531.2)	40 mL, Glass w/PTFE-lined septa	Sodium thiosulfate (12.5 mg)	Potassium dihydrogen citrate; adjust sample pH to ~3.8; Store at <4°C	28 days	HPLC-Fluorescence
Metals (Methods 200.7. 200.8, 200.9)	500 mL Plastic	None	Trace metal grade nitric acid to pH ≤2	6 months	AA, ICP-MS
Cyanide (Methods 335.2, 335.3, 335.4)	1 L, Plastic	Ascorbic acid (0.06 g)	Sodium hydroxide to pH ≥ 12; Store at ≤4°C	14 days	Titrimetric Spectrophotometric
Semi-Volatiles (Methods 525.2, 8270D/3535A)	1 L, Amber w/PTFE-lined screw cap	Sodium sulfite (40–50 mg)	6M HCl to pH <2 Store at <4°C	7 days to extraction, 28 days to analysis	SPE GC/MS
Water Quality Chemistry	1 L, Plastic	None	None—Mark samples Not Preserved	Immediate to 14 days	Conductivity, pH, alkalinity, hardness, turbidity

Source: U.S. EPA (2017).

Abbreviations: Atomic adsorption spectrophotometry (AA); hydrochloric acid (HCl); high-performance liquid chromatography (HPLC); inductively coupled plasma mass spectrometry (ICP-MS); polytetrafluoroethylene (PTFE); solid phase extraction gas chromatography/mass spectroscopy (SPE GC/MS).

TABLE 8.5
Radiochemical Sampling and Preservation Guidelines

Contaminant	Container Volume and Type	Disinfection Reducing Agent	Preservative	Holding Time	Analytical Technique
Radiochemical	1 5-L cubitainer or 4 1-L plastic containers	None	Trace metal grade nitric acid to pH ≤2	65 hours to 6 months (isotope specific)	Gross alpha, gross beta, gamma isotopes, specific radionuclides

Source: U.S. EPA (2017c).

8.8 SAMPLING STATISTICAL ANALYSIS

The words *sample* and *sampling* have appeared liberally throughout this chapter. Both play an important role in water monitoring and analysis. It is therefore important that the reader understand both the meaning and statistical implications of these two words. Shaefer and Theodore (2007) provide most of the following material.

When there is a set of n measurements, one may wish to use the values of these measurements in many instances to estimate certain characteristics of a larger number of measurements or observations that have not been made. In other words, the interest is not directly on a particular set of measurements but, rather, in using them to estimate something about a potentially larger set. One refers to the measurements as a *sample*. The larger "supply" of measurement, which one may have, is called a *population*. From the characteristics of a sample of measurements, one can estimate similar characteristics for the population.

When one calculates some measured value of a sample, it is referred to as a *statistic*. A statistic is any characteristic set of measurements calculated from a sample. The measure corresponding to a statistic obtained from a population is referred to as a *parameter*. In most cases of interest, the values of parameters are unknown and must be estimated by the statistics derived from a sample (Shaefer and Theodore 2007). *Thus, the sample value is an estimate of the population value.*

By definition, the mean, μ, and the variance, σ^2, of a random variable are constants characterizing the random variable's average value and its dispersion about its mean, respectively. The mean and variance can also be derived from the probability distribution function (PDF) of the random variable (Shaefer and Theodore 2007). If the PDF is unknown, however, the mean and variance can be estimated on the basis of a random sample of some, but not all, measurements on the random variable. Details on the mean and variance follow.

8.8.1 Estimation of the Mean

Let $X_1, X_2, \ldots, X_n$ denote a random sample of n measurements on X. The sample mean $\overline{X}$ is then defined by

$$\overline{X} = \sum_{i=1}^{n} \frac{X_i}{n} \tag{8.1}$$

If the numbers $\overline{X}_i$ have weighting factors W_i associated with them,

$$\overline{X} = \frac{W_1X_1 + W_2X_2 + \ldots + W_nX_n}{n} = \frac{\sum_{i=1}^{n} W_iX_i}{n} \tag{8.2}$$

Weighting factors are often normalized, i.e., $\sum W_i = 1.0$. For this condition,

$$\overline{X} = \sum_{i=1}^{n} W_i X_i \tag{8.3}$$

8.8.2 The Geometric Mean

The geometric mean, $\overline{X}_G$, is given by

$$\overline{X}_G = \left(X_1, X_2, \ldots X_n\right)^{\frac{1}{n}} = \left(\prod_{i=1}^{n} X_i\right)^{\frac{1}{n}} \tag{8.4}$$

8.8.3 The Median and Mode

The *median* is defined as the middle value (or arithmetic mean, of the two middle values) of a set of numbers. The *mode* is the value that occurs with the greatest frequency in a set of numbers. Thus, it is the typical or most common value in a set.

8.8.4 Estimation of Variance

Once again, let X_1, X_2, …, X_n denote a random sample of n measurements on X. Then the *sample* variance s^2 is defined by

$$s^2 = \sum_{i=1}^{n} \frac{\left(X_i - \overline{X}\right)^2}{n-1} \tag{8.5}$$

where $\overline{X}$ and s^2 are random variables in the sense that their values vary from sample to sample of the measurements on X. It can be shown that the *expected* value of X is the population mean, μ, and that the *expected* value of s^2 is the population variance of σ^2. Because of this, $\overline{X}$ and s^2 are called *unbiased estimators* of μ and σ^2, respectively. The calculation of s^2 can be facilitated by use of the computational formula

$$s^2 = \frac{n\sum_{i=1}^{n} X_i^2 - \left(\sum_{i=1}^{n} X_i\right)^2}{n(n-1)} \tag{8.6}$$

In the case of a random number of samples of observations of a continuous random variable assumed to have a so-called normal PDF, the graph of which is a bell-shaped curve, the following statements give a more precise interpretation of the sample standard deviation, s, as a measure of spread or dispersion.

1. $\overline{X} \pm s$ includes approximately 68% of the sample observations.
2. $\overline{X} \pm 2s$ includes approximately 95% of the sample observations.
3. $\overline{X} \pm 3s$ includes approximately 99.7% of the sample observations.

The source of these percentages is the normal probability distribution. Chebyshev's theorem provides an interpretation of the sample standard deviation (the positive square root of the sample variance) as a measure of the spread (dispersion) of the sample observations about their mean. Chebyshev's theorem states that for $k > 1$, at least $(1-\frac{1}{k^2})$ of the sample observations lie in the interval $(\overline{X} - ks, \overline{X} + ks)$. For $k = 2$, for example, this means that at least 75% of the sample observations lie in the interval $(\overline{X} - 2s, \overline{X}_0 + 2s)$. The smaller value of s, the greater the concentration of observations in the vicinity of $\overline{X}$ (Shaefer and Theodore 2007).

One clearly needs a measure of scatter, which can be used in samples of any size and in some sense, to make use of all the measurements in the sample. There are several measures of scatter that can be used for this purpose, and the most common of this is the aforementioned *standard deviation*. The standard deviation may be thought of as the "natural" measure of scatter. Three Illustrative Examples, one addressing the mean and two addressing the variance of a data set are presented in the next section.

8.9 APPLICATIONS

Six Illustrative Examples compliment the material presented in this chapter.

Illustrative Example 8.1

The U.S. EPA publishes drinking water standards that aim to control the quality of the nation's drinking water. The regulations include water monitoring schedules and methods to measure contaminates in water. Cadmium is one of these contaminates and must not exceed 0.005 mg/L in the finished water. Express this concentration in lb/1,000 gal.

Solution

The following is the conversion from mg/L to lb/1,000 gal.

$$\left(\frac{0.005\text{ mg}}{L}\right)\left(\frac{1\text{g}}{1{,}000\text{ mg}}\right)\left(\frac{1\text{ lb}}{454\text{g}}\right)\left(\frac{3785\text{ L}}{1{,}000\text{ gal}}\right)=\frac{4.17\times10^{-5}\text{ lb}}{1{,}000\text{ gal}}$$

Alternatively, the same result may be reached by using water conversion factor of $1\text{ mg/L}=1\text{ ppm}_{\text{water}}=8.34\times10^{-6}\text{ lb/gal}$ as follows

$$\frac{\left(\frac{0.005\text{ mg}}{\text{L}}\right)\left(\frac{8.34\times10^{-6}\text{lb}}{\text{gal}}\right)\left(\frac{1{,}000\text{ gal}}{1{,}000\text{ gal}}\right)}{\left(\frac{\text{mg}}{\text{L}}\right)}=\frac{4.17\times10^{-5}\text{ lb}}{1{,}000\text{ gal}}$$

Illustrative Example 8.2

What is biochemical oxygen demand (BOD)? What is chemical oxygen demand (COD)? Discuss the difference between them in laymen's terms.

Solution

The *biochemical oxygen demand (BOD)* is defined as the amount of oxygen used by microbes to oxidize biodegradable waste materials in a sample, usually measured for a standardized 5-day period, at 20°C, in the dark. BOD_5 is the designation used to express the 5-day incubation period, and has units of part per million (ppm) or milligrams per liter (mg/L). The *chemical oxygen demand (COD)* is defined as the amount of chemical oxygen equivalent (ppm or mg/L) consumed under high temperature, low pH, and strong oxidant (potassium permanganate) conditions in the rapid (3 hours) chemical reaction of organic material in a waste stream. Although not a biological property of a material, COD is relevant to industrial wastes where toxicity to biological system could be suspected.

In laymen's terms, BOD and COD are similar in the fact that they measure the amount of organic material in water. COD will be much larger than BOD, given COD will chemically oxidize many organic chemicals that are not biologically degradable.

Illustrative Example 8.3

A large variety of point sources using water in their manufacturing process (mining operations, coal-fired power plants, breweries, etc.) or treat wastewater from municipal, commercial, or industrial sources are required by law to obtain a National Pollutant Discharge Elimination System (NPDES) permit if the facility has a potential to discharge pollutants through a point source into a water of the United States. Go to the web at www.epa.gov/npdes to answer the following questions:

1. When was the NPDES program created? By which entity?
2. How many NPDES program areas exist and what are they?
3. How do NPDES permits protect the waters of the United States?

Solution

1. The NPDES program was created in 1972 by the Clean Water Act (CWA).
2. There are currently 12 NPDES program areas defined by the NPDES permit program. They are as follows:
   - Animal Feeding Operations (AFOs)
   - Aquaculture
   - Biosolids
   - Forest Roads
   - Industrial Wastewater
   - Municipal Wastewater (Combined Sewer Overflows [CSOs], Integrated Planning, Peak Flows, and Sanitary Sewer Overflows [SSOs])
   - National Pretreatment Program (Standards and Requirements, Roles and Responsibilities, and Events, Trainings, and Publications)
   - Pesticide Permitting
   - Stormwater (Discharges from Construction Activities, Discharges from Industrial Activities, Discharges from Municipal Sources, Discharges from Transportation Sources, Oil and Gas Stormwater Permitting, EPA's Residual Designation Authority, Stormwater Rules and Notices, and Stormwater Maintenance)
   - Vessels Incidental Discharge Permitting
   - Water Quality Trading
   - Whole Effluent Toxicity (WET)
3. The NPDES permit program helps protect the waters of the United States by specifying an acceptable level of pollutant discharge (e.g., maximum level of zinc). NPDES permits ensure that federal minimum technology based standards are met, as well as enforcing state water quality standards if these more stringent standards are required to ensure the protection of public health and the environment.

Illustrative Example 8.4

The average weekly industrial wastewater discharge from a small industrial facility for 6 consecutive weeks are 2.6, 3.4, 1.5, 2.4, 3.1, 2.7 ft^3/s. Determine the mean of the discharge rates.

Solution

Substituting these weekly values into Equation (8.1) yields:

$$\overline{X} = \frac{X_1 + X_2 + \ldots + X_n}{n} \tag{8.1}$$

$$\overline{X} = \frac{2.6 + 3.4 + 1.5 + 2.4 + 3.1 + 2.7}{6} = 2.1\,\text{ft}^3/\text{s}$$

Illustrative Example 8.5

Using the same data given in Illustrative Example 8.4, calculate the standard deviation of the measurements.

Solution

Recalling Equation 8.5:

$$s^2 = \sum_{i=1}^{n} \frac{\left(X_i - \overline{X}\right)^2}{n-1} \quad (8.5)$$

The *standard deviation* is simply the square root of *sample variance*, therefore,

$$\sqrt{s^2} = \sqrt{\sum_{i=1}^{n} \frac{\left(X_i - \overline{X}\right)^2}{n-1}};\ s = \sqrt{\sum_{i=1}^{n} \frac{\left(X_i - \overline{X}\right)^2}{n-1}} \quad (8.7)$$

The mean, $\overline{X}$, has been calculated in Illustrative Example 8.4.

The values of weekly average wastewater discharge from Illustrative Example 8.4 are then used to calculate $\sum_1^6 (X_i - \overline{X})^2$

$$\sum_1^6 (X_i - \overline{X})^2 = (2.6-2.61)^2 + (3.4-2.61)^2 + (1.5-2.61)^2 + (2.4-2.61)^2 + (3.1-2.61)^2 + (2.7-2.61)^2 = 2.15$$

Therefore,

$$s = \sqrt{\frac{(2.15)}{6-1}} = 0.66\ \text{ft}^3/\text{s}$$

Illustrative Example 8.6

Using the same data given in Illustrative Example 8.4, calculate the sample variance of the measurements.

Solution

Recalling Equation 8.5:

$$s^2 = \sum_{i=1}^{n} \frac{\left(X_i - \overline{X}\right)^2}{n-1} \quad (8.5)$$

The values of weekly average wastewater discharge from Illustrative Example 8.4 are then used to calculate $\sum_1^6 (X_i - \overline{X})^2$

$$\sum_1^6 (X_i - \overline{X})^2 = (2.6-2.61)^2 + (3.4-2.61)^2 + (1.5-2.61)^2 + (2.4-2.61)^2 + (3.1-2.61)^2 + (2.7-2.61)^2 = 2.15$$

Equation 8.5 becomes,

$$s^2 = \sum_{i=1}^{n=6} \frac{2.15}{6-1} = 0.43\ \text{ft}^3/\text{s}$$

Or, with $s = 0.66\ \text{ft}^3/\text{s}$ from Illustrative Example 8.5, $s^2 = (0.66)^2 = 0.43\ \text{ft}^3/\text{s}$.

REFERENCES

American Society of Testing and Materials. 1990. *1990 Annual Book of ASTM Standards,* Volume 11.04. ASTM D4448. Conshohocken, PA: ASTM.

City of East Williston. 2018. *East Williston Annual Water Report.* City of East Williston, NY.

Dupont, R.R., J. Richardson, and D. Willey. 2018. *Stormwater Impacts to an Urban River in the Intermountain West: the Use of Continuous Monitoring Datasets.* Paper #410122 presented at the 111th Air and Waste Management Association Annual Conference and Exhibit, Sustainability Metrics, Initiatives, and Analytics Session, June 25–28, Hartford, CT.

Horsburgh, J.S., A.S. Jones, D.K. Stevens, D.G. Tarboton, and N.O. Mesner. 2010. A sensor network for high frequency estimation of water quality constituent fluxes using surrogates. *Environmental Modelling & Software* 25:1031–1044. doi:10.1016/j.envsoft.2009.10.012.

Michael Baker International, LimnoTech, and MapTech, Inc. 2015. *Continuous Monitoring Data Sharing Strategy.* Report EP-C-12-052 submitted to U.S. EPA under Task Order No. 0005. Washington, DC: Michael Baker International. https://www.epa.gov/sites/production/files/2016-02/documents/.

New England Waterworks Association. 2004. *Pocket Sampling Guide for Operators of Small Water Systems.* Holliston, MA: New England Water Works Association. https://www3.epa.gov/region1/eco/drinkwater/pdfs/SamplingGuide.pdf.

Shaefer, S., and L. Theodore. 2007. *Probability and Statistics Applications for Environmental Science.* Boca Raton, FL: CRC Press/Taylor & Francis Group.

Schilling, K.E., S.W Kim, and C.S. Jones. 2017. Use of water quality surrogates to estimate total phosphorus concentrations in Iowa rivers. *Journal of Hydrology: Regional Studies* 12:111–121. doi:10.1016/j.ejrh.2017.04.006.

U.S. Environmental Protection Agency. 1981. *Manual of Ground-Water Quality Sampling Procedures.* EPA-600/2-81-160. Ada, OK: Robert S. Kerr Environmental Research Laboratory. https://nepis.epa.gov/Exe/ZyPDF.cgi?Dockey=2000AS86.PDF.

U.S. Environmental Protection Agency. 1982. *Handbook for Sampling and Sample Preservation of Water and Wastewater.* EPA-600/4-82-029. Cincinnati, OH: Environmental Monitoring and Support Laboratory. https://nepis.epa.gov/Exe/ZyPDF.cgi/30000QSA.PDF?Dockey=30000QSA.PDF.

U.S. Environmental Protection Agency. 1992. *NPDES Storm Water Sampling Guidance Document.* EPA 833-8-92-001. Washington, DC: Office of Water. https://www3.epa.gov/npdes/pubs/owm0093.pdf.

U.S. Environmental Protection Agency. 1993. *Subsurface Characterization and Monitoring Techniques, A Desk Reference Guide, Volume I: Solids and Ground Water Appendices A and B.* EP A/625/R-93/003a. Washington, DC: Office of Research and Development. https://nepis.epa.gov/Exe/ZyPDF.cgi/30004L8E.PDF?Dockey=30004L8E.PDF.

U.S. Environmental Protection Agency. 2003. *Index to EPA Test Methods.* Boston, MA: U.S. EPA New England Region 1 Library. https://www.epa.gov/sites/production/files/2015-03/documents/testmeth.pdf.

U.S. Environmental Protection Agency. 2004. *Response Protocol Toolbox: Planning for and Responding to Drinking Water Contamination Threats and Incidents, Interim Final – August 2004. Response Guidelines.* EPA 817-D-04-001. Washington, DC: Office of Ground Water and Drinking Water, Water Security Division. https://www.epa.gov/sites/production/files/2015-06/documents/2004_11_24_rptb_response_guidelines.pdf.

U.S. Environmental Protection Agency. 2016. *Quick Guide to Drinking Water Sample Collection.* 2nd edition update. Golden, CO: Region 8 Laboratory, U.S. EPA. https://www.epa.gov/sites/production/files/2015-11/documents/drinking_water_sample_collection.pdf.

U.S. Environmental Protection Agency. 2017a. *Sampling Guidance for Unknown Contaminants in Drinking Water.* EPA-817-R-08-003. Washington, DC: Office of Water. https://www.epa.gov/sites/production/files/2017-02/documents/sampling_guidance_for_unknown_contaminants_in_drinking_water_02152017_final.pdf.

U.S. Environmental Protection Agency. 2017b. *Sample Collection Information Document for Pathogens.* EPA/600/R-17/374. Washington, DC: Office of Research and Development.

U.S. Environmental Protection Agency. 2017c. *Sample Collection Information Document for Chemicals, Radiochemicals and Biotoxins.* EPA/600/R-17/389. Washington, DC: Office of Research and Development.

U.S. Geological Survey (USGS). 2006. *Collection of water samples (ver. 2.0).* U.S. Geological Survey Techniques of Water-Resources Investigations, Book 9, Chapter A4. Reston, VA: U.S. Geological Survey. http://pubs.water.usgs.gov/twri9A4/.

Section II

Water Resources

9 Water Resources of the United States

9.1 INTRODUCTION

The present outlook of water resources in the United States affects each and every citizen living today and each generation to come. As in any engineering undertaking, the concern associated with water resources must be identified and understood. The most basic question this chapter tries to address is the current and future projected discrepancy between water supply and water demand within the United States, and what practices will need to be implemented to ensure sustainable water resources in the future.

There are two major categories of water sources that are used in the United States. Surface water and groundwater both contribute to the national supply and are used in different proportions that varies by state. Trends in past usage and analysis of present issues creates a picture of what water resources will be available in the future.

One important metric to determine is if present water supplies will meet demands in the near future. Another equally important issue pertaining to present water resources is maintaining water quality standards. It could be argued that the quality issue is becoming increasingly more important in the face of increasing urbanization and population growth. Obviously, drinking water consumed by humans needs to be of the highest quality, ensuring health and vitality to all U.S. inhabitants.

In addition to studying water quality, water use and sustainable reuse methods are considered in this chapter. In the presence of increasing demands, more and sustainability efforts should be implemented in the United States. The feasibility and benefits of several types of reuse methods are discussed, and hopefully developing nations will continue to learn and emulate sustainable water practices in the developed world, making the overall future global water outlook a brighter one.

This chapter summarizes the state of water resources in the United States and addresses surface water, groundwater, quality of water resources, and water use and sustainable reuse methods. An application section provides three Illustrative Examples related to the general subject of water resources in the United States.

9.2 SURFACE WATER

Surface water consists of streams, lakes, and reservoirs. It should be noted that surface water contains both fresh and saline resources. Several coastal states resourcefully use saline surface water for industry and other uses. Two factors must be understood to establish what the most important surface water concerns are: the largest sectors of current water resource usage, and the trends in usage in these sectors.

The usage of fresh surface water has declined since 1980 and has actually decreased approximately 25% between 2005 and 2015. The decline in total withdrawals in 2015 was primarily caused by significant decreases (28,800 MGD) in thermoelectric power water use because of increased efficiency and closures of plants with once-through cooling systems and accounted for 89% of the decrease in total withdrawals. Between 2010 and 2015, withdrawals decreased in all categories except irrigation (2% increase), mining (1% increase), and livestock (no change). Both fresh surface-water withdrawals

Contributing Author: Jesse McLaury

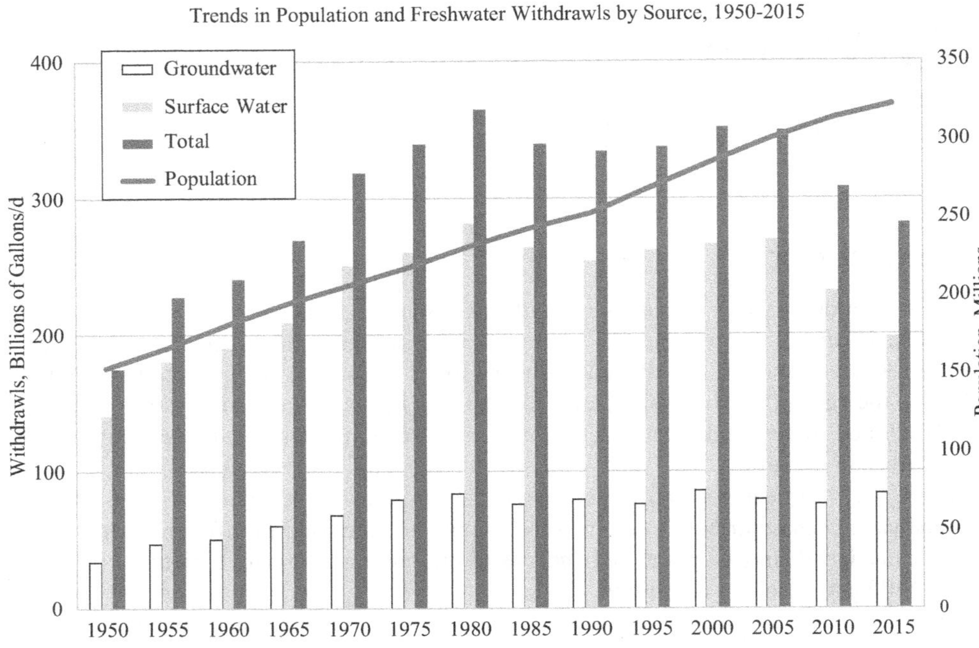

FIGURE 9.1 Trends in groundwater and surface water withdrawals and U.S. population, 1950–2015. (From Dieter, C.A. et al., *Estimated Use of Water in the United States in 2015*, Circular 1441, U.S. Geological Survey, Reston, VA, 65 p, 2018.)

(198,000 MGD) and saline surface-water withdrawals (38,600 MGD) were 14% less than corresponding values in 2010 (Dieter et al. 2018). Considering the continuing increase in the U.S. population over time, significant strides are being made in surface-water use efficiency (Figure 9.1). Total surface-water withdrawals were estimated to be 237,000 MGD, or 74%, of total withdrawals. About 84% (198,000 MGD) of total surface-water withdrawals were freshwater.

In 2015, more surface water than groundwater was withdrawn for all categories of use except domestic, livestock, and mining; however, irrigation was nearly evenly split between surface water and groundwater. Thermoelectric power accounted for about 48% of the total fresh surface-water withdrawals and irrigation accounted for about 31%. The largest surface-water withdrawals were in Texas, Idaho, Florida, California, New York, and North Carolina, cumulatively accounting for about 29% of total surface-water withdrawals for all categories nationwide. Large quantities of fresh surface water were withdrawn for thermoelectric power in Texas, Illinois, Michigan, and Alabama. Large saline surface-water withdrawals for thermoelectric power occurred in Florida, New York, Maryland, and New Jersey, which cumulatively accounted for 61% of the national total saline surface-water withdrawals.

Based on the present population distribution and large industrial uses of water; the usage rates discussed should be expected to at least be maintained, if not decrease in coming years. Efforts should be focused on sustainable methods that use water more efficiently, especially in the irrigation and thermoelectric power sectors. These two industries literally feed and power the nation and currently represent the largest water consumption sector in the United States, making up 78% of total water withdrawals in the United States in 2015. Significant reductions in water use in the thermoelectric power sector have recently been achieved as discussed. Irrigation withdrawals steadily increased from 1950 to 1980, when they peaked (150,000 MGD) and then have remained relatively steady from 2005 (127,000 MGD) through 2010 (116,000 MGD) to 2015 (118,000 MGD). The trend toward using more efficient irrigation systems has continued over time, with 10% more irrigated lands using sprinkler systems (including micro-irrigation) and 11% less land area using surface (flood) irrigation systems in

2015 than in 2010, but even with these improvements, water use for irrigation has increased over time. Given the large percentages of surface and groundwater consumed by both of these industries, they would be the largest leverage points in water conserving efforts and implementing reuse strategies.

With regard to trends in streamflow, there seems to be mixed results reported in the literature. Anderson and Woosley (2005) suggest that there is widespread perception that streamflow in the West is decreasing. Analysis of mean daily streamflow for selected major western rivers for 1950 to 2000, however, indicates either no trend, or a significant increasing trend in mean daily streamflow over time. These trends occur despite alterations in the flow regime because of reservoir operations (Anderson and Woosley 2005). Overall, the data from most of the river basins in the United States indicate constant mean daily streamflow rates over the recent past.

Reservoir resources are another facet of the surface water supply. The maximum amount of reservoir storage available in the United States is no longer changing because since 1985 a little reservoir capacity has been added (Figure 9.2), signaling the conclusion of the dam-building era (Anderson and Woosley 2005). Reservoir storage is dependent on the amount and timing of streamflow produced. A distinguishing characteristic of western rivers is the large proportion of reservoir storage compared to annual stream flows. In the case of the Rio Grande, total reservoir storage is about 28 times the annual mean flow of the river. Lake Powell stores about 3 times the annual discharge of the Colorado River at Lees Ferry, Arizona. Annual flows in western rivers between years, therefore, can reflect adjustments in reservoir storage, which is affected by precipitation in the contributing watershed and water demands downstream.

Total reservoir contents have declined in recent years for the Rio Grande and the Colorado and Missouri Rivers. Historical lows were set in 2003 for the Rio Grande and the Colorado Rivers at least since the reservoirs reached full operating potential. The 1996–2004 drought in the Southwest reduced inflows to most western reservoirs, lowering water levels and depleting reservoir storage as indicated in Figure 9.3 for the Rio Grande, Colorado, and Missouri River basins.

Not only has maximum reservoir capacity been unchanged in recent years, but the economics of reservoirs seems to be in conflict with the water budget. As freshwater is fully used, the evaporation losses from reservoirs, particularly in arid climates, may become a less affordable component of the water budget. Developing new reservoirs and dam building are no longer financially advantageous, but surface-water storage in reservoirs remains a vital component of a sustainable water resource strategy.

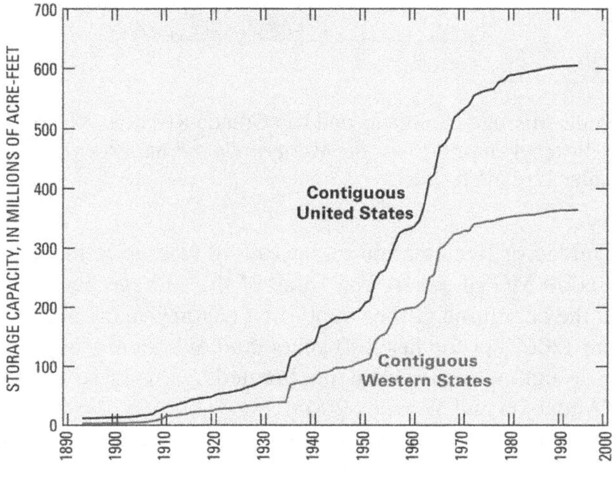

FIGURE 9.2 Total surface-water reservoir capacity in the contiguous western United States and the contiguous United States from 1894 to 1994. (From Anderson, M. and Woosley, L. Jr., *Water Availability for the Western United States—Key Scientific Challenges*, U.S. Geological Survey, Circular 1261, 85 p, 2005.)

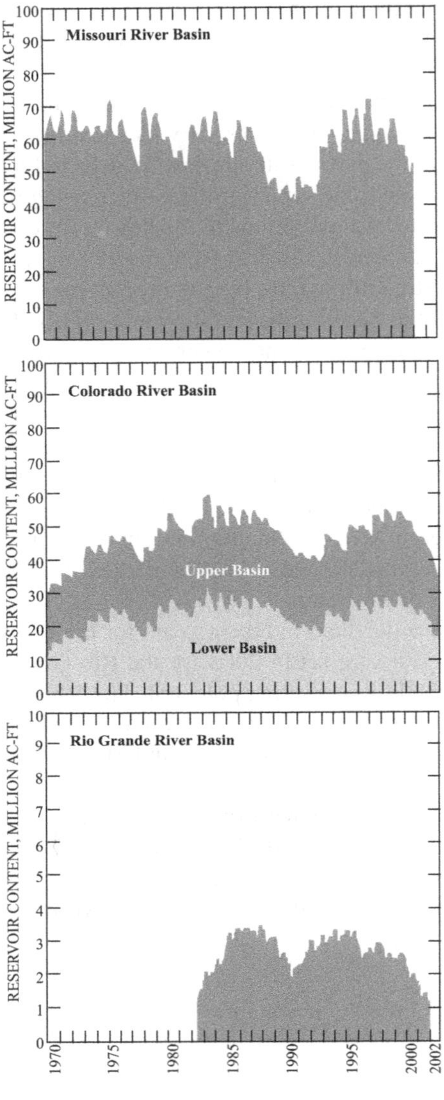

FIGURE 9.3 Trends in the Missouri, Colorado, and Rio Grande River basins 1982–2002. (From Anderson, M. and Woosley, L. Jr., *Water Availability for the Western United States—Key Scientific Challenges*, U.S. Geological Survey, Circular 1261, 85 p, 2005.)

For example, the Bureau of Reclamation manages 348 reservoirs in 17 western states that can store a total of 83,400,000 MG of water. The value of this storage becomes most evident during periods of drought. At the beginning of the twenty-first century, many areas of the west were experiencing the driest year (2002) in the last 100 years, and although reservoirs were low and being depleted, this western drought was somewhat ameliorated by this large volume of reservoir storage available at the time (Anderson and Woosley 2005).

There is another complicating factor associated with surface water supply, particularly in the western United States. That is the effect of climate change on seasonal precipitation patterns and implications regarding water storage the region has historically relied upon in the form of mountain snowpack. Several studies summarized in "Climate Change 2001: Synthesis Report" (Intergovernmental Panel on Climate Change, 2001) show seasonal shifts to increased winter runoff and reduced summer flows. Changes in snowfall and snowmelt dynamics for the west, even if annual amounts of precipitation change little, would be an important consequence of a warmer climate, especially given the importance of high-elevation precipitation to the available water supply. The consequences of these changes in streamflow timing trends may include: (i) a reduction in the amount of runoff that occurs in the high-demand spring and summer seasons; (ii) an increase in the amount of runoff during the winter storm season, thereby decreasing the available flood storage space in reservoirs in the spring; and (iii) reducing the water available to sustain soil and fuel moisture throughout the summer, which could initiate ecosystem changes (Dettinger et al. 2004).

It appears for the time being that the United States is keeping up with the increasing population in terms of surface water usage and water supply. However, in the coming years, population growth and climate change pressures will continue, making the continued monitoring and tracking of this invaluable resource critically important. New technological advances in water use infrastructure and water reuse methods that can improve efficiency of surface-water utilization will be discussed in a later section and should continue to be made in the future to improve the sustainability of surface water supplies.

9.3 GROUNDWATER

The U.S.'s second major water resource is groundwater. Groundwater systems consists of the water table, wells, and aquifers. More than 90% of the Earth's readily available freshwater is represented by groundwater. The total volume of ground water in storage is much larger than the volume of surface water in all the lakes, streams, and rivers in the United States combined. Because groundwater is the largest part of the available freshwater resources, it requires conscientious and intelligent analysis concerning its sustainable use. If groundwater is poorly managed in the coming years, there will be no simple solution to the looming water problem.

Groundwater sometimes presents interesting complications that large rivers also have (i.e., groundwater may lie beneath multiple state, national, or other political boundaries). The utilization of the resource is then subject to differing management strategies and often competing utilization goals. The wide range of interests that compete for groundwater resources illustrates the complex nature of groundwater systems, and their increasing importance as a source of water has led to broad consensus that an effective, integrated approach to groundwater management is essential (Jakeman et al. 2016).

It has already been stated that there is much more groundwater, compared to surface water, presently available. Interestingly, it has not been used nearly to the same degree as surface water has in the past. This is partly due to accessibility because some groundwater has historically been difficult to access. Technological advances in drilling and groundwater pumping is making more groundwater available than has been economically available before. According to Dennehy et al. (2015), much of the U.S. surface water supplies are fully apportioned for use; thus, in some areas, the only potential alternative freshwater source that can developed in the future is groundwater. Although frequently overlooked in some locations, groundwater serves as the principal reserve of freshwater in the United States and represents much of the potential supply during periods of drought (Dennehy et al. 2015).

Accessing groundwater however, must be done with caution and precision. When groundwater is pumped to the surface, proximal water moves to fill the void left by the water that was pumped. Groundwater depletion is a concern in some locations throughout the Midwest and California because it is used at an unsustainable rate in some areas. This observation highlights the need to manage groundwater effectively, both now, and in the future.

Each groundwater system is unique. The source and volume of water within an underground system without human development is dependent upon many external factors. These factors include the differing rates of precipitation, recharge, evaporation, and evapotranspiration. Also affecting the movement of water through an aquifer is the location and hydrological connection with streams, rivers, springs, reservoirs, and wetlands. The one common factor for all groundwater systems, however, is that the total volume of water entering, leaving, and being stored in the system must be conserved. An accounting of all the inflows, outflows, and changes in storage is called a *water budget*. Basin-wide water budgets are needed to properly manage groundwater resources. Under predevelopment conditions, a groundwater system is in long-term equilibrium. That is, averaged over some period of time, the volume of water entering or recharging the system is approximately equal to the volume of water leaving or discharging from the system. Because the system is in equilibrium, the quantity of water stored in the system is constant or varies about some average condition in response to annual or longer-term climatic variations. The equation that describes the water budget of the predevelopment groundwater system is:

$$\text{Recharge (water entering)} = \text{Discharge (water leaving)} \tag{9.1}$$

The groundwater being discharged to streams and rivers is called *base flow*. Humans change the natural or predevelopment groundwater system by withdrawing (pumping) water for use, increasing recharge by application of water to the land surface or by direct injection, and decreasing recharge by paving over permeable soils through urbanization. In all cases, the source of water for pumping will be supplied by (Anderson and Woosley 2005): (i) more water entering the groundwater system (increased recharge), (ii) removal of water that was stored in the system, (iii) less water leaving the system (decreased discharge), or (iv) some combination of the first three conditions. The effects of prolonged over-withdrawals on aquifers can be seen in Figure 9.4.

One major way that groundwater differs from surface water is the ease of measuring the resource's replenishment rate. Surface water is contained in reservoirs and lakes or flowing in rivers and streams and can be easily measured. Groundwater sources are not as easy to measure in most cases, necessitating more cautious groundwater withdrawals. Replenishment rates of groundwater cannot match past and current depletion rates in many parts of the world. In addition, declining quality of the remaining groundwater commonly cannot support all agricultural, industrial, urban, and ecosystem function beneficial uses, especially in the developed world. In the developing world, impacted groundwater can fail to even meet essential human needs (Jakeman et al. 2016). This brings up another point: Sometimes groundwater cannot be used directly in place of surface water. Groundwater in depleted aquifers often has poorer water quality than surface water sources.

The U.S. Geological Survey (USGS) is knowledgeable about the availability of U.S. groundwater resources. However, aquifer water levels are the least well defined and most critical part of the groundwater hydrologic budget. Quantitative regional groundwater systems analysis should estimate current conditions and trends in groundwater use, storage, recharge, and discharge making it possible to develop groundwater budgets leading to a greater understanding of the complete aquifer system. Once the effect of groundwater use and development are quantified, water managers can make informed decisions regarding the optimal use of the resource (Dennehy et al. 2015).

In California, where there have been two serious droughts (2006–2009 and 2013–2015) in the last decade, intelligent management of groundwater will be crucial to both their economy and their residents. As of 2015, California uses the most water of any state in the United States with 11,300 MGD from surface water and 17,400 MGD from groundwater (Dieter et al. 2018). California irrigation withdrawals declined 4,070 MGD (18%), whereas irrigated acres decreased by only 10% from 2010, likely as a result of the intense drought conditions in the region in 2015 and implementing more efficient irrigation systems. Groundwater became the primary source of irrigation water in California in 2015, and it appears that California has begun to develop a more sustainable mindset with many scientists, engineers, and other water experts working hard to make the future water outlook of their state a positive one.

There are many, often-interrelated, dimensions to managing groundwater effectively. Effective groundwater management is underpinned by sound science (biophysical and social) that actively

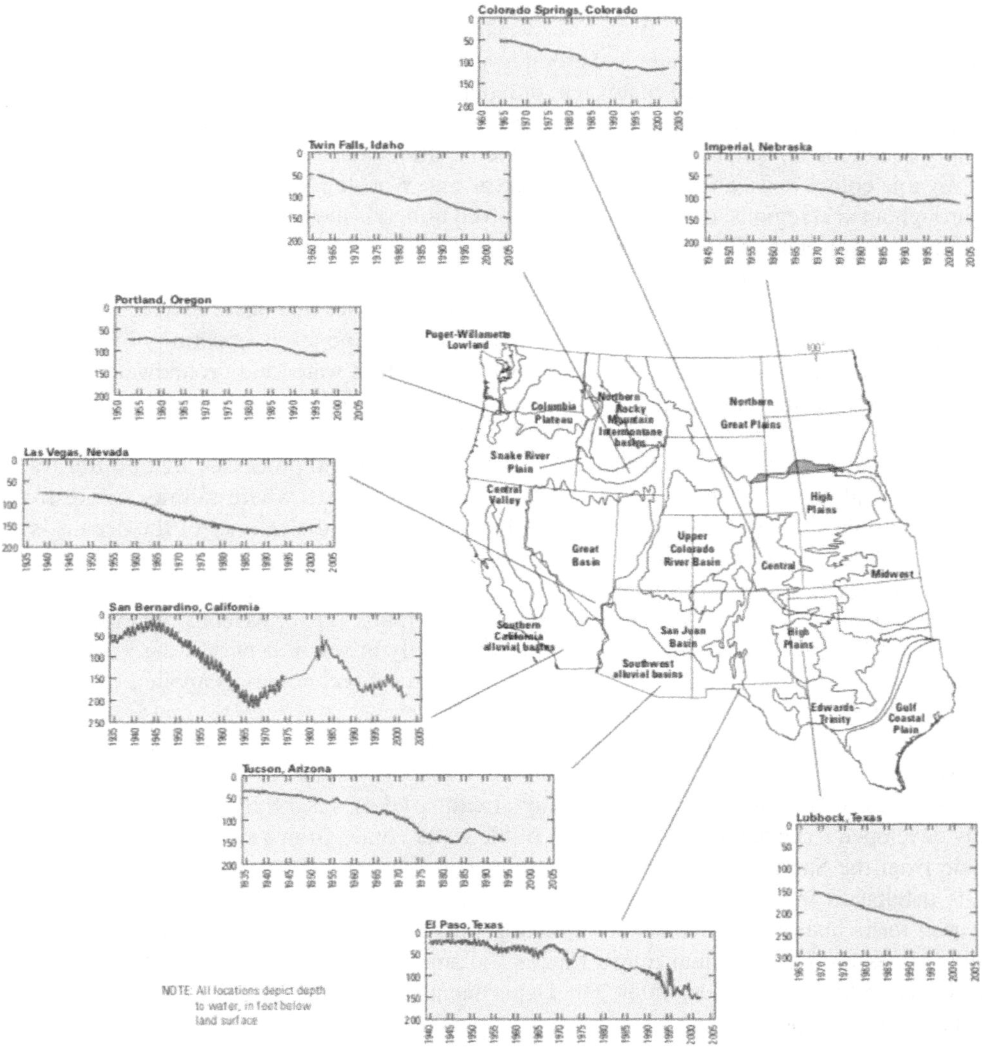

FIGURE 9.4 Trends in the depth of water below the surface in an aquifer in the Lower Colorado region. (From Anderson, M. and Woosley, L. Jr., *Water Availability for the Western United States—Key Scientific Challenges*, U.S. Geological Survey, Circular 1261, 85 p, 2005.)

engages the wider community and relevant stakeholders in the decision-making process. Generally, an integrated approach will mean thinking beyond the aquifer, a view that considers the wider context of surface water links, catchment management and cross-sector issues with economics, energy, climate, agriculture, and the environment (Jakeman et al. 2016).

Technological advancements are by far the greatest asset available for optimal utilization of groundwater in the United States. Integrated modeling is the common platform used for performing integrated assessments because it can support a systematic and transparent approach to water resource integration. Challenges of combining diverse tools and data have been the focus of work involving model and data standardizations and information exchange that is ongoing across the globe (Jakeman et al. 2016). Environmental conscientiousness and stewardship of current resources will ensure wise use of groundwater and effective delivery of groundwater to those who need it most.

9.4 QUALITY OF WATER RESOURCES

Information regarding surface water and groundwater resource volume is critical to understand present water concerns in the United States but, of equal importance, is the quality of those resources. The key water quality issues limiting water usability are the presence of elevated concentrations of naturally occurring constituents. Constituents of concern include salinity, nutrients, trace elements, trace organic compounds, and pesticides (Anderson and Woosley 2005).

Throughout arid regions, the presence of dissolved minerals most widely limits the use of surface water and groundwater. Water is an effective natural solvent that dissolves minerals as it interacts with geologic materials on the land surface and within an aquifer system. Deep aquifers represent significant untapped water resources. The geochemical processes that occur under extreme temperatures and pressures over centuries, however, have increased the levels of salinity in most of these deep aquifers to that of seawater. Interactions between surface water and groundwater also control natural levels of salinity in streams and reservoirs. In many parts of the arid west, saline aquifers and thermal springs are major sources of base flow. The region's characteristically high evaporation rates also increase surface and groundwater salinity. These natural processes are best illustrated by Utah's Great Salt Lake, the largest surface-water body in the west, where salinity ranges from 5% to 25% by weight or 50,000 to 250,000 mg/L. For comparison, the salinity of the oceans is about 3.5% or 35,000 mg/L of total dissolved solids. Saline lakes, such as the Salton Sea and Mono Lake in California, are found in several closed basins in the west. Many more closed basins contain dry lake beds or playas such as Death Valley, California; Willcox Basin, Arizona; and the Bonneville Salt Flats in Utah. Trace elements also can occur naturally in water as part of the weathering process at levels that limit water use unless expensive treatment technology is applied (Anderson and Woosley 2005).

Irrigation drainage can concentrate salts and trace elements in irrigation water by evaporation, leach salts and trace elements from soils, and can transport excess agrochemicals, such as nitrate and pesticides, to groundwater and receiving streams, lakes, reservoirs, and wetlands. In more highly developed irrigated areas where most of the water comes from a surface supply (i.e., the Rio Grande from the San Luis Valley in Colorado to Fort Quitman, Texas, and along the Gila River and its tributaries in Arizona), the irrigation return flow in the upper irrigation areas is repeatedly used for irrigation in the areas downstream. This cycle of use and reuse may be repeated six or more times until the residual return flow is too small in quantity and too saline in quality to have further value as irrigation water. The Department of the Interior's National Irrigation Water-Quality Program reported that selenium is the trace element most frequently found in elevated concentrations in irrigation return flows and has been responsible for wildlife deaths and deformities at several Department of the Interior irrigation project areas in the western United States (Anderson and Woosley 2005). Other constituents of concern include boron, arsenic, mercury, and some pesticides. Water-quality studies conducted by the USGS revealed that the highest concentrations of nitrogen occur in streams and groundwater in agricultural areas. Extensive herbicide use in agricultural areas (accounting for about 70% of the total national use of pesticides) has resulted in the widespread occurrence of herbicides in streams and shallow groundwater in agricultural areas (Anderson and Woosley 2005).

Mining remains an important industry supporting the economy of the west. The scars of past mining activities, however, remain visible on the western landscape. Mining for metals, coal, and other inorganic compounds, such as phosphates and limestone aggregates, can accentuate and accelerate natural geochemical processes. The development of underground mining works, open pits, ore piles, mill tailings, and spoil heaps and the extractive processing of ores enhance the likelihood of releasing trace elements to the surrounding area at elevated concentrations and at increased rates relative to unmined areas. Abandoned, inactive, and active mines can release mine drainage that is highly acidic and rich in trace elements that are toxic to aquatic life and humans. Tailings piles and mine roads are a source of sediment that can clog streambeds and adversely affect aquatic habitat in streams and reservoirs downstream. Groundwater quality also can be adversely affected by mine development. Restoration of affected streams and aquifers is costly, and residual mining waste ponds and piles require perpetual maintenance to prevent future damage to riparian habitats and degradation of water quality (Anderson and Woosley 2005).

Many areas in the United States are growing rapidly. Such development not only increases the quantity of wastewater return flows that are used downstream or used to recharge aquifers but also affects the quantity and quality of stormwater runoff. Traditional municipal wastewater treatment technology was not designed to effectively remove many of the pesticides, industrial and household chemicals, and pharmaceuticals that enter the collection system. Recent research also has shown that standard disinfection practices may be ineffective at killing some highly resistant pathogens, including protozoan pathogens, *Giardia* and *Cryptosporidium*, and viruses. Chlorination of water also can produce trihalomethanes, which are undesirable cancer-causing by-products. USGS studies revealed that some insecticides commonly used around homes and gardens and in commercial and public areas are widespread in occurrence in surface water throughout the United States. These insecticides occurred at higher frequencies, and usually in higher concentrations, in urban streams than in agricultural streams. The use and improper disposal of industrial chemicals has caused large groundwater contaminant plumes in some urban areas. Groundwater contamination is particularly difficult to remediate with present treatment technologies (Anderson and Woosley 2005).

Recent decades have brought increasing concerns for potential adverse human and ecological health effects resulting from the production, use, and disposal of numerous pharmaceuticals and personal care products (PPCPs). Research has shown that many such compounds can enter streams and groundwater, disperse, and persist to a greater extent than first anticipated. These PPCPs, including biogenic hormones, are released directly to the environment after passing through wastewater treatment processes or domestic septic systems, which often are not designed to remove them from the waste stream (Oelsner et al. 2017).

Treatment of the nation's municipal water supplies has played an important role in reducing disease and contributing to an overall higher standard of living and improved public health throughout the United States. Disease-causing organisms can be substantially reduced or even eliminated from the water supply by disinfection with chlorine or other methods. However, increasing amounts of PPCPs are present in city water supplies. As time progresses, treatment processes must address the contaminants discussed here. Water quality issues will continue to affect more of the population, and safe water supplies for all of U.S. urban areas will become more of a priority with increased urbanization and the aging of the U.S water infrastructure. Reevaluation of present treatment techniques are needed, and improvements to those processes will ensure safer municipal water systems in the United States moving forward.

9.5 WATER USE AND SUSTAINABLE REUSE METHODS

Major industrial water use is discussed in this section, as well as sustainable water reuse systems that will reduce water withdrawals in the future. As discussed previously, the industrial sectors currently using the most water in the United States are agriculture and thermoelectric power production. According to the USGS (Dieter et al. 2018), the total water withdrawn per day in the United States is approximately 322,000 MGD. The industries that consume the largest amounts of water are thermoelectric power (41%) and irrigation (37%). Figure 9.5 shows the trends in water withdrawals in the United States from 1950 to 2015 for these and other various beneficial uses. These total withdrawals include mostly freshwater from lakes, streams, reservoirs and aquifers, but there are several coastal states that use saline water from the ocean for some of their water needs.

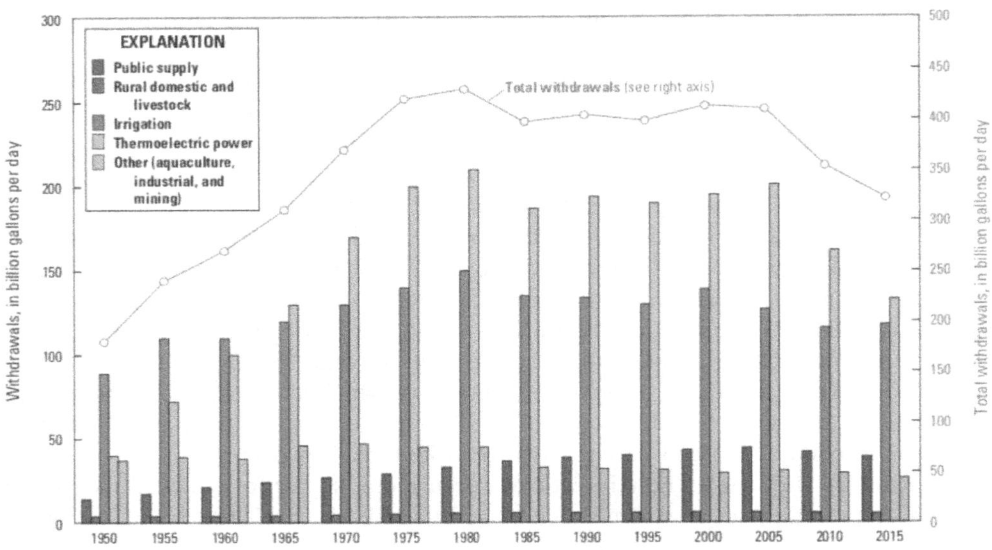

FIGURE 9.5 Trends in total water withdrawals by water-use category, 1950 through 2015. (From Dieter, C.A. et al., *Estimated Use of Water in the United States in 2015*, Circular 1441, U.S. Geological Survey, Reston, VA, 65 p, 2018.)

The total amount of groundwater withdrawn per day in the United States is approximately 84,500 MG. The largest withdrawals of fresh groundwater come from irrigation in five states: California, Arkansas, Nebraska, Idaho, and Texas, making up 46% of total groundwater use. Despite the continual rise in the U.S. population, water use has hit a plateau, with peak usage occurring in 1980, and decreasing continually since 2005 (Figure 9.5). Total surface withdrawals are decreasing, with thermoelectric power constituting the largest reduction. Public supply is also decreasing slightly, despite continued population growth.

Artificial recharge of aquifers is a familiar concept in the water industry, and experimentation dates back to the 1930s. One definition of artificial recharge is the practice of increasing by artificial means the amount of water that enters a groundwater reservoir. Today, artificial recharge is rapidly gaining acceptance as a conjunctive water-use strategy and as a mitigation measure for groundwater overdraft. Aquifer storage and recovery (ASR) is a simple concept whereby treated wastewater effluent, imported surface water, excess streamflow, or stormwater runoff is used to recharge an aquifer for subsequent withdrawal. Two principal methods of ASR are used—a high-rate method, whereby recharge water is directly injected into an underground source of drinking water using recharge wells (Class V injection wells as defined by the U.S. EPA under the Underground Injection Control Program), and a low-rate method, whereby the recharge water is diverted to recharge basins and allowed to slowly percolate into an underlying shallow aquifer. Artificial recharge is also being used to control the migration of seawater into coastal aquifers being extensively pumped for water supply. For example, in Orange County, California, freshwater is injected into this coastal aquifer to arrest the intrusion of salt water. Arizona is rapidly expanding its capability to recharge aquifers with Central Arizona Project (CAP) water (Figure 9.6). There were about 66 active or permitted sites in Arizona where about 162,925 MG of water per year was recharged, for a cumulative total through 1999 of about 782,042 MG (Anderson and Woosley 2005).

Arizona and Nevada have reached a novel agreement to "bank" Colorado River water in alluvial basins. Part of Nevada's allocation of river water will be stored in Arizona basins to be withdrawn later in a complicated exchange for future flows from the river. This ASR method of water management is relatively inexpensive with fewer environmental impacts compared to the construction and

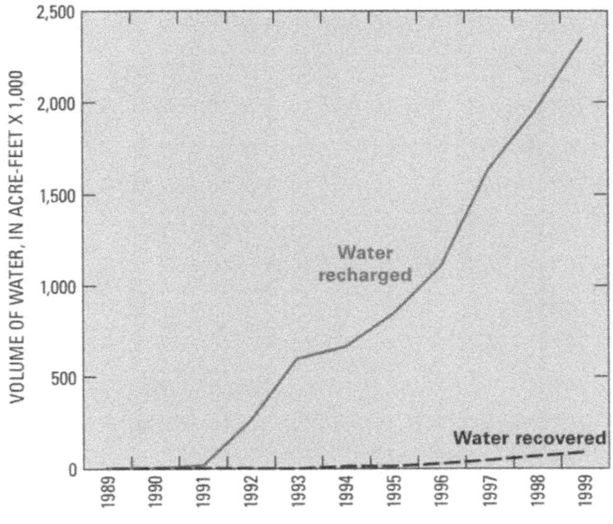

FIGURE 9.6 Cumulative volume of water artificially recharged in Arizona, 1989 through 1999. (From Anderson, M. and Woosley, L. Jr., *Water Availability for the Western United States—Key Scientific Challenges*, U.S. Geological Survey, Circular 1261, 85 p, 2005.)

operation of new reservoirs and can potentially accommodate a much greater volume of water with much less water lost due to evaporation.

Conservation as a water-use strategy is commonly proposed as a solution to meet increasing demand. Although seemingly a capitulation to shortage, water conservation holds promise for the future because in many cases new infrastructure is not required—only a change in user attitudes and practices. Several municipalities including Tucson (Arizona), Albuquerque (New Mexico), and Austin (Texas) support aggressive water conservation initiatives including public education highlighting the need for and benefits of water conservation (Anderson and Woosley 2005). Austin and Tucson provide literature on the use of native desert plants to xeriscape residential and business landscapes. Austin also has passed ordinances that limit lawn watering to specified days during periods of high water use and requires that all new business and residential structures be equipped with low-volume plumbing fixtures. The city of Albuquerque promotes a water-conservation program as well called "Water Watch." Tucson Water's efforts to encourage conservation seem to be working. The per-capita water use, as measured by the amount of potable water delivered, has declined since 1970 (Figure 9.7). Municipal water use, without lawns, gardens, swimming pools, or other landscaping, actually represents a nominal volume of water consumption. Most of this water is returned to the environment, whether a stream or a groundwater aquifer, but in a diminished state of water quality.

Because of the large consumptive loss of water used for irrigation, the U.S. Department of Agriculture (USDA) and state agricultural extension agencies actively support technology transfer and cost-sharing programs for farm water conservation practices and equipment. For example, low-energy precision-application irrigation systems are seeing widespread use by irrigators on the high plains. This system distributes irrigation water under low pressure through drop tubes rather than through overhead sprinklers. Other examples of water-conservation practices being supported by agriculture include soil moisture monitoring to more appropriately time water applications, installation of drip-irrigation systems, and dry farming (Anderson and Woosley 2005).

Conjunctive use is a sustainability strategy that considers surface water and groundwater to be a single resource (Anderson and Woosley 2005). As water development has evolved, surface water and groundwater are used in combination—the most commonly in the west today. Despite the

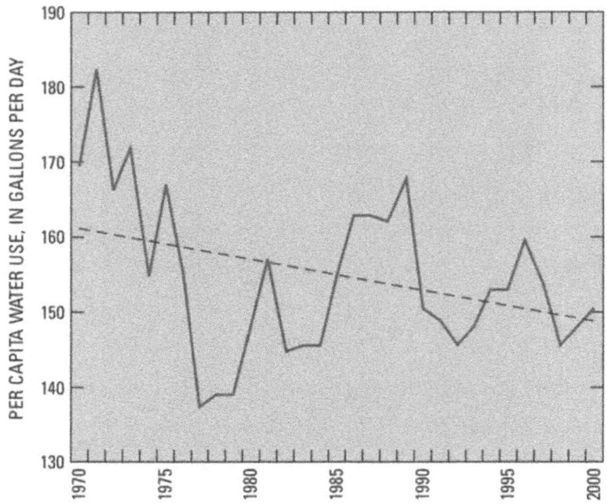

FIGURE 9.7 Trends in per-capita delivery of potable water in Tucson, Arizona, 1970 through 2000. (From Anderson, M. and Woosley, L. Jr., *Water Availability for the Western United States—Key Scientific Challenges*, U.S. Geological Survey, Circular 1261, 85 p, 2005.)

knowledge that groundwater and surface water resources are in fact a single resource, the body of law and administrative statutes often treat them as separate resources. The linkages between surface water and groundwater and the effects of development will need to be better defined to support the strategy of conjunctive use. The Salt River Project, the water purveyor for Phoenix, Arizona, delivers about 325,851 MG per year to water users (Anderson and Woosley 2005). It relies first upon the renewable portion of its water supply, which is surface water stored in reservoirs. Groundwater withdrawals from a network of 250 wells are increased when reservoir storage is depleted by seasonal demand or prolonged drought. Similarly, Rapid City, South Dakota, has a series of wells completed in the Madison and Minnelusa aquifers that are used to augment the supply during the winter, when surface-water flows are refilling upstream reservoirs, and especially during periods of drought (Anderson and Woosley 2005). Thus, strategies that rely first on surface water, which is the primary renewable portion and hold groundwater in reserve for drought periods, provide great flexibility in overcoming natural variability if surface water supplies.

Natural groundwater recharge can be enhanced by impounding stormwater flows along normally dry river channels and allowing water to infiltrate into riverbed sediments. In Las Vegas and Phoenix, urban runoff is captured and diverted into dry wells, which are large diameter (6 to 8 feet) wells up to 100 feet deep that are completed in the unsaturated zone. Runoff from parking lots and streets is diverted into the dry wells and allowed to infiltrate into the groundwater instead of producing surface runoff and evaporating back into the atmosphere. Similarly, recharge to the Edwards aquifer in south-central Texas is being enhanced within urban and rural settings by instream detention structures within the recharge zone and by shallow drainage wells or improved sinkholes. Artificial replenishment of surface-water supplies, either by interbasin transfer or return flow in the form of treated effluent, is becoming a common conjunctive-use strategy (Anderson and Woosley 2005). Other potential strategies include the construction of artificial aquifers, in which a membrane-lined depression is filled with an artificial medium, such as coarse gravel, and covered to eliminate evaporation. Again, stormwater runoff or imported water is allowed to percolate into the medium and to be withdrawn as needed. The issue of biofouling of the media is less of a concern when the strategy is used because the media at the point of recharge is normally dry (Anderson and Woosley 2005).

Water reuse is a simple sustainability strategy that redesignates water that has been used for one purpose for another use that does not require the same level of purity; essentially, it is direct water recycling. There is now wider interest in and economic benefits to reusing municipal wastewater rather than directly returning it to the environment. Some industries and power-generation facilities that use large amounts of water are now co-locating with municipal treatment plants to facilitate the reuse of treated effluent for cooling and process water. Other reuse practices include the use of wastewater effluent as irrigation water and a nutrient source on golf courses and right-of-ways. Some residential developments have encouraged gray-water systems, whereby new homes are equipped with a separate plumbing system to collect wash water from the kitchen and laundry drains. The gray water is then used for watering residential landscapes. The Los Angeles County Sanitation District, for example, is rapidly expanding the number of sites and volume of nonpotable wastewater effluent reused within their service area to reduce the demand of potable water for purposes for which treated effluent is suitable (Figure 9.8).

Desalination is a reuse strategy that removes dissolved solids from saline water sources. In 2010, 39% of the nation's population lived within 47 miles of a coast. This narrow fringe along the coasts constitutes 10% of the contiguous United States land area, but accounts for more than one-half of the nation's housing supply. The population of these coastal areas grew by more than 38 million people between 1960 and 1990. The tantalizing prospect of securing an unlimited, drought-proof supply of freshwater from the ocean has historically encountered the sobering reality of cost.

Dewvaporation is an atmospheric pressure desalination process that is a relatively new heat-efficient tower process. The process uses a humidification-dehumidification approach in which air is used as a carrier gas to evaporate water from saline source water and dew to form pure condensate at constant atmospheric pressure. Initial testing suggests that the average operating costs for brackish and seawater desalination for small plants of \$3.35 per 1,000 gallons can be reduced to \$1.55 per 1,000 gallons using dewvaporation. This lowering of cost and the difficulty of securing conventional freshwater supplies are making desalination more feasible.

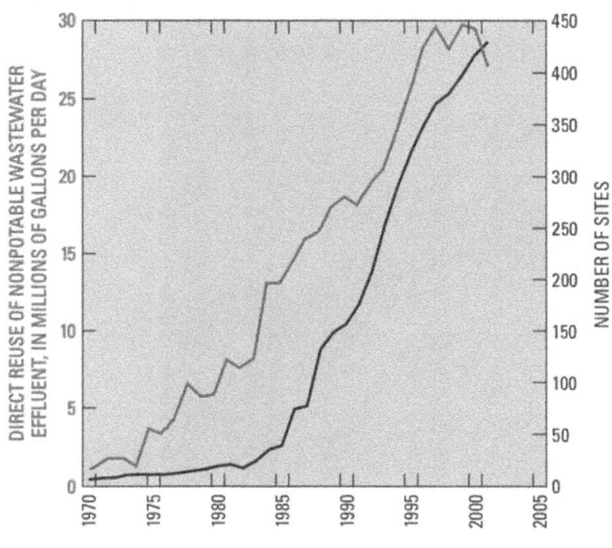

FIGURE 9.8 Increase in the number of sites using treated wastewater and volume of treated wastewater reuse in Los Angeles County, California, 1970 through 2001. (From Anderson, M. and Woosley, L. Jr., *Water Availability for the Western United States—Key Scientific Challenges*, U.S. Geological Survey, Circular 1261, 85 p, 2005.)

Ten communities in California and four in Texas are proposing or supporting desalination facilities. Brackish water is available in many inland areas of the west and with some treatment could relieve stress on limited freshwater supplies. Vast inland areas of the western United States are underlain by aquifers containing brackish or slightly saline water. The reduced cost of desalination for brackish water is offset, however, by the issue of waste-brine disposal. Disposal pipelines (brine lines) to deliver the waste stream from inland locations to an ocean are prohibitively expensive, making it impractical to secure permits, easements, and facility costs. A technological advance is needed to find economical ways to solidify the salts from the brine before desalination can be brought into common use for inland areas. For many municipalities, salts are enriched as they are cycled through the distribution system, collected again, treated in the wastewater-treatment process, and recharged back into an aquifer. Treated effluent is usually about one-third more saline than the influent water delivered to the water user as potable water (Anderson and Woosley 2005).

This cost of converting seawater to freshwater is currently five to six times that of treating freshwater. Advances in existing technologies (membrane filtration, electrodialysis reversal, and reverse osmosis) and development of new desalination systems (i.e., dewvaporation; Chapter 21 in this volume) continue to drive down the cost of desalination (Figure 9.9) so that it is expected that desalination will continue to increase as a source of water supply for various beneficial uses in the future.

Water managers in the western United States are increasingly seeking strategies to achieve a level of water use that can be sustained long term. Sustainability, as presently interpreted, goes beyond mere water availability for human water supply and includes water for ecosystems and even individual species. Sustainability by this definition will continue to be a significant challenge for science to translate into measurable water management strategies. A sustainable water supply for a community ideally would provide enough water to support population and economic growth, adjacent ecological system services, and be sufficient to endure protracted periods of drought without compromising the standard of living and environmental quality (Anderson and Woosley 2005). The overall outlook for U.S. water resources going forward is reasonably good.

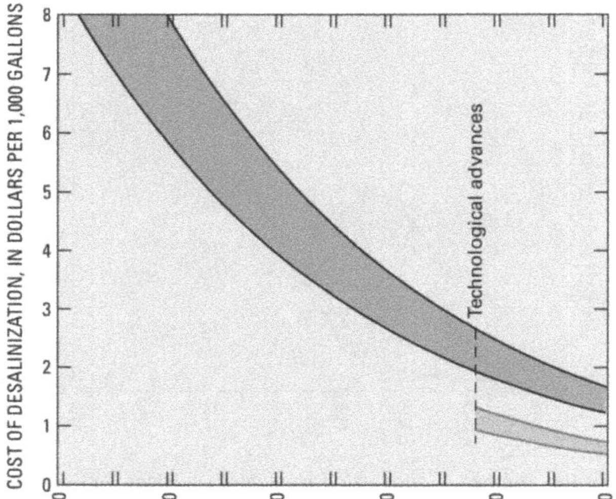

FIGURE 9.9 Declining costs of desalination, 1970 through 2020. (From Anderson, M. and Woosley, L. Jr., *Water Availability for the Western United States—Key Scientific Challenges*, U.S. Geological Survey, Circular 1261, 85 p, 2005.)

9.6 APPLICATIONS

Three Illustrative Examples complement the material presented in this chapter.

Illustrative Example 9.1

Texas and California are the two states that consume the most water in the United States. Compare and contrast the data for the two states, showing how they use their water resources differently.

SOLUTION

Figure 9.10 presents water use by major sector by state. This figure shows that Texas and California are in fact the largest water consumers in the United States, with the largest water use being irrigation in California, whereas thermoelectric power represents the largest water use in the state of Texas. Public water supplies represent the second-largest water use in California, and irrigation is the second-largest water user in Texas. Specific water use data for each sector in California and Texas in 2015 are shown in Table 9.1, reflecting the relative importance of the agriculture and energy production sectors in these two state economies and the relative population of each state. These differences highlight the importance of state-specific water use characteristics and the unique water conservation and reuse options that will be relevant on a state by state basis.

Illustrative Example 9.2

There are a wide range of units used to describe water use and it is important for a scientist or engineer to be fluent in these various unit conventions and conversions. A community in Florida has a potable water use of 15 MGD. If the state of Florida's 2015 total public supply water consumption is given as 2,670,000 ac-ft/yr (Dieter et al. 2018), what fraction of the total Florida water public supply consumption is represented by this community?

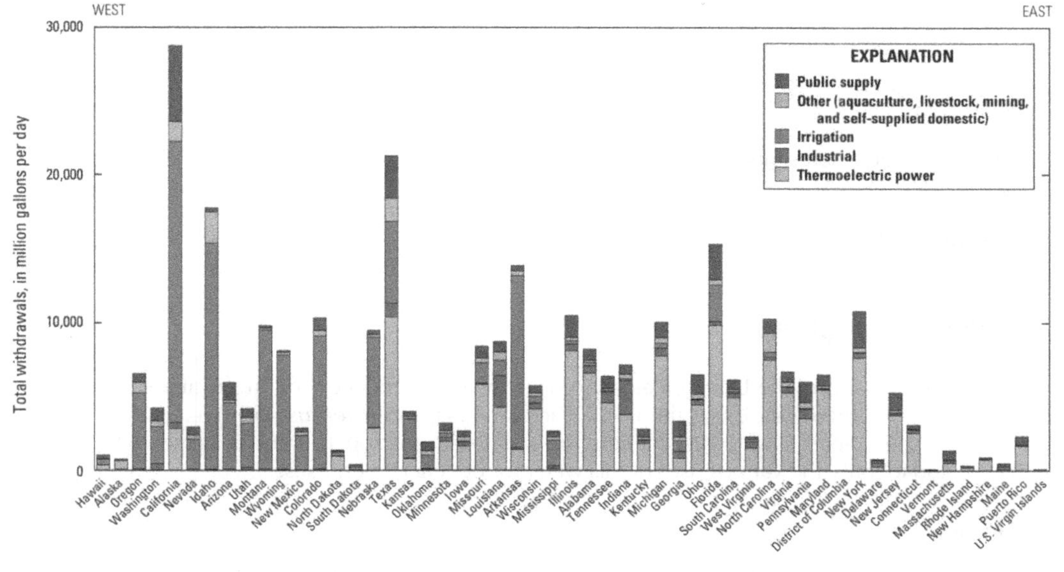

FIGURE 9.10 Total water consumption for major water use sectors by state in 2015. (From Dieter, C.A. et al., *Estimated Use of Water in the United States in 2015*, Circular 1441, U.S. Geological Survey, Reston, VA, 65 p, 2018.)

TABLE 9.1
Water Use Rate Data for Various Sectors in California and Texas, 2015

Quantity, ac-ft/yr	California	Texas
Public supply	5,150	2,890
Irrigation	19,000	5,490
Industrial	399	924
Thermoelectric power	2,876	10,397
Other	1,375	1,599
Total water consumption	28,800	21,300

Source: Dieter, C.A. et al., *Estimated Use of Water in the United States in 2015*, Circular 1441, U.S. Geological Survey, Reston, VA, 65 p, 2018.

Solution

First, the community's daily potable water use is converted to an annual use as follows:

$$(15\text{ MGD})(365\text{ d/yr}) = 5{,}475\text{ MG/yr}$$

This volume is then converted to ac-ft/yr using the following conversions:

$$1\text{ ac-ft} = \text{volume equivalent to 1 ft of water over 1 ac} = (1\text{ ft})(1\text{ ac})$$

$$= (1\text{ ft})(1\text{ ac})\left(43{,}560\text{ ft}^2/\text{ac}\right) = 43{,}560\text{ ft}^3$$

$$1\text{ ac-ft} = 43{,}560\text{ ft}^3\left(7.48\text{ gal/ft}^3\right) = 325{,}829\text{ gal} = 0.326\text{ MG}$$

Therefore,

$$(5{,}475\text{ MG/yr})\left(1\text{ ac-ft}/0.326\text{ MG}\right) = 16{,}800\text{ ac-ft/yr}$$

This community represents the following percentage of total public water supply in the state of Florida:

$$\%\text{ of Total Public Water Supply} = (16{,}800\text{ ac-ft/yr})/(2{,}670{,}000\text{ ac-ft/yr}) \times 100 = 0.63\%$$

Illustrative Example 9.3

Some coastal states in the United States use ocean water to reduce their freshwater surface and groundwater withdrawals. Using the USGS Circular 1441, *Estimated use of water in the United States in 2015* (Dieter et al. 2018), identify the states with the top 10 rates of saline water withdrawal and identify the various uses of this water for beneficial purposes.

Solution

Referring to Dieter et al. (2018), Table 9.2 is a summary of the states in the United States with the top 10 rates of saline water withdrawal and the various beneficial uses for this water. As indicated in Table 9.2, the majority of saline water withdrawals in all states are used in the thermoelectric power sector where the water is used for cooling purposes.

TABLE 9.2
States with the Top 10 Saline Water Withdrawal Rates (MGD) and the Distribution of Major Beneficial Uses for This Water in 2015

State	Industrial	Mining	Thermoelectric	Total
Florida	1.08	0	9,420	9,600
New York	8.43	0.95	5,470	5,480
Maryland	1.37	0	5,300	5,300
New Jersey	0	0	3,430	3,430
California	0	272	2,840	3,160
Connecticut	41.6	0	2,470	2,510
Virginia	5.15	0	2,400	2,420
Texas	323	1,000	757	2,390
North Carolina	0	0	1,360	1,360
New Hampshire	0	0	693	693

Source: Dieter, C.A. et al., *Estimated Use of Water in the United States in 2015*, Circular 1441, U.S. Geological Survey, Reston, VA, 65 p, 2018.

REFERENCES

Anderson, M. and L. Woosley Jr. 2005. *Water Availability for the Western United States—Key Scientific Challenges*. U.S. Geological Survey, Circular 1261, 85 p. https://pubs.usgs.gov/circ/2005/circ1261/pdf/C1261.pdf.

Dennehy, K., T. Reilly, and W. Cunningham. 2015. Groundwater availability in the United States: The value of quantitative regional assessments. *Hydrogeology Journal* 23(8):1629–1632. https://doi.org/10.1007/s10040-015-1307-5.

Dettinger, M.D., D.R. Cayan, M.K. Meyer, and A.E. Jeton. 2004. Simulated hydrologic responses to climate variations and change in the Merced, Carson, and American River basins, Sierra Nevada, California, 1900–2099. *Climatic Change*. 62(1):283–317.

Dieter, C.A., M.A. Maupin, R.R. Caldwell, M.A. Harris, T.I. Ivahnenko, J.K. Lovelace, N.L. Barber, and K.S. Linsey. 2018. *Estimated Use of Water in the United States in 2015*. Circular 1441. Reston, VA: U.S. Geological Survey. 65 p. https://doi.org/10.3133/cir1441.

Intergovernmental Panel on Climate Change. 2001. *Climate Change 2001 Synthesis Report. A Contribution of Working Groups I, II, and III to the Third Assessment Report of the Intergovernmental Panel on Climate Change*. New York: Cambridge University Press. https://www.ipcc.ch/report/ar3/syr/.

Jakeman, A., O. Barreteau, R.J. Hunt, J.-D. Rinaudo, A. Ross, M. Arshad, and S.Hamilton. 2016. Integrated groundwater management: An overview of concepts and challenges. In *Integrated Groundwater Management*, ed. A. Jakeman, O. Barreteau, R.J. Hunt, J.-D. Rinaudo, and A. Ross, pp. 3–20. New York: Springer. https://doi.org/10.1007/978-3-319-23576-9_1.

Oelsner, G.P., L.A. Sprague, J.C. Murphy, R.E. Zuellig, H.M. Johnson, K.R. Ryberg, J.A. Falcone et al. 2017. *Water-Quality Trends in the Nation's Rivers and Streams, 1972–2012—Data preparation, Statistical Methods, and Trend Results (version 2.0)*. U.S. Geological Survey Scientific Investigations Report 2017–5006, 136 p. https://doi.org/10.3133/sir20175006.

10 Global Water Resources

10.1 INTRODUCTION

This chapter on global water resources is a companion to Chapter 9, which dealt with water resources in the United States. The term *global*, refers to the entire world, whereas the term *water resources* refers to all sources of water that have the potential to be used in areas such as agriculture, industry, municipalities, and to support ecological function. Thus, this chapter addresses all potentially useful sources of water on the planet to support the human endeavor.

This topic is addressed through the topics of the history of global water resources, current global water resources, general global water resource issues, and human health issues related to global water resources. An application section provides three Illustrative Examples related to global water resources and water resource management issues.

10.2 HISTORY OF GLOBAL WATER RESOURCES

Water is the most sought after and prized resource on the planet. However, rarely do people take the time to examine how the planet came to be made up of roughly 71% water (Mullen 2012). As described in Chapter 2 water began forming on the surface of the Earth more than 4.5 billion years ago, from the collective impact of thousands of asteroids and comets alike (Fazekas 2014). The water that formed on the surface of the Earth began to separate into eight distinct compartments, which are shown in Table 10.1 along with how much of the Earth's total water budget each makes up.

The eight components shown in Table 10.1 are more commonly organized into three distinct groupings, which account for close to 100% of Earth's total water. First there is surface water, which includes oceans, lakes, rivers, seas, and so on. In short, surface water is all forms of liquid water that lay on the surface of the Earth. The second grouping is glacial and frozen water. These are resources that are located on the surface; however, these resources are in frozen, long-term storage. Long-term storage is defined as water that has historically taken thousands of years to move through the water cycle. The third and final grouping is groundwater. This is the water that is found below the surface of the Earth in conditions where the percentage of saturation in soil pore water is 100% (Mullen 2012).

The process known as the hydrologic, or *water cycle* has played a crucial role in the formation and distribution of all water resources on Earth today. The water cycle is used to describe the constant circulation of water through its three phases of solid, liquid, and gas. Figure 10.1 is a depiction of the water cycle by the U.S. Geologic Survey (USGS 2019). This depiction shows how water contained in its solid snow/ice phase will undergo melting, which involves a phase change from solid to liquid. This phase change results in the formation of rivers, lakes, oceans, etc. The next part of the cycle is a phase change of water from liquid to gas. This change is commonly known as *evaporation*. The water cycle is completed when the gaseous phase of water contained in the atmosphere begins condensing. As this water moves out of the gaseous phase, it will transform back into either liquid or solid water depending on where in the world the water is completing the water cycle.

Contributing Author: Wyatt L. Gilmour

TABLE 10.1
Breakdown of the Total Water Budget on Earth

Water Compartment	% of Total Earth Water Budget
Oceans	97.2
Frozen Glacier Water and Other Ice	2.15
Groundwater or Subsurface Water	0.61
Freshwater Lakes	0.009
Inland Seas	0.008
Soil Moisture	0.005
Atmospheric Water	0.001
Rivers and Streams	0.001

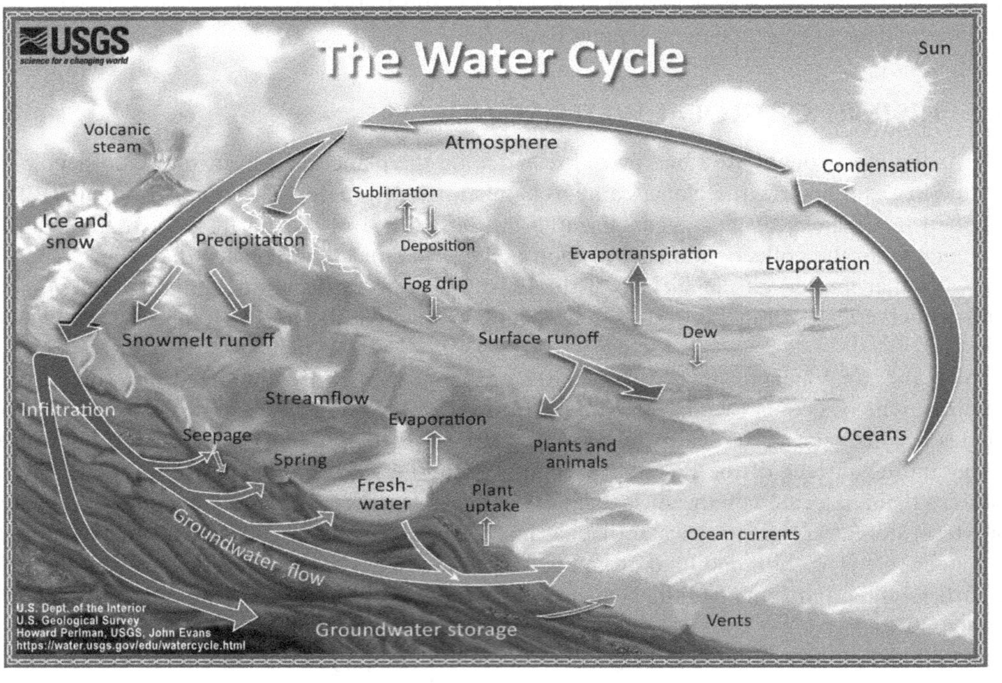

FIGURE 10.1 The water cycle.

10.3 GLOBAL WATER RESOURCES TODAY

Usable water is the key to sustainable life. However, out of all the water on Earth today, it is estimated that only 2.5% – 3.0% is usable. That is quite a large amount of water, considering the Earth's total global water supply is more than 1.38 billion km^3. However, when one delves deeper into this estimated usable supply it is found that a little more than 0.3% of the usable water on Earth is easily accessible (Gleick 1996). This usable water is found in rivers, streams, lakes, and swamps. With the ease of access to these forms of usable water being so high, it is estimated that more than 50% of the world's population lives within 3.0 km of these surface-water sources (Kummu et al. 2011). The majority of the available water that is not considered easily accessible is found in the subsurface in natural aquifers. Figure 10.2 is a depiction of the distance populations are from fresh surface water in different areas of the globe.

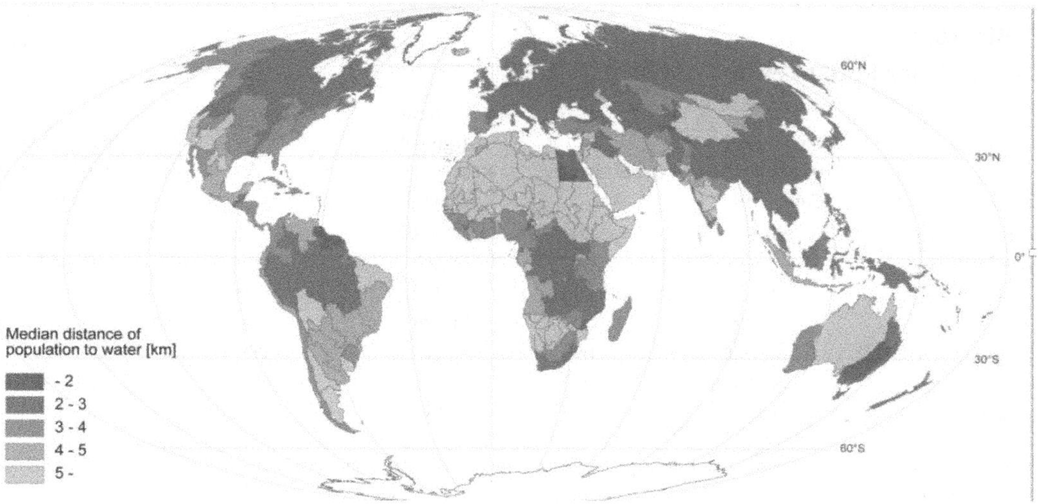

FIGURE 10.2 Median distance of population from fresh surface water in different global regions. (From Kummu, M. et al., *PLoS One*, 6, e20578, 2011.)

TABLE 10.2
Fresh Groundwater, Total Groundwater, and Total Global Water Resources

Water Source	Water Volume, mi³	Water Volume, km³	% of Total Water Resource	% of Total Freshwater Resource
Fresh Groundwater	2,526,000	10,530,000	0.80	30.1
Total Groundwater	5,614,000	23,400,000	1.7	
Total Global Water Resource	332,500,000	1,386,000,000		

Natural aquifers are the most common forms of groundwater storage. Today, these natural aquifers are thought to hold roughly 34.2 million km³ of total water. Out of these 34.2 million km³ of water, it is estimated that close to 10.4 million km³ of this water is usable (Gleick 1996). Because these aquifers are below the surface of the earth, they do require the creation of wells for water recovery. These wells can range from tens to hundreds of feet deep, depending on where the water table is located. It is also important to note that these wells will generally require the installation of a pumping system to bring the water to surface. Table 10.2 provides a comparison between the total water on Earth to the amount that is both groundwater and fresh groundwater.

10.3.1 Usable Water Resources by Continent

As small as the total volume of usable water to nonusable water is in the world, there are still considerable amounts of usable water resources present among the seven continents. The amount of usable water on each continent can be divided among the three water categories as described (i.e., groundwater; glacier/permanent ice caps; and wetlands, lakes, reservoirs, and rivers). These categories, along with the volumes of water in each for each continent across the globe are summarized in Table 10.3.

Examining Table 10.3 it becomes immediately obvious that Antarctica has by far the largest amount fresh water of any continent. However, this volume of fresh water is in the form of glaciers and permeant ice caps, thus making it non-usable. When comparing usable water resources, the top three continents are Asia, with roughly 7,830,622 km³ of water; Africa, with approximately

TABLE 10.3
Global Water Resources by Continent

Continent	Glacier and Permanent Ice Caps, km^3	Groundwater. km^3	Wetlands, Large Lakes, Reservoirs and Rivers, km^3	Totals, km^3	Total Usable Water Resource, km^3
North America	90,000	4,300,000	27,003	4,427,003	4,327,003
South America	900	3,000,000	N/A	3,00,900	3,000,000
Europe	18,216	1,60,000	2,239	1.620,745	1,602,529
Africa	0	5,500,000	31,776	5,531,776	5,531,776
Australia	180	1,200,000	221	1,200,401	1,200,221
Asia	60,984	7,800,000	30,622	7,891,606	7,830,622
Antarctica	30,109,800	N/A	N/A	30,109,800	0

Source: Rekacewicz, P., *Freshwater Resources: Volume by Continent*, GRID-Arendal, Arendal, Norway, http://www.grida.no/resources/5608, 2006.

5,531,776 km^3 of water; and North America with 4,327,003 km^3 of total usable water resources. Furthermore, these three continents contain more than 75% of the usable water on Earth. With this large amount of water being contained in three of the seven continents, mankind has adapted through the development of artificial reservoirs. These artificial reservoirs are most commonly constructed by the creation of a dam around an existing natural water source. Today, it is estimated that there are more than 515,000 of these man-made reservoirs of greater than 2.5 ac in size that cover roughly 335,000 km^2 of the Earth's surface (Osborn 2019).

10.3.2 Usable Water Resources by Country

Table 10.4 summarizes the annual renewable freshwater resources for the top 25 counties in the world. Renewable freshwater resources are an estimate of the average precipitation, recharged groundwater, and fresh surface inflows from surrounding countries that each of these countries receives in any given year. What is interesting to note from these data is that despite a country having a great deal of renewable fresh water resources, water shortages and unequal distribution of resources can still take place. Consider Brazil, which tops the list with more than 8,233 km^3 of renewable fresh water each year. Despite what would appear to be an abundance of fresh water, Brazil is experiencing a water crisis caused by poor water infrastructure and water resource management, which has left over 5 million, or more than 2% of its citizens without access to safe potable water (Water.org 2019).

Even in what is considered a highly advanced country like the United States, which receives 3,069 km^3 a year of renewable fresh water, and has historically invested in water infrastructure as well as the management of its water resources, it is estimated that as of 2014, 1.6 million people (0.5% of its population) were without a toilet, a tub or shower, or running water (UNC Environmental Finance Center 2017). Despite availability of renewable freshwater resources, equitable and affordable supply and access to these resources remain a significant challenge, particularly to minority and low-income communities, even in the most developed countries in the world.

10.3.3 Water Usage per Person by Country

Figure 10.3 summarizes the total per-capita water consumption (consumption of surface or groundwater for domestic, commercial, and industrial purposes; water for crop production; and water required for the assimilation of pollutants produced from domestic, commercial, and industrial

TABLE 10.4
Renewable Annual Freshwater Resources for Top 25 Countries

Rank	Country	Total Freshwater Renewable Resource, km^3
1	Brazil	8,233
2	Russia	4,498
3	Canada	3,300
4	United States	3,069
5	Indonesia	2,838
6	China	2,830
7	Colombia	2,132
8	Peru	1,913
9	India	1,908
10	Democratic Republic of the Congo	1,283
11	Venezuela	1,233
12	Bangladesh	1,211
13	Burma	1,046
14	Chile	922
15	Vietnam	891
16	Republic of the Congo	832
17	Argentina	814
18	Papua New Guinea	801
19	Bolivia	622
20	Malaysia	580
21	Philippines	479
22	Cambodia	476
23	Mexico	457
24	Ecuador	432
25	Japan	430

Source: Advameg, Inc, Encyclopedia of the Nations, Total renewable water resources – The World Factbook – CIA, http://www.nationsencyclopedia.com/WorldStats/CIA-Total-renewable-water-resources.html, 2010.

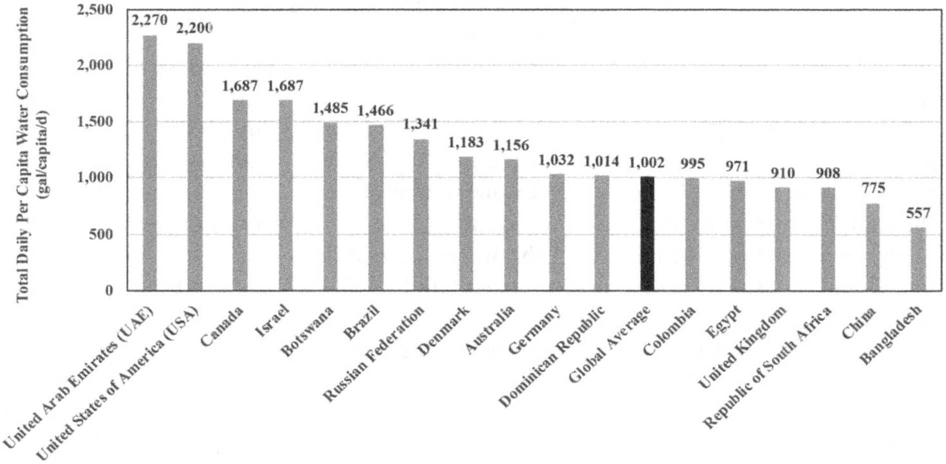

FIGURE 10.3 Total daily per-capita water consumption by selected country. (Data from Grace Communication Foundation, Water Footprint Comparisons by Country, https://www.watercalculator.org/footprints/water-footprints-by-country/, 2019.)

activities) for the select countries in the world whose daily consumption is above and below the global average (Mekonnen and Hoekstra 2011). The global average for total per-capita daily water usage is roughly 1,000 gal/capita/d (Grace Communication Foundation 2019). This is water that is used for domestic purposes such as laundry, cooking, drinking, washing, and, as well as water used for food production and the production of consumer goods, like cars, plastic bags, clothing, etc. At the top this list with 2,270 gal/capita/day is the United Arab Emirates (UAE), 76% of which is associated with products or ingredients of consumer goods imported annually into the UAE. In the United States, of the 2,200 gal/capita/day of total water use only 20% is associated with imported external water consumption. The significant differences seen in the daily internal versus external per-capita total water consumption are directly attributed to the consumption habits of each country's citizenry.

10.4 GENERAL GLOBAL WATER RESOURCE ISSUES

The root of all major issues pertaining to global water resources today stems from a lack of water accessibility. Today, it is estimated that more than 1 billion people live without access to safe water (WHO 2017). This number more than doubles to over 2.3 billion people when basic sanitation services are considered. It is this lack of access to safe drinking water and adequately treated wastewater that drives other major water related issues in the world today. As indicated previously, this access issue stems directly from the lack of necessary water infrastructure as well as the poor maintenance and management of the limited infrastructure that currently exists.

Improved water resource management and water storage capacity makes an economy more resilient to external shocks, such as rainfall variability and, thus, provides a stable and sustainable base for increased food and industrial productivity and production to maintain economic growth and development. Consider a heavily water dependent country like Kenya where frequent floods and droughts take a heavy toll on the economy (Stockholm International Water Institute [SIWI] 2005). These floods and droughts translate to a direct long-term fiscal liability and declines in economic growth of about 2.4% Kenya's annual gross domestic product (GDP; SIWI 2005). Much of this loss in economic output comes from lack of adequate water storage capacity, a deficit of more than $10 billion in 2004 (SIWI 2005). If Kenya, as for many other African countries, had invested in storage facilities for water, they would be able to store the excess water seen in heavy flood years for use in heavy drought years, thus, avoiding flood liabilities and water scarcity cost allowing the country to avoid the drops in economic growth that have occurred in the past.

The SIWI (2005) indicated that national economies are more resilient to rainfall variability, and economic growth is boosted when water storage capacity is improved. Investing in adequate water resource management in developing countries would be expected to produce significant economic gains of between two to four times the infrastructure investment. The SIWI (2005) concluded that an even greater return on investment of $8.00 to $200 would be saved in water treatment costs for every $1.00 invested in watershed protection. Investments in improved water and sanitation infrastructure in developing countries also has significant economic benefits when considering increased productivity from the improved health of workers in households and the commercial and industrial sectors, ranging from hundreds of million to billions of dollars annually (SIWI 2005).

In today's world, the term *hydro-politics* is becoming more and more known. Hydro-politics can be defined as debate or conflict amongst individuals, parties, or organizations within governments about water-related issues. These debates/conflicts may even take place between multiple counties and not just the individuals within those countries. Commonly, these types of conflicts are solved by establishing a system known as *Riparian Water Rights*. This is a system used to allocate water among different countries or jurisdictions within countries that are in control of the land through which the water travels. These types of rights are considered common law among many countries in the world such as the United States, Canada, Australian, and Great Britain.

Even though many of the largest counties in the world have using this system to manage water conflicts for years, there are many other counties, however, where these systems are relatively new. A prime example of this is conflict between Egypt and Ethiopia. For centuries, these two countries have been in constant conflict over who owns the developmental rights for the water in the River Nile. It was not until 2015 that these two counties set aside their differences and signed an agreement ensuring shared and equitable access to the river (Lufkin 2017).

Even with compromises such as the one between Egypt and Ethiopia described, some individuals, such as Ismail Serageldin (former Vice President of the World Bank), see water becoming the next so called "Oil War." Ismail Serageldin famously said, "The wars of this century were fought over oil, the wars of the next century will be fought over water" (Barnaby 2009). However, when one sees countries coming together to solve their water resources conflicts, forming compromises like Egypt and Ethiopia, the potential for water wars seem less likely, and the arguments by Barnaby (2009), Fröhlich (2012), and McCrary (2018) appear to have more validity. These later authors believe that physical conflict over water will not happen because established systems for international trade and international agreements leading to mutual benefits ease these conflicts, thus preventing the scale of these problems from becoming large enough to cause military style confrontations.

10.5 GLOBAL WATER HEALTH ISSUES

Waterborne diseases describe any form of illness that is caused by pathogens that originate in fecal material and are transmitted by ingestion (i.e., cholera and typhoid fever). *Water-washed* illnesses occur by organisms that originate in feces and are transmitted through contact because of inadequate sanitation or hygiene (i.e., trachoma). *Water-based* illnesses are caused by organisms that originate in water or spend a portion of their life cycle in aquatic organisms and come into direct contact with humans via water or through inhalation (i.e., schistosomiasis and legionellosis). Finally, *water-related* illnesses are caused by microorganisms that have life cycles associated with insects that live or breed in water (i.e., yellow fever; Pepper et al. 2014). A wide range of microorganisms (viruses, bacteria, protozoa, and helminth worms), causing a wide range of human diseases, are transmitted via contaminated water supplies or are in some way water related; examples of which are indicated in Table 10.5.

Many pathogenic organisms can persist for long periods of time in contaminated soils, sediments, and water, in addition to being viable in both human and animal hosts. Concentrations in fecal material can range from 10^4 to 10^5 cells of *Ascaris*/g feces to as high as 10^{11} Adenovirus or Norovirus/g feces (Pepper et al. 2014). With such high loadings of pathogens from contaminated individuals, both effective drinking water and wastewater treatment are essential to control disease transmission within susceptible populations. These waterborne and water-related illnesses are most common in developing countries because of the lack of sufficient water and sanitation infrastructure. In fact, it is estimated that out of all illnesses in these developing countries, roughly four out of five are waterborne or somehow water-related Bos (2005). The World Health Organization (WHO 2017) estimates that 2.3 billion people still lack basic sanitation services, and nearly 900 million people still practice open defecation.

A number of pathogens are of particular interest because of their prevalence, persistence, or virulence, and these are described briefly.

10.5.1 *Vibrio cholera*

Cholera, resulting in severe diarrhea, vomiting, and dehydration, is caused by the bacterium *Vibrio cholerae*. People become infected after eating food or drinking water that has been contaminated by the feces of infected persons. Raw or undercooked seafood may be a source of infection in areas where cholera is prevalent and sanitation is poor. Vegetables and fruit that have been washed with water contaminated by sewage may also transmit the infection if *V. cholerae* is present. A cholera

TABLE 10.5
Common Waterborne or Water-Related Human Pathogens Contained in Human Feces

Pathogen Group and Name	Associated Diseases	Transmissibility
Viruses:		
Adenoviruses	Respiratory, eye infections	Low infectious dose
Caliciviruses	Diarrhea	Low infectious dose
Enteroviruses		
Polioviruses	Aseptic meningitis, poliomyelitis	Low infectious dose
Echoviruses	Aseptic meningitis, diarrhea, respiratory illness	Low infectious dose
Coxsackie Viruses	Aseptic meningitis, herpangina, myocarditis	Low infectious dose
Hepatitis A Virus	Infectious hepatitis	Low infectious dose
Rotaviruses	Gastroenteritis	Low infectious dose
Norovirus	Gastroenteritis, diarrhea	Low infectious dose
Bacteria:		
Campylobacter	Gastroenteritis	Medium to high infections dose, moderate persistence
Legionella pneumophila	Pneumonia	Medium to high infections dose, moderate persistence
Salmonella typhi	Typhoid fever	Medium to high infections dose, moderate persistence
Other salmonellae	Gastroenteritis	Medium to high infections dose, moderate persistence
Shigella spp.	Bacillary dysentery	Medium to high infections dose, moderate persistence
Vibrio cholerae	Cholera	Medium to high infections dose, moderate persistence
Other vibrios	Diarrhea	Medium to high infections dose, moderate persistence
Yersinia enterocolitica	Gastroenteritis	Medium to high infections dose, moderate persistence
Protozoan:		
Giardia lamblia	Diarrhea	Low infectious dose
Crytosporidium spp.	Diarrhea	Low infectious dose
Entamoeba histolytica	Amoebic dysentery	Low infectious dose
Helminth:		
Ancylostoma duodenale (Hookworm)	Hookworm	Latent and persistent
Ascaris lumbricoides (Roundworm)	Ascariasis	Latent and persistent
Hymenolepis nana (Dwarf tapeworm)	Hymenolepiasis	Low infectious dose
Necator americanus (Hookworm)	Hookworm	Latent and persistent
Strongyloides stercoralis (Threadworm)	Strongyloidiasis	Latent and persistent
Trichuris trichiura (Whipworm)	Trichuriasis	Latent and persistent

Source: Viessman, W. et al., *Water Supply and Pollution Control,* 8th edition, Pearson Prentice-Hall, Upper Saddle River, NJ, 2009.

outbreak in 1991 in Peru spread throughout Central and South America resulting in more than 1 million cases of the disease and more than 9,600 deaths (Pepper et al. 2014). The risk of cholera remains high throughout the world as changing climate conditions resulting in increased temperatures and algal growth increase the survivability of *V. cholerae* in the environment, increasing risk of human exposure and disease transmission.

10.5.2 Pathogenic *Escherichia coli*

E. coli are gram-negative rods found in the intestinal track of warm-blooded animals. They are generally considered harmless and used as pathogen indicator organism, but some infectious strains of *E. coli* have been identified. *E. coli* are transmitted via fecal-oral route, and approximately 210 million cases of the infection and 380,000 deaths per year occur from infections *E. coli* worldwide (Pepper et al. 2014). The eterohemorrhagic (EHEC) O157:H7 produces shiga toxins like those produced by *Shigella dysenteriae* and is particularly harmful to elderly people, generating a mortality rate in this population of greater than 50%. *E. coli* O157:H7 is transmitted by contaminated raw hamburger, raw milk, unpasteurized fruit juice, and contaminated vegetables and is commonly detected in domestic wastewater as well as slaughterhouse wastewater.

10.5.3 *Shigella*

This organism is closely related to *E. coli* and is principally a disease of humans and occasionally of other primates. *Shigella* transmitted via the fecal-oral route and can have fatality rate high as 10% to 15%. It is a common cause of recreational disease outbreaks in the United States as well as food contamination. Approximately 440,000 cases of *Shigella* occur annually in United States alone (Pepper et al. 2014).

10.5.4 *Campylobacter jejuni*

This organism is transmitted via food and by direct contact with infected hosts. It is a leading cause of bacterial diarrheal illness in the world with approximately 2 million cases per year (Pepper et al. 2014). *C. jejuni* is somewhat unique in that its concentration is not well correlated to organisms normally used as indicators of fecal contamination (i.e., non-infectious *E. coli*) and is can survive for more than 120 days in the environment at temperatures below 4°C (Pepper et al. 2014).

10.5.5 *Salmonella*

Salmonella typhi is the cause of typhoid fever in humans, and with the advent of water treatment, filtration and chlorination the occurrence rate of typhoid fever has been reduced significantly. Those populations without this level of water treatment would still be susceptible to infection by *S. typhi*. Foodborne strains of *Salmonella* are also responsible for more than 1 million cases per year around the world of gastroenteritis.

10.5.6 Cyanobacterial Toxins

Cyanobacteria are a group of photosynthetic bacteria, some of which fix nitrogen from the atmosphere, and some of which produce a wide range of taste and odor compounds as well as human toxins including hepatotoxins, neurotoxins, and gastrointestinal irritants (Pepper et al. 2014). Cyanobacteria have increasingly been identified with harmful algal blooms (HABs) and elevated cyanotoxin levels that pose significant risks to humans and domestic animals coming in contact with HAB-impacted surface waters. HAB occurrence is related to excessive nutrient inputs and elevated temperatures of surface waters and are being stimulated as a result of changing climate conditions.

10.5.7 *Giardia lamblia*

G. lamblia is a human protozoan parasite that forms a nominal 10 μm oocyst as a survival response, and these cysts are resistant to chlorine disinfection. Infected individuals may shed 10^6 cysts/g feces, and the infective dose in humans less than 10 cysts/dose. Untreated infections may last from months to years, and the organism is most commonly spread person to person. *Giardia* is the leading cause of waterborne disease in United States.

10.5.8 *Cryptosporidium parvum*

C. parvum is another human protozoan parasite that forms a nominal 5 μm oocyst. The infectious dose in humans is approximately <1,000 oocysts. The disease is a self-limiting infection that lasts 1 to 10 days in healthy people but may persist for months and be fatal in immunocompromised persons. *C. parvum* is most commonly spread person to person and was responsible for a waterborne outbreak in Milwaukee, Wisconsin, in April 1993, which involved more than 400,000 people (25% of the service area) because of oocyst penetration of filters in the water treatment plant and the spread of infectious oocysts throughout the water distribution system.

10.5.9 *Ascaris lumbricoides*

A. lumbricoides or roundworm is a human parasitic worm which is the most prevalent parasite in the world, with more than 1 billion people (approximately 22% of the world's population) infected with the organism. The infection rate in the United States is 1% – 2% (Pepper et al. 2014). *Ascaris* eggs in contaminated water are swallowed, the larvae hatch in the small-intestines, then enter bloodstream, move to lungs, and immature worms move to the trachea and epiglottis where they are swallowed and move back to the intestines. *Ascaris* eggs shed in feces can survive in soil for long periods, 30% to 53% eggs in soil are reported to have survived more than 10 years, making them of great risk to humans that may come in contact with these viable eggs over time (Pepper et al. 2014). Treatment methods for this disease usually focuses on removal of *Ascaris* eggs from wastewater and sludge in those areas of the world where wastewater treatment is commonly practiced. Without adequate wastewater treatment, the risk of *Ascaris* infection remains high.

10.5.10 Viral Pathogens

There are approximately 140 known enteric viruses, and even respiratory viruses may also be found in wastewater (Pepper et al. 2014). Most virus transmission is person to person, and foodborne transmission is likely more significant than waterborne transmission. Also, the minimal infectious dose for viruses is, in general, lower than for bacterial pathogens. Table 10.5 lists a wide range of human viral pathogens responsible for a range of human afflictions including meningitis, poliomyelitis, hepatitis, gastroenteritis, and other respiratory and other intestinal illnesses (Viessman et al. 2009; Pepper et al. 2014). As indicated for other pathogens, a lack of adequate drinking water and wastewater treatment and basic sanitation practices are responsible for the transmission of viral pathogens in many parts of the developing world.

It was estimated that in 2016 more than 1 million people died worldwide from these waterborne illnesses, with the major cause of these deaths being diarrhea (WHO 2019). Due to this high prevalence of waterborne and water-related diseases and the large number of people infected and suffering from them, the World Bank is pushing countries to increase their $37.5 billion per year in water-related spending to $150 billion per year (Hares 2017). This significantly increased level of funding is what the World Bank believes necessary to universally deliver safe water and sanitation to all people in the world (Hares 2017).

10.6 ILLUSTRATIVE EXAMPLES

Three Illustrative Examples complement the material presented in this chapter related to global water resources issues.

Illustrative Example 10.1

SIWI (2005) and Hares (2017) have made strong cases for the economic benefit of water infrastructure investments around the world. In the United States, the U.S. Water Alliance (2017) outlined the way forward for ensuring all people have access to clean, affordable water, and all benefit from water infrastructure investments to make water supplies more reliable and resilient in the face of climate change pressures. Review the U.S. Water Alliance (2017) national briefing paper on water equity and summarize the "Three Pillars" of water equity put forth in this document.

Solution

This is an open-ended question exploring the national survey of water supply and system needs, and innovative approaches for addressing water stress and equity challenges in vulnerable communities across the United States that was conducted by the U.S. Water Alliance in 2017. Through discussions with stakeholders, surveys of utilities, discussions with research institutions, and a comprehensive review of the literature, the U.S. Water Alliance compiled findings and developed recommendations for implementing promising practices and strategies that organizations in the United States are implementing to improve water equity and resilience, the Three Pillars of Water Equity. These Three Pillars are:

1. *Pillar 1—Ensure all people have access to clean, safe, affordable water service.* Some items highlighted in this area were: *affordability* of water supplies and infrastructure improvements through expansion of affordability programs, funding support increases to smaller utilities; *increased access to infrastructure* through consolidation of small utilities, scaling up of successful community-centered direct service programs, and deploying technology solutions for communities without existing infrastructure; and *improvements in water quality* through support of waterfront revitalization for multiple benefits, advance water quality goals through citizen science, and supporting tribal government source water protection efforts.
2. *Pillar 2—Maximize the community and economic benefits of water infrastructure investment.* For this pillar, some items highlighted included: *workforce development* through building a water career pipeline for youth and adults, using proactive, inclusive hiring requirements for construction and nonconstruction water infrastructure careers, and aligning workforce training with employer needs at a regional level; *contracting and procurement changes* to break up large contracts to create opportunities for small, minority-, and women-owned businesses, partner on business accelerator and development programs, and create incentives within the contracting process for community benefit outcomes; and *neighborhood revitalization* that creates synergistic benefits among multiple infrastructure investments, channels green infrastructure to disinvested neighborhoods, and cultivates resident and community stewardship of water projects.
3. *Pillar 3—Foster community resilience in the face of a changing climate.* In this pillar, some items highlighted included: *planning and assessment improvements* to include community considerations in climate vulnerability assessments, incorporate community vulnerability assessments into climate planning tools, and connect community-based organizations to climate planning efforts; *funding* dedicated to adaptation, mitigation, and disaster relief for vulnerable communities, community development initiatives to build climate resilience, and building of partnerships with the flood insurance industry; and *project delivery approaches* that leverage climate adaptation projects to create economic opportunity for vulnerable communities, and that prioritize vulnerable communities needing physical adaptation for improving their resilience.

Illustrative Example 10.2

A comparison of global drinking water supply and sanitation services provided by WHO (2017) suggests that in 2015, North America has some of the highest rates of improved water supplies (96%) and basic sanitation (97%) and safely managed, improved sanitation facilities (78%) in the world.

The statistics for the United States are even higher at 99% of the population using improved water supplies, whereas 89% were using improved sanitation facilities. For those in the United States not either having access to improved water supplies or improved sanitation facilities, University of North Carolina (UNCE; 2017) has indicated that conditions for them are similar to individuals in developing countries. Review the UNC Environmental Finance Center (2017) document on *Clean Water Access Challenges in the United States* and summarize the six major clean water access challenges they identified in twenty-first century America.

Solution

This is an open-ended question exploring clean water access challenges in modern America. The six major clean water access challenges UNC identified in their project along with examples of the level and severity of impact follow:

1. *Lack of Indoor Plumbing.* An estimated 1.6 million individuals in the United States do not have a toilet, a tub or shower, or running water in their home. Some specific examples presented in the document included: Sandbranch (Texas), 14 miles from Dallas, that has never had running water in the 138 years since it was established. The residents, only 80 of which have remained, are 97% minority, and all live below the federal poverty threshold; 40% of Navajo Nation residents lack running water and instead get their water from monthly deliveries; and Texas Colonias, home to approximately 500,000 people in 2,294 communities. More than 40% live below the poverty line, and as of 2014, 337 of these colonias still lacked access to basic infrastructure.
2. *Unsanitary On-Site Waste Disposal.* Approximately 20% of the U.S. population are not connected to community sewer systems and rely on some form of decentralized wastewater treatment such as septic tanks. The last census of on-site wastewater treatment systems nationwide was conducted in 1990, and at the time, many systems were inadequately treating the wastewater they received. Examples of these issues nationwide include: in the state of Michigan where it is estimated that 10% of the state's 1.3 million on-site wastewater treatment systems have failed or are polluting the environment. Specifically, within Barry and Eaton counties, the health district discovered nearly 1,000 failed septic systems and 300 houses with no septic systems at all, where household sewage was running into the nearest farm drain, lake, or stream; Lowndes County, Alabama, a county that is 74% African American and where almost a third of the population lives below the poverty line. Hookworm, a pathogenic helminth, is thriving, largely because of a lack of sanitary waste disposal. Of the residents surveyed, 73% reported raw sewage washing into their homes from faulty septic or waste pipes, and 34% of residents who were tested were found positive for traces of hookworm.
3. *Contaminated and At-Risk Individual Wells.* The USGS estimates that 14% of the U.S. population receive their drinking water from self-supplied sources, of which 23% of 2,167 domestic wells sampled in 2010 were found to contain at least one contaminant exceeding recommended human health standards. In California's Central Valley, 700,000 people are currently being exposed to contaminated water at home or at school. In addition to those with contaminated water, another 3,511 California households reported having wells that are still dry according to state data released in January 2017; in May, 2017, residents of Airway Heights, Washington, were instructed not to drink their tap water after samples from private wells on and around the Fairchild Air Force Base indicated elevated levels of polyfluorinated alkyl substances (PFAS). Fairchild is one of 30 Air Force bases around the country where PFAS contamination was discovered in 2017.
4. *Community Systems with Contaminated or Depleted Water Supplies.* As of 2010, an estimated 105 million people receive water from public water systems that use groundwater, and based on USGS study mentioned previously, an estimated 22% of public wells for those systems have one or more contaminants at unsafe levels for human health. Some examples of contaminated community well situations around the United States include: the San Joaquin Valley in California has groundwater contaminated from agricultural practices that has exposed more than 1 million people to elevated nitrate levels via their drinking water; in Jim Hogg County, Texas, more than 5,000 people have been exposed

to arsenic concentrations in their water four times above the drinking water standard for a period of at least 5 years, and in 2016, 65 community water systems in the county have had arsenic levels above the drinking water standard for more than 2 years; according to the Minnesota Pollution Control Agency, approximately 40% of shallow wells in central Minnesota and most sand and gravel aquifers in southern Minnesota have nitrate levels that exceed safe drinking water standards.

5. *Community Systems Struggling to Pay for Service*. Water and wastewater utility service rates in certain locations in the country are rising at a rapid as a result of drought, to significant infrastructure investment, and to costs associated with lost water because of antiquated infrastructure. Even where there is customer financial assistance available at the state or utility level, it falls short of the needs of the increasing number of low-income customers nationwide. Actual or perceived legal barriers exist to the use of rate revenue for customer assistance programs. The following are examples of communities struggling with their customer base being unable to pay for services: in Detroit, Michigan, more than one in six households had their water cut off for unpaid bills in 2016, and in Philadelphia, Pennsylvania, more than 40% of their water utility customers are delinquent in paying their water bills.
6. *Community System Customers with Substandard Plumbing*. Customers served by even the most efficient public water systems but who may live in residences with vulnerable plumbing, especially affordable housing and rental markets, still may face serious access challenges. Leaky household plumbing with unaffordable repairs, or as in the Flint, Michigan, crisis where vulnerability households relying on service lines and indoor plumbing that contain lead are particularly vulnerable to these issues of substandard plumbing. Noteworthy cases of problematic plumbing across the country include: Baltimore, Maryland, that sent more than 315 owner-occupied properties to tax sale in 2016, due to unpaid water bills, many of which were the result of leaks of sinks or toilets; Pittsburg, Pennsylvania, where an estimated 17,000 homes served by the local utility have lead service lines; and Flint, Michigan, where an estimated 29,100 residents have lead or galvanized steel service lines that need to be replaced, at a cost of at least $140 million.

Illustrative Example 10.3

Mekonnen and Hoeskstra (2011) provide a summary of national water footprints of production and consumption for countries around the world. They provide a guideline for the determination of the representative water footprint of various commodities that may be both produced within a country as well as imported for consumption from elsewhere around the globe. An example that is provided determines the average water footprint for tomatoes consumed in Germany. Using the data presented in Table 10.6, determine what the water footprint for tomato consumption in Germany is assuming the quantity imported is consumed, and use a weighted average of each water footprint.

TABLE 10.6
Water Footprint and Quantities of German Tomato Consumption as a Function of Place of Origin

Place of Origin	Quantity Imported or Consumed, T/yr	Water Footprint, m^3/T
Germany	47,000	36
The Netherlands	252,000	10
Spain	244,000	83
Italy	72,000	109
France	52,000	110
Total	667,000	

Solution

Using a weighted average of the water footprint and Germany's import levels from the countries listed in Table 10.6 yields the following results using Equation 10.1 for the calculations.

$$\text{Weighted Average Water Footprint} = \frac{\int_1^n (\text{Tomato Consumption, T/yr})_i (\text{Water Footprint, m}^3/\text{T})_i}{\text{Total Tomato Consumption}} \quad (10.1)$$

where Tomato Consumption is the annual T of tomatoes produced or imported and consumed from each country, i; Water Footprint is the m^3 water/T tomatoes produced from each country, i; and n is the total number of countries, including itself, from which Germany imports tomatoes. Substituting values from Table 10.6 into Equation 10.1 yields:

Weighted Average Water Footprint

$$= \frac{(47{,}000)(36) + (252{,}000)(10) + (244{,}000)(83) + (72{,}000)(109) + (52{,}000)(110)\ (\text{T/yr})\ (\text{m}^3/\text{T})}{(667{,}000\ \text{T/yr})}$$

$$= \frac{38{,}032{,}000}{667{,}000} = 57\ \text{m}^3/\text{T}$$

REFERENCES

Advameg, Inc. 2010. Encyclopedia of the Nations. Total renewable water resources –: The World Factbook –: CIA. http://www.nationsencyclopedia.com/WorldStats/CIA-Total-renewable-water-resources.html.

Barnaby, W. 2009. Do nations go to war over water? *Nature* 458: 282–283.

Bos, R. 2005. A global picture of the diverse links between water and health. *Comptes Rendus Geoscience* 337(1–2): 277–278.

Fazekas, A. 2014. *Mystery of Earth's Water Origin Solved*. National Geographic. https://www.nationalgeographic.com/news/2014/10/141030-starstruck-earth-water-origin-vesta-science/.

Fröhlich, C.J. 2012. Water: Reason for conflict or catalyst for peace? The case of the Middle East. *L'Europe en Formation* 365: 139–161.

Gleick, P.H. 1996. Water resources. In *Encyclopedia of Climate and Weather*, ed. S.H. Schneider, Vol. 2 pp. 817–823. New York: Oxford University Press.

Grace Communication Foundation. 2019. Water Footprint Comparisons by Country. https://www.watercalculator.org/footprints/water-footprints-by-country/.

Hares, S. 2017. The cost of clean water: $150 billion a year, says World Bank. *Reuters*. London, UK: Thomson Reuters Corporation. https://www.reuters.com/article/us-global-water-health/the-cost-of-clean-water-150-billion-a-year-says-world-bank-idUSKCN1B812E.

Kummu, M., H. de Moel, P.J. Ward, and O. Varis. 2011. How close do we live to water? A global analysis of population distance to freshwater bodies. *PLoS One* 6(6): e20578.

Lufkin, B. 2017. Future – Why "hydro-politics" will shape the 21st Century. BBC Future Now. http://www.bbc.com/future/story/20170615-why-hydro-politics-will-shape-the-21st-century.

McCrary, K. 2018. The politics of water development: A source of conflict? *International Association for Political Science Students*. http://www.iapss.org/wp/2018/02/28/the-politics-of-water-development-a-source-of-conflict/.

Mekonnen, M.M., and A.Y. Hoekstra. 2011. *National Water Footprint Accounts: The Green, Blue and Grey Water Footprint of Production and Consumption. Volume 1: Main Report*. Value of Water Research Report Series No. 50. Delft, the Netherlands: UNESCO-IHE Institute for Water Education. https://waterfootprint.org/media/downloads/Report50-NationalWaterFootprints-Vol1.pdf.

Mullen, K. 2012. *Distribution of Earth's Water*. Westerville, OH: National Groundwater Association. https://www.ngwa.org/what-is-groundwater/About-groundwater/information-on-earths-water.

Osborn, L. 2019. Number of Reservoirs in the World. Current Results. Weather and Science Facts. https://www.currentresults.com/Environment-Facts/number-reservoirs-in-world.php.

Pepper, I., C. Gerba, and T. Gentry. 2014. *Environmental Microbiology*, 3rd edition. Cambridge, MA: Academic Press.

Rekacewicz, P. 2006. *Freshwater Resources: Volume by Continent*. Arendal, Norway: GRID-Arendal. http://www.grida.no/resources/5608.

Stockholm International Water Institute (SIWI). 2005. *Making Water a Part of Economic Development. The Economic Benefits of Improved Water Management and Services*. Stockholm, Sweden: SIWI and the World Health Organization. http://www.who.int/water_sanitation_health/waterandmacroecon.pdf.

University of North Carolina (UNC) Environmental Finance Center. 2017. *An Overview of Clean Water Access Challenges in the United States*. Chapel Hill, NC: University of North Carolina at Chapel Hill Environmental Finance Center. https://efc.sog.unc.edu/sites/default/files/2018/An%20Overview%20of%20Clean%20Water%20Access%20Challenges%20in%20the%20United%20States. Final_.pdf

U.S. Geological Survey. 2019. *The Water Cycle*. Art credit to H. Perlman, and J. Evans. Reston, VA: U.S. Geological Survey. https://www.usgs.gov/media/images/water-cycle-natural-water-cycle.

U.S. Water Alliance. 2017. *An Equitable Water Future. A National Briefing Paper*. Washington, DC: U.S. Water Alliance. http://uswateralliance.org/sites/uswateralliance.org/files/publications/uswa_waterequity_FINAL.pdf.

Viessman, W. Jr., M.J. Hammer, E.M. Perez, and P.A. Chadik. 2009. *Water Supply and Pollution Control*, 8th edition. Upper Saddle River, NJ: Pearson Prentice-Hall.

Water.org. 2019. *Brazil's Water and Sanitation Crisis*. Kansas City, MO: Water.org. https://water.org/our-impact/brazil/.

World Health Organization (WHO). 2017. *Progress on Drinking Water, Sanitation and Hygiene 2017. Update and SDG Baselines*. Geneva, Switzerland: World Health Organization (WHO) and the United Nations Children's Fund (UNICEF). https://apps.who.int/iris/bitstream/handle/10665/258617/9789241512893-eng.pdf;jsessionid=BFB5D93E1AC22CFE3A7EAB3C0F603BF9?sequence=1.

World Health Organization (WHO). 2019. *Global Health Observatory (GHO) Data. Mortality and Burden of Disease from Water and Sanitation*. Geneva, Switzerland: World Health Organization. https://www.who.int/gho/phe/water_sanitation/burden_text/en/.

Section III

Water Treatment Technologies

11 Drinking Water Treatment

11.1 INTRODUCTION

Millions of Americans receive high-quality drinking water every day from their potable water system, which may be publicly or privately owned. Drinking water safety cannot be taken for granted, however, because there are many chemical, biological, and physical threats to drinking water supplies. Chemical and biological threats include contamination from improper chemical handling and disposal, animal wastes, pesticides, human wastes, stormwater runoff, and naturally occurring substances. Drinking water that is not properly treated or disinfected or that travels through an improperly maintained distribution system can also become contaminated and pose a health risk. Physical threats include failing water supply infrastructure and threats posed by tampering or terrorist activity.

Total water withdrawal in the United States for public water supplies was approximately 12% (39,000 MGD) of the total withdrawals for all uses (322,000 MGD) in 2015 (Dieter et al. 2018), of which 61% was from fresh surface water. These public water supplies provide water to 87% (283 million) of the U.S. population (325 million) for domestic, industrial, and commercial uses, and treatment to meet strict drinking water standards is highly regulated and closely monitored to ensure safe and sustainable supplies. The balance of the population (42.5 million) relies on self-supplied water, 98% of which comes from fresh groundwater sources.

The bulk of drinking water treatment in the United States is directed to removal of contaminants commonly found in surface water supplies, which generally include suspended solids and turbidity and natural organic material responsible for taste and odor issues. Groundwater sources are susceptible to elevated levels of inorganic materials because of their dissolution into groundwater that is in intimate contact with aquifer sols. These pollutants can include dissolved constitutes (iron and manganese) causing hardness and arsenic as a result of natural weathering processes, as well as dissolved gases in the groundwater due to chemical dissolution or microbial degradation processes generating carbon dioxide (CO_2) and hydrogen sulfide (H_2S). Both surface and groundwater supplies must be treated for microbial contaminants that can be the source of disease transmission in contaminated water supplies, and both surface water and shallow groundwater systems are susceptible to synthetic organic chemical contamination from improper waste handling and disposal as well as nitrate contamination from fertilizer use in agricultural and domestic settings.

This chapter provides an overview of drinking water treatment processes for surface and groundwater through a discussion of conventional treatment systems, advanced treatment processes for organic and inorganic chemicals of concern, treatment process by-products and related health concerns for these materials, and concerns regarding water quality changes in water distribution systems. An applications section provides five Illustrative Examples related to the general subject of drinking water treatment.

11.2 CONVENTIONAL DRINKING WATER TREATMENT SYSTEMS

Water treatment originally focused on improving the aesthetic qualities of drinking water, and methods to improve the taste and odor of drinking water were recorded as early as 4000 BCE. Ancient Sanskrit and Greek writings recommended water treatment methods such as filtering through charcoal, exposing to sunlight, boiling, and straining (U.S. Environmental Protection Agency [EPA] 2000). To clarify water, the Egyptians reportedly used the chemical alum as a coagulant for the removal of turbidity as early as 1500 BCE. During the 1700s, filtration was established as an effective means of removing particles from water, although the degree of clarity achieved was not measurable at that time. By the early 1800s, slow sand filtration was beginning to be used regularly in Europe (U.S. EPA 2000).

In 1855, epidemiologist Dr. John Snow proved that cholera was a waterborne disease by linking an outbreak of illness in London to a public well that was contaminated by sewage; it has since been known as the Broad Street Well incident. In the late 1880s, Louis Pasteur demonstrated the "germ theory" of disease, which explained how microbes could transmit disease through media like water. During the late nineteenth and early twentieth centuries, concerns regarding drinking water quality continued to focus mostly on disease-causing microbes, or pathogens, in public water supplies. Scientists discovered that particles in source water, such as fecal matter, could harbor pathogens. As a result, the design of most drinking water treatment systems built in the United States during the early 1900s was driven by the need to reduce turbidity, thereby removing microbial contaminants that were causing typhoid, dysentery, and cholera in exposed populations. Although filtration was a fairly effective treatment method for reducing turbidity, it was disinfectants like chlorine that played the largest role in reducing the number of waterborne disease outbreaks in the early 1900s. In 1908, chlorine was used for the first time as a primary disinfectant of drinking water in Jersey City, New Jersey. The use of other disinfectants such as ozone also began in Europe around this time but were not employed in the United States until many decades later (U.S. EPA 2000).

By the late 1960s it became apparent that the aesthetic problems, pathogens, and chemicals identified by the Public Health Service were not the only drinking water quality concerns. The creation of new man-made chemicals for industrial and agricultural applications also had negative impacts on the environment and public health. Many of these new chemicals were finding their way into water supplies through factory discharges, street and farm field runoff, and leaking underground storage and disposal tanks (U.S. EPA 2000). Although treatment techniques such as aeration, flocculation, and granular activated carbon adsorption existed at the time, they were either underused or ineffective at removing some new contaminants (U.S. EPA 2000).

Today, filtration and chlorination remain effective treatment techniques for protecting U.S. water supplies from harmful microbes, although additional advances in disinfection have been made over the years. In the 1970s and 1980s, improvements were made in membrane development for reverse osmosis and other treatment techniques such as ozonation (U.S. EPA 2000). Some treatment advancements have been driven by the discovery of chlorine-resistant pathogens in drinking water that can cause illnesses like hepatitis, gastroenteritis, Legionnaire disease, and cryptosporidiosis. Other advancements resulted from the need to remove more and more chemicals found in sources of drinking water.

Many of the treatment techniques used today by drinking water plants include methods that have been used for hundreds and even thousands of years. However, newer treatment processes (e.g., reverse osmosis and granular activated carbon) are also being employed by some modern drinking water plants. The number and combinations of processes developed and the extent to which these advanced processes will be employed in drinking water treatment plants is expected to increase with time as more complex contaminants are discovered and regulated and as water treatment plants take advantage of the multi-billion-dollar state revolving loan fund that is designed to help water systems, especially those serving small and disadvantaged communities, upgrade or install new treatment facilities (U.S. EPA 2000).

A variety of sources are available to provide detail on the design and evaluation of water treatment systems (AWWA 2011; Hendricks 2011; AWWA and ASCE 2012; Howe et al. 2012; Droste and Gehr 2019). The U.S. EPA (2019a) and the U.S. Bureau of Reclamation (2010) have also developed guidelines for drinking water systems that are directed to identifying treatment processes that are effective for specific contaminants of interest. EPA's Drinking Water Treatability Database (EPA 2019a) is an online tool that allows the user to search for contaminants of interest to identify treatment technologies effective for their removal or to search for treatment technologies to identify removal effectiveness for contaminants that technology treats. The following discussion of conventional treatment technologies provides a summary of conventional process treatment trains that are employed for surface and groundwater sources and discusses the conventional pollutants these process trains are designed to remove. Additional treatment processes for nonconventional pollutants (i.e., synthetic organic compounds, heavy metals, chlorine resistant microbes, etc.) are discussed in the next section on advanced treatment processes.

11.2.1 Surface-Water Sources

Surface water is made up of direct runoff and base flow. Precipitation that does not enter the ground through infiltration or is not returned to the atmosphere by evaporation flows over the ground surface and is classified as direct runoff. Direct runoff is water that moves over saturated or impermeable surfaces and in stream channels or other natural or artificial storage sites. The dry weather (base) flow of streams is derived from groundwater. Once precipitation reaches the Earth's surface, many opportunities are presented for the introduction of mineral and organic substances, microorganisms, and other forms of pollution. When water runs over or through the ground surface it may pick up particles of soil. This is noticeable in the water as cloudiness or turbidity. Clay or other inert suspended particles in drinking water may not directly affect health, but water containing such particles can serve as a source of nutrients and food for any bacteria, viruses, or protozoa that may be in the water; these organisms can attach themselves to these particles, and these particles can interfere with the ability of disinfectants to eliminate pathogens (Pollution Probe 2002). Thus, surface water with high levels of turbidity will require treatment to make it safe for human consumption and suitable for its intended use. Surface water is also susceptible to taste and odor problems because of organic matter from domestic, agricultural, or natural sources and as a result of excessive growth of algae and other aquatic organisms in rivers, streams, and reservoirs.

Conventional drinking water systems for surface-water sources are subsequently focused on removal of suspended material and turbidity, organic materials resulting in taste and odor problems in the raw water, and removal of pathogenic organisms to prevent waterborne disease transmission. Conventional treatment for surface water sources typically consists of screening, coagulation, flocculation, clarification, and filtration, followed by disinfection. Figure 11.1 presents a typical conventional treatment process train for surface-water sources.

Screening consists of physical removal the large/heavy solids (e.g., wood, plastic materials, and rags) suspended in the raw surface water. The screens are usually mechanically cleaned and the screened solids are drained and typically disposed of in a sanitary landfill. Multiple screens are employed to minimize downtime for cleaning or maintenance. *Chemical coagulation* consists of rapidly mixing a positively charged chemical coagulant (aluminum sulfate, ferric sulfate, ferric

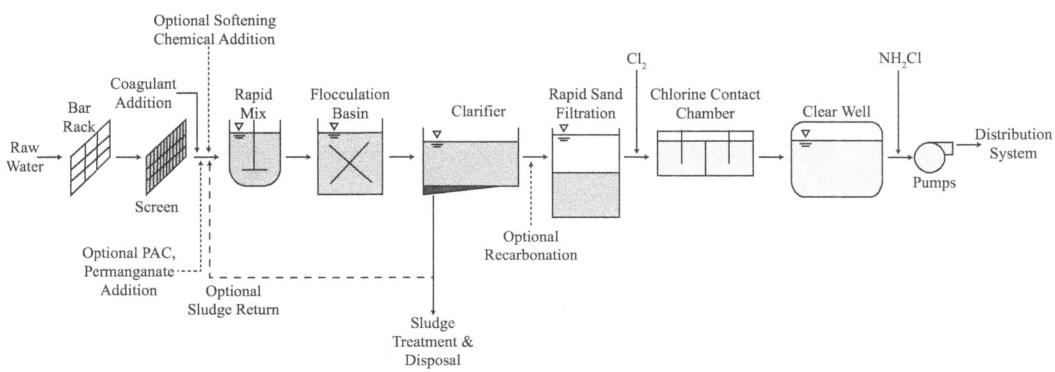

FIGURE 11.1 Typical water treatment process flow sheet for surface-water sources. Cl_2 = chlorine; NH_2Cl = chloramine; PAC = power activated carbon.

chloride, polymer, or coagulant aide) to neutralize and destabilize colloidal particles that are generally negatively charged. The optimal pH range for coagulation is 6 to 7 when using alum and 5.5 to 6.5 when using iron salts. There are two primary destabilization mechanisms in drinking water treatment: charge neutralization and sweep flocculation. The mechanism is dependent on the coagulant dose. Most drinking water treatment plants operate using sweep flocculation, which requires a higher coagulant dose than charge neutralization. In charge neutralization, the positively charged metal coagulant is attracted to the negatively charged colloids via electrostatic interaction. Floc starts to form during the neutralization step as particle collisions occur and is completed in flocculation tanks. Adding excess coagulant in the coagulation step beyond charge-neutralization results in the formation of metal coagulant precipitates (e.g., $Al(OH)_3$ or $Fe(OH)_3$) that are heavy, sticky, and larger in particle size. Sweep flocculation occurs when colloidal and suspended contaminants are entrained by the precipitates formed in the flocculation tank as they settle in a subsequent sedimentation tank (U.S. EPA 2019a).

During *flocculation*, gentle mixing accelerates the rate of particle collision, and the destabilized particles are aggregated and enmeshed into larger precipitates. Flocculation is affected by several parameters, including the mixing speed, mixing intensity (G), and mixing time. The product of the mixing intensity and mixing time (Gt) is frequently used to model and design the flocculation process (U.S. EPA 2019a).

Following flocculation, agglomerated particles enter a *clarification unit* where they are removed by sedimentation by gravity. In the sedimentation processes, the majority of the solids are removed by gravitational settling; particles that do not settle and are still suspended are removed during the filtration process. Sedimentation is generally accomplished in rectangular or circular basins and is often enhanced by the addition of inclined plates or tubes (*plate or tube settlers*), which increase the effectiveness of the process by effectively increasing the surface area of the sedimentation basin (Figure 11.2). Two parameters frequently used to describe the clarification process are the *surface overflow rate (SOR)* and the *hydraulic retention time (HRT)*. The SOR is a hydraulic

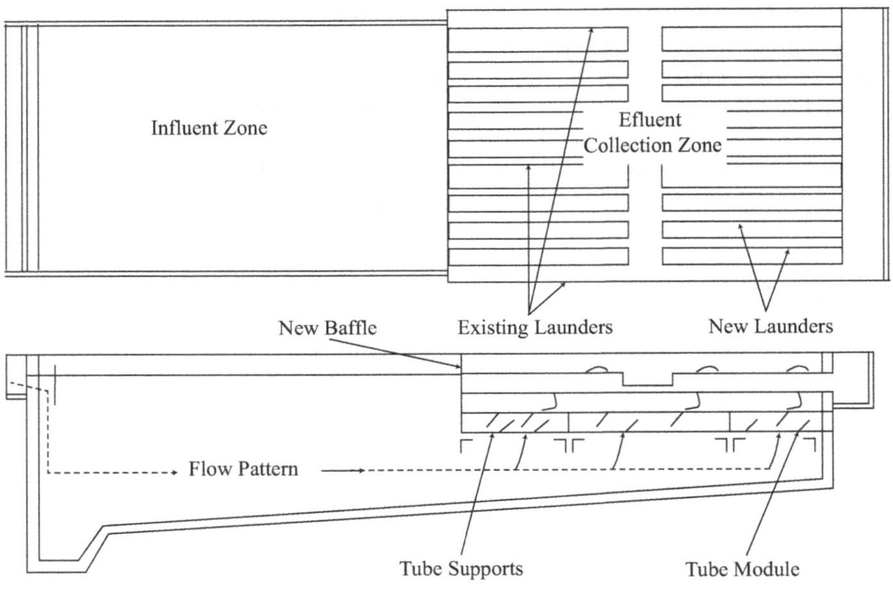

FIGURE 11.2 Typical tube settler installation in rectangular sedimentation basins. (From U.S. Environmental Protection Agency, *Technologies for Upgrading Existing or Designing New Drinking Water Treatment Facilities*, EPA/625/4-89/023, Center for Environmental Research Information, Office of Drinking Water, Cincinnati, OH, https://nepis.epa.gov/Exe/ZyPDF.cgi/300048WU.PDF?Dockey=300048WU.PDF, 1990.)

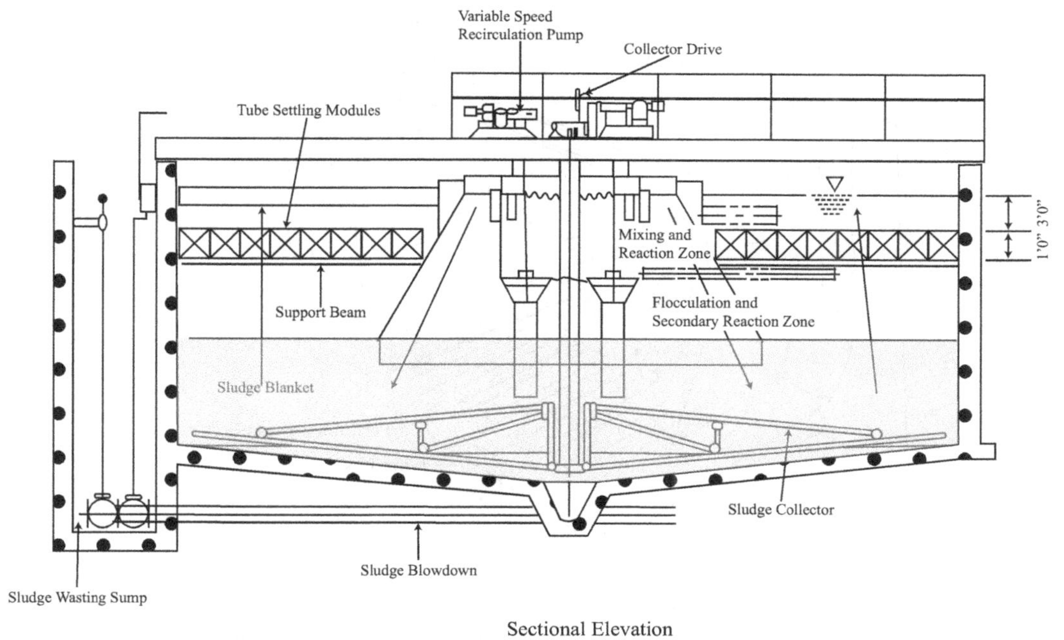

FIGURE 11.3 Solids contact clarifier with tube settlers. (From U.S. Environmental Protection Agency, *Technologies for Upgrading Existing or Designing New Drinking Water Treatment Facilities*, EPA/625/4-89/023, Center for Environmental Research Information, Office of Drinking Water, Cincinnati, OH, https://nepis.epa.gov/Exe/ZyPDF.cgi/300048WU.PDF?Dockey=300048WU.PDF, 1990.)

loading rate expressed in gpd/ft^2. SORs for conventional sedimentation generally range from 500 to 1500 gpd/ft^2. Typical HRTs range from 1 to 2 hours, although many states require up to 4 hours for full-scale surface water treatment plants (U.S. EPA 2019a).

An alternative to separate coagulation/flocculation/sedimentation basis is a *solids contact unit* that integrates these three unit processes together in a single tank. Figure 11.3 shows a cross section of a typical solids contact clarifier with tube settlers. As indicated in Figure 11.3, these units have an additional feature, that being a sludge blanket that entraps slowly settling small particles as the water moves upward vertically through the tank to the overflow weirs to achieve a high level of clarification. The sludge blanket is maintained at a specific level and concentration by the controlled removal of sludge from the bottom of the tank.

Filtration separates fine suspended solids, flocs, and pathogens not removed by gravity separation by passing the water through a porous medium (Figure 11.4). The most commonly used filter type in the conventional treatment process is a *dual-media filter* comprised of anthracite and sand; however, *single-media* (e.g., sand), *multimedia* (e.g., garnet, anthracite, and sand), and other media configurations, including the use of granular activated carbon, are also used in drinking water treatment. During filtration, the majority of suspended particles are removed in the top portion of the filter media. When filter head loss increases flow the filters is reversed and the filters are backwashed to dislodge and remove particles trapped within the filter bed. Once the filter media is cleaned flow is once again reversed and normal filtration resumes. Multiple filters are used in water treatment plants to allow continuous filter backwashing of clogged filters without disrupting a continuous water supply.

Filter hydraulic loading rate is used for filter design, and is a measure of the filter production per unit area expressed in units of gpm/ft^2. Typical filter loading rates range from 2 to 4 gpm/ft^2; however, higher filter loading rates, 4 to 6 gpm/ft^2, are becoming more common at full-scale.

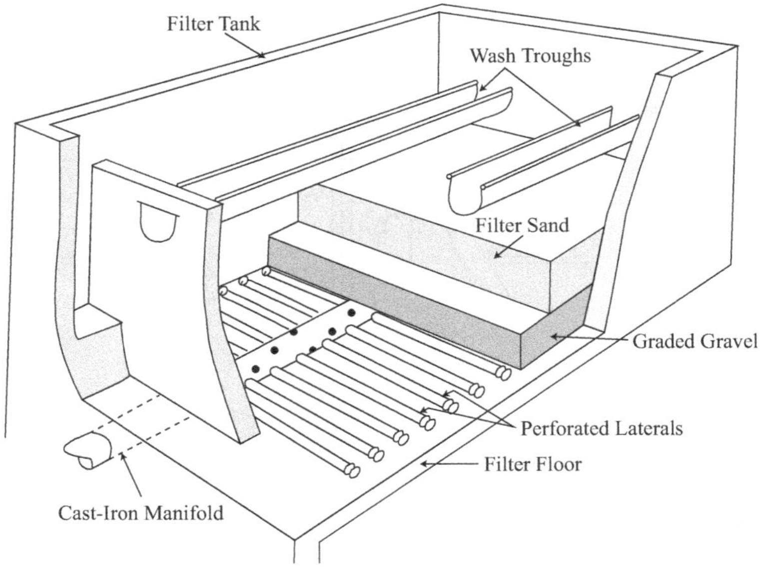

FIGURE 11.4 Cross section of typical single-media rapid sand filter. (From U.S. Environmental Protection Agency, *Technologies for Upgrading Existing or Designing New Drinking Water Treatment Facilities*, EPA/625/4-89/023, Center for Environmental Research Information, Office of Drinking Water, Cincinnati, OH, https://nepis.epa.gov/Exe/ZyPDF.cgi/300048WU.PDF?Dockey=300048WU.PDF, 1990.)

This can be a critical parameter because it determines the water velocity through the filter bed and can impact the depth to which particles pass through the media. The filter run time describes the length of time between filter backwashes during, which a filter is in production mode. The filter run time is not only an indicator of the effectiveness of prior treatment (i.e., the ability of the coagulation and clarification steps to remove suspended solids) but also plays a role in the effectiveness of the filter itself. Filter performance, particularly with regard to particulate contaminants, is often poorest immediately following a backwash. As the filter run time increases and the concentration of solids in the media increases, and the filtration process often performs better with regard to particulate contaminant removal (U.S. EPA 2019a).

Conventional treatment of surface water is often preceded by pre-oxidation by permanganate to control taste and odor, remove color, control biological growth in treatment plants, and control zebra mussels in intake structures and pipelines. Permanganate can also be used for controlling the formation of trihalomethanes and other disinfection by-products by oxidizing precursors and reducing the demand for other disinfectants. Permanganate has also shown to lower coagulant dose requirements and improve clarification (U.S. EPA 2019a).

At full-scale water treatment applications, potassium permanganate is most commonly added early in the treatment process at the raw water intake. Other possible points of application are at the rapid mix tank in conjunction with coagulants or at clarifiers upstream of filters. In all cases, potassium permanganate is added before filtration. Dose, contact time, pH, and temperature affect the effectiveness of permanganate application. Typical doses are approximately 1 mg/L but can vary depending on water quality and the application point (generally ranges between 0.1 and 5 mg/L). Higher potassium permanganate dose increases its oxidation capabilities. Dosage needs to be carefully controlled so that permanganate residual does not remain in the water because it tends to give water a pink color. Sufficient contact time must be provided to exhaust the permanganate residual before reaching the distribution system. Powdered activated carbon (PAC), iron salts, and hydrogen peroxide can be used for removing excess permanganate (U.S. EPA 2019a).

PAC is also commonly used at conventional surface-water treatment plants for the control of taste and odor causing compounds. PAC is made from organic materials with high carbon contents such as wood, lignite, and coal. PAC typically has a diameter less than 0.1 mm and an apparent density ranging between 23 and 46 lb/ft^3, depending on the material used and manufacturing process. PAC can be fed as a powder using dry feed equipment or as a slurry using metering pumps. Dry feed systems are typically used for smaller dosages and where PAC feed is infrequent. Dry feed systems typically include a bag-loading hopper, an extension hopper, a dust collector, a dissolving tank, and an eductor. PAC can also be mixed with water and fed as a slurry. Slurry systems are normally used when PAC is frequently added and the required dosages are high. Slurry systems usually include a storage tank, day tank, and a chemical feeder (either a diaphragm pumps or rotary feeders) (U.S. EPA 2019a).

PAC is normally added early in the treatment process and is subsequently removed either by sedimentation or by the filter beds during backwashing. The PAC application point should (i) allow for an adequate contact time between the PAC and organics and (ii) avoid coating PAC particles with other water treatment plant chemicals. A minimum contact time of about 15 minutes is required for most taste and odor compounds; however, significantly longer contact times may be required for methyl-iso-borneol (MIB) and geosmin removal. The PAC should not be coated with coagulants or other water treatment chemicals before the PAC has had sufficient contact time with the source water. Also, PAC should not be added concurrently with chlorine or potassium permanganate because these chemicals will adsorb to the PAC. PAC is usually added at the head of the plant to provide the longest contact time possible before applying other treatment chemicals. Higher PAC dosages may be necessary if PAC is added later in the treatment process to account for reduced contact times and interference with other treatment chemicals (U.S. EPA 2019a).

PAC dosages can range between 1 and 100 mg/L depending on the type and concentrations of organic compounds present. Dosages of 1 to 20 mg/L are typical for nominal taste and odor control. Application of PAC generates additional sludge that is not able to be regenerated as is spent granular activated carbon (GAC). PAC sludge will contain elevated concentrations of the contaminants removed and must be disposed in accordance with state and federal laws; however, it is not likely to be classified as a hazardous waste (U.S. EPA 2019a).

11.2.2 Groundwater Sources

Groundwater is made up of water that drains downward below the plant root zone, finally reaching a level at which all of the voids in the subsurface soil or rock are filled with water. This zone is known as the *zone of saturation* and water in the zone of saturation is referred to as *groundwater*. The upper surface of the zone of saturation, if not confined by impermeable material, is called the *water table*. Because of the irregularities in underground deposits and in surface topography, the water table occasionally intersects the surface of the ground or the bed of a stream, lake, or ocean. As a result, groundwater moves to these locations and out of the groundwater reservoir. Thus, groundwater is continually moving within the aquifer even though the movement may be slow. The water table surface slopes from areas of recharge to areas of discharge. The pressure differences represented by these slopes cause the flow of groundwater within the aquifer. At any point the slope is a reflection of the rate of flow and resistance to movement of water through the saturated formation. Seasonal variations in the supply of water to the underground reservoir cause considerable changes in the elevation and slope of the water table (Public Health Service 1962).

As surface water seeps downward into the soil and through the underlying material to the water table, most of the suspended particles are filtered out. This natural filtration may be partially effective in removing bacteria and other particulate materials; however, the chemical characteristics of the water may change and vary widely when it comes in contact with mineral deposits, and groundwater can often contain more dissolved minerals than surface water. Typical dissolved contaminants of concern are calcium and magnesium that cause *hardness*, leading to excessive use of soaps and detergents and scale in water appliances and hot water heaters. In addition, groundwater

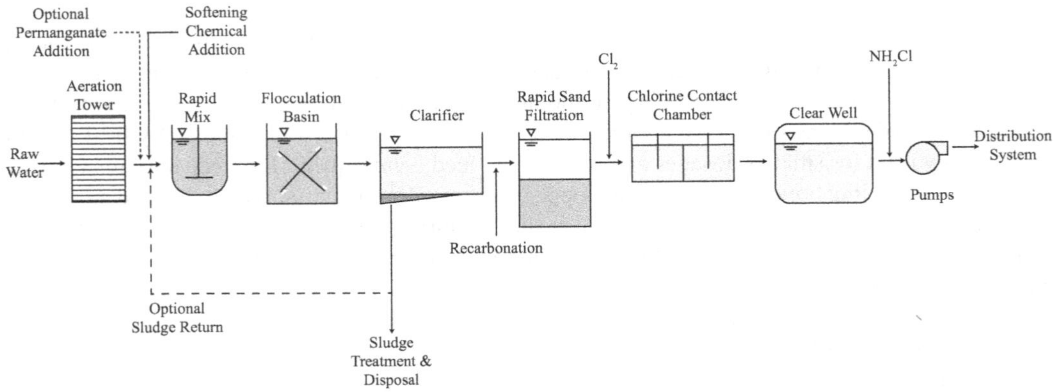

FIGURE 11.5 Typical water treatment process flow sheet for groundwater sources. Cl_2 = chlorine; NH_2Cl = chloramine.

may be susceptible to contamination from chemicals (i.e., pesticides, fertilizers notably nitrate, etc.) used in agricultural practices or from inorganic compounds released to groundwater in response to microbial activity stimulated from organic matter from man-made or natural sources that reach groundwater supplies. These inorganic compounds generated from microbial activity include dissolved iron, arsenic, and manganese, released when iron, arsenic, and manganese minerals are used as an electron acceptor during microbial metabolism of organic carbon, and hydrogen sulfide generated from the use of sulfate as an electron acceptor. These compounds are either potentially toxic (nitrate and arsenic) or cause taste, odor, or aesthetic problems (e.g., iron, manganese, hydrogen sulfide, hardness) and must be removed before distributing the finished water. Conventional treatment for groundwater sources typically consists of aeration, coagulation, flocculation, clarification, and filtration, followed by disinfection. Figure 11.5 presents a typical conventional treatment process train for surface-water sources.

Aeration processes are typically used in gas stripping (the process of removing dissolved gas from water) and aeration (the process of transferring oxygen to water). Major applications for aeration technologies include: (i) removal of carbon dioxide, taste- and odor-causing substances, volatile organic compounds (VOCs) and volatile synthetic organic compounds (SOCs), ammonia, trihalomethanes, pesticides, herbicides, and gases such as methane, hydrogen sulfide, and radon from the raw water; and (ii) oxidation of substances such as iron and manganese by oxygen addition to generate insoluble precipitates removed from the water in subsequent clarification and filtration steps (U.S. EPA 2019a).

Henry's law describes the tendency of a constituent to transfer from the liquid to the gas phase at equilibrium. The Henry's Law constant is the ratio of the equilibrium concentration of a particular contaminant in air to its concentration in water. Thus, a higher Henry's Law constant indicates a greater tendency of a species to volatilize into the air from the water. The efficiency of aeration depends almost entirely on the amount of surface contact between the air and water and increased with temperature and the level of turbulence in a system (U.S. EPA 2019a).

A commonly used aeration system in drinking water treatment is a tray aerator shown in Figure 11.6. These multitray or slat-tray aerators consist of series of trays equipped with slats or perforated or wire-meshed bottoms. Water is distributed over the staggered trays and is allowed to fall downward from one tray to another, eventually reaching a collection basin at the base. Air is forced or induced to flow upward countercurrent to the water path. These multiple trays are typically spaced 9 to 12 inches apart and are analogous to cooling towers and require adequate ventilation. They provide excellent oxygen absorption for preoxidation of iron and manganese, and effective gas stripping for carbon dioxide, ammonia, hydrogen sulfide, and VOC removal.

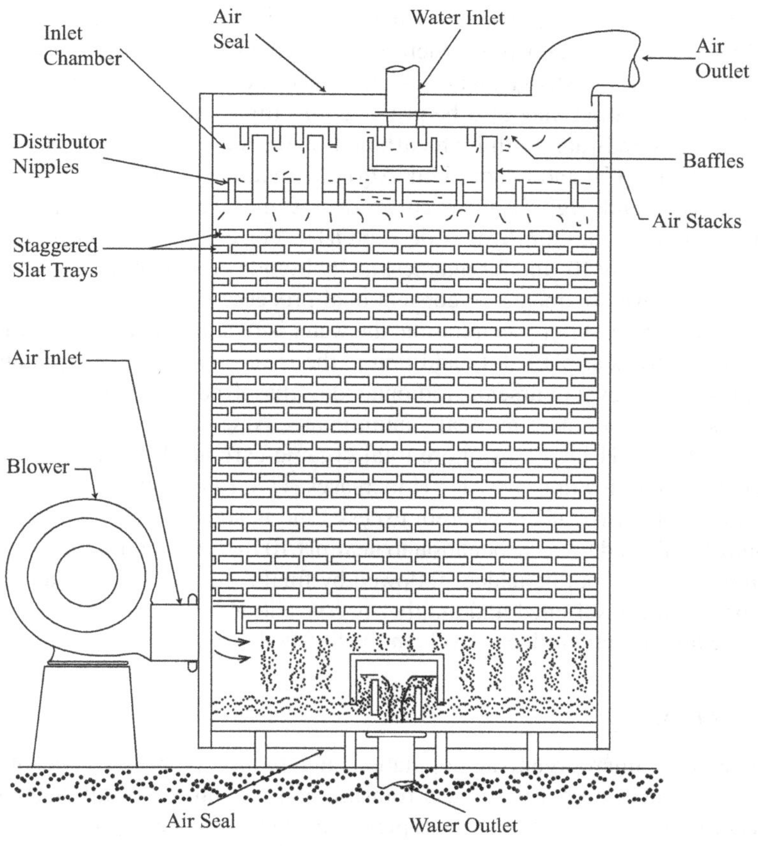

FIGURE 11.6 Typical tray aerator used for groundwater treatment.

Frequently, the off-gas from the column has to be treated, particularly if used for VOC removal. Typically, the gas is heated to reduce humidity and is then passed through an activated carbon adsorption chamber to trap the volatile compounds. Other process of treating off-gas includes thermal oxidation, which is expensive, and photooxidation, which is less feasible for continuous, full-scale operations. A common problem with tray aerators is fouling of trays with solids, resulting in loss of plant capacity, efficiency, and increased pressure drop.

Once aeration is completed various forms of chemical coagulation, flocculation, and sedimentation take place as described previously for surface water treatment. In treating groundwater, this chemical treatment is typically focused on iron and manganese removal and water softening rather than solids and colloidal removal. Permanganate is widely used for iron and manganese removal. With permanganate application, ferrous iron is oxidized to ferric iron (oxidation state +3), forming solid ferric hydroxide ($Fe(OH)_3$); and manganese (Mn^{+2}) is oxidized to manganese at a higher oxidation state (Mn^{+4}), forming solid manganese dioxide (MnO_2). These precipitates are removed by subsequent treatment steps, such as coagulation/flocculation/sedimentation, filtration, or GAC (U.S. EPA 2019a).

If softening is required to reduce hardness in source waters to below 120 mg/L as $CaCO_3$, lime softening, lime-soda ash softening, and caustic softening are employed. The selection of one of these alternatives is based on cost, water quality, and owner and operator preferences. Lime softening is typically used for water containing low concentrations of noncarbonate hardness. Lime-soda softening may be required with high concentrations of noncarbonate hardness. Caustic soda softening is typically used when the treated water has inadequate carbonate hardness to react with lime. Softening by lime and lime-soda ash is generally less expensive then by caustic soda. However, caustic softening produces less sludge and unlike lime does not deteriorate during storage. In some cases, pretreatment by aeration may also be practiced. In this context, aeration is used to remove carbon dioxide from groundwater with high carbon dioxide concentrations to facilitate pH adjustment to facilitate calcium (Ca) and magnesium (Mg) precipitation (U.S. EPA 2019a).

Precipitation of Ca^{2+} and Mg^{2+} is achieved by raising the pH of water to precipitate calcium carbonate ($CaCO_3$) and magnesium hydroxide $Mg(OH)_2$. For calcium carbonate precipitation the pH is raised to approximately 10, and for magnesium hydroxide precipitation, the pH is raised above 11. Precipitates are removed by means of conventional processes such as coagulation-flocculation, sedimentation, and filtration. After precipitation, the water is recarbonated to lower the pH to reduce scale formation, typically near pH 8.4. In addition to the removal of hardness, precipitative softening can be used for the removal of arsenic, radionuclides, dissolved organics (including disinfection by-product precursors), color, and microbial contaminants. Arsenic removal is greatly increased at pH greater than 10.5. As pH increases, more total organic carbon (TOC) and color are removed. Removal of radionuclides also improves as pH increases (U.S. EPA 2019a).

Softening may involve two stages wherein excess lime is added to the first stage to pH 11 for magnesium control, followed by recarbonation to near pH 10 in the second stage for calcium control where soda ash may or may not be added, followed by final recarbonation. Softening may also be split wherein one stream is softened and another is conventionally treated. These waters are then blended to achieve target final hardness levels in the finished water.

11.2.3 Disinfection

Disinfection to control microorganism population and reduce pathogen concentrations to low enough levels to prevent waterborne disease transmission is required in all water treatment systems regardless of their source water and is typically placed as the last unit operation in the treatment plant. The most common types of disinfection are chlorine, and chloramines, with alternative disinfectants being developed and applied to reduce potential health effects from disinfection by-products from chlorine disinfection (see Chapter 7). The choice of disinfectant depends on cost, disinfection by-product formation potential, log inactivation required, cost, and whether it is used for primary disinfection or disinfectant residual. The disinfectant dose depends on log inactivation required, which may be influenced by temperature, pH, and turbidity.

Federal regulations state that all treated water leaving treatment plants must maintain a minimum free chlorine residual of 0.2 mg/L (U.S. Bureau of Reclamation 2010); therefore, some form of chlorination is required in water treatment regardless of disinfections technology used, if that disinfected water is to be used by the consumer. Free chlorine is the reactive chlorine portion and is not combined with organic nitrogen or ammonia. There are times when the water in the treatment plant requires disinfection for biofouling control or oxidation; however, it is not distributed to the public without further treatment. For example, some feedwaters to treatment plants require disinfection prior to treatment and ultraviolet (UV) can be used to disinfect, even though it does not provide chlorine residual.

As disinfection is typically the last treatment step in most water treatment facilities, most suspended solids should have been removed prior to disinfection. Dissolved organic matter (DOM) presence can exert a substantial chlorine demand along with disinfection by product generation and, therefore, should be removed as much as possible before chlorine disinfection. Tests to determine optimum pH and chlorine dose are required depending on the level of disinfection required.

Chlorine is available as compressed elemental gas, sodium hypochlorite solution (NaOCl), or solid calcium hypochlorite ($Ca(OCl)_2$). All forms of chlorine, when applied to water, form hypochlorous acid (HOCl). Gaseous chlorine acidifies the water and reduces the alkalinity, whereas the liquid and solid forms of chlorine increase the pH and the alkalinity at the application point.

The pH of the water will affect the dominating chlorine species such that HOCl dominates at lower pH, whereas the hypochlorite ion (OCl^-) dominates at higher pH. Of the two species, HOCl is the stronger oxidant. Therefore, chlorine is more effective as an oxidant and a disinfectant at lower pH. Both forms, HOCl and OCl^-, are referred to as free chlorine. Chlorine gas is typically used for large utilities over 200 gpm. For utilities under that flow rate, chlorine in the form of sodium hypochlorite and calcium hypochlorite dehydrate liquids are recommended because their handling and storage are much easier than chlorine gas (U.S. Bureau of Reclamation 2010; U.S. EPA 2019a).

The concentration (C), contact time (T), pH, and temperature affect the effectiveness of chlorine application. The product of concentration and time (CT) is the most important operational parameter in disinfection and inactivation. Although increasing the dose increases the ability of chlorine to oxidize and disinfect, it may also lead to taste and odor issues and to the formation of disinfection by-products by chlorine's reaction with DOM. The dose is also affected by the application point, chlorine demand of the water, and desired residual concentration. Total organic carbon (TOC) and UV absorbance are two measures of disinfection by-product–reactive DOM and of chlorine demand.

Chloramines are a family of oxidants formed by the reaction of chlorine and ammonia. In water treatment, chloramines are primarily used as a secondary disinfectant to provide a residual in the distribution system; however, chloramines are occasionally used as a primary disinfectant. In full-scale water treatment applications, ammonia is most frequently added to the water after chlorine (after achieving primary disinfection) and prior to entering the distribution system. Where alternative primary disinfectants, such as ozone or chlorine dioxide are used, ammonia may be added before chlorine or chlorine and ammonia may be added simultaneously. Simultaneous application is sometimes referred to as *chloramination*. Chloramine residuals are common during filtration to inhibit microbial (biofilm) growth on filter media that could increase filter head loss (pressure) build up (U.S. EPA 2019a).

The ratio at which chlorine and ammonia are fed control the species of chloramines present. At chlorine-to-ammonia-nitrogen (Cl_2/NH_3-N) mass ratios less than 5:1, monochloramine (NH_2Cl) is the predominant species. Monochloramine is also the preferred species because it is a more powerful oxidant and is less likely to cause taste and odor problems in the distribution system than the other chloroamine species. At ratios between 5:1 and 7.6:1, dichloramine ($NHCl_2$) and trichloramine ($NHCl_3$) are the dominant species. At mass ratios greater than 7.6:1, the breakpoint reaction occurs in which all of the available ammonia is oxidized to nitrogen gas, and free chlorine is the dominant chlorine species present. The pH of chloraminated water is often at 8 and above because monochloramine stability is enhanced in that range.

Although weaker than chlorine and chlorine dioxide, monochloramine oxidizes precursors of disinfection by-products, inactivates microorganisms, and controls biofilm. Chloramination is often an attractive alternative to chlorine for secondary disinfection because it minimizes the formation of disinfection by-products. The effectiveness of chloramination is dependent on chloramine concentration (C), contact time (T), pH, and temperature (U.S. EPA 2019a).

The downside of chloramination is it has been demonstrated to form currently unregulated disinfection by-products such as cyanogen chloride and N-nitrosodimethylamine and impact the stability of existing pipe scale in previously chlorinated distribution systems (U.S. EPA 2019a).

11.3 ADVANCED TREATMENT PROCESSES

Advanced drinking water treatment processes are used for the removal of nontraditional or emerging contaminants (Section 11.6) that are not removed by conventional water treatment technologies described previously. These advanced technologies include GAC treatment, advanced oxidation processes (AOPs), and membrane separation systems such as nanofiltration, reverse osmosis, and electrodialysis. These later membrane-based systems are the subject of Chapters 15 and 18. The discussion provided here highlights the use of GAS and AOPs for advance drinking water treatment.

11.3.1 GAC

Activated carbon is commonly used to adsorb natural organic compounds, taste and odor compounds, and SOCs not removed by other conventional treatment processes in drinking water treatment. Adsorption is both the physical and chemical process of accumulating a substance at the interface between liquid and solids phases. Activated carbon is an effective adsorbent because it is a highly porous material and provides a large surface area to which contaminants may adsorb. GAC is made from organic materials with high carbon contents such as wood, lignite, and coal. GAC typically has a diameter ranging between 1.2 and 1.6 mm and an apparent density ranging between 25 and 31 lb/ft^3, depending on the material used and manufacturing process. The bed density is about 10% less than the apparent density and is used to determine the amount of GAC required to fill a given size filter. The uniformity coefficient of GAC is quite large, typically about 1.9, to promote stratification after backwashing and minimize desorption and premature breakthrough that can result from mixing activated carbon particles with adsorbed compounds with activated carbon particles with smaller amounts of adsorbed compounds (U.S. EPA 2019a).

The most common option for SOC removal by GAC is postfiltration adsorption, where the GAC unit is located after the conventional filtration process (postfilter contactors or adsorbers). In postfiltration applications, the GAC contactor receives the highest quality water and, thus, has as its only objective the removal of dissolved organic compounds. Backwashing of these adsorbers is usually unnecessary, unless excessive biological growth occurs. This option provides the most flexibility for handling GAC and for designing specific adsorption conditions by providing long contact times.

Primary factors in determining the required GAC contactor volume are the (i) breakthrough, (ii) empty bed contact time (EBCT), and (iii) design flow rate. The breakthrough time is the time when the concentration of a contaminant in the effluent of the GAC unit exceeds the treatment requirement. As a rule of thumb, if the GAC effluent concentration is greater than the performance standard for over three consecutive days, the GAC is exhausted and must be replaced or regenerated. The EBCT is calculated as the empty bed volume divided by the flow rate through the carbon. Longer EBCTs can be achieved by increasing the bed volume or reducing the flow rate through the filter. The EBCT and the design flow rate define the amount of carbon to be contained in the adsorption units. A longer EBCT can delay breakthrough and reduce the GAC replacement or regeneration frequency. The carbon depth and adsorber volume can be determined once the optimum EBCT is established. Typical EBCTs for water treatment applications range between 5 and 25 minutes (U.S. EPA 2019a).

The surface loading rate for GAC filters is the flow rate through a given area of GAC filter bed and is expressed in units of gpm/ft^2. Surface loading rates for GAC filters typically range between 2 and 10 gpm/ft^2. High surface loading rates can be used when highly adsorbable compounds are targeted for removal. The surface loading rate is not important when mass transfer is controlled by the rate of adsorption as is the case for less-adsorbable compounds.

The carbon usage rate (CUR) determines the rate at which carbon will be exhausted and how often carbon must be replaced or regenerated. Carbon treatment effectiveness improves with increasing contact times. Deeper beds will increase the percentage of carbon that is exhausted at breakthrough. The optimum bed depth and volume are typically selected after carefully evaluating capital and operating costs associated with reactivation frequency and contactor construction costs (U.S. EPA 2019a).

GAC contactors can be configured as either: (i) downflow fixed beds, (ii) upflow fixed or expanded beds, or (iii) pulsed beds; with single or multiple adsorbers operated in series or in parallel. In downflow fixed beds in series, each unit is connected in series with the first adsorber receiving the highest contaminant loading and the last unit receiving the lightest contaminant load. Carbon is removed for reactivation from the first unit, with the next adsorber becoming the lead unit. For downflow fixed beds in parallel, each unit receives the same flow and contaminant load. To maximize carbon usage, multiple contactors are frequently operated in parallel-staggered mode in which each contactor is at a different stage of carbon exhaustion. Because effluent from each contactor is blended, individual contactors can be operated beyond breakthrough such that the blended flow still meets the treatment goal. Upflow expanded beds permit removal of suspended solids by periodic bed expansion and allow using smaller carbon particles without significantly increasing head loss. In pulsed bed adsorbers, removal of spent carbon occurs from the bottom of the bed, whereas fresh carbon is added at the top without system shutdown. A pulsed bed cannot be completely exhausted, which prevents contaminant breakthrough in the effluent (U.S. EPA 2019a).

Depending on the economics, facilities may have on-site or off-site regeneration systems or may waste spent carbon and replace it with new. The three most common GAC regeneration methods are steam, thermal, and chemical; of which, thermal regeneration is the most common method used. Available thermal regeneration technologies used to remove adsorbed organics from activated carbon include: (i) electric infrared ovens, (ii) fluidized bed furnaces, (iii) multiple hearth furnaces, and (iv) rotary kilns (U.S. EPA 2019a).

GAC treatment is particularly effective for nonvolatile, low water-soluble organic contaminants that have low acceptable drinking water concentrations and cannot be removed via air stripping and have high adsorption rates to the activated carbon. Once removed from the treated water, these compounds would be destroyed in the carbon regeneration process through thermal destruction.

11.3.2 AOPs

A number of organic micropollutants cannot be removed by conventional drinking water treatment alone. Chemical oxidants such as ozone (O_3) are able to oxidize a wide range of these substances; however even these technologies are ineffective in removing a number of emerging contaminants. AOPs such as O_3/H_2O_2, UV/H_2O_2, and UV/O_3 can be used to remove persistent organic contaminants in drinking water treatment. Some of these contaminants include pesticides, gasoline additives, taste and odor compounds, pharmaceutical compounds, and the per- and polyfloroalkyl (PFOA, PSOS) family of compounds of significant recent human health and environmental interest (U.S. EPA 2019b).

In the UV/O_3 process, ozone is applied ahead of UV so that the UV irradiates water with dissolved ozone, and the dissolved ozone absorbs UV light. The photolysis reaction of ozone reacting with UV light forms hydroxyl radicals (·OH), which are very strong oxidants. The oxidation potential of a hydroxyl radical (2.8 V) is much greater than other oxidizing agents such as ozone (2.07 V) and chlorine (1.39 V) and, thus, has the capability of oxidizing a wide variety of organic and inorganic contaminants. DOC removal by UV light or ozone alone is negligible, but substantial removal can be expected for the combined UV/O_3 process. In the presence of UV radiation, the initial ozonation rate of organic substances is increased 10 to 10,000 times. This process has been shown to be capable of mineralizing up to 50% of the total organic matter and, thus, is capable of oxidizing disinfection by-products precursors as well as SOCs of human health concern such as pesticides and other organic refractory compounds. This technology can be considered viable complement to conventional treatment in drinking water and is still being optimized for engineering applications (U.S. EPA 2019a).

The UV/H_2O_2 system is an AOP in which hydrogen peroxide (H_2O_2) is added in the presence of UV light to catalyze the dissociation of hydrogen peroxide into hydroxyl radicals (·OH) through chain reactions that then react with the contaminants of interest in the water. In the UV/H_2O_2 process, H_2O_2 is applied ahead of UV so that H_2O_2-treated water is irradiated.

This technology has been shown to be effective in destroying many micropollutants present in the groundwater (e.g., methyl tert-butyl ether [MTBE], perchlorate, pesticides, 1,4-Dioxane, etc.) and surface water (pharmaceutical and personal care products, taste and odor compounds such as MIB and Geosmin) through direct chemical oxidation. The hydroxyl radical can be generated using either low-pressure (LP) or medium-pressure (MP) continuous UV light in the presence of H_2O_2. Low pressure high output (LP-HO) lamps are preferred when continuous contaminant removal is required (e.g., MTBE and perchlorate removal). MP lamps are recommended for periodic events (such as taste and odor [T&O] or pesticides). LP-HO lamps use less power and have lower operating temperatures (i.e., lower temperatures produce less scaling of sleeves) than MP lamps (U.S. EPA 2019a).

The Electrical Energy per Order (EEO) is used to determine the removal efficiency of organic contaminants in a UV/H_2O_2 and O_3 system. EEO is defined as the number of kilowatt-hours (kWh) of electrical energy required to reduce the concentration of a pollutant by one order of magnitude (90%) in 1 cubic meter of contaminated water. The EEO has units of kWh/1,000 gal/order of removal. The EEO is dependent on water quality and is measured at the optimum H_2O_2 or O_3 dose. It enables a direct comparison of the effectiveness of removing different organic compounds using UV radiation. Contaminants with greater EEO values are more difficult and costly to remove in comparison to those with lower EEOs. For example, the EEO values for MTBE and benzene are 10 and 2, respectively, indicating that MTBE is more difficult to treat than benzene. An EEO of 10 for MTBE means it would take 10 kWh of energy to reduce MTBE from 1,000 to 100 μg/L (90% reduction) in 1,000 gallons of water. It would take another 10 kWh to reduce the MTBE from 100 to 10 μg/L, and so on (U.S. EPA 2019a).

The key design and operating parameters for UV/H_2O_2 systems include peroxide dose, UV lamp type and intensity, reactor contact time, and control system (pH and temperature). The hydroxyl radicals generated in these AOPs is nonselective and, thus. can be "used up" by organic and inorganic scavenging compounds. Water quality parameters, such as organic matter, alkalinity, and nitrite, play an important role for the AOPs because they are considered hydroxyl radical "scavengers," which reduce system effectiveness during oxidation of contaminants (U.S. EPA 2019a).

The most common point of application for an AOP is in the last step of the conventional treatment process, typically after filtration (i.e., lower turbidity, reduced obstruction/shielding of UV light, and fewer scavengers). Closed vessel UV reactors are generally preferred for drinking water applications with hydrogen peroxide and ozone feed due to their smaller footprints, reduced airborne material pollution, reduced exposure to UV light, and installation simplicity. Some concerns of AOP systems include lamp fouling and subsequent lamp sleeve cleaning. Lamp breakage and aging are other potential problems because UV intensity output decreases with time (U.S. EPA 2019a).

11.4 TREATMENT PROCESS BY-PRODUCTS

Residuals generated by the conventional treatment process include coagulated and settled solids (sludge), spent backwash, and spend filter or GAC material and must be managed properly to prevent adverse environmental impacts (U.S. EPA 1996). Spent backwash is often returned to the treatment process as a means to minimize water loss (see Backwash Rule Chapter 7). Recycled filter backwash may concentrate oocyst/cysts and result in a significant source of increased turbidity and crypto/giardia. As a result, a period of filter-to-waste flow may be required after postbackwash or cleaning periods (U.S. Bureau of Reclamation 2010).

Sludge may also be recycled to minimize coagulant and coagulant aid doses and improve process performance. Process solids (i.e., coagulation sludge and filtered solids) will contain elevated concentrations of contaminants removed during the treatment process. Depending on the source water concentration of a particular contaminant and any disposal limitations, it may be necessary evaluate the disposal of process solids with respect to state and local hazardous waste regulations (U.S. EPA 2019a).

Softening generates large quantities of sludge that can be very expensive to dispose of. The amount of sludge produced depends on the water's hardness. Depending on the hardness of water, the average water treatment plant produces 1,000–8,000 lb dry solids/MG water treated. Lime sludge is frequently recycled to the clarification process to improve precipitation, reduce chemical usage, and improve process performance. Sludge generated by softening can also be disposed by discharge to a sanitary sewer, drying lagoons, and land application. If the sludge contains high concentrations of metals or toxic substances, it may be a requirement to dispose of it in a hazardous waste landfill (U.S. EPA 2019a).

Spent GAC must be disposed of recognizing that adsorbed pollutants can be desorbed from the spent GAC when exposed to percolating water. These leached pollutants can potentially result in contaminating soils or groundwater, and consequently, spent GAC regeneration is typically favored over disposal.

11.5 WATER DISTRIBUTION SYSTEMS

Once raw water is adequately treated to remove chemical and microbiological contaminants of health and aesthetic concern, it is essential that water distribution systems are adequately inspected and managed to ensure that this high-quality water reaches the consumers without deterioration in quality or recontamination during transmission. Proper techniques for protection of finished water quality and elimination of cross-connections within water distribution systems have been available for many years as federal guidelines (U.S. Public Health Service 1962) and in local plumbing codes. The sanitary protection of new or repaired pipelines can be ensured by proper attention to certain details of construction. All connections should be made under dry conditions, either in a dry trench or, if it is not possible to completely dewater the trench, above the ground surface. Contaminated piping should be thoroughly cleaned and disinfected before connections are made. Flush valves or cleanouts should be provided at low points to provide pipe section draining where there is no possibility of flooding.

When not properly designed or installed, frost-proof hydrants may permit contamination to enter the water system. Such hydrants should be provided with suitable drainage to a free atmosphere outlet where possible. The drainage from the base of the hydrant should not be connected to a seepage pit or to a sewer that is subject to contamination. The water supply inlet to water tanks should be placed with an air gap (twice pipe diameter) above the flooding level of the fixtures to prevent danger of back siphoning. There should be no cross connection, auxiliary intake, bypass, or other piping arrangements by which polluted water or water of questionable quality can be discharged or drawn into the domestic water supply system (U.S. Public Health Service 1962).

Before a distribution system is placed in service it should be completely flushed and disinfected. The entire system, including tank or standpipe, should be thoroughly flushed with water to remove any sediment that may have collected during operation with raw water. Following flushing, the system should be filled with a disinfecting solution of calcium hypochlorite and treated water. This solution is prepared by adding 1.2 pounds of 70% calcium hypochlorite to each 1,000 gallons of water. A mixture of this kind provides a solution having not less than 100 mg/L of available chlorine. The disinfectant should be retained in the system, tank, and standpipe for not less than 24 hours and then examined for residual chlorine and drained out. If no residual chlorine is found, the process should be repeated. The system is next flushed with treated water and put into operation (U.S. Public Health Service 1962).

Elevated storage tanks and ground-level reservoirs and cisterns are typically used in drinking water system distribution and should be completely covered and constructed to prevent potential contamination by outside water or other foreign matter. All tanks require adequate screening of any openings to protect against the entrance of small animals, mosquitoes, flies, and other small insects. Disinfection of storage facilities subsequent to construction or repair should be carried out in accordance with the recommendations presented for water distribution systems listed previously.

Once pipelines and storage tanks are disinfected and free of contamination, water age is a major factor contributing to water quality deterioration within the distribution system. Water age is primarily controlled by system design and system demands and can vary significantly within a given system at a particular time. Increased temperatures typically associated with increased water age can cause reactions to proceed faster and go further than if water recently released to the distribution system. Water quality problems that can be exacerbated by increased water age include disinfection by-product formation, corrosion control effectiveness, taste and odor problems, water discoloration, nitrification, and microbial growth/regrowth. The Stage 1 and proposed Stage 2 DBP Rules (Chapter 7) recognize the relationship between DBP occurrence and water age and have established monitoring requirements based on this link. The Lead and Copper Rule (Chapter 7) requires utilities to maintain optimal water quality parameters within the distribution system to ensure effective corrosion control treatment. For poorly buffered waters, increased water age can impact compliance with water quality parameter requirements. Tools for evaluating water age include hydraulic models, tracer studies, and monitoring programs (U.S. EPA 2002a). A study conducted by Black and Veatch for the Water Research Foundation (Brandt et al. 2004) evaluated the feasibility and effectiveness of using distribution system retention time (or water age) as a tool for managing distribution system water quality. This study generated a guidance manual for distribution system operators (Brandt 2006) identifying retention time management strategies (e.g., altering hydraulic boundaries, flushing, abandoning or downsizing mains, and reconfiguring storage or mode of operation) that can be implemented to control changes in water quality that occur as a result of stagnant water in distribution systems.

Microbiological, chemical, and physical water quality problems can also occur in finished water reservoirs that are underused or poorly mixed. Poor mixing can be a result of design or operational practices. Several guidance manuals have been developed to address design, operations, and maintenance of finished water reservoirs. Water quality issues that have the potential for impacting public health include disinfection by-product formation, nitrification, pathogen contamination, and increases in dissolved organic carbon concentrations. Elevated disinfection by-products' levels within storage facilities could result in a maximum contaminant level (MCL) violation under the proposed Stage 2 Disinfectants and Disinfection Byproduct Rule (Chapter 7), based on a locational running annual average approach. Nitrite or nitrate levels are unlikely to approach MCL concentrations within the distribution system as a result of nitrification unless finished water nitrate/nitrite levels are near their respective MCLs (U.S. EPA 2002b). Pathogen contamination from floating covers or unprotected hatches is possible. Recommended tank cleaning and inspection procedures have been developed by the American Water Works Association (AWWA 1992, 1995) to address these issues. Elevated levels of dissolved organic carbon have been measured in finished water storage facilities (Alben et al. 1989). AWWA and National Sanitation Foundation (NSF) standards have been developed to ensure that approved coatings function as intended (AWWA 1996, 1997). Additional data and evaluation would be required to determine if there is a significant potential for coatings and other products used in distribution system construction and maintenance to cause an MCL violation based on sampling within the distribution system rather than the currently required monitoring at the point of entry (U.S. EPA 2002b).

One final issue of concern in water distribution systems is the protection of them against corrosion both internally and externally. Internal corrosion is associated with the aggressiveness of the finished water they are transporting, and external corrosion is associated with the aggressiveness of the soil into which they are placed. External corrosion protection can include various external surface coatings (bituminous enamels, asphalt mastic, liquid epoxies, extruded plastic coatings, etc.) or cathodic protection (Beavers and Thompson 2006). Internal mechanical coatings are of concern because of potential leaching and water quality impacts from the coatings, so chemical coating approaches integrated into softening or solids removal steps are normally used.

The chemical composition of water determines the stability of the water (i.e., the likelihood of precipitation of sparingly soluble salts or the tendency to cause corrosion by scavenging minerals from surrounding materials such as tanks and pipes). Scaling or corrosion potential indices have been developed to predict the likelihood of calcium carbonate precipitation or dissolution. These indices are convenient, easy to use, and provide a reasonable estimate of the likelihood of scaling or aggressive water in distribution systems. Scaling potential indices are only capable of predicting the potential of a water to form scale and are generally conservative estimates of precipitation formation as they do not account for the kinetics of the precipitation reactions, the retention time required for precipitate formation, and assume adequate buffer capacity to support precipitate formation. The most widely used index in the water treatment industry is the Langelier Saturation Index (LSI; U.S. Bureau of Reclamation 2013). The presence of a thin layer of scale is usually for prevention of corrosion of metallic piping. For maximum benefit, one needs neither too much nor too little scale deposition. The LSI was initially developed to quantify this water conditioning operation.

The LSI is used to determine the need for control of $CaCO_3$ precipitation to either control excess scaling or minimize corrosion of metallic pipe by aggressive water. The LSI is computed by comparing the pH of the water to that of the saturation pH of $CaCO_3$, pH_{sat}, in a given water. The value of pH_{sat} is calculated as:

$$pH_{sat} = pCa + pAlk + C \tag{11.1}$$

where pCa is the negative log of the molar Ca concentration, pAlk is the negative log of the molar bicarbonate concentration, and C is a function of the TDS (mg/L) and temperature, T, in °F of the water being evaluated. C is quantitatively described as (U.S. Bureau of Reclamation 2013):

$$C = 3.26\,e^{-0.005T} - 0.0116\log_{10}\left(TDS^3\right) + 0.0905\log_{10}\left(TDS^2\right) - 0.133\log_{10}\left(TDS\right) - 0.02 \tag{11.2}$$

The LSI is then calculated as shown in Equation 11.3:

$$LSI = pH - pH_{sat} \tag{11.3}$$

Water with an LSI value more negative than −0.5 is described as "aggressive," and will not deposit a protective film or layer of $CaCO_3$, whereas a water with a positive LSI would be expected to form a film of $CaCO_3$ on the pipe interior. Aggressive water will dissolve a $CaCO_3$ film where one exists leaving any exposed metallic surfaces unprotected and open to corrosion resulting from phenomena like galvanic differences, low pH, or significant concentrations of sulfate or chloride ions (U.S. Bureau of Reclamation 2013). An aggressive water will also attack a concrete lining on a metallic pipe. Carrier Air Conditioning Company (1965) presented an interpretation of LSI values as follows: −2.0 = Serious Corrosion; −0.5 = Slightly Corrosive, Not Scale Forming; 0 Balanced, Pitting Corrosion Possible; +0.5 Slightly Scale Forming, Slightly Corrosive; +2.0 Scale Forming, Non-Corrosive.

If the LSI is unacceptable, it can be lowered by reducing the calcium concentration through softening, or the pH can be lowered by acid addition or via recarbonation. The LSI can be increased with adjustment to the softening process to increase the calcium concentration or by pH adjustment through changes in the postsoftening recarbonation process.

Finally, the use of orthophosphates has been a successful practice for minimizing corrosion of piping and materials containing lead and copper. For utilities with hard water and high levels of dissolved inorganic carbonate, blended ortho- and polyphosphates have been used (U.S. EPA 2002a). Corrosion protection and pH stability can be difficult to achieve for soft, poorly buffered waters common in surface water supplies of the Pacific Northwest, New England, and the Southeastern United States. Interaction with cement linings can significantly increase pH, but unstable, soft, low-mineralized waters can revert to untreated water pH conditions by the time the water reaches the customer's tap. For these waters, corrosion must be controlled with pH and alkalinity adjustment or addition of polyphosphate or calcium (U.S. EPA 2002a).

11.6 APPLICATIONS

Five Illustrative Examples complement the material presented in this chapter on the drinking water treatment.

Illustrative Example 11.1

It is common practice in drinking water treatment to "overtreat" a portion of the flow and blend the balance of the untreated water to produce the maximum allowable level of a primary drinking water contaminant to reduce the overall water treatment cost and still ensure a safe drinking water supply is delivered to the customer. With this in mind, answer the following problem.

A water supply contains a primary drinking water compound at a concentration of 600 mg/L. The primary drinking water standard for this compound is 50 mg/L. If a portion of the flow can be treated to a level of 6 mg/L of the compound, what fraction of the raw water can bypass the treatment step so that the blended water has a final concentration at the primary drinking water standard?

Solution

Let B = the fraction of the raw water bypassing the treatment step with a compound concentration of 200 mg/L, and let $1 - B$ = the fraction of the raw water passing through the treatment step to yield a final compound concentration of 5 mg/L. A mass balance around the treatment step can be written as:

$$B(200 \text{ mg/L}) + (1 - \text{B})(5 \text{ mg/L}) = (1)(50 \text{ mg/L})$$

Rearranging and solving for B yields:

$$195\ B = 45;\ B = 45/195 = 0.23$$

Therefore, 23% of the raw flow can bypass the treatment step and be blended with the treated water to yield a finished water concentration for the compound that meets the primary drinking water standard.

Illustrative Example 11.2

Lake Pristine is a small, shallow lake in the Adirondack mountains and is a source of drinking water and home to thousands of fish, having escaped the ravages of acid rain. Unfortunately, a waste dump is sited only 0.5 miles upstream from the lake. On January 1, 2012, polychlorinated biphenyls (PCBs) began to enter the lake and began to accumulate there. The lake has a surface area of 9×10^5 m^2 and an average depth of 3.0 m. The average annual flow rate into the lake from the surrounding drainage area is 80 m^3/h, while the annual loss due to evaporation averages 20% of its depth.

1. Calculate the average annual outflow of the lake and the annual PCB influx (in units pf kg), given that the PCB concentration in the contaminated water is 0.10 mg/L, but that only 10%; of the incoming water is contaminated.
2. If the lake volume is assumed constant, calculate the concentration of PCBs (mg/L) in the lake water after 1 year if there is no outflow.
3. Derive an equation that describes the mass and concentration of PCBs in the lake with respect to time. Assume perfect mixing.
4. In what year will the concentration of PCBs become 10 times that of Year 1?
5. In what year will the lake become saturated with PCBs? (Solubility of PCBs in water ≈ 0.20 mg/L.)
6. How many kg of PCBs will have been lost from the dump (into the lake) when the lake reaches its maximum concentration?

Solution

First calculate the volume of the lake as:

$$\text{Volume} = (\text{surface area})(\text{depth}) = \left(9\times10^5\,\text{m}^2\right)(3.0\,\text{m}) = 2.7\times10^6\,\text{m}^3$$

Next, calculate the volume of water lost annually due to evaporation:

$$\text{Evaporation Loss} = (0.2)\left(2.7\times10^6\,\text{m}^3\right) = 5.4\times10^5\,\text{m}^3/\text{yr}$$

Next, calculate the volume of water flowing into the lake annually:

$$\text{Volume In} = (\text{water rate})(\text{time}) = \left(80\ \text{m}^3/\text{h}\right)(24\ \text{h/d})(365\ \text{d/yr}) = 7.0\times10^5\,\text{m}^3/\text{yr}$$

Next, calculate the volume of water flowing out of the lake annually:
Volume In = Volume Out because the lake volume is assumed to remain constant, therefore,

$$7.0\times10^5\text{m}^3/\text{yr} = \text{Evaporation} + \text{Outflow} = 5.4\times10^5\,\text{m}^3/\text{yr} + \text{Outflow}$$

$$\text{Outflow} = 1.6\times10^5\,\text{m}^3/\text{yr}$$

Calculate the volume of contaminated water entering the lake annually and then, using the inlet concentration of PCBs, the annual kg of PCBs in the inflow.

$$\text{Volume of Contaminated Water} = 10\%\text{ of Total Inflow} = 7.0\times10^4\,\text{m}^3/\text{yr}$$

$$\text{Concentration of PCBs in Inflow} = 0.10\ \text{mg/L} = 1.0\times10^{-4}\,\text{kg/m}^3$$

$$\text{Annual Inflow of PCBs, kg} = (\text{Volume})(\text{Concentration})$$

$$= \left(7.0\times10^4\,\text{m}^3/\text{yr}\right)\left(1.0\times10^{-4}\,\text{kg/m}^3\right) = 7.0\ \text{kg/yr}$$

An equation can be derived by applying the conservation law for the mass of PBCs on a time rate basis. The following equation can be written:

$$(\text{mass PCBin}) - (\text{mass PCBout}) = (\text{rate of PCB accumulation}) \tag{11.4}$$

Substituting (in kg/yr) gives:

$$(7.0\ \text{kg/yr}) - \left(1.6\times10^5\text{m}^3/\text{yr}\right)(\text{C}) = \left(2.7\times10^6\text{m}^3\right)\frac{\text{dC}}{\text{dt}}$$

where C is the concentration of PCB in the lake at time t, kg/m³.
This equation reduces to:

$$2.59\times10^{-6} - 0.059\,\text{C} = \frac{\text{dC}}{\text{dt}};\ \text{C} = 0\ \text{at}\ \text{t} = 0$$

This is an equation of the form $\frac{dx}{dt} = a + bx$ which may be integrated to give:

$$\text{C} = 4.38\times10^{-5}\left(1 - e^{-0.059t}\right) \tag{11.5}$$

The mass, M, of PCB in the lake at any time is then given as:

$$M = C\,(\text{Volume of the lake}) = \left(2.7 \times 10^6 m^3\right) C$$

After Year 1, the concentration of PCBs in the lake, assuming complete mixing is:

$$\text{Concentration Year 1} = 4.38 \times 10^{-5}\left(1 - e^{-0.059(1)}\right) = 2.51 \times 10^{-6}\ kg/m^3 = 2.51\ \mu g/L$$

Ten times the Year 1 concentration = 25.1 μg/L = $25.1 \times 10^{-6}\ kg/m^3$. Equation 11.5 can be solved for t to determine when this concentration will occur as:

$$C = 25.1 \times 10^{-6} kg/m^3 = 4.38 \times 10^{-5}\left(1 - e^{-0.059t}\right)$$

$$0.573 = \left(1 - e^{-0.059t}\right);\ 0.427 = e^{-0.059t};\ \ln(0.427) = -0.059t;\ 0.851 = 0.059t$$

$$t = 0.851/0.059 = 14.4\ yr$$

For $C = 0.2\ mg/L = 0.2 \times 10^{-3} kg/m^3$, Equation 11.5 can be solved again for t as:

$$C = 0.2 \times 10^{-3} kg/m^3 = 4.38 \times 10^{-5}\left(1 - e^{-0.059t}\right)$$

$$4.57 = \left(1 - e^{-0.059t}\right);\ -3.57 = e^{-0.059t};\ \ln\left(-3.57\right) = -0.059t$$

There is no solution for this problem indicating that the lake will never reach saturation with respect to PCB concentrations. At long time periods, the maximum concentration that the lake will reach is $4.38 \times 10^{-5} kg/m^3$ = 0.0438 mg/L = 43.8 μg/L.

When the lake reaches its maximum concentration of $4.38 \times 10^{-5}\ kg/m^3$, the mass of PCBs in the lake is:

$$M = C\ (\text{Volume of the lake}) = C(2.7 \times 10^6\, m^3) = \left(4.38 \times 10^{-5}\, kg/m^3\right)\left(2.7 \times 10^6\, m^3\right)$$

$$= 118.3\ kg$$

Illustrative Example 11.3

Assume the lake described in Illustrative Example 11.2 is the source of water supply for the community of Upstate, New York. Assuming the lake will not require softening, propose a water treatment process schematic that would provide safe drinking water for the town.

Solution

The EPA's goal for drinking water's maximum contaminant level of PCPs is zero, and the enforceable MCL for PCBs in public water systems is 0.0005 mg/L = 0.5 μg/L. As indicated in Illustrative Example 11.2, the PCB concentration 1 year after release of the PCBs into the lake is estimated to be 2.51 μg/L, with a maximum concentration expected of 43.8 μg/L, both far exceed the MCL for PCBs. This means that in addition to conventional treatment for a surface water source for turbidity and pathogen removal, this lake water will require advanced treatment for PCB removal that would entail the use of GAC or perhaps AOP to ensure the PCB MCL is met in the finished water. The selection of a final process train would be based on an economic and performance evaluation of the advance treatment components of the surface-water treatment Process flow sheets for such treatment trains are shown in Figure 11.7.

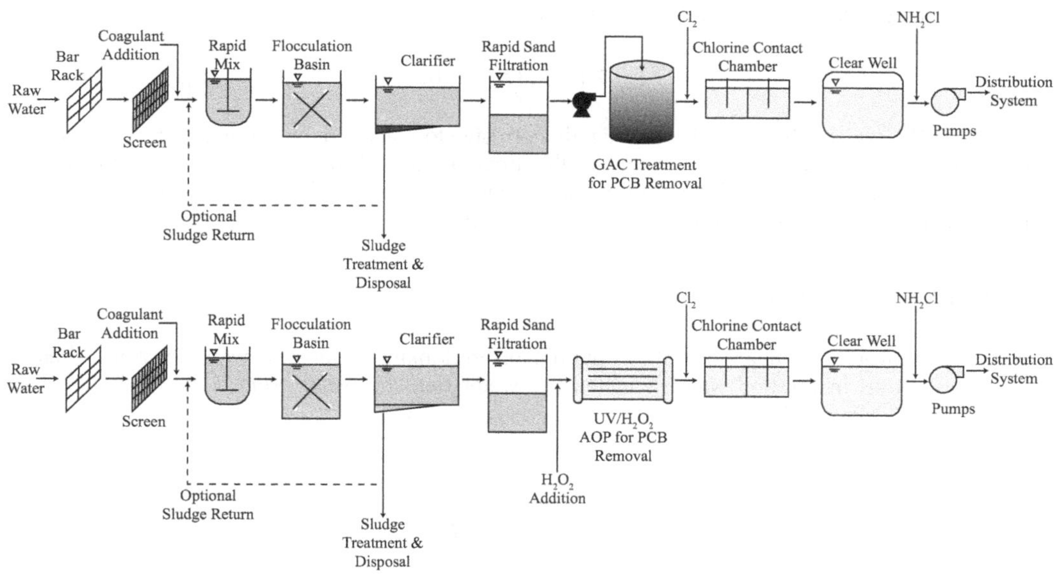

FIGURE 11.7 Process flow sheet alternatives for treatment of PCB contaminated water from Lake Pristine for potable use. AOP = advanced oxidation process; Cl_2 = chlorine; GAC = granular activated carbon; H_2O_2 = hydrogen peroxide; NH_2Cl = chloramine; PCB = polychlorinated biphenyl; UV = ultraviolet.

Illustrative Example 11.4

If the water in Lake Pristine has the following water quality, determine its LSI, interpret the result in terms of corrosion and precipitation potential, and discuss the result in terms of the need for any additional water conditioning that might need to be added to the water treatment process flow sheet proposed in Illustrative Example 11.3.

$$\text{Mean temperature} = 45°\text{F, TDS} = 250\text{ mg/L, }[\text{Ca}] = 100\text{ mg/L, pH} = 7.8, \left[\text{HCO}_3^-\right] = 150\text{mg/L}$$

Solution

The molecular weight of $\text{Ca} = 40.1\text{ g/gmol}; \text{HCO}_3^- = 61\text{ g/gmol}$. The pCa and pAlk values are calculated as:

$$\text{pCa} = -\log(\text{molar Ca concentration}) = -\log\left[(100\text{ mg/L})/(40.1\text{ g/gmol})\right] = -\log(0.0025)$$

$$\text{pCa} = 2.60$$

$$\text{pAlk} = -\log\left(\text{molar HCO}_3^-\text{concentration}\right) = -\log\left[(150\text{ mg/L})/(61\text{ g/gmol})\right] = -\log(0.00245)$$

$$\text{pAlk} = 2.61$$

The value of C is then calculated using Equation 11.2 as:

$$C = 3.26\,e^{-0.005(45)} - 0.0116\log_{10}\left(250^3\right) + 0.0905\log_{10}\left(250^2\right)$$

$$-\,0.133\log_{10}(250) - 0.02 = 2.60 - 0.083 + 0.434 - 0.319 - 0.02 = 2.61$$

pH_{sat} is then calculated using Equation 11.1 as:

$$pH_{sat} = 2.60 + 2.61 + 2.61 = 7.82$$

The LSI is then calculated from Equation 11.3 as:

$$\text{LSI} = 7.8 - 7.82 = -0.02$$

With an LSI = −0.02, this water is very slightly corrosive to balanced, but pitting corrosion is still possible in the distribution pipes delivering this treated water. A water conditioning step using calcium or orthophosphate addition, or pH adjustment upward would help to provide some protection against corrosion within this distribution system.

Illustrative Example 11.5

A term employed for noncarcinogenic contaminants in drinking water systems to determine their acceptable level in finished water is called the water *drinking water equivalent level* (*DWEL*). The DWEL is calculated as:

$$\text{DWEL} = \frac{(\text{NOAEL})(w)}{(Q_w)(UF)} \tag{11.6}$$

where NOAEL is the no observed adverse effects level, mg/kg-d; w is the average adult body weight, 70 kg; Q_w is the average adult daily water consumption rate, 2 L/d; and UF is the uncertainty factor of toxicity conversion to adult humans, 100. Calculate the DWEL for fluorides if their NAOEL = 11.4 mg/kg-d.

Solution

Applying the equation for fluorides yields:

$$\text{DWEL} = \frac{(11.4\ \text{mg/kg-d})(70\ \text{kg})}{(2\ \text{L/d})(100)} = \frac{798\ \text{mg/d}}{200\ \text{L/d}} = 3.99\ \text{mg/L}$$

REFERENCES

Alben, K., A. Bruchet, and E. Shpirt. 1989. *Leachate from Organic Coating Materials Used in Potable Water Distribution Systems*. Report submitted to AWWA and AwwaRR. Denver, CO: AWWA and AwwaRF.

American Water Works Association (AWWA). 1992. *AWWA Standard for Disinfection of Water Storage Facilities*. Report # AWWA C652. Denver, CO: American Water Works Association.

American Water Works Association (AWWA). 1995. *Water Transmission and Distribution*. 2nd edition. Denver, CO: American Water Works Association.

American Water Works Association (AWWA). 1996. *AWWA Standard for Flexible-Membrane Lining and Floating-Cover Materials for Potable-Water Storage*. Report # AWWA D130. Denver, CO: American Water Works Association.

American Water Works Association (AWWA). 1997. *AWWA Standard for Coating Steel Water Storage Tanks*. Report # AWWA D102-97. Denver, CO: American Water Works Association.

American Water Works Association (AWWA). 2011. *Water Quality & Treatment: A Handbook on Drinking Water,* 6th edition. Chicago, IL: McGraw-Hill. https://www.accessengineeringlibrary.com/content/book/9780071630115.

American Water Works Association (AWWA) and American Society of Civil Engineers (ASCE). 2012. *Water Treatment Plant Design,* 5th edition. Chicago, IL: McGraw-Hill. https://www.accessengineeringlibrary.com/content/book/9780071745727.

Beavers, J.A., and N.G. Thompson. 2006. External Corrosion of Oil and Natural Gas Pipelines. In *ASM Handbook, Volume 13C, Corrosion: Environments and Industries* (#05145). Materials Park, OH: AMS International. https://www.asminternational.org/documents/10192/1849770/ACFAB96.pdf.

Brandt, M.J. 2006. *Managing Distribution Retention Time to Improve Water Quality–Phase II: Guidance Manual.* Final Report # 91121 and CD for Project #2769b. Denver, CO: Water Research Foundation. http://www.waterrf.org/Pages/Projects.aspx?PID=2769.

Brandt, M.J., J. Clement, J. Powell, R. Casey, D. Holt, N. Harris, and C.T. Ta. 2004. *Managing Distribution Retention Time to Improve Water Quality–Phase I.* Final Report # 91006F. Denver, CO: Water Research Foundation. http://www.waterrf.org/Pages/Projects.aspx?PID=2769.

Carrier Air Conditioning Company. 1965. *Handbook of Air Conditioning System Design.* New York: McGraw-Hill.

Dieter, C.A., M.A. Maupin, R.R. Caldwell, M.A. Harris, T.I. Ivahnenko, J.K. Lovelace, N.L. Barber, and K.S. Linsey. 2018. *Estimated Use of Water in the United States in 2015.* Circular 1441. Reston, VA: U.S. Geological Survey. 65 p. https//doi.org/10.3133/cir1441.

Droste, R.L., and R.L. Gehr. 2019. *Theory and Practice of Water and Wastewater Treatment*, 2nd edition. Hoboken, NJ: John Wiley & Sons.

Hendricks, D. 2011. *Fundamentals of Water Treatment Unit Processes.* Boca Raton, FL: CRC Press. https://doi.org/10.1201/9781439895092.

Howe, K.J., D.W. Hand, J.C. Crittenden, R.R. Trussell, and G. Tchobanoglous. 2012. *Principles of Water Treatment*, 3rd edition. Hoboken, NJ: John Wiley & Sons.

Pollution Probe. 2002. *The Drinking Water Primer.* Toronto, Canada: Pollution Probe. http://www.nesc.wvu.edu/ecommerce/products/DW_PublicEducation/DWBKPE428DL.pdf.

U.S. Bureau of Reclamation. 2010. *Water Treatment Primer for Communities in Need.* Desalination Series Report No. 68, Reclamation, Managing Water in the West. Denver, CO: Technical Services Center, Bureau of Reclamation. https://www.usbr.gov/research/dwpr/reportpdfs/report068.pdf.

U.S. Bureau of Reclamation. 2013. *Water Chemistry Analysis for Water Conveyance, Storage, and Desalination Projects.* Manuals and Standards Program. Denver, CO: Technical Service Center, Civil Engineering Services Division, Water Treatment Group, Bureau of Reclamation. https://www.usbr.gov/tsc/techreferences/mands/mands-pdfs/WQeval_documentation.pdf.

U.S. Environmental Protection Agency (U.S. EPA). 1990. *Technologies for Upgrading Existing or Designing New Drinking Water Treatment Facilities.* EPA/625/4-89/023. Cincinnati, OH: Center for Environmental Research Information, Office of Drinking Water. https://nepis.epa.gov/Exe/ZyPDF.cgi/300048WU.PDF?Dockey=300048WU.PDF.

U.S. Environmental Protection Agency (U.S. EPA). 1996. *Technology Transfer Handbook: Management of Water Treatment Plant Residuals.* EPA/625/R-95/008. Cincinnati, OH: Office of Research and Development, National Risk Management Research Laboratory. https://nepis.epa.gov/Exe/ZyPDF.cgi/30004PI0.PDF?Dockey=30004PI0.PDF.

U.S. Environmental Protection Agency (U.S. EPA). 2000. *The History of Drinking Water Treatment.* EPA-816-F-00-006. Washington, DC: Office of Water. https://nepis.epa.gov/Exe/ZyPDF.cgi/P1002SMN.PDF?Dockey=P1002SMN.PDF.

U.S. Environmental Protection Agency (U.S. EPA). 2002a. *Effects of Water Age on Distribution System Water Quality.* Distribution System Issue Paper. Washington, DC: Office of Water, Office of Ground Water and Drinking Water. https://www.epa.gov/sites/production/files/2015-09/documents/2007_05_18_disinfection_tcr_whitepaper_tcr_waterdistribution.pdf.

U.S. Environmental Protection Agency (U.S. EPA). 2002b. *Finished Water Storage Facilities.* Distribution System Issue Paper. Washington, DC: Office of Water, Office of Ground Water and Drinking Water. https://www.epa.gov/sites/production/files/2015-09/documents/2007_05_18_disinfection_tcr_whitepaper_tcr_storage.pdf.

U.S. Environmental Protection Agency (U.S. EPA). 2019a. *Drinking Water Treatability Database.* Cincinnati, OH: Office of Research and Development. https://iaspub.epa.gov/tdb/pages/general/home.do. (Accessed July 2019).

U.S. Environmental Protection Agency (U.S. EPA). 2019b. *EPA's Per- and Polyfluoroalkyl Substances (PFAS) Action Plan.* EPA 823R18004. Washington, DC: Office of Ground Water and Drinking Water. https://www.epa.gov/sites/production/files/2019-02/documents/pfas_action_plan_021319_508compliant_1.pdf.

U.S. Public Health Service. 1962. *Manual of Individual Water Supply Systems.* Public Health Service Publication No. 24. Washington, DC: U.S. Department of Health, Education, and Welfare, Public Health Service. https://nepis.epa.gov/Exe/ZyPDF.cgi/9101ODLR.PDF?Dockey=9101ODLR.PDF.

12 Municipal Wastewater Treatment

12.1 INTRODUCTION

In the United States, during the early 1900s, nuisance and health conditions resulting from the improper disposal of large quantities of untreated municipal wastewater from large cities created the need for more effective means of wastewater management. Large expanses of land that had previously been used for wastewater disposal were no longer readily available, especially in larger cities. This led to the planning, design, construction, and operation of more advanced wastewater treatment facilities.

If untreated wastewater is allowed to be discharged, the decomposition of the organic material it contains can lead to the production of offensive odors and gases. In addition, untreated wastewater may contain numerous pathogenic (i.e., disease causing) microorganisms that are released in very large numbers from the human intestinal system of infected individuals. Many of these microorganisms can persist in the environment for long periods of time, making the potential for disease transmission highly likely. Domestic wastewater also contains nutrients (nitrogen and phosphorous), which can stimulate the growth of aquatic life that may lead to eutrophication and significantly impacted surface water quality. Domestic wastewater may also contain toxic compounds from commonly used consumer products and from commercial and industrial wastewater entering the municipal system. For these reasons, the immediate removal of wastewater from its source of generation, followed by proper treatment and safe disposal, is imperative to ensure protection of public health and the environment.

Today, not only must a wastewater treatment plant satisfy effluent quality limits, but it must also satisfy many other environmental requirements. Some of these conditions are to satisfy requirements for aesthetics and minimization of obnoxious odors at treatment and disposal sites; to prevent contamination of water supplies from physical, chemical, or biological agents; to prevent adverse impacts to fish, shellfish, and other aquatic life; to prevent degradation of the quality of receiving waters from overfertilization; to prevent the impairment of beneficial uses of receiving waters (for recreational, agricultural, commercial, drinking water, or industrial purposes); to protect against the spread of disease from crops grown on reclaimed wastewater or treatment plant biosolids; to prevent the decline in property values so as to not restrict community growth and development; and to encourage other beneficial uses of the effluent.

The purpose of any wastewater treatment plant is to convert the components in raw wastewater into a harmless final effluent for discharge to a receiving body of water, essentially to return this municipal wastewater safely back into the water cycle and to safely dispose of the solids (biosolids) produced in the process. The primary means of accomplishing this for municipal wastewater is through the use of a range of biological processes. The planning, design, construction, and operation of wastewater treatment facilities are a complex problem. The reader is referred to the literature (Water Environment Federation 2012; Metcalf & Eddy et al. 2014; Viessman et al. 2014) for texts entirely or in large part devoted to the topic of municipal wastewater treatment. This chapter presents a brief insight into the various factors affecting the design, operation, and management of municipal wastewater treatment systems by addressing the following topics: overview of wastewater regulations, municipal wastewater characteristics, wastewater plant design considerations, wastewater treatment options, overview of advanced wastewater treatment technologies, sludge disposal considerations, and wastewater reuse options. An application section completes the chapter with six Illustrative Examples related to the general subject of municipal wastewater treatment and wastewater management.

12.2 OVERVIEW OF WASTEWATER REGULATIONS

The reader is referred to Chapter 6 for a detailed discussion of the historical development and implementation of wastewater regulations in the United States. An overview of these topics is provided here as an introduction to municipal wastewater management requirements.

The Federal Water Pollution Control Act Amendments of 1972, which was amended in 1977 and relabeled the Clean Water Act, established the basic structure for regulating discharges of pollutants into the waters of the United States. It gave the Environmental Protection Agency (EPA) the authority to implement pollution control programs such as setting wastewater standards for industry and established minimum levels of wastewater treatment for municipal and industrial treatment plants, deadlines for meeting these levels, and penalties for violators. The act mandated that requirements for treatment performance that had historically been based on percentage of solids removed could no longer be used as primary design standards. Receiving water quality criteria established by the states were to also be used to define limitations of pollutant concentrations resulting from the comingled effluent and receiving water (*Stream Standards*), and the level of treatment required to meet these Stream Standards.

Through this act the federal government established certain minimum effluent criteria as a first step in upgrading water quality throughout the United States. These minimum *effluent limits*, regardless of a treatment plant's location or surface water it discharges to are listed in Table 12.1. If these minimum effluent limits are not sufficient to maintain acceptable water quality for the designated *beneficial use* of the receiving water, the Stream Standards, the water quality standards of the receiving waters, will dictate the level of treatment required for a specific plant. Individual states also reserve the right to impose more stringent effluent requirements than the federal standards.

For wastewater treatment plant discharge to surface waters, effluent requirements are defined in permits issued by the National Pollutant Discharge Elimination System (NPDES). The NPDES permits have generally mandated the upgrading of existing treatment plants or the construction of new plants to provide higher levels of treatment and reliability.

The Clean Water Act of 1977 contains two major provisions for wastewater solids (those biosolids produced during treatment): utilization and disposal. Guidelines and regulations limit the quantity and type of toxic materials reaching the general public based on limits on the quantity and quality of biosolids from wastewater treatment plants that can be distributed for public use or applied to lands where crops are grown for human consumption. The methods by which sludge is applied to land are also designed to aesthetic requirements. The Resource Conservation and Recovery Act (RCRA) of 1976

TABLE 12.1
Federal Secondary Effluent Standards for Municipal Wastewater Treatment Plants

Parameter[a]	Monthly Average	Weekly Average
BOD_5, mg/L	30	45
$CBOD_5$, mg/L	25	40
TSS, mg/L	30	45
BOD and TSS Removal Efficiency	≥85%	
pH	Within the limits of 6.0 to 9.0	

[a] $CBOD_5$ = carbonaceous BOD_5 = the measure of oxygen consumed in a 5-day period to remove the biodegradable organic material in a sample; TSS = total suspended solids.

requires that solid wastes, including sludge from wastewater treatment plants, be used or disposed of in a safe and environmentally acceptable manner. The Marine Protection, Research and Sanctuaries Act 1977 amendments prohibited disposal of "sewage sludge" by barging to the ocean after December 31, 1981. State and municipal requirements, which are more stringent than federal regulations, govern in many areas. This is commonly noted in regulations regarding sludge disposal. Many municipalities that apply sludge to land on which food crops are grown analyze their sludges more frequently than required by federal guidelines or limit sludge application rates more severely to ensure long-term protection of public health and safety. More discussion of sludge disposal issues is provided in Section 12.7.

12.3 MUNICIPAL WASTEWATER CHARACTERISTICS

Municipal wastewater is composed of a mixture of dissolved, colloidal, and particulate organic and inorganic materials. Wastewater characteristics are described in terms of the chemical composition of a wide range of contaminants in the wastewater that represent biodegradable organic material, suspended and dissolved solids, nutrients, specific organic compounds, and gross properties of the wastewater such as pH, dissolved oxygen, and conductivity. The characteristics depend largely on the types of water usage in the community and industrial and commercial contributions to the wastewater treatment plant. During wet weather, a significant quantity of infiltration or inflow may also enter the municipal wastewater collection system. This will significantly dilute the concentration of pollutants in the wastewater and potentially hydraulically overload the treatment plant.

The characteristics of a wastewater may be obtained from the plant operating records that should contain flow and laboratory data at the municipal wastewater treatment plant. The data describing the wastewater characteristics should include minimum, average, and maximum dry weather flows, peak wet weather flows, sustained maximum flows, and chemical parameters such as biochemical oxygen demand (BOD), total suspended solids (TSS), total dissolved solids (TDS), pH, total nitrogen (Total N), phosphorus (Total P), and toxic chemicals. It is important that reliable estimates of the wastewater characteristics be made because this is what the municipal wastewater treatment plant will be responsible for treating.

Municipal wastewater contains 99.99% water. The remaining materials include suspended and dissolved organic and inorganic matter as well as microorganisms. These materials make up the physical, chemical, and biological properties that are characteristic of municipal wastewater.

The physical quality of municipal wastewater is generally reported in terms of temperature, color, odor, and solids content. The temperature of wastewater is slightly higher than that of the water supply. It is an important parameter because of its effect upon aquatic life and the solubility of gases. Temperature varies slightly with the seasons, normally higher than air temperature during most of the year and lower only during the hot summer months. The color of a wastewater is usually indicative of age. Raw wastewater is usually gray; septic wastes impart a black appearance. Odors in wastewater are caused by the decomposition of organic matter that produces offensive smelling gases such as hydrogen sulfide. TSS is defined as the dry weight of matter that can be removed from water by filtration through 0.45-μm filters. Volatile suspended solids (VSS) represent the organic portion of TSS and is determined based on the weight of TSS lost on combustion of TSS at 500°C. TSS may cause undesirable conditions of increased turbidity and silt load in the receiving water.

Chemical characteristics of wastewater are expressed in terms of various organic and inorganic constituents. Different chemical analyses furnish useful and specific information with respect to the quality and strength of wastewater. Organic components in wastewater are the most significant factor in the pollution of many natural waters. The principal groups of organic substances found in municipal wastewater are proteins (40%–60%), carbohydrates (25%–50%), and fats and oils (10%). Carbohydrates, proteins, and fats and oils are easily biodegradable by microorganisms in the wastewater treatment system and represent the major oxygen demand associated with domestic wastewater. In addition, wastewater may also contain small fractions of synthetic detergents, phenolic compounds, and pesticides and herbicides from households and industrial sources that are biodegradable to varying degrees depending on specific compounds being considered. These compounds may create problems such as foaming or may be toxic to treatment plant microorganisms depending on their concentration.

The inorganic compounds most often found in wastewater are chloride, hydrogen ions, alkalinity-causing compounds, nitrogen, phosphorus, sulfur, and heavy metals. Trace concentrations of heavy metal compounds can significantly affect microorganisms in the wastewater treatment plant as wells as in the receiving waters because of their toxicity. Nitrogen and phosphorous released from wastewater treatment plants can also be detrimental to receiving waters due to their stimulation of algal growth in areas where these nutrients would otherwise be growth-limiting.

Gases commonly found in raw wastewater include nitrogen, oxygen, carbon dioxide, hydrogen sulfide, ammonia, and methane. Of all these gases mentioned, the ones that are of most concern in the design of a treatment facility are oxygen, hydrogen sulfide, and methane. Oxygen is required for all aerobic life forms either within the treatment facility or in the receiving water. In the absence of oxygen, microorganisms will continue to metabolize using alternative electron acceptors, which results in the reduction of inorganic salts such as nitrate, sulfate, or carbon dioxide to generate the end products nitrogen, hydrogen sulfide, or methane, respectively, that are obnoxious, corrosive, and a potential human health hazard in the case of hydrogen sulfide or that may be explosive in the case of methane. To avoid such conditions when they are not desired, it is important aerobic conditions (i.e., dissolved oxygen [DO] $\geq$1 mg/L) be maintained.

The quantity and type of microorganisms that make up the biological community in wastewater treatment plants can significantly affect a systems performance. Within a wastewater treatment facility, the wastewater provides the perfect medium for microbial growth, whether under aerobic or anaerobic conditions. Bacteria are the key to the biological treatment process used at most treatment facilities and to the natural biological cycle in receiving waters. In the presence of sufficient DO, bacteria convert the soluble, biodegradable organic matter into new cells and stabilized end products. This conversion of biodegradable organics (i.e., BOD) under controlled conditions within a wastewater treatment plant reduces the loading of this oxygen demanding material to the receiving water.

Within a wastewater treatment plant handling domestic wastewater, bacteria (with concentrations ranging from 10^5 to 10^8/mL) will be the dominant organism, with other organisms (e.g., protozoa, rotifers, algae) achieving varying degrees of importance depending on the wastewater treatment plant design. Consequently, municipal wastewater treatment is directed toward using bacteria to remove BOD from the wastewater and removing the excess bacteria from the treatment plant for further processing and disposal. Biological wastewater treatment processes do not remove phosphorus and nitrogen to any substantial extent, and advanced treatment technologies (Section 12.6) must be employed to remove these nutrients from the wastewater. In fact, biological wastewater treatment processes convert the organic forms of these substances into mineral forms, making them more usable by plant life.

The BOD test, highlighted in Illustrative Example 6.2 in Chapter 6, was developed in an attempt to reflect the depletion of oxygen that would occur in a stream as a result of microorganisms, primarily bacteria, because they use oxygen to metabolize biodegradable organic matter. The BOD is commonly used as the sole basis for determining the efficiency of the treatment plant in stabilizing organic matter and is the parameter used to verify a plant's compliance with regulatory treatment requirements. Wastewater ammonia-nitrogen poses an analytical problem in measuring BOD because at 20°C (68°F), nitrifying bacteria in domestic wastewater usually are significant in number and can exert a measurable oxygen demand during a 5-day BOD test, especially in effluent samples. To obtain a true measure of the treatment plant performance in removing organic matter, the BOD test generally requires inhibition of nitrification so the BOD that is measured is due only to the carbonaceous oxygen demanding material (CBOD) in a sample.

The COD analysis is more reproducible and less time consuming than the BOD test. The COD test measures both the nonbiodegradable as well as biodegradable organics in a sample through the use of a very strong oxidizing chemical, usually potassium dichromate, added to samples of different dilution. To ensure full oxidation of the various compounds found in the samples, a strong acid and a chemical catalyst are added. One of the products of this oxidation-reduction reaction is the chromate ion that causes a very sharp color change that can be easily detected via a spectrophotometer. The COD and BOD tests can be correlated so an estimate of the concentration of biodegradable organic material can be rapidly made using the COD test. Such correlations are specific to a particular wastewater, but for a given plant, COD measurements can be useful for plant operational monitoring. It should be noted, however, that the only acceptable parameter to monitor and report for regulatory purposes remains BOD.

12.4 WASTEWATER PLANT DESIGN CONSIDERATIONS

A significant amount of time is involved in the planning and design of a wastewater treatment facility. The initial phase consists of a facility plan that first must be prepared before grants from the federal and state governments can be obtained (Water Environment Federation 2012; Metcalf & Eddy et al. 2014).

A facility plan is prepared to identify the water pollution problems in a specific area, develop design data, evaluate alternatives, and recommend a solution. Most of the data developed in a facility plan are used in the preparation of design plans, specifications, and cost estimates of the wastewater treatment facility.

Steps in the facility plan include:

1. List effluent limitations: The plan should list the effluent limitations applicable to the facility being planned. Publicly owned wastewater treatment works (POTWs) built after June 30, 1974, must achieve "best practicable waste treatment technology" (equivalent to secondary treatment).
2. Assess current situation: Describe briefly the existing conditions to be considered when examining alternatives during planning. The following conditions should be described:
    a. Planning area
    b. Organizational context
    c. Demographic data
    d. Water quality
    e. Other existing environmental conditions
    f. Existing wastewater flows and treatment systems, including system performance
    g. Infiltration and inflow
3. Assess future conditions:
    a. Twenty-five year planning period beyond start-up of the facility. Phased construction should be considered.
    b. Land use must be carefully coordinated with the state, municipal, and regional regulations, policies, and plans.
    c. Demographic and economic projections.
    d. Forecasts of flow and waste loads; includes projection of population growth, infiltration/inflow estimates, analysis of pollutant content and flows in the existing system, sewer overflow data, industrial waste load data projections, and pollution reduction possibilities.
    e. Future environment of the planning area without the project.
4. Develop and evaluate alternatives: Develop alternative treatment systems and their impact on the environment, long-range sewer plans for the planning area, sludge utilization or disposal, and select a facility location.

5. Select plan: The public is provided with alternative proposals and hearings are held to explain each proposal.
6. Preliminary design of treatment works: The following items are included: a schematic flow diagram; unit processes; plant site plans; sewer pipe plans and profiles; design data regarding detention items, flow rates, and sizing of units; operation and maintenance summary; cost estimates; and a completion schedule.
7. Arrangements for implementation: Following selection of plan and design, existing institutional arrangements should he reviewed and a financial program developed, including preliminary allocation of costs among various classes of users of the system.

Wastewater treatment plants use a number of unit operations and processes to achieve the desired degree of treatment. The collective treatment schematic is called a *process flowsheet*. Many different flow schemes can be developed from various unit operations and processes for the desired degree of treatment. However, the most desirable flow scheme is the one that is the most cost effective.

Wastewater treatment plants are designed to process the liquid and solid portions of the wastewater. The liquid treatment systems and solids treatment and disposal systems must be put together so as to assure the most efficient utilization of resources such as money, materials, energy, and work force in meeting treatment requirements. Logic dictates what the process elements must be and the order in which they go together.

A methodical process of selection must he followed in choosing a resource efficient and environmentally sound system from the numerous treatment and disposal options available. The basic selection mechanism used is the "principle of successive elimination," an iterative procedure in which less effective options arc progressively eliminated until only the most suitable system or systems for the particular site remain.

The concept of a "treatment train" is a result of a systems approach to problem solving. However, this concept is useful only if all components of the train are considered. This includes not only sludge treatment and disposal components but also wastewater treatment options and other critical linkages such as sludge transportation, storage, and side stream treatment.

The general sequence of events in system selection is:

1. Selecting relevant criteria
2. Identifying options
3. Narrowing the list of candidate systems
4. Selecting a system

The basis of unit process design is the initial and future volume and characteristics of the wastewater, anticipated variations in flow and wastewater composition, and statutory requirements of regulatory agencies. Data acquisition required will be, in part, determined by the treatment processes considered and largely determined by the treatment requirements. The plant process design is usually based on average loading conditions, with provisions required for some system components to have capacity for peak and minimum loading conditions (i.e., aeration equipment operating capacity).

For updating an existing wastewater facility, existing flow data, presented on a per-capita basis, should provide a reliable estimate of future flows. In the case of a new facility, utilization of per-capita potable water demands and adjustment to per-capita wastewater production volumes based on data from nearby communities of similar demographic make-up is typically used for flow predictions.

Several flow rates are used for design of various elements in a wastewater treatment plant. The average daily flow is used to determine such items as reactor size, pumping and chemical costs, sludge solids production rates, and mass loading of treatment units. Peak design rate, usually 2 to 3 times the design average flow rate, is used for hydraulic sizing.

The method of solids disposal usually controls the selection of solids treatment systems and not vice versa. Thus, the system selection procedure normally begins when the solids disposal option is specified. The process selection procedure consists of developing treatment/disposal systems that are compatible with one another and appear to satisfy local relevant criteria and choosing the best system or systems by progressive elimination. Proper selection of the sludge processing equipment is important for trouble-free operation of a wastewater treatment plant. Sludge is quite odorous and may cause environmental concerns. Therefore, such factors as solids captured, chemical quality of return flows, ability to handle variable quality of sludge, ease of operation, and odors potentially generated from the process are often given serious considerations.

12.5 WASTEWATER TREATMENT OPTIONS

Wastewater treatment plants use a number of individual or unit operations and processes to achieve the desired degree of treatment. The collective treatment schematic is called a *flow diagram*, a *flow sheet*, a *process train*, or a *process flow sheet*. Many different process trains can be developed from various unit operations and processes for the desired level of treatment. Unit operations and processes are grouped together to provide what is known as preliminary, primary, secondary, and tertiary (or advanced) treatment. *Preliminary* treatment processes refer to those used to physically pretreat the incoming wastewater to remove nonbiodegradable components of the incoming wastewater and protect subsequent treatment units from damage. These preliminary treatment units include bar racks and screens, grit removal, particle size reduction via comminutors, and flow measurement. The term *primary* refers to physical and biological unit operations focused on settleable solids removal and stabilization, *secondary* refers primarily to biological unit processes for the removal of soluble and colloidal biodegradable organic material and solids, and *tertiary* refers to combinations of all three for the removal of pollutants not effectively removed through primary or secondary treatment (i.e., nutrients, metals, synthetic organic compounds, etc.).

Treatment methods in which the application of physical forces predominates are known as *physical unit operations*. These were the first methods to be used for wastewater treatment. Screening, grit removal, comminution, sedimentation, floatation, and filtration are typical physical unit operations used for wastewater treatment. Treatment methods in which the removal or conversion of contaminants is brought about by the addition of chemicals or by other chemical reactions are known as *chemical unit processes*. Precipitation, aeration, adsorption, and disinfection are the most common examples uses in wastewater treatment. Treatment methods in which the removal of contaminants is brought about by biological activity are known as *biological unit processes*. Using biological treatment, colloidal or dissolved biodegradable organic substances and nitrogen, in some cases, can be converted into biological cell mass that can be removed by settling and into gases (i.e., N_2, CO_2, CH_4) that can escape to the atmosphere or, in the case of methane, be captured and used as an alternative source of energy within the plant. With proper environmental control of inhibitory or toxic pollutants in a waste stream, wastewater can in most cases be cost-effectively treated biologically.

Municipal wastewater treatment plants are mandated by the federal government to provide secondary treatment to comply with the final effluent discharge to the receiving waters. As a result of this mandate, secondary treatment is now considered the minimum acceptable design criteria for a wastewater treatment plant.

Secondary treatment implies that biological processes are used in the overall treatment process in addition to physical treatment. An overview of a typical secondary wastewater treatment plant would consist of the following components (Figures 12.1 and 12.2):

1. Preliminary treatment
2. Primary treatment

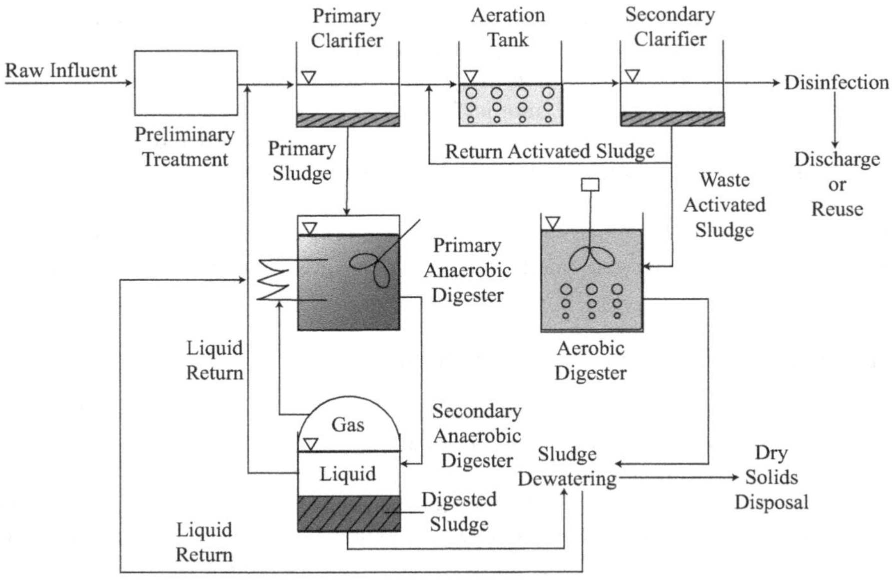

FIGURE 12.1 Layout for a suspended growth wastewater treatment plant.

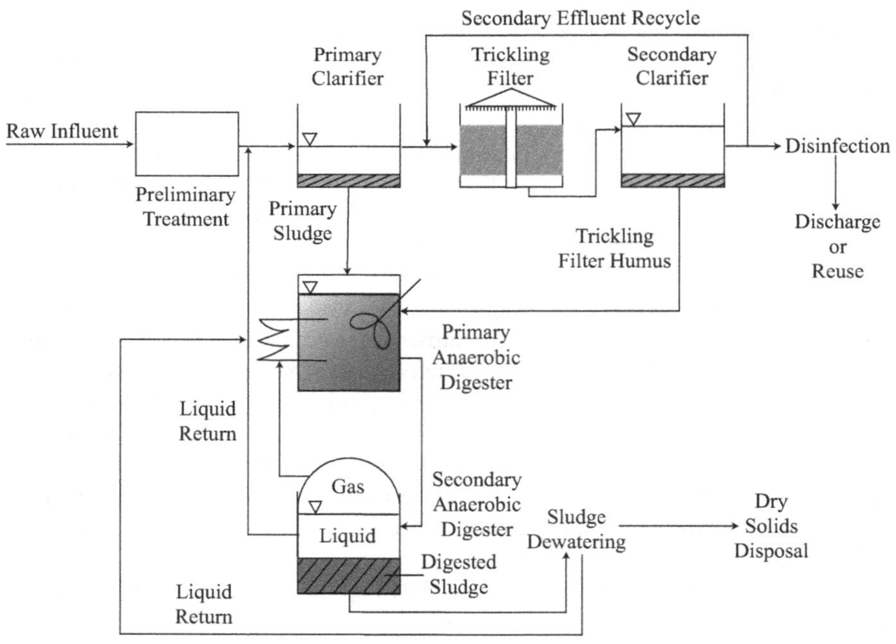

FIGURE 12.2 Layout for a fixed-film wastewater treatment plant.

3. Secondary treatment consisting of suspended growth (i.e., activated sludge or fixed-film, i.e., trickling filter units)
4. Sludge treatment, storage, and disposal facilities designed to provide primary sludge stabilization and dewatering as well as secondary solids handling and disposal (highlighted in Section 12.7).

12.5.1 Preliminary Treatment

Raw sewage enters the plant by means of gravity sewer systems and then enters preliminary treatment where screens or bar racks physically separate large objects such as rags and sticks from the raw sewage flow. Grinders and shredders called *comminutors* are also frequently incorporated into preliminary treatment facilities. The next phase of processing involves grit removal where materials such as sand, coffee grounds, and cigarette filter tips are separated in grit tanks or chambers. Grit removal can be accomplished in shallow rectangular tanks equipped with velocity control weirs to maintain horizontal velocities of 1 ft/s at varying flow rates to allow dense grit to settle by gravity from the bulk wastewater flow. Raking devices in these *horizontal grit chambers* allow the removal of settled grit from the bottom of the grit chambers. Grit removal in larger facilities can also be accomplished in *aerated grit chambers* in which air is introduced at the bottom of a 6 to 8 ft deep tank producing spiraling flow within the tank and allowing grit separation through centrifugal force. Finally, *vortex grit chambers* are designed to hydraulically generate centrifugal force for grit separation by directing flow to form a vortex within the treatment device. Grit tanks of any design are used for the following purposes:

1. Protect moving mechanical equipment from abrasion and abnormal wear
2. Reduce clogging in pipes and sludge hoppers
3. Prevent accumulations in aeration tanks and sludge digesters and the consequent loss of usable volume.

From the grit chambers, the raw sewage enters the primary clarifiers. The purpose of the primary clarifiers is to provide the detention time needed to separate settleable and floatable solids from the wastewater for subsequent stabilization and disposal. The effluent from the primary clarifiers, *primary effluent*, consists of dissolve and colloidal organic and inorganic materials not removable by gravity separation. Primary treatment provides approximately 50% TSS removal and 30% BOD removal. Further treatment by secondary treatment is required to reach the minimum secondary treatment standards required for modern wastewater treatment plants.

12.5.2 Secondary Treatment

The most common secondary treatment process is the suspended growth activated sludge process. The aeration tank is the heart of the activated sludge process. Here, oxygen is mixed with the effluent from the primary treatment process to satisfy the oxygen demand of the microbes (the activated sludge) in the aeration tank and to keep the sludge completely mixed in the aeration tank. These organisms use the carbon in the waste material remaining in the primary effluent as a source of energy and carbon for cell reproduction, and in this way, soluble and nonsettleable compounds are converted to settleable solids in the form of microbial biomass.

The biological solids grown in the aeration tank are settled out in secondary clarifiers. A portion of the settled solids is returned to the aeration tank, *return activated sludge*, to ensure sufficient biomass for the efficient removal of the incoming primary effluent, whereas excess settled solids are wasted, *waste activated sludge*, to maintain the required level of total active biomass in the system (Figure 12.1).

Aerobic lagoons, also labeled stabilization ponds or oxidation ponds, are another type of suspended growth system without biosolids recycle that is used for municipal wastewater treatment for small towns. They consist of large, shallow (3 to 5 ft deep) earthen basins that provide wastewater treatment by using natural processes involving both algae and bacteria. The objective is the microbial conversion of organic wastes into biological solids and stable end products. Aerobic conditions prevail throughout the process through oxygen transfer across a large water surface and through algal production of oxygen via photosynthesis.

In photosynthesis, oxygen produced by the algae is used by the bacteria in the biochemical oxidation and degradation of organic waste. Carbon dioxide, ammonia, phosphate, and other nutrients released in the biochemical oxidation reactions by the bacterial are, in turn, used by the algae, forming a symbiotic relationship between the algae and bacteria in the lagoon.

These suspended growth systems can be aerated using either *diffused* or *mechanical aerati*on. In the *diffused aeration* system, compressed air is introduced at the bottom of the aeration tank. This causes the tank's contents to be circulated by the air-lift effect. Some mechanical aerators use a blade to agitate the surface of the tank to create fine droplets that are oxygenated as they move through the atmosphere above the aeration tank. The continuous mechanical mixing of the aeration tank liquid, *mixed liquor*, disperses oxygenated droplets into the aeration tank, providing oxygen to microbes in the mixed liquor. Others circulate the mixed liquor by an updraft or downdraft pump or turbine that produces surface and subsurface turbulence, while at the same time injecting air through the circulating liquid.

Fixed-film or attached growth systems are a second commonly used secondary treatment system for municipal wastewater (Figure 12.2). These systems are designed to use microbial biomass that is attached to some type of media (i.e., rock, plastic sheets, plastic rings, etc.). In these attached growth systems, either the wastewater trickles over the surface of the media as in *trickling filters* or *biotowers*, or the medium is rotated through the wastewater as in *rotating biological contactors*. The process is related to the aerobic suspended growth process in that both depend on biochemical oxidation of organic matter in the wastewater to provide energy and a carbon source to promote the growth of bacteria and remove the oxygen demanding material from the wastewater.

There are several advantages of attached growth processes over other biological processes. First, due to dense microorganism growth in the biofilms that grow on the fixed-film media, these systems are inherently resistant to upsets from accidental spikes of toxic materials. Second, because oxygen is supplied naturally by convection of air through the unsaturated media, the need for aeration equipment is eliminated. As in activated sludge systems, secondary clarifiers follow the attached growth reactors, and a portion of the clarified effluent liquid is recycled back to the bioreactor to dilute the incoming wastewater, to provide additional liquid to help shear biosolids from the media to prevent clogging, and to provide additional contact time between the wastewater and attached biomass to improve overall BOD removal.

One disadvantage of attached growth treatment processes is that they experience operating difficulty in cold climates. Enclosing the units for temperature protection can lead to other problems such as excessive condensation and potentially corrosive conditions. These attached growth systems are also susceptible to clogging if dense media is used or high solid loadings are applied, and all of these attached growth systems require primary treatment before they can safely be used.

In general, aerobic attached growth processes are not as efficient nor as flexible as aerobic suspended growth processes in removing BOD, suspended solids, and toxic pollutants.

12.5.3 Disinfection

Secondary treatment as described can remove greater than 85% removal of BOD and TSS and up to 99% of the bacteria from raw wastewater. After treatment, however, the number of coliform bacteria in the effluent from trickling filters, activated sludge, and wastewater treatment lagoons can still be in the range of 100,000–10,000,000/100 mL. The final treatment step to ensure the minimization of disease transmission via treated wastewater is the use of disinfection for the destruction or inactivation of pathogenic organisms that may remain in secondary effluent. Disinfection of secondary effluent is required to achieve a maximum 30-day and 7-day log mean average of fecal coliform bacteria of 200 and 400/100 mL, respectively, to ensure the elimination of bacterial pathogens and inactivation of enteric viruses and other infectious pathogens below levels that pose risks of disease transmission to subsequently exposed human populations.

Chlorine is the disinfectant that has historically been used at wastewater treatment plants. Gaseous chlorine is the most common form used for wastewater disinfection and is applied in a rapid mix chamber at concentrations from 8 to 15 mg/L prior to entering a plug flow chlorine contact chamber to provide contact times of 20–30 minutes so the required disinfection performance can be produced. Chlorine can be toxic to aquatic life, so it is common that a de-chlorination step involving sodium metabisulfite be required to reduce chlorine residual to 0–0.5 mg/L before discharge.

Chlorine is effective and reliable but concerns regarding the storage of large quantities of hazardous materials (i.e., chlorine gas) at treatment facilities, and the potential for chlorine storage being targeted for terrorist activities (see Chapter 19) has made it a less desirable choice for modern wastewater treatment plant adoption. Chlorine also generates disinfection by-products (see Chapter 11), particularly in wastewater effluents that can be toxic to biota in receiving streams and that are known to be carcinogenic to humans. Because of these negative aspects of chlorine disinfection for wastewater, many new and upgraded facilities are moving to ultraviolet (UV) disinfection for final effluent treatment.

In UV disinfection systems, medium pressure UV lamps are placed parallel or perpendicular to the wastewater flow and expose the secondary effluent to high intensity UV radiation for 2–4 seconds. The DNA of microorganisms absorbs UV light in the 250–265 nm range, causing changes in this DNA, rendering potential microbial pathogens inactive and unable to reproduce and, subsequently, to transmit disease. These UV disinfection systems are applicable for large and small plants alike, and although they are expensive relative to chlorination systems due to power demand and lamp replacement costs, these systems have lower life-cycle costs because they do not produce a residual nor require a dechlorination step to protect receiving water ecosystems as chlorination does. The threat of foul play is also completely eliminated with the use of UV disinfection systems.

12.6 OVERVIEW OF ADVANCED WASTEWATER TREATMENT TECHNOLOGIES

Advanced wastewater treatment technology is designed to remove those constituents that are not adequately removed in the secondary treatment process. These include nitrogen, phosphorus, many synthetic organic compounds, and heavy metals. Nitrogen and phosphorus are nutrients that accelerate the growth of plants in the receiving waters. Ammonia is toxic to fish, exerts nitrogenous oxygen demand, and increases chlorine demand. Heavy metals, halogenated hydrocarbons, and phenolic compounds are toxic to fish and other aquatic life, concentrate in the food chain, and may create taste and odor problems in water supplies. Many of these constituents must be removed to meet stringent water quality standards and to allow reuse of the effluent for municipal, industrial, irrigation, recreation, and other beneficial water needs.

The most commonly used advanced wastewater treatment processes are chemical precipitation of phosphorus, nitrification, denitrification, ammonia stripping, filtration, carbon adsorption, ion exchange, and reverse osmosis. Filtration and carbon adsorption are highlighted in Chapter 11, ion exchange is highlighted in Chapter 13, and reverse osmosis is highlighted in Chapters 15 and 18. The rest of these unit operations and processes are discussed in this chapter.

12.6.1 Chemical Phosphorous Removal

As indicated in Chapter 13, a wide range of dissolved inorganic compounds can be removed from water via *chemical precipitation*. The overall process of chemical precipitation involves rapid mix of a chemical coagulant followed by flocculation of precipitated species to form larger and larger particles that can be easily removed from the water via sedimentation and filtration. Coagulation involves the reduction of surface charges and the formation of complex hydrous oxides. Flocculation involves combining the coagulated particles to form settleable floe. The coagulant (alum, ferric chloride, ferrous sulfate, ferric sulfate, etc.) is typically mixed rapidly to the plant's final effluent and then stirred to encourage formation of floc prior to settling. The objective of chemical precipitation in a wastewater context is to improve the removal of phosphorus, but the process also increases removal efficiency of residual particulate BOD and TSS as well. The chemical dosage is adjusted to give the desired amount of floc formation and phosphorus removal. The approximate average alum and ferric chloride dosages in municipal wastewater are 170 and 80 mg/L, respectively. Lime dosages of about 180–250 mg/L as CaO may also be used for chemical phosphorous removal. The effluent from the sedimentation tank is then carbonated by adding carbon dioxide (CO_2) to lower the pH to meet effluent standards prior to final discharge.

Major equipment for a chemical phosphorous removal system includes chemical storage; chemical feeders, piping, and control systems; flash mixer; flocculator; sedimentation basin; and recarbonation unit. Proper mixing of chemicals at the point of addition, along with proper flocculation prior to clarification, is essential for maximum effectiveness. Flocculation may be accomplished in a few minutes to half an hour in basins equipped with mixers, paddles, or baffles. The coagulants react with alkalinity to produce insoluble metal hydroxide for floc formation. If sufficient alkalinity is not present, lime or soda ash (sodium carbonate) is added in the required dosages.

In biological treatment units, the addition of coagulants has a marked influence on biota. Population of protozoans and higher animals is adversely affected. However, the BOD, TSS, and phosphorus removal is significantly improved. Overdosing of chemicals may cause toxicity to microorganisms necessary to the treatment process, so separate chemical addition to the final secondary effluent is recommended to prevent adverse impacts to the biological reactors that are essential for proper secondary treatment. Large quantities of chemical sludge will be produced from the chemical precipitation of phosphorous. The chemical sludges may cause serious handling and disposal problems, particularly if mixed with biological sludges generated from the secondary treatment system. Therefore, separate chemical sludge handling and dewatering is recommended.

12.6.2 Nitrification

Nitrification is the aerobic biological process that converts ammonia to nitrate, thus eliminating toxicity to fish and other aquatic life from ammonia and reducing the nitrogenous oxygen demand of the wastewater effluent. Ammonia oxidation to nitrite and then to nitrate is performed by autotrophic bacteria. The reactions are shown by Equations 12.1 and 12.2.

$$NH_3 + {}^{2}\!/_{3}\,O_2 \rightarrow NO_2 + H^+ + H_2O + \text{biomass} \tag{12.1}$$

$$NO_2 + {}^{1}\!/_{2}\,O_2 \rightarrow NO_3 + \text{biomass} \tag{12.2}$$

Temperature, pH, dissolved oxygen, and the ratio of BOD to total Kjeldahl nitrogen (TKN) are important factors affecting the extent and rate of nitrification.

12.6.3 Denitrification

Nitrite and nitrate are reduced to gaseous nitrogen by a variety of facultative heterotrophs in an anaerobic environment. A carbon source, such as acetic acid, acetone, ethanol, methanol, or sugar, is needed to act as an electron donor and to supply carbon for synthesis. Methanol is preferred because it is least expensive. Equations 12.3 through 12.5 express the basic reactions involved in denitrification. Sequenced nitrification-denitrification provides near complete biological nitrogen removal from municipal wastewater.

$$3\,O_2 + 2\,CH_3OH \rightarrow 2\,CO_2 + 4\,H_2O \tag{12.3}$$

$$6\,NO_3 + 5\,CH_3OH \rightarrow 3\,N_2 + 5\,CO_2 + 7\,H_2O + 6\,OH \tag{12.4}$$

$$2\,NO_2 + CH_3OH \rightarrow N_2 + CO_2 + H_2O + 2\,OH \tag{12.5}$$

12.6.4 Biological Phosphorous Removal

Biological phosphorus removal has also received considerable attention in recent years. Basic benefits reported for biological phosphorous removal include monetary savings through the obviated expense for chemical treatment. Biological phosphorous removal involves various sequential anaerobic, aerobic, and anoxic reactors in place of conventional aeration tanks. Based on the anaerobic, anoxic, and aerobic treatment sequence and internal recycling, a wide range of treatment systems have been developed that can be further reviewed in Water Environment Federation (2012), Metcalf & Eddy et al. (2014), and Viessman et al. (2014). More than 90% phosphorus and near-complete nitrogen removal (by nitrification and denitrification) has been reported using these biological nutrient removal approaches.

12.6.5 Ammonia Stripping

Ammonia gas can be removed from an alkaline solution by air stripping as expressed by Equation 12.6:

$$NH_4^{\,+} + OH^- \rightarrow NH_3\uparrow + H_2O \tag{12.6}$$

The basic equipment for an ammonia stripping system includes chemical feed, a stripping tower, a pump and liquid spray system, forced air draft, and a recarbonation system. This process requires raising the pH of the wastewater to approximately 11, forming droplets in the stripping tower, and providing air-water contact and droplet agitation by countercurrent circulation of large quantities of air through the tower. Ammonia stripping towers are simple to operate an can be effective in ammonia removal, but their efficiency is highly dependent on air temperature. As the air temperature decreases, the ammonia removal efficiency drops significantly. This process, therefore, is not recommended for use in a cold climate. A major operational disadvantage of stripping is the need for neutralization and prevention of calcium carbonate scaling on the tower. Also, there is some concern over discharge of ammonia into the atmosphere rather than its conversion to N_2 as is done via nitrification-denitrification.

12.7 SLUDGE DISPOSAL CONSIDERATIONS

The principal sources of sludge at a municipal wastewater treatment plant are the primary clarifiers that generate primary solids and the final clarifiers that produce secondary solids. Additional sludge may come from chemical precipitation, nitrification facilities, screening, and grinder devices. Sludge contains large volumes of water. The small fraction of solids in the sludge is highly offensive, however, thus requiring stabilization before final disposal. Common sludge management processes include thickening, stabilization, dewatering, and disposal.

Most wastewater treatment plants use primary sedimentation to remove readily settleable raw organic solids from the influent wastewater. In a typical plant, the dry weight of primary solids is roughly 50% of the total sludge solids. Primary sludge is usually easier to manage than biological and chemical sludges, which are produced in the secondary or advanced stages of treatment, for several reasons. Most importantly, primary sludge is readily thickened by gravity, either within a primary sedimentation tank or within a separate gravity thickener.

Primary sludge always contains some grit, even when the wastewater has been processed through degritting. Primary sludge production is typically in the range of 800–1,600 lb/MG (100–200 mg/L) of wastewater. A basic approach to estimating primary sludge production for a particular plant is by computing the quantity of TSS entering the primary sedimentation tanks and assuming a removal efficiency of 50%.

Biological sludges are produced by secondary treatment processes such as activated sludge. trickling filters, and rotating biological contactors. Quantities and characteristics of biological sludges vary with the metabolic and growth rates of the various microorganisms present in the sludge. Biological sludge containing debris, such as grit, plastics, paper, and fibers will be produced at plants lacking primary treatment. Plants with primary sedimentation normally produce a fairly pure biological sludge. Biological sludges are generally more difficult to thicken and dewater than are primary sludge and most chemical sludges due to the high water content of the microbial biomass generated in the secondary treatment process. This is particularly true for biosolids generated from suspended growth systems compared to those from attached growth systems such as trickling filters.

Sludge thickening is used to concentrate solids and reduce sludge volume. Thickened sludge requires less tank capacity and for stabilization and smaller piping and pumping equipment for transport. Common methods of sludge thickening used at medium-to-large plants are gravity thickening, dissolved air flotation, and centrifugation.

The main purpose of sludge stabilization is to reduce pathogens, eliminate offensive odors, and control the potential for putrification of organic matter in the sludge. Sludge stabilization can be accomplished by biological, chemical, or physical means. Selection of a stabilization method depends largely on the nature of the sludge being stabilized and the potential for capture and recovery of by-products generated in the stabilization process. Various methods of sludge stabilization include anaerobic or aerobic digestion (biological); chemical oxidation or lime stabilization (chemical); and thermal conditioning (physical). The anaerobic digestion process is used in most plants for primary solids and mixtures of primary and attached growth solids stabilization. Aerobic digestion it typically employed for the stabilization of secondary solids generated from suspended growth activated sludge systems.

The principal purpose of biological sludge digestion is twofold:

1. To reduce in volume the solids from the treatment process
2. To decompose highly putrescible organic matter to relatively stable or inert organic and inorganic compounds

In anaerobic digestion, tanks receiving the primary solids are covered to exclude oxygen, to prevent release of offensive odors, and to collect digester gases generated in the process of anaerobic decomposition of raw organic solids in the primary sludge. These gases are comprised mostly of methane (≈60% by volume) and carbon dioxide (≈40% by volume). This methane gas is routinely used by the treatment plant to preheat the primary solids entering the process and as an energy source to meet plant needs or to sell to the power grid if the energy generation capacity exceeds that needed by the facility.

Anaerobic digesters are most commonly operated at 30°C–35°C. Typically a two-stage, high-rate digestion system is used, with the first digester being heated and completely mixed. It is in this primary digester that the anaerobic conversion of high strength organic solids to product gases is accomplished over a typical detention time of 10–20 days. The second-stage digester is unheated and unmixed, typically with a floating cover, and is used as an anaerobic clarifier to separate the digested solids from the supernatant liquor and provide storage for gas generated in the primary-stage digester. A small amount of additional digestion and gas recovery may also be achieved in this second-stage digester.

Aerobic digesters are used for the stabilization of secondary solids. In this process, wasted sludge is pumped to an unheated and uncovered 15–20 ft deep tank and aerated for 10–20 days to encourage the endogenous respiration (use of internally stored substrate) of viable microorganisms remaining in the sludge. At the end of this aeration period, the very low levels of viable organism remain and the sludge is ready for final processing and disposal.

Ensuring the safe disposal of municipal sludge and other residues, such as screenings, grit, and skimmings, is considered an integral part of good planning, design, and management of municipal wastewater treatment facilities. Once biological solids are stabilized, they are processed for final disposal depending on the final disposal method being used. Stabilized sludge is increasingly being used as a fertilizer and soil conditioner for agricultural benefits and for restoring disturbed lands. Under this sludge management approach, digested sludge in semiliquid form is transported to the application areas. The slurry, containing nutrients from the waste, is subsurface injected for crop production or spread over the land to support revegetation of barren and disturbed land. If the stabilized solids are to be disposed of via land application by dry solids spreading or are to be landfilled or incinerated, some form of sludge dewatering will be required.

Sludge dewatering is designed to bring high water content stabilized solids at 8%–10% solids to a solids content of 18%–20% or more, so the sludge may be handled as a moist solid rather than a viscous liquid. Sludge dewatering at small facilities in arid climates is often carried out in sand drying beds where the stabilized sludge is deposited and the solids area allowed to freely drain and the water evaporate to produce a high solids content sludge that can be disposed of appropriately. For larger facilities, ≥1 million gallon/day (MGD), mechanical dewatering via vacuum filtration, pressure filtration, or centrifugation are commonly employed. These mechanical dewatering devices require the use of coagulants to aid in flocculating the stabilized solids to avoid fouling of the various belts, plates, or interior surfaces of the equipment and to generate dewatered solids cakes of 20%–30% solids. More details regarding these systems and their design can be found in the literature (Water Environment Federation 2012; Metcalf & Eddy et al. 2014; Viessman et al. 2014).

12.8 WASTEWATER REUSE OPTIONS

As the Earth's population continues to increase and the demand for all natural resources, including water, increases, interest is growing in the safe and cost-effective reuse of treated wastewater for both municipal and commercial/industrial/agricultural uses. The U.S. EPA (2012, 2017) has published guidelines for water reuse and is facilitating the development of a national water reuse action plan. Their *2012 Guidelines for Water Reuse* document (U.S. EPA 2012) contains an appendix that links to state reuse regulations for virtually every state in the United States and provides a compendium of water reuse case studies from the United States and across the globe. The EPA's *2017 Potable Reuse Compendium* (U.S. EPA 2017) supplements the *2012 Guidelines for Water Reuse* to inform current practices and approaches in wastewater reuse, including those related to direct potable water reuse and provides an updated compilation of technical information on potable reuse practices to provide planners and decision makers with a summary of the current state of the practice. The reader is referred to these documents for expansive detail on a range of water reuse practices both nationally and internationally. The following discussion provides a summary of the main points presented in these documents, highlighting the various reuse applications implemented across the globe.

A significant concept related to modern wastewater reuse practices is that of treatment "fit for purpose." This concept suggests that the appropriate level of treatment should be dictated by the end application of the reclaimed water for achieving economic efficiency and environmental sustainability. This concept suggests that for many nonpotable uses of reclaimed water (i.e., urban landscape and garden use) secondary treatment will be adequate to protect public health and the environment. Other applications, such as for cooling or industrial process water needs, may require additional tertiary treatment to reach required water quality levels. Direct potable water reuse would require advanced treatment to ensure near complete removal of microbial and trace metals and synthetic organic compounds and minimal risk to the public. Table 12.2 provides a summary of the defined categories of water reuse applications as per the U.S. EPA (2012).

TABLE 12.2
Categories of Water Reuse Applications as Defined by U.S. EPA (2012)

Category of Reuse		Description
Urban Reuse	Unrestricted	The use of reclaimed water for nonpotable applications in municipal settings where public access is not restricted
	Restricted	The use of reclaimed water for nonpotable applications in municipal settings where public access is controlled or restricted by physical or institutional barriers, such as fencing, advisory signage, or temporal access restriction
Agricultural Reuse	Food Crops	The use of reclaimed water to irrigate food crops that are intended for human consumption
	Processed Food Crops and Non-food Crops	The use of reclaimed water to irrigate crops that are either processed before human consumption or not consumed by humans
Impoundments	Unrestricted	The use of reclaimed water in an impoundment in which no limitations are imposed on body-contact water recreation activities
	Restricted	The use of reclaimed water in an impoundment where body contact is restricted
Environmental Reuse		The use of reclaimed water to create, enhance, sustain, or augment water bodies including wetlands, aquatic habitats, or stream flow
Industrial Reuse		The use of reclaimed water in industrial applications and facilities, power production, and extraction of fossil fuels
Groundwater Recharge	Nonpotable Reuse	The use of reclaimed water to recharge aquifers that are not used as a potable water source
Potable Reuse	Indirect Potable Reuse	Augmentation of a drinking water source (surface or groundwater) with reclaimed water followed by an environmental buffer that precedes normal drinking water treatment
	Direct Potable Reuse	The introduction of reclaimed water (with or without retention in an engineered storage buffer) directly into a water treatment plant, either collocated or remote from the advanced wastewater treatment system

The required level of treatment for these specific reuse applications can vary widely among the states, as summarized in detail by the U.S. EPA (2012). As an example, for unrestricted urban reuse, the state of Arizona requires secondary treatment followed by effluent filtration and disinfection, and the state of Nevada requires only secondary treatment with disinfection. For the same use, Arizona requires a 24-hourr average turbidity not to exceed 2 NTU; Nevada requires only a final TSS to be no more than 30 mg/L on a 30-day average; North Carolina specifies only filtration or equivalent and requires a 10 nephelometric turbidity unit (NTU) maximum turbidity and 5 mg/L monthly average TSS; and Texas has no minimum treatment process requirement, no TSS limit and a 3 NTU standard. The level of treatment required for a specific end use is highly dependent on the state requirements in which the reuse application is to take place, and reference to current state regulations is essential for the evaluation and design of a water reuse project. Despite these wide-ranging state-level requirements, Table 12.3 summarizes a general guideline regarding the minimum level of treatment and corresponding treatment technologies that are appropriate for the range of water reuse options defined in Table 12.2. As this table indicates, as the level of treatment increases, so do the costs and the acceptable levels of human exposure to the treated water that can be tolerated.

TABLE 12.3
Type of Water Reuse Appropriate for Increasing Levels of Treatment

	Increasing Level of Treatment →			
Treatment Level	**Primary**	**Secondary**	**Filtration and Disinfection**	**Advanced (Tertiary)**
Processes	Sedimentation	Biological oxidation and disinfection	Chemical coagulation, biological or chemical nutrient removal, filtration, and disinfection	Activated carbon, reverse osmosis, advanced oxidation processes, soil aquifer treatment, etc.
End Use	No uses recommended	Surface irrigation of orchards and vineyards Nonfood crop irrigation Restricted landscape impoundments Groundwater recharge of nonpotable aquifer Wetlands, wildlife habitat, stream augmentation Industrial cooling processes	Landscape and golf course irrigation Toilet flushing Vehicle washing Food crop irrigation Unrestricted recreational impoundment Industrial systems	Indirect potable reuse including groundwater recharge of potable aquifer and surface water reservoir augmentation and potable reuse
Human Exposure	Increasing Acceptable Levels of Human Exposure →			
Cost	Increasing Levels of Cost →			

Source: U.S. Environmental Protection Agency, *2012 Guidelines for Water Reuse*, EPA/600/R-12/618, U.S. Environmental Protection Agency, Office of Wastewater Management, Office of Water, U.S. Agency for International Development, Washington, DC, https://nepis.epa.gov/Adobe/PDF/P100FS7K.pdf, 2012.

12.9 APPLICATIONS

Six Illustrative Examples complement the material presented in this chapter related to the concepts and practices of municipal wastewater treatment and reuse.

Illustrative Example 12.1

A municipal wastewater contains 100 mg/L of organic material composed of protein, carbohydrates, and fats and oils representative of domestic wastewater with the molecular formula, $C_{10}H_{19}O_3N$. What is the theoretical oxygen demand (ThOD) of this wastewater? The molecular weight of this representative domestic wastewater compound = 201 mg/mgmol. Note that the ThOD is a calculated oxygen demand that assumes oxidation of all species to their most highly oxidized stable forms such as CO_2, H_2O, NO_3, etc., and is roughly equivalent to the ultimate BOD of the wastewater.

Solution

The number of moles, n, of this representative organic compound in 1 L is as follows:

$$n = (100 \text{ mg/L})/(201 \text{ mg/mgmol}) = 0.50 \text{ mgmol}$$

Note that the units are milligram moles because the concentration is given in mg/L. Convert all of the species to their most highly oxidized stable form and balance the equation as:

$$C_{10}H_{19}O_3N + 14.75\ O_2 \rightarrow 10\ CO_2 + 9.5\ H_2O + NO_3 \qquad (12.7)$$

The ThOD for 0.50 mgmol of municipal wastewater is calculated employing Equation 12.7 as:

$$\text{ThOD} = (14.75\ \text{mgmol})\ O_2/(1\ \text{mgmol}\ C_{10}H_{19}O_3N)\ (0.50\ \text{mgmol}\ C_{10}H_{19}O_3N)$$

$$\text{ThOD} = 7.38\ \text{mgmol}\ O_2 = 7.38\ \text{mgmol}\ (32\ \text{g}\ O_2/\text{mgmol}) = 236\ \text{mg/L}\ O_2$$

The ThOD for this wastewater is 236 mg/L.

Illustrative Example 12.2

A municipal wastewater treatment plant treats a daily average flow of 10 MGD. BOD and TSS coming into the plant average 200 mg/L each. The primary clarifier of this plant achieves a 50% TSS and 30% BOD removal. Calculate the pounds of TSS and BOD removed in the primary clarifier on a daily basis.

Solution

The pounds of any pollutant removed by a unit process can be calculated using the following equation:

$$\text{lb removed/d} = (\text{Flow, MGD})\ (\Delta\text{Conc, mg/L})(8.34\ \text{lb/MG/mg/L}) \qquad (12.8)$$

For this example, the pounds of TSS and BOD removed on a daily basis are found using Equation 12.8 to be as follows:

$$\text{lb TSS removed/d} = (10\ \text{MGD})(0.5 \times 200\ \text{mg/L})(8.34\ \text{lb/MG/mg/L}) = 8{,}340\ \text{lb/d}$$

$$\text{lb BOD removed/d} = (10\ \text{MGD})\ (0.3 \times 200\ \text{mg/L})(8.34\ \text{lb/MG/mg/L}) = 5{,}004\ \text{lb/d}$$

Illustrative Example 12.3

The municipal wastewater treatment plant in Illustrative Example 12.2 uses anaerobic digesters to stabilize the primary solids generated from the primary clarifier. The hydraulic retention time to be used for sizing the anaerobic digester is to be no less than 20 days, and the volatile solids loading is to be no more than 0.2 lb VS/ft^3/d. Assume the primary solids are 70% volatile, that the specific gravity of wet sludge, SG_{ws} = 1.0 for solids contents less than 10%, and use Equation 12.9 to determine the volume of primary solids generated on a daily basis if the solids content at the bottom of the primary clarifier is 6% solids. Based on meeting both the hydraulic retention time and volatile solids loading design criteria, determine the size of the primary anaerobic digester for this plant.

$$\text{Volume of wet sludge} = \frac{\text{Pounds of Dry Solids}}{(\%\ \text{Solids Content})(\gamma_{\text{water}})(SG_{ws})} \qquad (12.9)$$

Solution

The volume of the wet sludge generated from the primary clarifier on a daily basis using Equation 12.9 is determined to be:

$$\text{Volume} = \frac{8{,}340\ \text{lb/d}}{(0.06)\ (62.4\ \text{lb/ft}^3)(1.0)} = \frac{8{,}340\ \text{lb/d}}{3.74\ \text{lb/ft}^3} = 2{,}230\ \text{ft}^3/\text{d}$$

The daily loading of volatile solids from the primary clarifier to the primary anaerobic digester is:

$$\text{lb VS/d} = \text{lb TSS/d (\%VS)} = (8{,}340 \text{ lb TSS/d})\ (0.70) = 5{,}838 \text{ lb VS/d}$$

The primary digester size based on the required hydraulic retention time for the sludge is:

$$\text{Digester Volume, ft}^3 = (2{,}230 \text{ ft}^3\text{/d})(20 \text{ d}) = 44{,}600\text{ft}^3$$

The primary digester size based on the maximum VS loading is:

$$\text{Digester Volume, ft}^3 = (5{,}838 \text{ lb VS / d}) / (0.2 \text{ lb VS / ft}^3 \text{ / d}) = 29{,}190 \text{ ft}^3$$

The hydraulic retention time criteria controls the size of the primary digester, which must be 44,600 ft^3. Based on this digester volume, the actual VS loading to the digester will be:

$$\text{Actual digester VS Loading} = (5{,}838 \text{ lb VS/d})/(44{,}600 \text{ ft}^3) = 0.13 \text{ lbVS/ft}^3\text{/d}$$

Illustrative Example 12.4

A municipal wastewater treatment plant in northern Utah is considering developing a water reuse program that would pump treated secondary effluent from a membrane bioreactor plant into a pressurized secondary water system. This secondary water system would blend water from a surface reservoir with treated effluent and provides outdoor water taps to each home in the community for lawn and garden watering, car washing, and other outdoor uses around the home. The municipal membrane bioreactor plant incorporates preliminary treatment, modified secondary treatment with anoxic and aerobic basis plus alum addition for phosphorous removal, membrane separation of solids and secondary effluent, and UV disinfection prior to effluent discharge. During the irrigation season, the plant would not be required to provide chemical addition for phosphorous removal, saving the municipality approximately $60,000 a year in chemical costs. Based on the recommended treatment level for this form of water use from Table 12.2, is this a wise move on the part of the municipality?

Solution

The recommended level of treatment in Table 12.2 for these types of water uses and levels of human exposure would be "Filtration & Disinfection," which includes chemical coagulation, biological or chemical nutrient removal, filtration, and disinfection. Based on the description of the municipal wastewater treatment process it appears that these basic steps are included, especially with the membrane solids separation that is provided, and the treated effluent would be a good candidate for augmentation of surface water sources currently being used for this secondary irrigation use. This appears to be a particularly good option because the pressurized secondary water system is already in place and effluent can be directly pump from the treatment plant to the secondary system with little additional infrastructure needed.

Illustrative Example 12.5

The municipal wastewater treatment plant in Illustrative Example 12.4 has data that suggests some regrowth of microbial pathogens and evidence of part per trillion levels of some pharmaceuticals and personal care products (PPCPs) including carbamazepine, ibuprofen, sulfamethoxazole, triclosan, and progesterone within the pressurized secondary water distribution system in the zone near the wastewater treatment plant. Review data in Chapter 6 of the EPA (2012) water reuse guidelines, specifically Table 6.5 in this document and recommend an approach to deal with both bacterial regrowth and further reduction of these PPCPs.

Solution

The removal efficiency of a range of PPCPs as affected by a range of treatment technologies from conventional secondary treatment to advanced oxidation processes is summarized in Table 6.5 of EPA (2012). Only a few of these technologies are appropriate for both microbial pathogen reduction and PPCP removal, and only one technology, chlorination, provides a disinfection residual that would provide protection against microbial regrowth in the secondary water distribution system. Table 6.5 indicates that in addition to providing the disinfection residual, it can also provide >80% removal of sulfamethoxazole, triclosan, and progesterone. It provides less than 20% removal of carbamazepine and ibuprofen, but both compounds have high clinical dose levels thousands of times higher than doses that might occur from exposure to the reuse water, and application of a posttreatment chlorine residual of 0.5–1.0 mg/L is recommended to reduce the potential for microbial regrowth and provide additional removal of the antibiotics and hormones in the treated effluent.

Illustrative Example 12.6

The municipal wastewater treatment plant in Illustrative Example 12.3 uses a belt filter press for the dewatering of their stabilized primary solids. Assume that the VS destruction in the primary digester is 60%, that the solids removed from the secondary digester as feed to the belt filter press is 8% solids, that the coagulant is added to the press at a dose of 0.05 lb/lb dry solids, and that the final solids content of the cake produced from the belt press is 18%. Determine the volume of the dewatered solids, the volume of water from the filter press returned to the front of the plant, and the volume reduction of digested sludge feed to the filter press. Use Equations 12.10 and 12.11 to determine the specific gravity of the wet sludge for input to Equation 12.9 for the final filter press cake.

$$\text{Specific gravity of dried sludge, } SG_{ds} = \frac{250}{100 + 1.5(\%VS)} \tag{12.10}$$

$$\text{Specific gravity of wet sludge, } SG_{ws} = \frac{100(SG_{ds})}{100 + \%\text{Water}(SG_{ds} - 1)} \tag{12.11}$$

where %VS is the numeric value of the percentage of volatile solids in the digested sludge, and % water is the numeric value of the percentage of water in the dewatered solids.

Solution

The %VS in the digested sludge from Illustrative Example 12.3 is based on a 0% reduction of the nonvolatile solids (NVS) in the sludge and 70% reduction of the volatile solids in the sludge. The NVS and % VS after digestion are found by:

$$\text{lb NVS} = \text{lb TSS} - \text{lb VS} = \text{lb TSS}(1 - \%\text{VS}) = 8{,}340\ \text{lb/d}(1 - 0.70) = 2{,}502\ \text{lb/d}$$

$$\text{lb VS remaining after digestion} = \text{lb VS in }(1 - \%\text{VS Destruction}) = 5{,}838\ \text{lb/d}(1 - 0.60)$$

$$= 5{,}838\ \text{lb/d}(0.40) = 2{,}335\ \text{lb/d}$$

$$\text{Total solids after digestion} = \text{lb/d NVS} + \text{lb/d VS} = 2{,}502\ \text{lb/d} + 2{,}335\ \text{lb/d} = 4{,}837\ \text{lb/d}$$

$$\%\text{VS after digestion} = (\text{lb/d VS})/(\text{lb/d TS})(100) = (2{,}335\ \text{lb/d})/(4{,}837\ \text{lb/d})(100) = 48.3\%$$

The SG_{ds} from Equation 12.10 is found to be:

$$SG_{ds} = \frac{250}{100 + 1.5(48.3)} = \frac{250}{172.45} = 1.45$$

The SG_{ws} is then found from Equation 12.11 as:

$$SG_{ws} = \frac{100\ (1.45)}{100 + 82\ (1.45 - 1)} = \frac{145}{136.9} = 1.06$$

The volume of the dried sludge is then determined from Equation 12.9. The pounds of dry solids must account for the digested solids plus the dry solids added as a coagulant. The total pounds of dry solids loaded to the belt press daily is:

$$\text{Total lb dry solids / d} = 4{,}847\ \text{lb / d}\ (1 + 0.05) = 5{,}089\ \text{lb dry solids / d}$$

The volume of the dewatered, stabilized sludge is then:

$$\text{Volume of dewatered sludge} = \frac{5{,}089\ \text{lb/d}}{(0.18)(62.4\ \text{lb/ft}^3)(1.06)} = \frac{5{,}089\ \text{lb/d}}{11.9\ \text{lb/ft}^3} = 428\ \text{ft}^3/\text{d}$$

The volume of water returned to the front of the plant from the belt press is the difference between the sludge volume loaded to the belt press and the volume of dried cake produced. The volume of the stabilized sludge fed to the belt press is:

$$\text{Volume to the belt press} = \frac{4{,}837\ \text{lb/d}}{(0.08)(62.4\ \text{lb/ft}^3)(1.0)} = \frac{4{,}837\ \text{lb/d}}{4.99\ \text{lb/ft}^3} = 969\ \text{ft}^3/\text{d}$$

Therefore,

$$\text{Volume returned to the front of the plant} = 969\ \text{ft}^3/\text{d} - 428\ \text{ft}^3/\text{d} = 541\ \text{ft}^3/\text{d}$$

The overall volume reduction of sludge through the belt filter press is found to be:

$$\%\ \text{Volume Reduction} = \frac{(969 - 428\ \text{ft}^3/\text{d})}{969\ \text{ft}^3/\text{d}}(100) = 55.8\%$$

The overall volume reduction of sludge through stabilization and dewatering based on 2,230 ft^3/d of sludge loaded to the primary anaerobic digester as indicated in Illustrative Example 12.3 is found to be:

$$\%\ \text{Volume Reduction} = \frac{(2{,}330 - 428\ \text{ft}^3/\text{d})}{2{,}330\ \text{ft}^3/\text{d}}(100) = \frac{1{,}902}{2{,}330} = 81.6\%$$

REFERENCES

Metcalf & Eddy, Inc., G. Tchobanoglous, F. Burton, and H. D. Stensel. 2014. *Wastewater Engineering: Treatment and Reuse*, 5th edition. New York: McGraw-Hill.

U.S. Environmental Protection Agency. 2012. *2012 Guidelines for Water Reuse*. EPA/600/R-12/618. Washington, DC: U.S. Environmental Protection Agency, Office of Wastewater Management, Office of Water, U.S. Agency for International Development. https://nepis.epa.gov/Adobe/PDF/P100FS7K.pdf.

U.S. Environmental Protection Agency. 2017. *2017 Potable Reuse Compendium*. EPA/810/R-17/002. Washington, DC: Office of Ground Water and Drinking Water, Office of Water. https://www.epa.gov/sites/production/files/2018-01/documents/potablereusecompendium_3.pdf.

Viessman, W. Jr., M.J. Hammer, E.M. Perez, and P.A. Chadik. 2014. *Water Supply and Pollution Control. Pearson New International Edition*. Harlow, UK: Pearson Education Limited.

Water Environment Federation. 2012. *Design of Municipal Wastewater Treatment Plants MOP 8*, 5th edition. Alexandria, VA: Water Environment Federation.

13 Industrial Wastewater Treatment

13.1 INTRODUCTION

This chapter on industrial wastewater treatment is a companion to Chapter 12, which dealt with municipal wastewater treatment systems. Although municipal wastewater is similar from communities across the country, industrial wastewater, and consequently, the processes required to treat it can differ markedly both within and among industries. The impact of industrial discharges depends not only on their collective characteristics such as biochemical oxygen demand and the concentration of suspended solids but also on their content of specific inorganic and organic constituents. Three options, which are not mutually exclusive, are available in treating industrial wastewater: (i) treatment at the point of generation (at the process level) within the plant, (ii) pretreatment of the industrial waste prior to discharge to a municipal wastewater treatment plant, and (iii) treatment of mixed industrial wastewater streams at an industrial wastewater treatment plant to the point that it can be discharged directly to a receiving stream or so that it can be reused within the plant.

To meet industrial wastewater management objectives, four major areas of concern must be addressed (Metcalf and Eddy et al. 2013):

1. Sources and characteristics of wastes and wastewaters must be ascertained.
2. The particular type of wastewater treatment process, or combination of processes must be determined.
3. Wastewater effluent must be effectively controlled.
4. Sludge and other solid wastes must be properly managed.

This chapter provides a general overview of these four areas of concern through topics that include sources and characterization of industrial wastewater; a description of physical, chemical, and biological industrial wastewater treatment processes; options for management of industrial wastewater effluent; sludge management; and future developments in industrial wastewater treatment. An application section provides six Illustrative Examples of industrial wastewater treatment concepts and treatment process trains.

13.2 SOURCES AND CHARACTERIZATION OF INDUSTRIAL WASTEWATER

13.2.1 Types of Pollutants

The various types of pollutants found in industrial wastewaters can be classified into eight categories: common sewage and other oxygen-demanding wastes; disease-causing agents; nutrients; synthetic organic chemicals; inorganic chemicals and other mineral substances; sediments; radioactive substances; and heat (U.S. EPA 1980).

Oxygen-demanding wastes are the traditional organic wastes, ammonia, iron, or any other oxidizable compound contributed by industrial wastes. Such wastes result from food processing, paper mill production, tanning, and other manufacturing processes. These wastes are usually destroyed by bacteria if there is sufficient oxygen present in the water. Because fish and other aquatic life depend on oxygen for life, the oxygen-demanding wastes must be controlled otherwise this aquatic life will die.

The disease-causing agents include infectious organisms that are carried into surface water and groundwater by certain kinds of industrial wastes, such as tanning and meat-packing plants. Humans or animals may come in contact with these pathogens either by drinking the water or through swimming, fishing, or other activities. Although modern disinfection techniques have greatly reduced the danger of this type of pollutant, these effluents must be carefully monitored. Nutrients are the substances in the food chain of aquatic life (such as algae and water weeds) that support and stimulate plant growth. Carbon, nitrogen, and phosphorus are the three chief nutrients present in natural waters. Large amounts of these nutrients are present in sewage, certain industrial wastes, and drainage from fertilized land. Biological wastewater treatment processes do not remove phosphorus and nitrogen to any substantial extent. In fact, biological wastewater treatment processes convert the organic forms of these substances into mineral forms, making them more usable by plant life. The problem starts when an excess of these nutrients overstimulates the growth of water plants (particularly algae). When the plants and algae eventually die, the decomposition of their organic matter can lead to a severe depletion of dissolved oxygen in the body of water, a process called *eutrophication.*

Synthetic organic chemicals include pesticides, synthetic industrial chemicals, and wastes from their manufacture. Many of these substances are toxic to aquatic life and are possibly harmful to humans as well. They can cause taste and odor problems in drinking water, and they may resist water and wastewater treatment. Some are known to be highly toxic at very low concentrations.

A vast array of metal salts, acids, solid matter, and many other chemical compounds are included in the category of inorganic pollutants. Their sources include mining and manufacturing processes, oil field operations, and agricultural practices. Although a wide variety of acids are discharged as waste by industry, the largest single source comes from mining operations and mines that have been abandoned.

Many of these types of chemicals are being created each year. If untreated, they can interfere with natural stream purification, destroy fish and other aquatic life, and cause excessive hardness of water supplies. Moreover, such chemicals can corrode expensive wastewater treatment equipment and increase the cost of wastewater control. The Environmental Protection Agency (EPA) has developed a list of 129 priority pollutants that include both toxic organic and inorganic chemicals. The heightened danger of these chemicals has necessitated the drafting of specific limitations for these chemicals in most wastewater treatment plant permits.

Although not as dangerous as some other types of pollutants, sediments arc a major problem because of the sheer magnitude of the quantities reaching waterways. Sediments fill stream channels and harbors, requiring expensive dredging. They also fill reservoirs, reducing their capacity and useful life. They erode power turbines and pumping equipment, and they reduce fish and shellfish populations by blanketing fish spawning areas and food supplies.

Radioactive pollution results from the mining and processing of radioactive ores, from the use of refined radioactive materials in power reactors, and from industrial, medical, and research purposes. Because radiation bioaccumulates, control of this type of pollution must take into consideration total exposure in the environment.

Although not directly toxic, heat reduces the capacity of water to absorb oxygen. Tremendous volumes of water are used by power plants and industry for cooling. Most of the water, with the added heat, is returned to natural bodies of water, raising their temperatures. With less oxygen, the water is not as efficient in assimilating oxygen-consuming wastes and in supporting fish and aquatic life. Unchecked discharges of waste heat can seriously alter the ecology of a lake, a stream, or even tidal areas.

To complicate matters, most industrial wastes are a mixture of these eight types of pollution. Such mixtures make the problems of treatment and control that much more difficult and costly.

13.2.2 Characterization of Wastewater

The volume and strength of industrial wastewaters are usually defined in terms of units of production. For example, for a pulp-and-paper mill, waste is often measured in gal/ton of pulp. However, because of variations in the production process as well as differences in housekeeping and water reuse, considerable variations may occur in the flow and characteristics of wastewaters of similar industries. In fact, very few plant within a particular industry use identical process operations. Thus, in the development of an effective strategy for wastewater control, an industrial waste survey is usually required to determine the character of a particular industry's waste load (Eckenfelder et al. 2009).

The industrial waste survey involves a definite procedure designed to develop a flow and material balance of all processes using water and producing wastes and to establish the variation in waste characteristics from specific process operations as well as from the plant as a whole. The results of the survey should establish possibilities for water conservation and reuse, and the variation in flow and strength to undergo wastewater treatment.

The general procedure to be followed in developing the necessary information with a minimum of effort can be summarized in four steps:

1. Develop a sewer map from consultation with the plant engineer and an inspection of the various process operations. This map should indicate possible sampling stations and a rough order of magnitude of the anticipated flow.
2. Establish sampling and analysis schedules. Continuous samples with composites weighted according to flow are the most desirable, but these either are not always possible or do not lend themselves to the physical sampling location. The period of sample composite and the frequency of sampling must be established according to the nature of the process being investigated.
3. Develop a flow-and-material-balance diagram. After the survey data are collected and the samples analyzed, a flow-and-material-balance diagram should be developed that considers all significant sources of waste discharge. How closely the summation of the individual sources checks the measured total effluent provides a check on the accuracy of the survey.
4. Establish statistical variation in significant waste characteristics. The variability of certain waste characteristics is significant for wastewater treatment plant design. These data should be prepared as a probability plot showing frequency of occurrence.

Data from industrial waste surveys are highly variable and are usually subjected to statistical analysis. Statistical analysis of variable data provides the basis for wastewater treatment plant process design.

13.3 DETERMINATION OF WASTEWATER CONTAMINANTS

As implied in the industrial waste survey, the objective of water quality management is to control the discharge of pollutants so that water quality in the receiving stream is not degraded to an unacceptable extent below the natural background level. Direct measurement of the pollutants must be made. The impact of the pollutants on water quality must be predicted and compared to the background water quality, which would be present without human intervention. Based on this, a decision can be made as to the pollutant levels acceptable for the intended uses of the water (see Section 6.4 Water Quality Standards and Section 6.5 Water Quality Criteria), and the level of treatment required by an industrial discharger to meet these acceptable pollutant levels (see Section 6.6 Total Maximum Daily Loads).

The *biochemical oxygen demand* (BOD) test is used to determine the amount of biodegradable organic material in a sample of wastewater. As discussed in Chapter 6, the BOD test measures the amount of oxygen consumed by living organisms (mainly bacteria) as they metabolize the organic matter present in a waste. As is true for any bioassay, the success of a BOD test depends on the control of such environmental and nutritional factors as pH and osmotic conditions, essential nutrients, constant temperature, and population of organisms representative of natural conditions.

Although it theoretically takes an infinite amount of time for all the oxidizable organic material in a sample of water to be consumed, it has been empirically determined that a period of 20 days is required for near completion. Because the 20-day waiting period is too long to wait in most cases, a 5-day incubation period (the time in which ≈68% of available material is usually oxidized) has been adopted as standard procedure. This shorter test is referred to as BOD_5. The $BODS_5$ test also serves the purpose of avoiding the contribution of nitrifying bacteria to the overall dissolved oxygen (DO) measurement because these nitrifying bacterial populations do not become large enough to express a significant oxygen demand until about 7–10 days from the start of the BOD test (Metcalf & Eddy et al. 2013).

Although the BOD test is the only test presently available that gives a measure of the amount of biologically oxidizable organic material present in a body of water, it does not provide an accurate picture of the total amount of overall oxidizable material and, thus, the total oxygen demand. For such measurements, the chemical oxygen demand (COD) test is used. In this test, a very strong oxidizing chemical, usually potassium dichromate, is added to samples of different dilution. To ensure full oxidation of the various compounds found in the samples, a strong acid and a chemical catalyst are added. One of the products of this oxidation-reduction reaction is the chromate ion, which causes a very sharp color change that can be easily detected via a spectrophotometer. Thus, the greater the absorbance measured (inverse of transmittance), the greater the amount of oxidation that has taken place; also, the more the oxidizable material originally present, and the greater the chemical oxygen demand of that material. To find the corresponding concentration of DO in the sample, the absorbance measured is related to an absorbance-concentration graph of a standard whose oxygen concentration is known.

A third method for measuring the organic matter present in a wastewater is the total organic carbon (TOC) test. The test is performed by placing a sample into a high-temperature furnace or chemically oxidizing environment. The organic carbon is oxidized to carbon dioxide. The carbon dioxide that is produced can then be measured. Although the TOC test does directly measure the concentration of organic compounds, it does not provide a direct measurement of the rate of reaction or the degree of biodegradability. For this reason, the TOC test has been accepted as a monitoring technique but has not been used in the establishment of treatment regulations.

The BOD, COD, and TOC tests provide estimates of the general organic content of a wastewater. However, because the particular composition of the organics remains unknown, these tests do not reflect the response of the wastewater to various types of biological treatment technologies. It is therefore necessary to separate the wastewater into its specific components.

The solids analysis focuses on the quantitative investigation and measurement of the specific content of solid materials present in a wastewater. The total solids are the materials that remain after the water from the solution has evaporated. For any solution, the total solids consist of suspended and dissolved particles. Dissolved particles are classified according to size as either soluble or colloidal.

When a solution is poured through a filter (usually 0.45 μm in size), those solids that are too large to pass through the filter openings will be retained. These solids are called *suspended solids*. The filter containing the suspended solids is heated to 103°C to remove the water and to quantify the total suspended solids (TSS). The TSS are further heated in a muffle furnace at 550°C. At such high temperatures, the volatile suspended solids (VSS; i.e., organics) will be vaporized, leaving behind the remaining fixed or nonvolatile suspended solids (NVSS). Those solids that are small enough to pass through the filter are called dissolved (soluble) solids. By heating this solution to 103°C to remove the water, the total dissolved solids (TDS) are obtained. As with the suspended solids, these solids can be further heated to 550°C to combust the volatile dissolved solids (VDS), leaving the fixed or nonvolatile dissolved solids (NVDS).

The assumption is made that all volatile solids (VSS and VDS) are organic compounds. A distinction is made as to whether such solids may or may not be adsorbed by a specific treatment media. The soluble organics that are not sorbed can be further separated into degradable and nondegradable constituents.

The EPA has also established guidelines and developed standard methods for the analysis of the many and varied disease-causing agents, nutrients, synthetic organic chemicals, inorganic chemicals, and other mineral and radioactive substances that can exist in industrial wastewaters. The reader is directed to these EPA-approved Clean Water Act methods at the following URL: www.epa.gov/cwa-methods.

13.4 INDUSTRIAL WASTEWATER TREATMENT PROCESSES

Numerous technologies exist for treating industrial wastewater. These technologies range from simple clarification in settling ponds to a complex system of advanced technologies requiring sophisticated equipment and skilled operators. Finding the proper technology or combination of technologies to treat a particular wastewater to meet federal and local requirements and still be cost effective can be a challenging task. To facilitate this effort, the EPA (2017) has recently released the Industrial Wastewater Treatment Technology (IWTT) Database as a publicly accessible web application at www.epa.gov/iwtt. IWTT provides industrial wastewater treatment technology performance data from peer-reviewed journal articles, conference proceedings, industry trade organizations, and other government data sources as a web-based database application searchable by industry, treatment technology, or pollutant. The database can also be downloaded for offline searching and analysis through the main IWIT webpage. The information here provides a summary of the most widespread physical, chemical, and biological wastewater treatment technologies that are used in the industrial field (Figure 13.1). The reader is directed to the this IWIT database for a more comprehensive presentation of relevant treatment technologies.

13.4.1 Physical Treatment Processes

13.4.1.1 Clarification (Sedimentation)

When an industrial wastewater containing a suspension of solid particles that have a higher specific gravity than the transporting liquid is in a relatively calm state, the particles will settle out because of the effects of gravity. This process of separating the settleable solids from the liquid is called *clarification* or *sedimentation*. In some treatment systems employing two or more stages of treatment and clarification, the terms *primary*, *secondary*, and *final* clarification are

CANDIDATE TREATMENT TECHNOLOGY / INDUSTRY TYPE	ADHESIVES AND SEALANTS	ALUMINUM FORMING	AUTO & OTHER LAUNDRIES	BATTERY MANUFACTURING	COAL MINING	COIL COATING	COPPER FORMING	ELECTRICAL PRODUCTS	ELECTROPLATING	EXPLOSIVES MANUFACTURING	FOUNDRIES	GUM & WOOD CHEMICALS MANUFACTURING	INORGANIC CHEMICALS MANUFACTURING	IRON & STEEL MANUFACTURING	LEATHER TANNING & FINISHING	MECHANICAL PRODUCTS	NONFERROUS METALS MANUFACTURING	ORE MINING & DRESSING	ORGANIC CHEMICALS	PAINT & INK FORMATION	PESTICITES MANUFACTURING	PETROLEUM REFINING	PHARMACEUTICALS MANUFACTURING	PHOTOGRAPHIC PROCESSES	PLASTIC & SYNTHETIC MATERIALS	PLASTIC PROCESSING	PORCELAIN ENAMELLING	PRINTING & PUBLISHING	PULP, PAPER, & PAPERBOARD	RUBBER	SOAP & DETERGENT MANUFACTURING	STEAM ELECRIC	TEXTILE MILLS	TIMBER PRODUCTS PROCESSING
BIOLOGICAL TREATMENT	●					●				●		●		●	●				●	●	●	●	●		●	●			●	●	●		●	●
CLARIFICATION			●		●	●	●				●			●		●	●	●		●					●		●		●					●
FLOTATION	●		●			●			●							●										●	●	●		●	●			
OIL-WATER SEPARATION		●									●					●										●								●
COAGULATION-PRECIPITATION	●	●		●	●	●		●	●		●	●	●	●	●	●	●	●	●	●	●	●		●		●	●	●		●	●	●	●	
NEUTRALIZATION					●	●	●							●			●				●			●			●							●

FIGURE 13.1 Candidate treatment technologies for various industrial categories. Taken from the IWTT Database.

used. *Primary clarification* is the term normally used for the first clarification step in a system. This process is used to remove the readily settleable solids prior to subsequent treatment, particularly in biological treatment systems. This treatment step results in significantly lower pollutant loadings to downstream processes and is appropriate for industrial wastewaters containing a high suspended solids content.

The actual physical sizing of the clarifier (surface area, depth, inlet structure, etc.) is highly dependent on the quantity and composition of the wastewater flow to be treated (i.e., the settling rate of the particles being removed). Because these criteria (maximum surface overflow rate, gal/d/ft^2; hydraulic retention time, hour; etc.) will vary substantially among industries, sizing must be done following a detailed study on a site-by-site basis.

Clarification units can be either circular or rectangular and are normally designed to operate as continuous flow reactors. Circular units are generally called *clarifiers*, whereas rectangular units are commonly referred to as *sedimentation tanks*.

Generally, clarifiers are designed on the basis of the type of suspended solids to be removed from the waste stream. There are three types of suspended solids. Type I solids are discrete particles that will not readily flocculate and are typical of raw influents. Type II solids are characterized by solids that flocculate and interact with one another, increasing in size and particle settling velocity as they settle through the water column. These types of particles are usually found in wastewaters that have been subjected to chemical addition or that are generated in biological wastewater treatment plants. Type I particles are typical of physical manufacturing operations as inorganic particles and grit, whereas Type II particles are more common to chemical manufacturing operations and biological wastewater treatment plants. The last type are Type III particles that describes the solids generated from a biological treatment process at the bottom of clarifiers as the flocculated biosolids begin to thicken and settle with *zone settling* characteristics. Under zone settling, high-concentration flocculated solids settle as a mass with a clear, well-defined interface between the settling mass of solids and the clarified water above. The upward velocity of the displaced water reduces the velocity of the settling solids. Because of their poor settling properties, special care must be given in the design of the removal of these Type III solids.

Clarification is effective in removing a substantial portion of the suspended solids and BOD in an industrial wastewater. Not only does this aid in the performance of other downstream processes, but it also protects them from negative effects, such as fouling from excessive solids loading. This technology is relatively adaptable to modifications, such as chemical addition, which allows for response to changes in wastewater characteristics. Finally, clarification is a relatively simple process, which requires neither sophisticated equipment nor highly skilled operators. The disadvantages of clarification as a component of an industrial wastewater treatment system include high surface area requirements and limitations on specific pollutant loadings and flows.

13.4.1.2 Flotation

As opposed to clarification, which separates suspended particles from liquids by gravitational forces, flotation separates particles by reducing their density with the introduction of air into the system. In dissolved air flotation (DAF), air is dispersed through a liquid in a closed pressure tank, supersaturating it. When the supersaturated liquid is mixed with the incoming sludge at the bottom of the DAF and exposed to atmospheric pressure, fine gas bubbles are generated as a result of the release of a gas from the supersaturated mixture. These fine bubbles adhere to the solids, which are then lifted to the surface and skimmed from the tank via a float collector.

DAF tanks can be either rectangular or circular in shape and constructed of either concrete or steel. Tanks are constructed with equipment to provide uniform flow distribution at the inlet, to provide pressurized gas, and to skim off float material. In designing a DAF separator system, the following variables are typically considered: full, partial, or recycled pressurization; feed characteristics; surface area; float characteristics; hydraulic loading; chemical usage; type of pressurization equipment; and operating pressure.

DAF is an appropriate technology for treating suspended solids and oil and grease in industrial wastewaters. Generally, the process will achieve 40%–65% suspended solids removal and 60%–80% oil and grease removal. The reduction of BOD will also be realized through the removal of biodegradable suspended solids. The addition of chemicals prior to actual flotation can significantly improve the performance of this technology.

In addition to the pollutant removals mentioned, flotation has other advantages. The resulting float material may be used as an auxiliary fuel source. Minimized construction costs may result because the higher overflow rates and shorter detention times for DAF compared to clarification resulting in smaller tank volume requirements. Also, because of the short detention periods and the presence of residual DO in the wastewater, odor problems are kept at a minimum. Some disadvantages associated with this type of treatment process are high operations and maintenance costs; inability to remove "heavy" suspended solids; and the requirement of more skilled personnel than for conventional settling processes.

13.4.1.3 Oil-Water Separation

In practically all manufacturing industries, oil and grease can be found in a plant's wastewater. Oil and grease generally results from equipment lubrication, accidental spills, and leaks, etc. However, for some industries, such as petroleum and edible oil refining, oil compounds can represent a significant constituent of their flows. This is due to the use of these oils as the raw materials in production. Not only can the oil and grease have a detrimental effect on the performance of a wastewater treatment system, but it also represents a valuable raw material being wasted. Thus, specific treatment processes have been developed to separate the major portion of the oil from the waste stream. These are called *oil-water separators*.

The configuration of the oil-water separators is that of a flow-through tank. The basic principle by which oil-water separators work is the differential between the specific gravities of water and the oils to be removed. Because the oils have lower specific gravities, they will rise to the top of the unit, and the heavier water sinks to the bottom. An important consideration in the separator design is the oil globule size because Stokes' law for terminal velocity of spheres in a liquid medium will determine the rate at which the oil rises. Retaining baffles and skimmers capture the oil compounds as the separated water leaves the tank. Sludge collectors can be used to scrape the bottom of the tank to remove any settled solids.

The major advantage of oil-water separators is their ability to treat wastewater, which is heavily laden with oil. Because of their design, they represent a simple treatment operation, which minimizes personnel requirements. Because this technology relies on natural forces, as does clarification, rather than relying on chemicals or aeration, its operating costs are minimized. Also, it results in a "purer" oil, which can make recycling much easier.

The disadvantages of oil-water separators are as follows. The ability to remove only specific size oil globules can result in very small oil droplets passing through the system. Also, if the wastewater has a high solids content and the oil and solids become mixed, the resultant globules may be too heavy to float to the surface and, thus, remain suspended. Finally, improperly maintained units are subject to odor problems.

13.4.2 Chemical Treatment Processes

13.4.2.1 Coagulation-Flocculation-Sedimentation

Often the nature of an industrial wastewater is such that conventional physical treatment methods will not provide an adequate level of treatment. Particularly, ordinary settling or flotation processes will not remove ultrafine colloidal particles and metal ions. In these instances, stabilizing forces (such as electrostatic repulsion and physical separation) predominate over the aggregating forces and mechanisms, namely, Van der Waals' forces and Brownian motion, which tend to cause particle contact. Therefore, to adequately treat these particles in industrial wastewaters, coagulation-precipitation may be warranted.

The first and most important step of this technology is coagulation, which involves rapid mixing of destabilizing/charge neutralizing chemicals (coagulants) to thoroughly disperse them throughout the wastewater for uniform reaction. Next, the wastewater undergoes flocculation, which provides slow mixing to encourage destabilized/charge neutralized particle contact, so that the particles can agglomerate to a size large enough for gravity separation. Chemical precipitation is carried out in a similar manner, with the addition of hydroxide salts to raise the wastewater pH and form insoluble metal hydroxide precipitates during the coagulation/flocculation steps instead of destabilized flocculated solids.

Coagulation-flocculation-sedimentation is capable of removing colloidal BOD, COD, and TSS from industrial wastewater. In addition, depending on the characteristics of the wastewater being treated, coagulation-flocculation-sedimentation via chemical precipitation can remove additional pollutants such as phosphorus, hardness, and heavy metals. This technology is attractive to industry because a high degree of both conventional and toxic pollutant removal can be realized in a single treatment process. A disadvantage of this process is the substantial quantity of chemical sludge generated, potentially of both high organic and heavy metal content, which can present a sludge processing and disposal problem.

13.4.2.2 Neutralization

In virtually every type of manufacturing industry, chemicals play a major role. Whether they result from the raw materials or from the various processing agents used in the production process, some residual compounds will ultimately end up in a process wastewater. Thus, it can generally be expected that most industrial waste streams will deviate from the neutral state (i.e., will be acidic or basic in nature).

Highly acidic or basic wastewaters are undesirable for two reasons. First, they can adversely impact aquatic life in receiving waters; second, they might significantly affect the performance of downstream treatment processes at the plant site or at a publicly owned treatment works. Therefore, to rectify these potential problems, one of the most fundamental treatment technologies, neutralization, is often employed at industrial facilities. Neutralization involves adding an acid or a base to a wastewater to neutralize its counterpart in the wastewater flow, namely, adding acids to alkaline wastewaters and bases to acidic wastewaters.

The most important considerations in neutralization are ensuring that wastewater constituents are thoroughly known so that the proper neutralizing chemicals are used, and that proper monitoring is carried out to ensure that the required quantities of these chemicals are used so that the effluent is in fact neutralized. For acid waste streams, lime, soda ash, and caustic soda are the most common base chemicals used in neutralization. In the case of alkaline waste streams, sulfuric, hydrochloric, and nitric acid are generally used for neutralization. Some industries have operations that separate acid and alkaline waste streams. If properly controlled, these waste streams can be mixed to produce a neutralized effluent without the need for adding additional neutralizing chemicals.

Although most people do not think of pH as a pollutant, it is in fact designated by the EPA as such. Because many subsequent treatment processes are pH-dependent, neutralization can be considered a pretreatment step in the treatment for many pollutants.

Eliminating adverse impacts of extreme pH on water quality and wastewater treatment system performance is not the only benefit of neutralization. Acidic or alkaline wastewaters can be highly corrosive. Thus, by neutralizing its wastewaters, a plant can protect its treatment units and associated piping. The major disadvantage of neutralization is that the chemicals used in the neutralization process are corrosive themselves and can be dangerous to store and handle.

13.4.3 Biological Treatment Processes

For many industrial wastewaters, one of the most important concerns has to do with those constituents that can exert an oxygen demand and have an impact on oxygen status in receiving waters. Although most industries are discussed in terms of the toxics they discharge, many are also significant sources of BOD and COD. In many instances, the most appropriate industrial treatment technology for removing this oxygen-demanding material is biological treatment. Discussion in this section will be limited to those processes frequently used in the industrial field: aerobic suspended growth processes (activated sludge), aerobic attached growth processes (trickling filters, rotating biological contactors, and biotowers), aerated lagoons (stabilization ponds), and anaerobic lagoons.

13.4.3.1 Aerobic Suspended Growth Processes (Activated Sludge)

An aerobic suspended growth process is one in which the biological growth products (microorganisms, mostly bacteria) are kept in suspension in a liquid medium (*mixed liquor*) consisting of entrapped and suspended colloidal and dissolved organic and inorganic materials. This biological process encourages the growth of microorganisms on the soluble and colloidal biodegradable organic material in the wastewater to produce an acceptable effluent quality by converting those substances exerting an oxygen demand into biosolids in the aeration tank and subsequently removing them in secondary clarifiers. Depending on the type of material in the raw wastewater, this process may be preceded by one or more other treatment technologies (i.e., screening, grit removal, primary clarification, oil and grease removal, etc.) to improve overall treatment process pollutant removal efficiencies.

In the suspended growth processes, wastewater enters an aeration tank of concrete, steel, or an earthen basin, where microorganisms are brought into contact with the incoming biodegradable organic components of the wastewater by some type of mixing device. This mixing device not only maintains this mixed liquor solids in suspension but also promotes transfer of oxygen to the wastewater, thus providing oxygen necessary for sustaining aerobic biological activity in the aeration tank. The biodegradable organic matter in the wastewater serves as a carbon and energy source for microbial growth and is converted into microbial cell tissue and oxidized end products during aeration.

A portion of the mixed liquor solids (MLSS) concentrated in the secondary clarifier are recycled back to the aeration basin, the return activated sludge (RAS), to maintain an appropriate level of microbial solids based on the concentration of BOD entering the aeration tank. This ratio of BOD loading to microbial biomass within the aeration tank is called the F/M ratio [lb BOD/d/(lb MLSS in aeration tank)] and varies based on the type of activated sludge system being utilized. The concentrated MLSS in the secondary clarifier not returned to the aeration tank is labeled waste activated sludge (WAS) and is removed from the activated sludge system, further processed and dewatered prior to disposal. The quantity of WAS, lb/d, is determined by the solids retention time (SRT) with units of days, that is also specified based on the type of activated sludge system being used.

The first suspended growth process, now called the *conventional activated sludge* process, was developed in the early 1900s to achieve carbonaceous BOD removal. However, since its inception, many modifications to the basic process have taken place. The variations in the activated sludge process are too numerous to be discussed in detail here. However, some of these are step aeration, tapered aeration, contact stabilization, complete mix, and the oxidation ditch (Metcalf and Eddy et al. 2013).

Aerobic suspended growth systems can be adapted to treat a wide range of industrial wastewaters. The process can be easily expanded to accommodate increased flows. However, these systems have drawbacks. Suspended growth systems generally perform best under uniform hydraulic and pollutant loadings. For some industries it is extremely difficult to maintain these conditions because of their manufacturing operations. A common event may be a "shock" loading of high strength waste entering the treatment process, with the result being poor pollutant treatment. Also, certain toxic pollutants can kill microorganisms in the aeration basin, causing a loss of treatment and a potentially lengthy recovery of the MLSS to shock loading levels. Operational skills and controls required to effectively operate an aerobic suspended growth system are higher than for most other biological processes. Finally, high-energy costs can be expected because of the need to provide mixing and oxygen in the process to sustain microbial growth.

Excluding those waste streams with very high concentrations of toxic pollutants, aerobic suspended growth systems can be used to treat any wastewater containing biodegradable matter. Removal efficiencies in excess of 90% for carbonaceous BOD been achieved through this process.

13.4.3.2 Aerobic Attached Growth Processes

An aerobic attached growth process is one in which the microbial biomass is attached to some type of media (i.e., rock, plastic sheets, plastic rings, etc.) and where either the wastewater trickles over the surface of the media or the medium is rotated through the wastewater. The process is related to the aerobic suspended growth process in that both depend on biochemical oxidation of organic matter in the wastewater to provide energy and a carbon source to promote the growth of bacteria and remove the oxygen demanding material from the wastewater. There are three general types of aerobic attached growth systems: conventional trickling filters, roughing filters or biotowers, and rotating biological contactors (RBCs).

There are several advantages of attached growth processes over other biological processes. First, due to dense microorganism growth in the biofilms that grow on the fixed-film media, these systems are inherently resistant to upsets from accidental spikes of toxic materials. Second, because oxygen is supplied naturally by convection of air through the unsaturated media, the need for aeration equipment is eliminated. This, along with a simpler operation, lowers the requirements for highly skilled treatment plant personnel and can result in substantial cost savings over suspended growth treatment systems. Secondary clarifiers follow the attached growth reactors, and a portion of the clarified effluent liquid is recycled back to the bioreactor to dilute the incoming wastewater to provide additional liquid to help shear biosolids from the media to prevent clogging and to provide additional contact time between the wastewater and attached biomass to improve overall BOD removal.

One disadvantages of attached growth treatment processes is they experience operating difficulty in cold climates. Enclosing the units for temperature protection can lead to other problems such as excessive condensation and potentially corrosive conditions. These attached growth systems are also susceptible to clogging if dense media is used or high solid loadings are applied, and all of these attached growth systems require primary treatment before they can safely be used.

Excluding those waste streams with very high concentrations of toxic pollutants, aerobic attached growth processes can be used to treat any wastewater containing biodegradable organic material. In general, aerobic attached growth processes are not as efficient nor as flexible as aerobic suspended growth processes in removing BOD, suspended solids, and toxic pollutants.

13.4.3.3 Aerobic Lagoons (Stabilization Ponds or Oxidation Ponds)

Aerobic lagoons are large, shallow (3–5 ft deep) earthen basins that provide wastewater treatment by using natural processes involving both algae and bacteria. The objective is the microbial conversion of organic wastes into biological solids and stable end products. Aerobic conditions prevail throughout the process through oxygen transfer across a large water surface and through algal production of oxygen via photosynthesis.

In photosynthesis, oxygen produced by the algae is used by the bacteria in the biochemical oxidation and degradation of organic waste. Carbon dioxide, ammonia, phosphate, and other nutrients released in the biochemical oxidation reactions by the bacterial are, in turn, used by the algae, forming a symbiotic relationship between the algae and bacteria in the lagoon. Aerobic lagoons are used for treatment of weak industrial wastewater containing negligible amounts of toxic or nonbiodegradable substances.

13.4.3.4 Anaerobic Lagoons

Anaerobic lagoons are earthen ponds built with a small surface area and a deep liquid depth of 8–20 ft. These lagoons are anaerobic throughout their depth, except for an extremely thin surface zone. Once greases form an impervious layer, complete anaerobic conditions develop. In a typical anaerobic lagoon, raw wastewater enters near the bottom of the lagoon (often at the center) and mixes with the active microbial mass in the sludge blanket, which is usually about 6 ft deep. The discharge is located near one of the sides of the lagoon, submerged below the liquid surface. Excess, undigested grease floats to the top, forming a heat-retaining and airtight cover. Excess sludge is washed out with the effluent. Anaerobic lagoons are effective prior to aerobic treatment of high-strength organic wastewater that also contains a high concentration of solids. Under optimal operating conditions, BOD removal efficiencies of up to 85% are possible.

The advantage in using either aerobic or anaerobic lagoons are low cost (excluding land if not readily available); simplicity of operation; low operation and maintenance cost (particularly for anaerobic lagoons that require no aeration); and when designed properly, high reliability. The disadvantages in using any lagoon process are: high land requirements; possible odor emissions; and the potential for seepage of wastewater into groundwater unless the lagoon is adequately lined. In addition, in most locales of the United States there are seasonal changes in both available light and temperature that reduced the systems treatment performance. In the winter, biological activity decreases because of a reduction in temperature, with rates of reaction typically decreasing by a factor of 2 for every 10°C reduction in water temperature.

13.5 TREATED EFFLUENT MANAGEMENT

The development of advanced treatment technologies, along with an increasing scarcity of freshwater, has led to marked changes in industrial wastewater treatment plant effluent management. Numerous strategies for treated wastewater reuse are presently being employed in ways appropriate to the particular industrial operation. The combination of scarcity of water with increasingly stringent regulations has made effluent disposal into natural receiving bodies the option of last resort.

13.5.1 Water Reuse and In-Plant Wastewater Segregation

Although it is theoretically possible to reuse ("closeup") all the water of many industrial processes, inevitable limits will be encountered as a result of quality control. In a paper mill, for example, a closed system will result in increasing concentrations of dissolved organic solids. Such a build-up will lead to an increase in the costs of slime control, greater downtime on the paper machines, and possible discoloration of the paper stock. The goal is to maximize reuse while maintaining the integrity of the production process.

In the design of water reuse strategies, the particular water requirements of the process must be addressed. For example, water use by hydropulpers in a paper mill does not have to be treated for suspended solid removal. However, if these solids are not removed from shower water used on the paper machines, clogging of the shower nozzles will result. In the production of produce, such as tomatoes, protection against microbial contamination through chlorination is of greater importance than overall purification.

In the case where two or more waste streams are generated, it is often necessary that they be separated. This is all the more important when mixing can result in health hazards. In a plating plant where both acidic metal rinses and cyanide streams are generated, the lack of separation can lead to the production of toxic hydrogen cyanide and undesirable exposure to plant personnel.

It is often the case that only a portion of the waste flow contains the majority of the suspended solids loading. Treatment for solids removal would then only be necessary on this part of the waste stream. Such is the case in a tannery where nearly 60% of the flow from the beam house yields 90% of the suspended solids.

Where little, if any, contamination of waste streams has occurred, such as in cooling water, the segregation of these effluents can allow for direct reuse or discharge into a receiving water with little or no pretreatment.

13.5.2 Stormwater Management

At industrial facilities, containment and control of pollutant discharges from stormwater is also regulated. Present strategies include adequate diking around process sewers and diversion of stormwater away from chemical process units to reduce the volume of contaminated stormwater requiring treatment prior to discharge. The design of holding basins for temporary storage of contaminated stormwater is based on storm frequency for a particular region. Once collected, the stormwater is passed through the treatment plant at a controlled rate following storm events to prevent hydraulic overloading of wastewater treatment facilities and subsequent dilution of wastewater requiring treatment.

13.5.3 Effluent Disposal

That portion of wastewater that is not reused must be disposed of to the environment, where it reenters the hydrologic cycle. Disposal thus may be considered a first step in the long-term, global process of reuse.

The most common strategy for disposal of treated wastewater effluent is release to local surface waters. Where streams or rivers of sufficient volume or flow are not available, it may be necessary to directly discharge treated wastewater into lakes or reservoirs. Because of their tremendous assimilative capacity, oceans and large lakes, such as the Great Lakes, are used by many communities for wastewater disposal. Discharge under these conditions is accomplished with the use of pipes or tunnels laid on, or buried in, the floor of the water body.

In very dry regions, land application is often used. Where successful, effluent will seep into the ground and replenish underlying aquifers. At the same time, a portion of the wastewater will return to the hydrologic cycle through evaporation. Of principle importance in wastewater disposal is the consideration of environmental impact. Environmental regulations, criteria, policies, and reviews protect ambient waters from negative impacts of treated wastewater discharges. When selecting discharge locations and outfall structures, ambient water quality standards developed by the states, as well as effluent regulations developed by the EPA, will govern the level of treatment required (see Chapter 6).

13.6 SOLIDS MANAGEMENT

The utilization and disposal of sludge is jointly addressed under the Clean Water Act and the Resource Conservation and Recovery Act. Both of these federal laws emphasize the need to employ environmentally sound sludge management techniques and to beneficially use sludge whenever possible. At the same time, the national requirements for improved wastewater treatment will result in the production of a greater quantity of residuals, and possibly more concentrated forms of contaminants will be present in these residuals than ever before. The permits required for effluent discharge from wastewater treatment plants are affected by, and contain, provisions related to sludge management techniques employed by facilities.

Prior to utilization or disposal, sludge must be stabilized to control odors and reduce the number of disease-causing organisms. The sludge is also routinely dewatered to reduce the volume to be transported or to prepare it for final processing. The liquid sludge, which contains 90%–98% water, can be partially dewatered from 20% to 30% solids (80% to 70% water) by a number of processes. Vacuum filtration, pressure filtration, and centrifugation are three of the more common dewatering technologies presently being employed.

Digestion of sludge is accomplished in heated tanks where the material can decompose anaerobically and the odors can be controlled. Anaerobic sludge digestion has the added benefit of producing methane gas, which may be used by the same treatment plant as an energy source.

Until recently, the majority of generated sludge had been incinerated, landfilled, or dumped into the ocean. Currently, much more attention is being given to sludge utilization by application to land as a soil conditioner or fertilizer and combustion in facilities to recover energy.

Liquid digested sludge has been used successfully as a fertilizer and for restoring areas disrupted by strip mining. Under this sludge management approach, digested sludge in semiliquid form is transported to the spoiled areas. The slurry, containing nutrients from the waste, is spread over the land to support revegetation of barren and disturbed land. Restoration of disturbed lands also can help control the flow of acids that drain from mines into streams and rivers, endangering fish and other aquatic life and adding to the difficulty of reusing the water.

13.7 DEVELOPMENTS IN INDUSTRIAL WASTEWATER TREATMENT

With the tremendous increase in recent years in the overall generation of industrial pollutants, along with the escalating costs of pollution control, increasing efforts are being made with regard to pollution prevention. The policies of the EPA now give the highest priority to source reduction, followed by recycling, with pollution control as the final alternative.

One example of this is the National Pretreatment Program, which is intended to eliminate the discharge of pollutants into the nation's waters. The major objective of the program is to protect publicly owned treatment works (POTWs) and the environment from the hazards of various toxic and other dangerous pollutants. Water quality based toxics limits and monitoring requirements are becoming a common provision in the National Pollutant Discharge Elimination System (NPDES) permits of POTWs. Pretreatment of toxic pollutants is necessary to comply with these water quality based toxics requirements.

Industry is responding in new, innovative ways to these mandates. A Union Carbide plant in South Charleston, West Virginia, is an example of what is happening in the organic chemicals industry. Through various in-plant changes, the average waste flow of 11.1 million gallons/day (MGD) and a waste load of 55,700 lb BOD/d was reduced to a flow of 8.3 MGD and load of 37,000 lb BOD/d through aggressive pollution prevention efforts. These changes were accomplished principally by replacement of scrubbers, waste segregation, collection and incineration of high strength wastewater streams for energy recovery, and additional processing of collected tail streams.

Responding to the difficulty of many dairy industries to meet present federal wastewater standards, Search, Inc., of Norman, Oklahoma, has developed the vertically integrated environmental waste (VIEW) treatment system. The VIEW system uses nothing other than basic wastewater treatment technologies. In this system, an integrated system has been constructed through the combination of an aeration basin, a trickling filter, and a clarifier. BOD levels are reduced to EPA standards through the continuous recirculation of plant effluent.

The largest water reclamation effort in California was initiated by the Metropolitan Water District of Southern California and the West Basin Municipal Water District to produce five types of recycled water from its Hyperion Water Reclamation Plant. The recycled water is provided to municipal, commercial, and industrial customers for the following purposes: irrigation, for a seawater barrier and groundwater replenishment, industrial cooling towers, low-pressure boiler feed water, and high-pressure boiler feed water.

13.8 APPLICATIONS

Six Illustrative Examples complement the material presented in this chapter related to the concepts and practices of industrial wastewater treatment.

Illustrative Example 13.1

Describe the mechanisms of ion exchange used to remove heavy metals from industrial wastewater.

Solution

Ion exchange involves the displacement of ions of given species from insoluble exchange materials by ions of different species when solutions of the latter are brought into contact with the exchange materials. The exchange can be expressed by the general equilibrium equations as:

$$B_1^+ + R^-B_2^+ \leftrightarrow B_2^+ + R^-B_1^+ \quad (13.1)$$

$$A_1^- + R^+A_2^- \leftrightarrow A_2^- + R^+BA_1^- \quad (13.2)$$

where B_1^+, B_2^+ are cations of two different species; A_1^-, A_2^- are anions of two different species; and R^-, R^+ are cationic and anionic exchange media, respectively.

The exchange behaves as a chemically reversible interaction between a fixed, ionized exchange site on the exchange material and ions in solution. The equilibrium expression for the first equation is:

$$K = \frac{\left[B_2^+\right]\left[R^-B_1^+\right]}{\left[B_1^+\right]\left[R^-B_2^+\right]} \quad (13.3)$$

where K is an equilibrium constant sometimes referred to as the equilibrium selectivity coefficient. A similar expression can be written for the second equation.

Although ion exchange has been used to concentrate valuable ionized materials in dilute solutions and for the selective fractionation of certain ionized solutes in solution, its widest application is found in the removal of objectionable ions from water used for domestic (primarily) and industrial purposes.

When the objective in water treatment is the removal of hardness ions Ca^{2+} and Mg^{2+} only, exchange materials are used which operate on a sodium cycle, that is, which exchange Na^+ for Ca^{2+} and Mg^{2+}. The equilibrium equation for the removal of calcium with such exchange materials is:

$$Ca^{2+} + R^-\left(Na^+\right)_2 \leftrightarrow 2\ Na^+ + R^-Ca^{2+} \quad (13.4)$$

When the exchange material is "spent," that is, when equilibrium is reached, the exchange material in the form of R^-Ca^{2+} is regenerated to $R^-\ (Na^+)_2$ so that it can be reused. Regeneration is accomplished by contacting the exchange material with a concentrated solution of sodium ions. Because the equilibrium selectivity coefficients for equilibriums such as the preceding equation have large values, at equilibrium almost all of the exchange material is in the R^-Ca^{2+} form and solutions with high concentrations of Na^+ ions are required for regenerations. Normally, regenerations is accomplished using solutions of 5%–10% NaCl.

Illustrative Example 13.2

Air strippers are commonly used for the removal of volatile organic compounds in industrial wastewater. An atmospheric packed tower air stripper is used to clean a contaminated waste stream with a concentration of 100 ppm trichloroethylene (TCE). The stripper was designed such that the packing height, Z, is 13 ft; the diameter, D, is 5 ft; and the height of the transfer unit (HTU) is 3.25 ft. Assume Henry's Law applies with a value, H, of 324 atm at 68°F. Also at these conditions, the air-to-water molar flow-rate ratio, G/L, is related to the air-to-water volume ratio, G''/L'', through $G''/L'' - 130\ G/L$, where units of G'' and L'' are $ft^3/(s\text{-}ft^2)$. With this input information, answer the following questions:

1. If the stripping factor, R, used in the design is 5, what is the tower removal efficiency?
2. If the air blower produces a maximum flow, q, of 106 acfm, what is the maximum water flow in gpm that can be treated by the stripper?

For purposes of this problem, the height of a packed tower can be calculated by:

$$Z = (N_{OG})(H_{OG}) = (\text{NTU})(\text{HTU}) \tag{13.5}$$

In addition, the following equation has been employed for the calculation of the number of transfer units (NTU) for an air-water stripping system, and is based on the stripping factor, R, and the inlet/outlet concentrations.

$$\text{NTU} = \frac{R}{R-1} ln\left(\frac{C_{in}/C_{out}(R-1)+1}{R}\right) \tag{13.6}$$

where C_{in} is the inlet contaminant concentration, mg/L; and C_{out} is the outlet contaminant concentration, mg/L.

Note also that

$$R = \frac{HG}{PL} \tag{13.7}$$

where P is the system pressure, atm; and G is the gas (air) loading rate (or flux), lbmol/ft^2-s; and L is the liquid loading rate (or flux), lbmol/ft^2-s.

Solution

The NTU are first calculated from rearrangement of Equation 13.5 as:

$$\text{NTU} = Z/\text{HTU} = (13\text{ ft})/(3.25\text{ ft}) = 4 \tag{13.8}$$

Rearranging Equation 13.6 to solve for C_{out} yields:

$$C_{out} = \frac{C_{in}(R-1)}{R\exp[(\text{NTU})(R-1)/R]-1} \tag{13.9}$$

Substituting known values into Equation 13.9 yields:

$$C_{out} = \frac{(100\text{ mg/L})(5-1)}{5\exp[(4)(5-1)/5]-1} = \frac{400\text{ mg/L}}{5[\exp(16/5)]-1} = \frac{400\text{ mg/L}}{120.66} = 3.3\text{ mg/L}$$

The removal efficiency of the stripping tower is then:

$$\text{Removal efficiency} = \frac{100\text{ mg/L} - 3.3.\text{ mg/L}}{100\text{ mg/L}} \times 100 = 96.7\%$$

The air-to-water molar ratio, G/L, is calculated from rearrangement of Equation 13.7 as:

$$G/L = (P)(R)/H \tag{13.10}$$

Substituting known values into Equation 13.10 yields:

$$G/L = (1 \text{ atm})(5)/(324 \text{ atm}) = 0.154$$

With this value of *G/L*, the air-water volume ratio, *G″/L″*, can be determined as:

$$G''/L'' = 130\, G/L = 130\ (0.154) = 20$$

The cross-sectional area, *A*, of the stripping tower can be determined as:

$$A = \pi D^2/4 = \pi (5 \text{ ft})^2/4 = 19.63 \text{ ft}^2$$

The air volumetric loading rate, *L″*, is:

$$G'' = (106 \text{ ft}^3/\text{min})(1 \text{ min}/60 \text{ s})/(19.63 \text{ ft}^2) = 0.09 \text{ ft}^3/(\text{s-ft}^2)$$

L″ is then determined to be:

$$G''/L'' = 20;\ L'' = G''/20 = \left(0.09 \text{ ft}^3/\left(\text{s-ft}^2\right)\right)/20 = 0.0045 \text{ ft}^3/\left(\text{s-ft}^2\right)$$

Converting this liquid loading rate to a maximum water flow rate that can be treated yields:

$$\text{Water flowrate} = 0.0045 \text{ ft}^3/\left(\text{s-ft}^2\right)\left(19.63 \text{ ft}^2\right)\left(60 \text{ s/min}\right)\left(7.48 \text{ gal/ft}^3\right) = 39.6 \text{ gpm}$$

Illustrative Example 13.3

Land treatment of industrial wastewater is a process in which wastewater is applied directly to the land. This type of treatment is most common for food-processing waste including meat, poultry, dairy, brewery, and winery wastes. The principal rationale of this practice is that the soil is a highly efficient biological treatment reactor, and food-processing wastewater is highly degradable. This treatment practice is usually carried out by distributing the wastewater through spray nozzles onto the land or letting the water run through irrigation channels.

Suppose that the rate of the wastewater flowing to a land application site is 178 gal/acre/min and the irrigated land area is 5.63 acres. The entire irrigation process lasts 7.5 hr/d. If the wastewater has a BOD_5 concentration of 50 mg/L, what is the mass of BOD_5 remaining in the soil after the land treatment process is complete (assume a BOD_5 removal efficiency by the land treatment process is 95%)?

Solution

The quantity of the wastewater, *q*, flowing to the land is:

$$q = (178 \text{ gal/acre/min})\ (5.63 \text{ acre})\ (60 \text{ min/hr})(7.5 \text{ hr/d}) = 450{,}963 \text{ gal/d}$$

The mass of BOD_5 remaining in the soil is:

$$\text{Mass of BOD}_5 \text{ remaining} = 0.05(\text{BOD}_5 \text{ Loading}) = 0.05(q, \text{MGD})\ (50 \text{ mg/L})$$

$$(8.34 \text{ lb/MG/ mg/L}) = 0.05(0.451 \text{ MGD})\ (50 \text{ mg/L})(8.34 \text{ lb/MG/ mg/L}) = 9.4 \text{ lb/d}$$

Illustrative Example 13.4

A 100,000 gal/d protein-rich wastewater with a COD of 4,000 mg/L is generated by a meat-processing plant. This waste is to be treated by an anaerobic contact process using a loading of 0.15 lb COD/ft^3-d and a solids retention time (SRT) of 20 d. Assuming 80% efficiency, you have been asked to determine or provide the following as part of a permit to be submitted to the state.

1. A flow diagram of the process
2. Hydraulic retention time, hours
3. Daily solids accumulation, lb/d (assume 6% net solids production from BOD removal)

Solution

A simple flow diagram of the process is provided in Figure 13.2. Daily COD Loading is calculated as follows:

$$\text{COD Loading} = (q, \text{MGD})\,(4{,}000\ \text{mg/L})(8.34\ \text{lb/MG/mg/L}) = (0.1\ \text{MGD})(4{,}000\ \text{mg/L})$$

$$(8.34\ \text{lb/MG/mg/L}) = 3{,}336\ \text{lb COD/d}$$

The volume of the anaerobic contact reactor is determined based on the COD Loading and the design COD Loading rate as:

$$\text{Volume} = \text{COD Loading/COD Loading Rate} = (3{,}336\ \text{lb COD/d})/\left(0.15\ \text{lb COD/ft}^3\text{-d}\right)$$

$$\text{Volume} = 22{,}240\ \text{ft}^3\left(7.48\ \text{gal/ft}^3\right) = 166{,}355\ \text{gal}$$

The contact reactor hydraulic retention time, HRT, is:

$$\text{HRT} = \text{Volume}/q = (166{,}355\ \text{gal})/(100{,}000\ \text{gal/d}) = 1.66\ \text{d} = 39.8\ \text{hr}$$

The solids growth, *SG*, may be calculated using the 6% synthesis value and 80% COD removal efficiency as:

$$SG = (3{,}336\ \text{lb COD/d})(0.8)(0.06\ \text{lb cell mass/lb COD removed}) = 160\ \text{lb cell mass/day}$$

SG represents the mass of volatile suspended solids and is the net active cell population generated from COD removal on a daily basis.

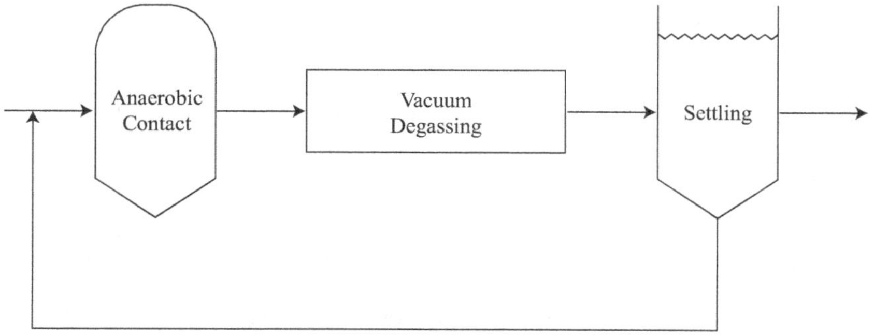

FIGURE 13.2 Flow diagram for anaerobic contact process for Illustrative Example 13.4.

Illustrative Example 13.5

Refer to Illustrative Example 13.4. Determine the N and P requirements for the anaerobic contact process assuming a 10% and 1.5% content in the generated cells, respectively. Also calculate the mixed liquor suspended solids in the reactor at steady state represented by the solids generated in the process each day.

Solution

Refer to Illustrative Example 13.4. Because 160 lb/d of cell mass is generated in the removal of COD, and the N requirement is stated as 10% of the biosolids content, the N requirement for cell mass production is:

$$\text{N required} = 10\% \text{ of cell mass produced} = 0.1\ (160\ \text{lb cell mass/day}) = 16\ \text{lb N/d}$$

Because 1.5% of the biosolids generated is P, the P requirement for cell mass production is:

$$\text{P required} = 1.5\% \text{ of cell mass produced} = 0.015\ (160\ \text{lb cell mass/day}) = 2.4\ \text{lb P/d}$$

Since the wastewater is from a meat-processing plant, it is proteinaceous (rich in N and P) and N and P will not be required.

The equivalent concentration of MLVSS generated in the process can be determined from the equation used to convert concentration to mass, lb/d, as:

$$\text{lb/d} = (\text{Volume, MG})(\Delta\text{MLVSS, mg/L})\ (8.34\ \text{lb/MG/mg/L}), \text{ or}$$

$$\Delta\text{MLVSS, mg/L/d} = (\text{lb/d})/\left[(\text{Volume, MG})(\text{Volume, MG})\right] = (160\ \text{lb/d})/(0.1663\ \text{MG})(8.34)$$

$$\Delta\text{MLVSS, mg/L/d} = 115\ \text{mg/L/d}$$

Because the *SRT* is 20 d, MLVSS concentrations will accumulate over this time before being wasted. The total MLSS concentration at steady state will then be:

$$\text{Steady} - \text{State MLVSS} = \Delta\text{MLVSS (SRT)} = (115\ \text{mg/L/d})\ (20\ \text{d}) = 2{,}300\ \text{mg/L}$$

Illustrative Example 13.6

The Perfect Plating Company is planning to build a plant on the Rugged River, discharging its wastewater into the river. The waste stream is produced 24 hr/d, has a cyanide concentration of 3.0 mg/L, and has a flow rate of 0.3 ft^3/s. The maximum allowable concentration of cyanide in the river below the outfall is 3.5 μg/L. The river flow is seasonal and the monthly average flow rates from the previous 5-year record are as follows:

Month	Oct	Nov	Dec	Jan	Feb	Mar	Apr	May	Jun	Jul	Aug	Sept
Flow, ft^3/s	146	254	300	395	690	607	333	386	321	177	185	154

1. For an average day in the month of October, show that discharging all of the waste would produce a downstream concentration of cyanide above the maximum allowable value. Assume no background cyanide concentration in the river.
2. Show that over the entire year, the discharge of the waste would not violate the permit.
3. Establish a discharge schedule assuming that, in the months of low river flow, the maximum amount of waste is discharged, with any excess flow diverted to storage, and that waste from storage is discharged with process waste during months of high flow.
4. Calculate the amount of storage required. Assume storage is covered so that there is no change in the concentration or volume of the stored waste.

Solution

Assume perfect mixing. For an average day in the month of October, the waste stream is diluted by the river so that the river concentration is the following mass fraction of the original waste stream concentration:

$$\text{Waste flow/(Waste + River flow)} = (0.3\ \text{ft}^3/\text{s})/(146 + 0.3\ \text{ft}^3/\text{s}) = 0.00205\ \text{ft}^3/\text{s}$$

The cyanide concentration in the river is then:

$$\text{River concentration} = 0.00205(3\ \text{mg/L}) = 6.1\times10^{-3}\text{mg/L} = 6.1\ \mu\text{g/L}$$

This is above the permitted level of 3.5 μg/L.

The yearly flow of the river and of waste stream are calculated as:

$$\text{River Annual Flow} = \sum \text{Monthly Flows} = 102.72\times10^{8}\text{ft}^3/\text{yr (see Table 13.1 below)}$$

$$\text{Waste stream Annual Flow} = (0.3\ \text{ft}^3/\text{s})(86{,}400\ \text{s/d})(365\ \text{d/yr}) = 9.46\times10^{6}\text{ft}^3/\text{yr}$$

$$\text{River} + \text{Waste Annual Flow} = 102.72\times10^{8}\text{ft}^3/\text{yr} + 9.46\times10^{6}\text{ft}^3/\text{yr} = 102.81\times10^{8}\text{ft}^3/\text{yr}$$

Yearly average dilution is:

$$\text{Annual Average Dilution} = \text{Waste Annual Flow/(Total Annual Flow)}$$

$$\text{Annual Average Dilution} = (9.46\times10^{6}\text{ft}^3/\text{yr})/(102.81\times10^{8}\text{ft}^3/\text{yr}) = 9.2\times10^{-4}$$

On an annual basis, the effective river cyanide concentration is then:

$$\text{River concentration} = (9.20\times10^{-4})(3.0\ \text{mg/L}) = 2.76\times10^{-3}\ \text{mg/L} = 2.76\ \mu\text{g/L}$$

Therefore, this discharge level is acceptable on an average annual basis because it is below 3.5 μg/L.

A proposed disposal schedule is based on the monthly average river discharge and the dilution provided to the waste stream by the river. As indicated in Table 13.1, excess waste flow that would result in monthly cyanide levels exceeding the 3.5 μg/L standard would be diverted to a covered storage tank and held until river flow is high enough to allow its discharge without impact to the Rugged River.

The allowable cyanide discharge rate in each month is calculated as follows:

$$\text{Allowable Waste Discharge, g/s} = \text{River Flow, ft}^3/\text{s (Discharge Limit, g/L)(l L/0.0353 ft}^3)$$

For October this becomes:

$$\text{Allowable Waste Discharge, g/s} = (146\ \text{ft}^3/\text{s})(3.5\times10^{-6}\text{g/L})(1\ \text{L}/0.0353\ \text{ft}^3) = 0.0145\ \text{g/s}$$

The waste flow discharge rate of Cyanide is calculated as:

$$\text{Waste Cyanide Discharge Rate} = (0.3\ \text{ft}^3/\text{s})(3.0\times10^{-3}\text{g/L})\ (1\ \text{L}/0.0353\ \text{ft}^3) = 0.0255\ \text{g/s}$$

The discharge schedule shown in Table 13.1 is based on the relationship between monthly allowable cyanide discharge rates and the waste stream cyanide discharge rate. Discharge is allowed from the waste stream only if the allowable rate is less than the waste discharge rate. If the

TABLE 13.1
Calculations for Storage and Dilution Requirements to Maintain Allowable Cyanide Levels in Rugged River for Illustrative Example 13.6

Month	Flow, ft^3/s	Days/Mo	Volume × 10^{-8}, ft^3	Allowable Cyanide Discharge, g/s	Storage Volume, ft^3	Waste Stream Discharge, ft^3/s	Flow from Storage, ft^3/s
Oct	146	31	3.91	0.0145	347,000	0.171	
Nov	254	30	6.68	0.0252	9,000	0.296	
Dec	300	31	8.04	0.0297		0.300	0.049
Jan	395	31	10.58	0.0392		0.300	0.161
Feb	690	28	16.69	0.0684		0.300	0.505
Mar	607	31	16.26	0.0602		0.300	0.408
Apr	333	30	8.31	0.0330		0.300	0.088
May	386	31	10.34	0.0383		0.300	0.150
Jun	321	30	8.32	0.0318		0.300	0.074
Jul	177	31	4.74	0.0175	252,000	0.206	
Aug	185	31	4.96	0.0183	227,000	0.215	
Sept	154	30	3.99	0.0153	311,000	0.180	
			102.72		1,146,000		

allowable discharge rate is less than the waste stream discharge rate, as in October, only a portion of the waste stream may be discharged to the Rugged River, the balance must be diverted to storage. For October this allowable discharge rate is:

$$\text{Allowable Monthly Waste Stream Discharge Rate} = (\text{Waste flow rate}) \times \frac{\text{Allowable Monthly Mass Discharge Rate, g/s}}{\text{Waste Stream Mass Discharge Rate, g/s}} \tag{13.11}$$

$$\text{Allowable October Waste Stream Discharge Rate} = (0.3\ ft^3/s)(0.0145\ g/s)/(0.0255\ g/s)$$

$$\text{Allowable October Waste Stream Discharge Rate} = 0.171\ ft^3/s$$

If the allowable mass rate of discharge of cyanide is greater than the waste stream mass discharge, the discharge of the waste stream is supplemented with discharge from storage, and the storage discharge rate is calculated as follows:

$$\text{Storage discharge rate} = (\text{Waste flow rate}) \times \frac{(\text{Allowable mass discharge rate} - \text{Waste mass discharge rate})}{\text{Waste mass discharge rate}} \tag{13.12}$$

This situation occurs in December due to high river flows, so for December, the discharge from the storage reservoir is calculated to be:

$$\text{December Storage Discharge Rate, } ft^3/s = (0.30\ ft^3/s)\ (0.0297 - 0.0255)/(0.0255)$$

$$\text{December Storage Discharge Rate, } ft^3/s = 0.049\ ft^3/s$$

The accumulation period runs from July through November, with discharge occurring from December through June.

If the allowable rate of cyanide mass discharge is less than the waste stream mass discharge rate, then waste flow must be diverted to storage. The required storage volume is calculated as:

$$\text{Monthly Storage Volume} = (\text{Waste flow rate}) \times (\text{d/mo}) \times (86{,}400\ \text{s/d})$$

$$\frac{(\text{Waste mass discharge rate} - \text{Allowable mass discharge rate})}{\text{Waste mass discharge rate}} \tag{13.13}$$

October required storage volume is calculated as:

$$\text{October Storage Volume} = (0.3\ \text{ft}^3/\text{s})\ [(0.0255 - 0.0145)/(0.0255)]\ (31\ \text{d})(86{,}400\ \text{s/d})$$

$$\text{October Storage Volume} = 347{,}000\ \text{ft}^3$$

Since the months in which the storage is required are consecutive, the total storage requirement is the sum for these months or 1,146,000 ft^3. A volume of 1,250,000 ft^3 would provide approximately 10% excess storage and some level of safety to ensure adequate storage to protect against permit violations and impacts to the Rugged River.

REFERENCES

Eckenfelder, W.W., D.L. Ford, and A.J. Englande, Jr. 2009. *Industrial Water Quality*, 4th edition. New York: McGraw-Hill.

Metcalf & Eddy, Inc., G. Tchobanoglous, F. Burton, and H.D. Stensel. 2013. *Wastewater Engineering: Treatment and Reuse*, 5th edition. New York: McGraw-Hill.

U.S. Environmental Protection Agency (U.S. EPA). 1980. *Primer for Wastewater Treatment*. 832-K-80-100. Washington, DC: Office of Water Programs Operations. https://nepis.epa.gov/Exe/ZyPDF.cgi/200044GQ.PDF?Dockey=200044GQ.PDF.

U.S. Environmental Protection Agency (U.S. EPA). 2017. Industrial Wastewater Treatment Technology Database

(IWTT) Web Application. EPA 821-F-17-005. Washington, DC: Office of Water. https://www.epa.gov/sites/production/files/2018-02/documents/iwtt_fact-sheet_dec-2017.pdf

14 Evaporation

14.1 INTRODUCTION

The vaporization of a liquid for the purpose of concentrating a solution has long been in the domain of the chemical engineer and is a common operation in the chemical process industry. The concentration of solutions of nonvolatile solutes through heat-induced vaporization of the solvent (generally water) is called *evaporation*. Evaporation is a relatively expensive operation, and its application in waste treatment is limited primarily to the recovery of valuable by-products from waste liquids and to the treatment of wastes where no alternate methods are available.

Purifying water via the evaporation route is not nearly as common as concentrating liquid solutions. The ocean/seawater salt concentration is approximately 30 to 40 g/L. Desalination by evaporation is a viable option, however, for two reasons. First, the operation is not affected by the salt concentration, and secondly, the salt concentration can easily be reduced via evaporation to concentrations below drinking water standards.

The overall desalination process requires the salt water to be heated to the boiling point where water vapor is formed. The "pure" vapor is then condensed to produce potable water. The thermal operating efficiency can be increased by employing a multi-stage evaporator, a type of unit that is addressed later in the chapter (Kern 1950; Farag and Reynolds 1996; Flynn et al. 2019).

The simplest device is an open pan or kettle that receives heat from a coil or jacket or by direct firing underneath the pan. Perhaps the traditional unit is the horizontal-tube evaporator in which a liquid (to be concentrated) in the shell side of a closed vertical cylindrical vessel is evaporated by passing steam or another hot gas through a bundle of horizontal tubes contained in the lower part of the vessel. The liquid level in the evaporator is usually less than half the height of the vessel; the empty space permits disengagement of entrained liquid from the vapor passing overhead (Kern 1950; Farag and Reynolds 1996; Flynn et al. 2019).

This chapter provides a review of the basis and fundamentals of evaporation through classification of vaporizing equipment, describing equations, multiple-effect evaporation, thermocompression, other evaporation units, and desalination via the evaporation route. An application section provides three Illustrative Examples of evaporation concepts and implementation for desalination.

14.2 CLASSIFICATION OF VAPORIZING EQUIPMENT

There are two principal types of tubular vaporizing equipment used in industry: *boilers* and *vaporizing exchangers* (Kern 1950; Flynn et al. 2019). Boilers are directly fired tubular apparatus that primarily convert fuel energy into latent heat of vaporization. Vaporizing exchangers are unfired and convert the latent or sensible heat of one fluid into the latent, heat of vaporization of another. If a vaporizing exchanger is used for the evaporation of water or an aqueous solution, it is now fairly conventional to call it an *evaporator*. If it is used to supply the heat requirements at the bottom of a distilling column, whether the vapor formed be steam or not, it is labeled a *reboiler*. When not used for the formation of steam and not a part of a distillation process, a vaporizing exchanger is simply called a *vaporizer*. When an evaporator is used in connection with a power-generating system for the production of pure water or for any of the evaporative processes associated with power generation, it is a *power-plant evaporator*. When an evaporator is used to concentrate a chemical solution by the evaporation of solvent water, it is a *chemical evaporator*. Both classes differ in design. Unlike evaporators, it is the object of reboilers to supply part of the heat required for distillation and not a change in solution concentration, although a change generally cannot be avoided. Very often the term *evaporator* is also applied to a combination of several pieces of equipment, each of which can also be defined as an evaporator (Flynn et al. 2019).

Unfortunately, certain classes of evaporators are still designed as part of an art rather than the rational summation of the individual resistances to heat flow (to be discussed shortly) that is the basis of the science. This is because of the high heat transfer coefficients with which certain classes of evaporators operate and the difficulty in identifying each of the small individual resistances that make up the overall resistance terms. As in the case of surface condensers (Kern 1950; Farag and Reynolds 1996; Flynn et al. 2019), numerous classes of evaporators are designed on the basis of *accepted* overall heat transfer coefficients, and it is these classes and their processes that are discussed in this chapter. Generally, they involve vaporization from pools, as compared with vaporization in the shell or tubes of a heat exchanger (Farag and Reynolds 1996; Theodore 2011; Flynn et al. 2019).

14.3 DESCRIBING EQUATIONS

The describing equation of an evaporator, like that of any heat exchanger given by:

$$Q = UA\Delta T \tag{14.1}$$

where Q is the rate of heat transfer, U is the overall heat transfer coefficient (the resistance to heat transfer), A is the available area for heat exchange, and ΔT is the temperature difference driving force across the heat transfer area. The term $(UA)^{-1}$ is equal to the sum of the individual resistances of the steam, walls of the tubes, the boiling liquid, and any fouling that may be present.

Consider Figure 14.1. Assume F lb of feed to the evaporator per hour, whose solid content is x_F. (The symbol x is employed to represent weight fraction.) Also, assume the enthalpy of the feed

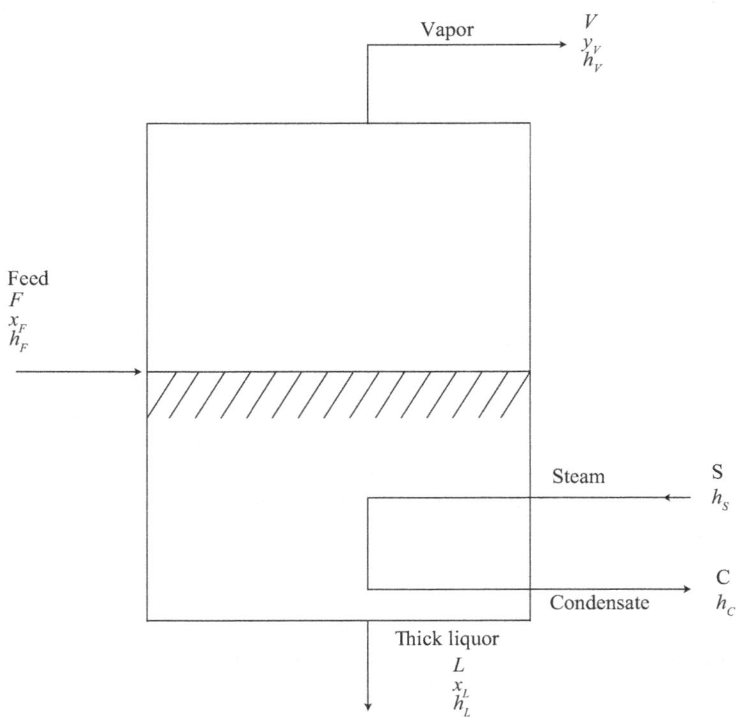

FIGURE 14.1 Material and enthalpy balance for a single-effect evaporator.

per lb to be h_F. L lb of thick liquor, whose composition in weight fraction of solute is x_L and whose enthalpy is h_L leaves from the bottom of the evaporator. V lb of vapor, having a solute concentration of y_V and an enthalpy of h_V Btu/lb, leaves the unit. In most evaporators, the vapor is pure water, and therefore, y_V is zero.

The material balance equations for this evaporator are relatively simple. A total material balance gives:

$$F = L + V \tag{14.2}$$

A componential balance leads to:

$$F\ x_F = L\ x_L + V\ y_V \tag{14.3}$$

To furnish the heat necessary for evaporation, S lb of stream is supplied to the heating surface with an enthalpy of h_S Btu/lb; S lb of condensate with an enthalpy of h_C Btu/lb is leaving as condensate. One simplifying assumption usually made is that, in an evaporator, there is little cooling of the condensate. This is usually less than a few degrees in practice; the sensible heat recovered from cooling the condensate is so small compared to the latent heat of the steam supplied to the heating surface that the condensate will leave at the condensing temperature of the steam. The enthalpy balance equation is therefore:

$$F\ h_F + S\ h_S = V\ h_V + L\ h_L + S\ h_C \tag{14.4}$$

Both Equations 14.1 and 14.4 are applied in tandem when designing or predicting the performance of an evaporator.

The calculations for an evaporator can be complicated because of an adiabatic solution temperature change. When two or more pure substances are mixed to form a solution, a heat effect usually results (Theodore et al. 2009). Many have experienced this effect on mixing concentrated sulfuric acid with water. This heat of mixing is defined as the enthalpy change that occurs when two or more pure substances are mixed at constant temperature and pressure to form a solution. Enthalpy concentration diagrams offer a convenient way to calculate enthalpy of mixing effects and temperature changes associated with this type of process. These diagrams, for a two-component mixture, are graphs of the enthalpy of a binary solution plotted as a function of composition (mole fraction or weight fraction of one component), with the temperature as a parameter. For an ideal solution, isotherms or an enthalpy-concentration diagram would be straight lines. For a real solution, the actual isotherm is displaced vertically from the ideal solution isotherm at a given point by the value Δh at that point, where Δh is the enthalpy of mixing (Farag and Reynolds 1996; Theodore 2011). If heat is evolved whenever the same pure components at a given temperature are mixed to form a solution at the same temperature, the system is said to be exothermic. An endothermic system is one for which the heats of solution are positive (i.e., the solution at constant temperature is accompanied by the absorption of heat). Organic mixtures often fit this description.

14.4 MULTIPLE-EFFECTS EVAPORATORS

One of the principal operating expenses of evaporators is the cost of steam for heating. A considerable reduction in those costs can be achieved by operating a battery of evaporators in which the overhead vapor from one evaporator (or "effect") becomes the heating medium in the steam chest of the next evaporator, thus saving both the cost of condensing the vapor from the first unit and supplying heat for the second. Several evaporators may operate in a battery in this fashion.

Basically, a multiple-effect evaporator may be thought of as a series of resistances to the flow of heat. The main resistances are those associated with heat transfer across the heating surface of each effect in the evaporator and across the final condenser if a surface condenser is used. The resistances of the heating surfaces are equivalent to the reciprocal of the product of the area and overall heat transfer coefficient (1/UA) for each effect (Kern 1950; Farag and Reynolds 1996; Theodore 2011; Flynn et al. 2019).

Neglecting all resistances but those noted and assuming they are equal, it can be seen that if the number of resistances (effects) is doubled, the flow of heat (steam consumption) will be cut in half. With half as much to heat, each effect will evaporate about half as much water. But because there are twice as many effects, the total evaporation will be the same. Thus, under these simplifying assumptions, the same output would be obtained regardless of the number of effects (each of equal resistances), and the steam consumption would be inversely proportional to that number of effects or the total heating surface installed.

This resistance concept is also useful in understanding the design and operation of such units. The designer places as many resistances (effects) in series as can be economically afforded to reduce steam consumption. One can also show that the lowest total area required arises when the ratio of temperature drop to area is the same for each effect.

In the production of distilled water, the vapor formed in the evaporator is useful steam as well as relatively pure water. If a pound of steam is supplied to an evaporator, it can be used to produce approximately 0.9 lb of steam from a pound of water. The remaining 0.1 lb of water contains the bulk of the impurities, and it is removed from the evaporator as evaporator *blowdown*. The 0.9 lb of vapor from the evaporator can be condensed by partially preheating the evaporator feed, or in a power plant application, it can be mixed directly with cold return condensate before being fed to a boiler.

If however, the original pound of steam was supplied to a multiple-effect evaporator, the vapor produced in the first evaporator would serve as the heat source in a second evaporator operating at a lower pressure than the first. Thus, an additional utilization could be made of most of the heat. If both evaporators were fed in *parallel* with raw water, about 0.85 lb of pure water would be formed in the first effect and about 0. 75 lb would be formed in the second effect. For each pound of steam supplied, about 1.6 lb of pure water would be produced, more than the 0.9 lb of vapor noted in the previous paragraph. The original pound of steam can also be considered a pound of pure water. When the vapor formed in the first effect is reused as the heating medium in a second effect, it is called a *double-effect* evaporator. When applied to three effects it is a *triple-effect* evaporator, and the original pound of steam will produce approximately 2.25 lb of pure water.

To maintain temperature differences for heat transfer between the vapor from one effect and the boiling liquid of the next effect, the pressure on each succeeding evaporator must be *lower* than its predecessor. The quantity of blowdown, which is arrived at from experience with water of different impurities, carries with it considerable sensible heat and thereby reduces the amount of vaporization that can be realized in succeeding effects. Due in part to this escape of heat from the system, there is a limit to the number of effects that are economically justifiable. The fixed charges for additional effects ultimately dissipate the savings in energy that result from the use of a 1arge number of effects (Theodore 2014).

In any event, the greater the number of effects the larger the amount of evaporation per pound of steam admitted to the first effect. The operating costs will be less the larger the number of effects used. This is offset, however, by the increased capital cost of the added equipment, and the increased maintenance costs for cleaning and replacement, both of which are considered fixed charges. Supervisory labor will be the same for the operation of any number of effects. The cost of condenser water must also be included, and it, too, will decrease the greater the number of affects employed. The optimum number of effects may be obtained by computing the process requirements with two, three, four, or up to six or eight effects and determining the fixed charges and operating cost resulting from each arrangement. When the total cost is plotted against the number of effects, a minimum will occur corresponding to the optimum number of effects for a given application.

Evaporators occasionally operate with the last effect under vacuum, and one of the important considerations for this option is the method used for continuously maintaining a vacuum. Kern (1950) has reported that the use of a mechanical compressor on the last stage is usually prohibitive because of the energy required for compression. The thermocompression option is discussed in the next section.

14.5 THERMOCOMPRESSION

The principle of thermocompression is continuously finding wider application in industry. Consider a single-effect evaporator. It is fed with steam and generates vapors that have nearly as much heat content as was originally present in the steam. These vapors are condensed with water as a convenient means of removal, but this is a significant waste of both energy and water. Were it not for the fact that the temperature of the generated vapor is lower than the steam, it would be possible to circulate the vapor back into the steam chest and evaporate continuously without supplying additional steam. But apart from the heat (enthalpy) balance, a temperature difference must exist between the steam and the generated vapors or no heat would have been transferred originally. If the vapors from the evaporator were compressed to the saturation pressure of the steam, however, the temperature of the vapors could be raised to that of the original steam. The cost of supplying the necessary amount of compression is usually small compared with the value of the latent heat in the vapors.

The practice of recompressing a vapor to increase its temperature and permit its reuse is thermocompression. Where fuel is costly, the compression may be accomplished with a centrifugal compressor (Theodore 2014). Where steam is available at a pressure higher than that required in the evaporator, the recompression can be effected in a steam jet booster. The latter operates on the principle of an ejector and is used to compress vapors instead of a noncondensable gas.

The pressure at which steam is discharged from a thermocompressor depends on pressure and proportion at which the live steam and the vapors are supplied. Kern (1950) and Flynn et al. (2019) have shown that the higher the desired discharge pressure, the greater the percentage of live steam that is required.

14.6 OTHER EVAPORATOR UNITS

Four classes of evaporator units are discussed: power plant evaporators, chemical process evaporators, heat transformer evaporators, and saltwater distillers.

14.6.1 Power Plant Evaporators

Power plant evaporation processes fall into four categories:

1. Make-up evaporators for boiler feed
2. Process evaporators for the production of purified water
3. Heat-transformer evaporators
4. Saltwater distillers

These processes are reviewed by Kern (1950) with typical flow sheets that contain all the necessary information computed from simple heat balances. Because only enthalpy differences are involved, it has not been felt necessary to include the individual heat balances for each case.

The purpose of the majority of power plant evaporators is the separation of pure water from raw or treated water. The impurities in the unevaporated residue are continuously withdrawn from the system as blowdown.

14.6.2 Chemical Process Evaporators

There are a number of industries that continuously require large quantities of distilled water. This type of plant employs double, triple, or quadruple effects evaporators and receives heat either from a bleed point or from the boiler directly. The selection of the number of effects is closely related to the relationship among the fixed charges and the steam operating cost. Multiple-effect evaporators with parallel feed need not have all effects simultaneously in operation and can, therefore, be adjusted if the demand for distilled water varies. Evaporators for this type of service are generally of medium size of approximately 500 to 2,000 ft^3 per shell.

In the chemical industry, the manufacture of heavy chemicals such as caustic soda, table salt, and sugar starts with dilute aqueous solutions from which large quantities of water must be removed before final crystallization can take place in suitable equipment. In these chemical evaporators, the unevaporated portion of the feed is the product being sought after.

14.6.3 Heat Transformer Evaporators

The heat-transformer evaporator is a single-effect, system of one or more shells *in parallel* receiving steam from the exhaust of a high-pressure turbine or high-pressure engine. The purpose of this type of evaporator is to condense steam from a high-pressure boiler that has passed through a high-pressure turbine and into the evaporator. The condensate is then returned directly to the high-pressure boiler by a pressure booster, thereby keeping the high-pressure circuit closed and continuously supplied with high-pressure boiler water and steam. Obviously high-pressure boiler and turbine installations are favorably affected by this system. By the condensation of the exhaust steam from the high-pressure turbine or engine, the heat transfer in the evaporator is used to produce large quantities of process steam, all or a large part of which is never returned to the evaporator system.

14.6.4 Saltwater Distillers

A pound of fuel can normally yield about 10 lb of steam, and in a double-effect evaporator for use with saltwater, 10 lb of steam would yield a total of approximately 18.5 lb of pure water. It is little wonder, then, that ships at sea commonly produce their own potable water requirements from seawater. Seawater contains about 3.5% dissolved solids by weight corresponding to about 35,000 ppm compared with < 500 ppm in fresh water. Instead of about 90% vaporization it is customary to vaporize only about one-third of the feed in a saltwater distiller. The remainder, containing about 5% dissolved solids, is disposed of as reject. Because of the degree of blowdown, the use of a vacuum system with low evaporation temperatures is desirable because low temperatures are also produce low scaling rates. Unlike evaporators in power plants, saltwater evaporators operate for 600 to 700 hours before having to descale.

14.7 DESALINATION VIA EVAPORATION

Evaporation is a common process used in water desalination. It may also be used to treat a variety of waste streams, including liquids, slurries, sludges, organic and inorganic streams, streams containing suspended or dissolved solids, and streams containing nonvolatile dissolved liquids. In waste treatment facilities, it has been used in processing radioactive wastes, wastes from paper mills and molasses distilleries, and wastes from trinitrotoluene (TNT) manufacturing.

The process and equipment used in evaporation are similar to that of distillation, except the vapor is not collected and condensed unless organic components are present. Usually the waste stream flows through metal pipes that are heated by low-pressure steam outside the pipe walls. Other modes of operation that have been used are the solar evaporation from ponds or the heating of open vessels. The process concentrates the wastes of the original feed stream and reduces its volume. Although evaporation is an energy intensive process and utility and equipment costs may be high, its use for desalination is certainly a viable option to be promoted in the future.

Details on this as well as other desalination processes are provided in Chapter 19 of this text.

14.8 APPLICATIONS

Three Illustrative Examples complement the material presented in this chapter related to the concepts and practices of industrial evaporation processes.

Illustrative Example 14.1

Qualitatively describe what an evaporator does.

Solution

This is an open-ended question. Standiford (1963) provides one answer to this example. The requirements for the correct functioning of any evaporator are:

1. It must transfer a great deal of heat, on the order of 1,000 Btu/lb of water evaporated. This, more than anything else, determines the type, size, and cost of each effect of the evaporator.
2. It must efficiently separate the vapor from the residual liquid. What is efficient in one evaporator may be many orders of magnitude different from what is efficient in another (e.g., from a loss of salt value at only a few dollars a ton, to lithium chemicals valued at a dollar a pound, to radioactive waste). Separation may to be important solely because of the value of product lost, pollution problems, or because of fouling or corrosion in the equipment in which the vapor is condensed or in which the condensate is subsequently used.
3. It must make as efficient use of the available heat or mechanical energy as is economically feasible. This means using the vapor evaporated in one part (effect) of the evaporator as the heating steam in another effect that is operating at a lower temperature (as in a multiple-effect evaporator), or compressing the vapor evolved so that it can be used as the heating medium in the same evaporator, or by employing a combination of these. As in most other cases, efficiency is usually gained only as capital cost is increased, and for evaporators, the designer has a wide range to choose from.
4. It must meet conditions dictated by the characteristics of either the liquid being evaporated or the product. In a crystallizing evaporator, crystal size, shape, and purity may be of the utmost importance. If a salting or scaling liquid is to be handled, the evaporator type selected must be capable of dealing with it. Product quality characteristics that may also be important are corrosiveness and degradation at high temperature, long holdup time, or contact with certain metals.

Other considerations are the size of the operation, any foaming characteristics of the liquor, the need for special types of materials of construction such as polished stainless steel (required for many food products), and easy access for cleaning.

Illustrative Example 14.2

An evaporator is to be fed 5,000 lb of solution containing 2% dissolved solids by weight The feed, F, is at a temperature of 100°F. It is to be concentrated to a solution of 5% solute by weight in an evaporator operating at a pressure of 1 atm in the vapor space. To carry out the evaporation, the heating surface is supplied with steam at 5 psig (227°F), and the overall heat transfer coefficient of the evaporator, U, is 280 Btu/hr-ft^2-°F. What is the mass of vapor produced, the total mass of steam required, and the surface area required? Neglect enthalpy of solution effects adapted from (Badger and Banchero 1955).

Solution

To simplify the solution, it will be assumed that the solution is so dilute that its boiling point is the same as the boiling point of water and that its heat capacity and latent enthalpy are the same as that of water. Under these circumstances, the thermal properties of the solution (both feed and product) and of the steam can be taken from the steam tables. This results in the following values for pertinent quantities on a per hour basis (refer to Figure 14.1):

$$F = 5{,}000 \text{ lb}$$

$$x_F = 0.02$$

Total solids in feed = (5,000) (0:02) = 100 lb (componential balance

Total water in feed = 5,000 − 100 = 4,900 lb (componential balance)

$$T_F = 100°\text{F}$$

h_F = 68 Btu/lb (estimated from steam tables at 100°F)

$$x_L = 0.05$$

Total solids in liquor = (5,000)(0.02) = 100 lb

$$L = 100/(0.05) = 2{,}000 \text{ lb}$$

T_L = 212°F (at 1 atm)

h_L = 180Btu/lb (estimated from steam tables)

V = 5,000 − 2,000 = 3,000 (overall balance)

$$T_V = 212°\text{F}$$

h_V = 1,156 Btu/lb (estimated from steam tables at 1.0 atm)

$$T_S = 227°\text{F}$$

h_S = 1,156 Btu/lb (estimated from steam tables at 227°F and 5 psig)

T_C = 227°F (condensate)

h_C = 195 Btu/lb (estimated from steam tables)

The following enthalpy (energy) balance (Equation 14.4) results (S representing steam):

$$F\ h_F + S\ h_S = V\ h_V + L\ h_L + S\ h_C \tag{14.4}$$

Substituting and solving yields:

(5,000 lb)(68 Btu/lb) + (S) (l,156 Btu/lb) = (3,000 lb) (1,150 Btu/lb) + (2000 lb) (180 Btu/lb) + (S)(l95 Btu/lb); S = 3,611 lb

The total heat requirement is:

$$Q = (3{,}611 \text{ lb})(1156 - 195 \text{ Btu/lb}) = 3{,}470{,}000 \text{ Btu}$$

The required area is (assuming all of the above is based on 1 hour) is determined from rearrangement of Equation 14.1 as:

$$A = Q/U\Delta T = 3{,}470{,}000 \text{ Btu/hr}/\left[(280 \text{ Btu/hr-ft}^2 - {}^\circ\text{F})(227 - 212\ {}^\circ\text{F})\right] = 826 \text{ ft}^2$$

Illustrative Example 14.3

An evaporator is fed with 5,000 lb/hr of a 20% solution of sodium hydroxide at 100°F. This is to be concentrated to a 40% solution. The evaporator is supplied with saturated steam at 5 psig. Although the unit operates with the vapor space at a pressure of 4 in Hg absolute, the boiling temperature of the solution in the evaporator is 198°F (due to the superheat created by the exposed heating element). The overall heat transfer coefficient is 400 Btu/hr-ft^2-°F. Calculate the steam rate and the required heat transfer area.

Solution

Assume a basis of 1 hour. The following data are known:

$$F = 5{,}000 \text{ lb/hr}$$

$$x_F = 0.20$$

$$T_F = 100^\circ\text{F}$$

$$x_L = 0.40$$

$$x_V = y_V = 0$$

From the steam tables at 228°F and 5 psig,

$$h_S = 1{,}156 \text{ Btu/lb}$$

$$h_C = 196 \text{ Btu/lb}$$

$$h_{vap} = 1{,}156 - 196 \text{ Btu/lb} = 960 \text{ Btu/lb}$$

In addition,

Boiling point of water at 4 in Hg absolute = 125.4°F

Enthalpy of saturated steam at 125°F = 1,116 Btu/lb (estimated from steam tables)

$h_F(100^\circ\text{F}) \approx 55$ Btu/lb (estimated from Theodore et al. 2009)

$h_L(198^\circ\text{F}) \approx 177$ Btu/lb (estimated from Theodore et al. 2009)

A componential balance from Equation 14.3 yields:

$$F\,x_F = L\,x_L + V\,y_V \tag{14.3}$$

$$(5{,}000 \text{ lb/hr})(0.20) = (0.40)\,(L);\ L = 2{,}500 \text{ lb/hr}$$

From the material balance from Equation 14.2:

$$F = L + V \tag{14.2}$$

$$V = 5{,}000 \text{ lb/hr} - 2{,}500 \text{ lb/hr} = 2{,}500 \text{ lb/hr}$$

An enthalpy balance from rearrangement of Equation 14.4 yields:

$$F\ h_F + S\ (h_S - h_C) = V\ h_V + L\ h_L \tag{14.4}$$

In calculating h_V, the enthalpy of vapor leaving the solution, it should be remembered that this vapor is in equilibrium with the boiling solution at a pressure of 4 in Hg absolute and, therefore, is superheated in comparison with vapor in equilibrium with water at the same pressure. Because the heat capacity of superheated steam in this range may be assumed to be 0.46 Btu/lb-°F, then

$$h_V = 1{,}116 \text{ Btu/lb} + (0.46 \text{ Btu/lb-°F})(198 - 125\text{°F}) = 1{,}150 \text{ Btu/lb}$$

Substituting into the enthalpy balance gives:

$$(5000 \text{ lb/hr})(55.0 \text{ Btu/lb}) + S(1{,}156 - 196 \text{ Btu/lb}) = (2{,}500 \text{ lb/hr})\ (1{,}150 \text{ Btu/lb}) +$$
$$(2{,}500 \text{ lb/hr})(177 \text{ Btu/lb})$$

Solving for S, the steam rate, yields:

$$S = 3{,}170 \text{ lb/hr}$$

The area may now be calculated using the standard heat exchanger equation, Equation 14.1:

$$Q = UA\Delta T \tag{14.1}$$

$$(3{,}170 \text{ lb/hr})(1{,}156 - 196 \text{ Btu/lb}) = (400 \text{ Btu/hr-ft}^2\text{-°F})\ A(228\text{–}198\text{°F})$$

$$A = (3{,}043{,}200 \text{ Btu/hr})/(12{,}000 \text{ Btu/hr-ft}^2) = 253.6 = 254 \text{ ft}^2$$

REFERENCES

Badger, W., and J. Banchero. 1955. *Introduction to Chemical Engineering*. New York: McGraw-Hill.

Farag, I., and J. Reynolds. 1996. *Heat Transfer. A Theodore Tutorial*. East Williston, NY: Theodore Tutorials. Originally published by the U.S. EPA/APTI, RTP, NC.

Flynn, A.M., T. Akashige, and L. Theodore. 2019. *Kern, Process Heat Transfer*, 2nd edition. Salem, MA: Schrivener-Wiley.

Kern, D. 1950. *Process Heat Transfer*. New York: McGraw-Hill.

Standiford, F.C., Jr. 1963. Evaporation is a Unit. *Chemical Engineering* 79(25):158–176.

Theodore, L. 2011. *Heat Transfer for the Practicing Engineer*. Hoboken, NJ: John Wiley & Sons.

Theodore, L. 2014. *Chemical Engineering: The Essential Reference*. New York: McGraw-Hill.

Theodore, L., F. Ricci, and T. VanVleet. 2009. *Thermodynamics for the Practicing Engineer*. Hoboken, NJ: John Wiley & Sons.

15 Membrane Separation Processes

15.1 OVERVIEW

There is virtually an infinite amount of water in the oceans, but because of an average salt content of 3.5 g/mL, it is essentially unusable for any beneficial use to humans without significant treatment. There are several processes that are available, however, that can be used to purify this salt water to render it usable even for drinking water purposes. These processes remove or separate the salt from the water and include the three principle processes of:

1. Evaporation
2. Distillation
3. Membrane separation

Evaporation is reviewed in the previous chapter and one of the authors (Theodore 2015) believes that this will become the desalination process of the future. Distillation is the backbone of the chemical engineering curriculum in academia, and details of the process and design considerations are available in the literature (Theodore and Ricci 2010; Theodore 2016). Membrane separation processes have come to the forefront in the last three decades and remain the principle method of purifying salt water. This chapter summarizes a wide range of membrane separation process topics by addressing membrane separation principles, reverse osmosis, ultrafiltration, microfiltration, gas permeation, pervaporation, and electrodialysis. An application section provides three Illustrative Examples related to the general subject of membrane separation and membrane separation processes.

15.2 MEMBRANE PROCESSES

Membrane processes today are state-of-the-art separation technologies that have shown promise for future technical growth and wide-scale industrial commercialization. They are used in many industries for process stream and product concentration, purification, separation, and fractionation, as well as for potable water production for marginal water sources. The use of membranes in wastewater treatment systems to replace conventional sedimentation tanks is also expanding. The need for membrane research and development is important because of the increasing use of membrane technology in both traditional and emerging chemical and environmental engineering fields. Membrane processes are increasingly finding their way into the growing engineering application areas of biotechnology, green engineering, specialty chemical manufacture, and biomedical engineering, as well as in the traditional chemical and environmental process industries. Membrane technology is also being considered as either a replacement for or supplement to traditional separations such as sedimentation, distillation, or extraction. Membrane processes can be more efficient and effective because they can simultaneously concentrate and purify and can also perform separations at ambient conditions in small or existing reactor volumes.

Membranes create a boundary between different bulk gas or liquid mixtures. Different solutes and solvents flow through membranes at different rates; this enables the use of membranes in separation processes. Membrane processes can be operated at moderate temperatures for sensitive components (food, pharmaceuticals, etc.). Membrane processes also tend to have low relative capital and energy costs. Their modular format permits simple and reliable scale-up or retrofitting in existing process tanks.

Key membrane properties include their size rating, selectivity, permeability, mechanical strength, chemical resistance, low fouling characteristics, high capacity, low cost, and consistency. Vendors characterize their equipment with ratings indicating the approximate size (or corresponding molecular weight) of components retained by their membranes. Commercial membranes consist of polymers or ceramics. Other membrane types include sintered metal glass and liquid films.

15.3 MEMBRANE SEPARATION PRINCIPLES

Membrane unit operations are often characterized by the following parameters: (i) driving force used, (ii) membrane type and structure, and (iii) species being separated. Reverse osmosis (RO), nanofiltration (Theodore 2006), ultrafiltration (UF), and microfiltration (MF) all use a pressure difference as the driving force to separate a liquid feed into a liquid *permeate* and *retentate*. They are listed in ascending order in their ability to separate a liquid feed on the basis of solute size. RO uses nonporous membranes and can separate down to the ionic level (e.g., seawater in the rejection of dissolved salt). Nanofiltration performs separations at the nanometer range. UF uses porous membranes and separates components of molecular size ranging from the low thousand to several hundred thousand atoms; an example includes components in biomedical processing. MF uses much more porous membranes and is typically employed in the micro- or macromolecular range to remove particulate or larger biological matter from a feed stream (e.g., in the range of 0.05 to 2.5 μm) (Schweitzer 1979; Applegate 1984).

Gas separation processes can be divided into two categories: gas permeation through nonporous membranes and gas diffusion through porous membranes. Both of these processes use a concentration difference across the membranes as the driving force for separation. The gas permeation processes are used extensively in industry to separate air into purified nitrogen and enriched oxygen. Another commercial application is hydrogen recovery in petroleum refineries. As noted previously, dialysis membrane processes use a concentration difference as a driving force for separation of liquid feed across a semipermeable membrane, with the major application in the medical field of hemodialysis. *Electrodialysis* separates a liquid employing an electric charge difference as the driving force and is widely used in water purification and industrial processing.

As noted, the main difference between RO and UF is that the overall size and diameter of the particles or molecules in solution to be separated is smaller in RO. In MF, the particles to be separated or concentrated are generally solids or colloids rather than molecules in solution. Figure 15.1 illustrates the different among the processes.

Naturally, the heart of the membrane process is the membrane itself; it is an ultrathin semipermeable barrier separating two fluids that permits the transport of certain species through the membrane barrier from one fluid to the other. As noted previously, a membrane is typically produced from various *polymers* such as cellulose acetate or polysulfone, but ceramic and metallic membranes are also used in some applications. The membrane is referred to as *selective* because it permits the transport of certain species while rejecting others. The term *semipermeable* is frequently used to describe this selective action.

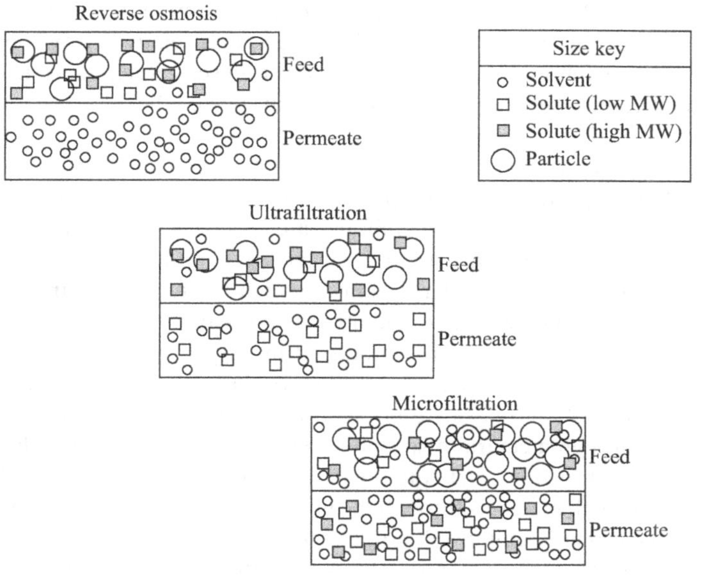

FIGURE 15.1 Three common membrane separation processes.

15.4 REVERSE OSMOSIS (RO)

The most widely commercialized membrane process by far is RO. It belongs to a family of pressure-driven separation operations for liquids that includes not only RO, but also UF and MF.

RO is an advanced separation technique that may be used when low molecular weight (MW) solutes such as inorganic salts or small organic molecules (e.g., glucose) are to be separated from a solvent (usually but not always water). In normal (as opposed to reverse) osmosis, water flows from a less concentrated salt solution to a more concentrated salt solution as a result of driving forces. As a result of the migration of water, an osmotic pressure is created on the side of the membrane to which water flows. In reverse (as opposed to normal) osmosis, the membrane is permeable to the solvent or water and relatively impermeable to the solute or salt. To make water pass through an RO membrane in the desired direction (i.e., away from the concentrated salt solution), a pressure must be applied that is higher than that of the osmotic pressure.

RO is widely used today by a host of chemical and environmental process industries for a surprisingly large number of operations. Aside from the classic example of RO for seawater desalination (Dupont et al. 1982), it has found a niche in the food industry for concentration of various fruit juices, in the galvanization industry for concentration of waste streams, and in the dairy industry for concentration of milk prior to cheese manufacturing (Parkinson 1983; Applegate 1984; Brooks 1985).

RO processes are classified into the following two basic categories:

1. *Purification* of the solvent such as in desalination where the permeate or purified water is the product.
2. *Concentration* of the solute such as in concentration of fruit juices where the retentate is the product.

The membranes used for RO processes are characterized by a high degree of semi-permeability, high water flux, mechanical strength, chemical stability, and relatively low operating and high capital costs. Early RO membranes were composed of cellulose acetate, but restrictions on process stream pressure, temperature, and organic solute rejection spurred the development of noncellulosic and composite materials.

RO membranes may be configured or designed into certain geometries for system operation: plate and frame, tubular, spiral wound (composite), and hollow fiber (Brooks 1985). In the *plate-and-frame* configuration, flat sheets of membrane are placed between spacers with heights of approximately 0.5 to 1.0 mm. These are, in turn, stacked in parallel groups. *Tubular* units are also commonly used for RO. This is a simpler design in which the feed flows inside of a tube whose walls contain the membrane. These types of membranes are usually produced with inside diameters ranging from 12.5 to 25 mm and generally produced in lengths of 150 to 610 cm. There is also the *hollow fine-fiber* (HFF) arrangement used in approximately 70% of worldwide desalination applications. Millions of hollow fibers are oriented in parallel and fixed in epoxy at both ends. A feed stream is sent through a central distributor where it is forced out radially through the fiber bundle. As the pressurized feed contacts the fibers, the permeate is forced into the center of each hollow fiber. The permeate then travels through the hollow bore until it exits the permeator. A *spiral-wound* cartridge is occasionally employed in this configuration. Here, the solvent is forced inward toward the product tube while the concentrate remains in the space between the membranes. A flat film membrane is made into a "leaf." Each leaf consists of two sheets of membrane with a sheet of polyester tricot in between that serves as a collection channel for the water product. Plastic netting is placed between each leaf to serve as a feed channel. Each leaf is then wrapped around the product tube in a spiral fashion.

It is no secret that water covers around 70% of Earth's surface, but 97.5% of it is unfit for human consumption. With the world facing a growing freshwater shortage from which the United States may not be spared, one method of producing fresh water is desalination. The major application of RO is water desalination. Some areas of the world that do not have a ready supply of fresh water have chosen to desalinate seawater or brackish water using RO to generate potable drinking water. Because no heating or phase change is required, the RO process is both a relatively low energy and economical water purification process. A typical saltwater RO system consists of an intake, a pretreatment component, a high-pressure pump, membrane apparatus, remineralization, and pH adjustment components, as well as a disinfection step. A pressure difference driving force of about 1.0 to 6.5 mPA is generally required to overcome the osmotic pressure of salt water.

Another important membrane application is *dialysis*. This technique is used in patients who suffer from kidney failure and can no longer filter waste products (urea) from the blood. In general, RO equipment used for dialysis can reduce ionic contaminants by up to 90%. In this process, the patient's blood flows in a tubular membrane while a dialysate flows countercurrently on the outside of the feed tube. The concentrations of undesirable salts (e.g., potassium, calcium, urea) are high in the blood (while low or absent in the dialysate). This treatment successfully mimics the filtration capabilities of the kidneys.

RO membranes are designed for high salt retention, high permeability, mechanical robustness (to allow module fabrication and withstand operating conditions), chemical robustness (for fabrication materials, process fluids, cleaners, and sanitizers), low extractables, low fouling characteristics, high capacity, low cost, and consistency. The predominant RO membranes used in water applications include cellulose polymers, thin composites consisting of aromatic polyamides, and cross-linked polyetherurea.

The membrane operation for the purification of seawater that incorporates a selective barrier can be simply described using the line diagram provided in Figure 15.2. This membrane operation typifies the case in which a feed stream (seawater) is separated by a semipermeable membrane that rejects salt but selectively transports water. A purified stream (the permeate) is, therefore, produced, while at the same time, a concentrated salt stream (the retentate) is discharged. With reference to

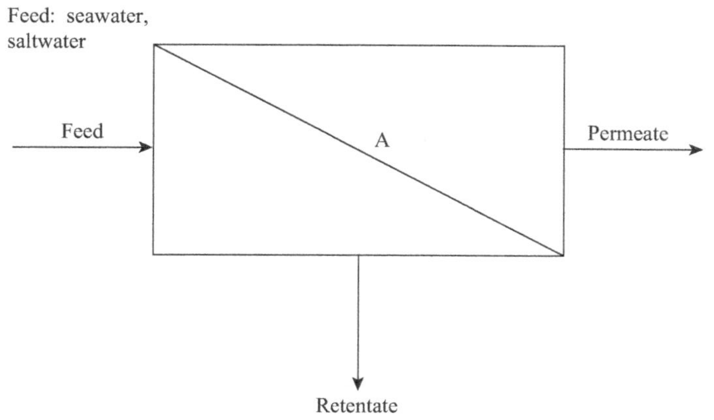

FIGURE 15.2 Desalination of seawater by reverse osmosis (RO).

Figure 15.2, a simple material balance can be written on the overall process flows and for that of the solute as:

$$q_f = q_r + q_p \tag{15.1}$$

$$C_f q_f = C_r q_r + C_p q_p \tag{15.2}$$

where q is the volumetric flow rates, C is the solute concentration, and subscripts f, r, and p refer to the feed, retentate, and permeate, respectively.

Osmosis occurs when a concentrated solution is partitioned from a pure solute or relatively lower concentration solution by a semipermeable membrane. The semipermeable membrane allows only the solvent to flow through it freely. Equilibrium is achieved when the solvent from the lower-concentration side ceases to flow through the membrane to the higher-concentration side (thus reducing the concentration) because the mass transfer concentration difference driving force has been reduced. This is shown in Figures 15.3 and 15.4. *Osmotic pressure* is the pressure needed to terminate the flux of solvent through the membrane or the force that pushes on the concentrated side of the membrane (see Figure 15.5). Applying a pressure on the concentrated side halts the solvent flux. *Reverse osmosis* (see Figure 15.6) takes place when an applied force (pressure) overcomes the osmotic pressure and forces the solvent from the concentrated side through the membrane, leaving the solute on the concentrated side.

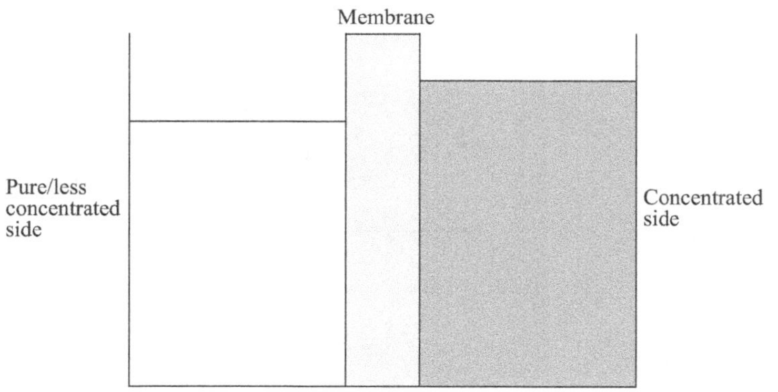

FIGURE 15.3 Pre-osmosis equilibrium.

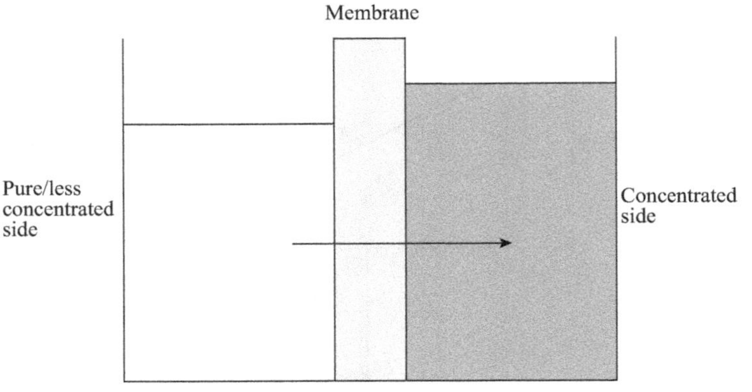

FIGURE 15.4 Osmosis of a solvent.

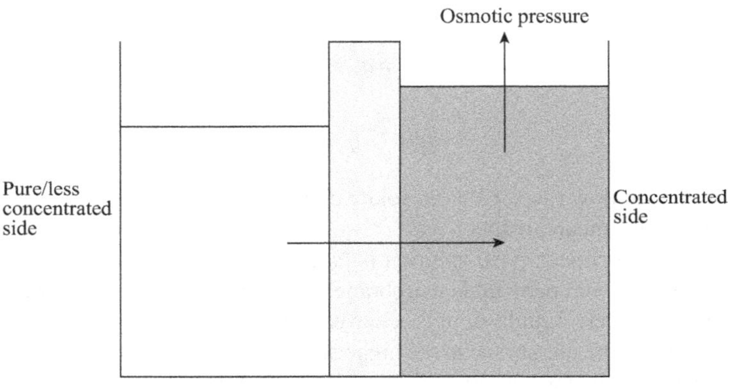

FIGURE 15.5 Osmotic pressure.

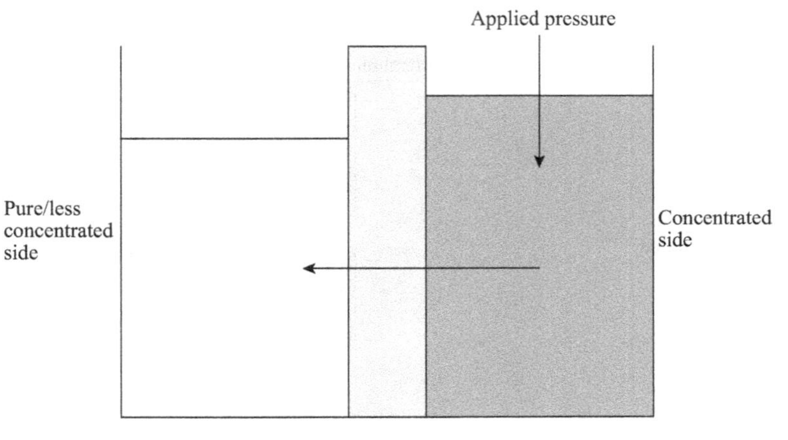

FIGURE 15.6 Reverse osmosis.

Osmotic pressure is related to both the solute concentration and the temperature of the solution as described in the Van't Hoff equation:

$$\pi = iC_sRT \tag{15.3}$$

where π is the osmotic pressure, psi; i is the Van't Hoff factor, dimensionless; C_s is the solute concentration mol/L; R is the universal gas constant (L-atm)/(mol-K); T is the absolute temperature, K.

The Van't Hoff factor, i, in Equation 15.3 factors in the number of ions in solution. For example, NaCl separates into two ions, Na^+ and Cl^-, therefore making the Van't Hoff factor equal to 2. Closer inspection of Equation 15.3 reveals that the Van't Hoff equation is analogous to the ideal gas law.

The change in osmotic pressure, $\Delta\pi$, across the membrane in this operation must be overcome to achieve RO. This is described in the following equation:

$$\Delta\pi = \pi_f - \pi_p \tag{15.4}$$

where π_f is the osmotic pressure in the feed, psi; and π_p is the osmotic pressure in the permeate, psi.

This change in osmotic pressure, $\Delta\pi$, can also be calculated using the concentrations of both the feed and the permeate as well as an empirical coefficient denoted as ψ. This formula is

$$\Delta\pi = \psi(C_f - C_p) \tag{15.5}$$

where ψ is a constant, L-psi/g; C_f is the feed concentration, g/L; and C_p is the permeate concentration, g/L.

The permeate flux, J_p, is an important characteristic of an RO system. It is related to the permeate flow as well as to the area of the membrane. This is represented in the following equation:

$$J_p = \frac{q_p}{A_m} \tag{15.6}$$

where A_m is the membrane surface area, ft^2; q_p is the permeate flow, gal/day; and A_m is the membrane surface area, ft^2.

The flux can be determined experimentally by measuring each incremental volume of permeate, ΔV, collected in time period Δt, and dividing by the surface area of the membrane. In water-based processes such as desalination, the permeate naturally consists mostly of water. Therefore, the permeate flux can be considered to be equal to *water flux*. Equation 15.7 defines this water flux:

$$J_p = J_w = A_w(\Delta P - \Delta\pi) = \frac{K_s}{t_m}(\Delta P - \Delta\pi) \tag{15.7}$$

where J_w is the water flux, gal/(day-ft^2); A_w is the water permeability coefficient, gal/(day-ft^2-psi); ΔP is the pressure drop, psi; K_s is the permeability constant, gal/(day-ft-psi); and t_m is the membrane thickness, ft.

The water permeability coefficient can, therefore, be experimentally determined by obtaining data on the unit with pure water; this eliminates the $\Delta\pi$ term because both sides of the membrane contain pure water. Water permeability coefficients supplied by a manufacturer is generally provided at 25°C. As temperature varies water diffusivity and viscosity vary, affecting the water flux rate through these membranes. Required membrane area correction factors, A_{wT}/A_{w25}, the area required at temperature, T, compared with that at 25°C, have been reported by Kaup (1973) to vary as follows: T = 10°C, 1.58; T = 15°C, 1.34; T = 20°C, 1.15; T = 25°C, 1.0; T = 30°C, 0.84.

Another important factor is the *solute flux*, J_s, which can be determined by:

$$J_s = \frac{q_p C_p}{A_m} = J_p C_p = \frac{\dot{m}_p}{A_m} \tag{15.8}$$

where J_s, is the solute flux, g/(day-ft^2); and $\dot{m}_p$ is the permeate mass flow rate, g/d.

The solute flux can also be related to the solute concentration by using the solute permeability factor. This relationship is given by:

$$J_s = B_s(\Delta C_s) \tag{15.9}$$

where B_s is the solute permeability coefficient, gal/(day-ft^2); and ΔC_s is the change in solute concentration, g/L, with appropriate unit conversions to yield proper units for J_s.

The selectivity of a membrane for a given solute can be expressed as the *percent rejection* (*%R*). Percent rejection represents a membrane's effectiveness and is a measure of the membrane's ability to selectively allow certain species to permeate and others to be rejected. This is an important characteristic or consideration when selecting a membrane for a separation process. The percent rejection represents the percentage of solute that is *not* allowed to pass into the permeate stream and is given by:

$$\%R = \left(\frac{C_f - C_p}{C_f}\right) \times 100 \tag{15.10}$$

Rejections are sometimes written in terms of the average concentration of the feed and the concentrate because the feed becomes the concentrate as water moves through the membrane unit. The percent rejection of a given solute will vary from the entry point of the system to the final membrane unit. Percent rejection values are high (i.e., >95%) for most solutes except for gases and small organic molecules.

Finally, the solvent recovery, *Y*, is a measure of the quantity of solvent that is allowed to pass through the membrane. This is defined as the quotient of the permeate flow divided by the feed flow, as shown by Equation 15.11.

$$Y = \frac{q_p}{q_f} \tag{15.11}$$

where q_f is the feed flow rate, gal/d.

The concentration of solute in the permeate is a function of the water flux, which is dependent on the osmotic pressure (i.e., the concentration of the solute on the feed-concentrate side of the membrane). Solute concentration is also a function of the solute flux, which is dependent on the solute concentration difference across the membrane.

The concentrate concentration is important with respect to fouling and scaling, which must be considered in both the design and operation of an NF or RO system. If the solubility of salts are exceeded there is a potential for precipitation on the membrane surface. In some cases, as with calcium carbonate, pH adjustment with an acid dosage to the feed can control the precipitation. With other compounds such as calcium sulfate and silica, polymeric anti-scalants are needed to control scaling. Fouling can be caused by the accumulation of natural organic matter on the membrane surface, and periodic offline cleaning of the membranes is needed.

The actual concentrate concentration, C_c, can be predicted by combining Equations 15.1, 15.2, 15.10, and 15.11, and expressing C_c in terms of *Y* and *R* as shown in Equation 15.12.

$$C_C = \frac{[1 - Y(1 - R)]C_f}{1 - Y} \tag{15.12}$$

15.5 ULTRAFILTRATION (UF)

UF is a membrane separation process that can be used to concentrate single solutes or mixtures of solutes. Transmembrane pressure (the membrane pressure drop) is the main driving force in UF operations, and separation is achieved via a sieving mechanism. The UF process can be used for the treatment or concentration of oily wastewater, for pretreatment of water *prior* to RO, and for the removal of bacterial contamination (pyrogens). In the food industry, UF is used to separate lactose and salt from cheese whey proteins, to clarify apple juice, and to concentrate milk for ice cream and cheese production (Maubois 1980; Kosikowski 1986; Garcia et al. 1989). The most energy-intensive step in ice cream production is concerned with the concentration of skimmed milk, where membrane processes are more economical for this step than vacuum evaporation (Garcia et al. 1989). UF processes are also used for concentrating or dewatering fermentation products as well as purifying blood fractions and vaccines.

UF may be regarded as a membrane separation technique where a solution is introduced on one side of a membrane barrier, while water, salts, or other low-molecular-weight materials pass through the unit under an applied pressure. As noted, these membrane separation processes can be used to concentrate single solutes or mixtures of solutes. The variety in the different membrane materials allows for a wide temperature-pH processing range.

The main economic advantage of UF is a reduction in both design complexity and energy usage because UF processes can simultaneously concentrate and purify process streams. The fact that no phase change is required leads to highly desirable energy savings. A major disadvantage is the high capital cost that might be required if low flux rates for purification demand a large system. However, UF processes are usually economically sound in comparison to other traditional separation techniques.

In addition to the applications described, UF membrane processes are used in various commercial applications. They are found in the treatment of industrial effluents and process water; in the concentration, purification, and separation of macromolecular solutions in the chemical, food, and drug industries; in the sterilization, clarification, and pre-filtering of biological solutions and beverages; and in the production of ultrapure water and pretreating of seawater in RO processes. The most promising area for the expansion of UF process applications is the biochemical industry. Some of its usage in this area includes purifying vaccines and blood fractions; concentrating gelatins, albumin, and egg solids; and recovering proteins and starches.

The rejection of a solute is a function of the size, size distribution, shape, and surface binding characteristics of the hydrated molecule. It is also a function of the pore size distribution of the membrane. Therefore, MW cut-off values can be used only as a rough guide for membrane selection. The retention efficiency of the solutes depends to a large extent on the proper selection and condition of the membrane. Replacement of highly used membranes and regular inspections of the separation units averts many problems that might otherwise occur because of membrane clogging and gel formation.

Ultrafiltration processes use driving forces of 0.2 to 1.0 MPa to drive liquid solvent (usually water) and small solutes through membranes while retaining solutes of 10 to 1,000 Å diameter. The membranes consist primarily of polymeric structures, such as polyethersulfone, regenerated cellulose, polysulfone, polyamide, polyacrylonitrile, or various fluoropolymers. Hydrophobic polymers are surface modified to render them hydrophilic.

The general design factors for any membrane system (including UF), as reported by Wankat (1990), are:

1. Thin, active layer of membrane
2. High permeability for Species A and low permeability for Species B
3. Stable membrane with long service life
4. Mechanical strength

5. Large surface area of membrane in a small volume
6. Elimination or control of concentration polarization
7. Ease in cleaning
8. Low construction costs
9. Low operating costs

UF system performance is usually defined in terms of the permeate flux, J_p, with dimensions of volume/(unit area-time) as was seen with RO systems. Typical units as shown in Equation 15.6 are gal/(day-ft^2). As with RO, permeate flux can be determined experimentally by measuring each incremental volume of permeate, ΔV, collected in time period, Δt, and dividing by the surface area of the membrane.

Transmembrane pressure is the main driving force in UF operations, and separation is achieved through the sieving mechanism mentioned above. Because UF is a pressure driven separation process, it is appropriate to examine the effects of pressure on flux. Equation 15.7 applies to UF operations and indicates that the water flux through a UF membrane is directly proportional to the applied pressure gradient across it and inversely proportional to the membrane thickness.

Such factors as the membrane porosity, pore size distribution, and viscosity of the solvent are accounted for by the permeability constant, K_s, in Equation 15.7 When membrane thickness is not available or is not known, the water permeability coefficient, A_w, in Equation 15.7 may be used in place of K_s. This water permeability coefficient is a function of the distribution coefficient (solubility), diffusion coefficient, membrane thickness, and temperature. The value of A_w can be determined by conducting ultrapure water flux experiments at varying operating pressures, whereas permeate collection occurs at or near atmospheric pressure.

Because the pressure is relatively low for macromolecular solutions, which are typically the ones recommended for UF processes, the $\Delta\pi$ term in Equation 15.7 can be neglected. This is because the molar concentration of the high-MW molecules separated by UF is low, even when the mass concentrations are high. When the $\Delta\pi$ term is neglected (Maubois 1980; Theodore and Ricci 2010) and A_w is used, Equation 15.7 simplifies to:

$$J_w = A_w \, \Delta P \tag{15.13}$$

When a solute, such as solids dissolved in water, flows through a typical UF process, some of the solute usually passes through the membrane because real membranes are *partially* permeable. The *apparent rejection* on a percentage basis, R_{app}, is then calculated in a manner similar to that described in Equation 15.10:

$$R_{app} = \left(\frac{C_r - C_p}{C_r}\right) \times 100 \tag{15.14}$$

where C_r is the retentate concentration, g/L.

Three additional factors need to be considered in UF separations: concentration polarization, gel formation, and fouling. A concentration gradient or boundary layer typically forms during UF operation. This concentration difference driving force appears near the membrane surface and is referred to as *concentration polarization*. It results from the counteracting effects of the convective flow of solute toward the membrane and diffusion of the solute toward the bulk fluid in the reverse direction. Although concentration polarization is regarded as a reversible boundary-layer phenomenon that causes a rapid initial drop in flux to a steady-state value, fouling is considered as an irreversible process that leads to a flux decline over the long term. The process of *gel formation*, however, may be reversible or irreversible. When the gel is difficult to remove, the membrane is said to be *fouled* and this gel formation is irreversible. Concentration polarization may occur with or without gel formation.

Concentration polarization occurs in many separations and for large solutes where the osmotic pressure can be safely neglected. Concentration polarization without gelling is predicted to have no effect on the flux. Therefore, if a flux decline is observed, it is usually attributed to the formation of a gel layer. The gel layer, once formed, usually controls mass transfer so that Equation 15.7 is no longer applicable (Porter 1990).

15.6 MICROFILTRATION (MF)

MF can separate particles from liquid or gas phase solutions. The usual materials retained by an MF membrane range in size from several micrometers down to 0.2 μm. Very large soluble molecules are retained by a microfilter at the low end of this spectrum. MF is employed in modern industrial biochemical and biological separation processes. For example, MF can be used instead of centrifugation or precoat rotary vacuum filtration to remove yeast, bacteria, or mycelia organisms from fermentation broth in cell harvesting. Both MF and UF are used for cell harvesting. Microfiltration is used to retain cells and colloids, while allowing passage of macromolecules into the permeate stream. UF is also used to concentrate macromolecules, cells, and colloidal material, while allowing small organic molecules and inorganic salts to pass into the permeate stream. As stated previously, pore sizes in MF are around 0.10 to 10 μm in diameter as compared with 0.001 to 0.02 μm for UF (ranges vary slightly depending on the manufacturer and specific application) (Porter 1990; Hollein et al. 1995; Mulder 1996).

Similar types of equipment are used for MF and UF; however, membranes with larger pore sizes are generally employed in MF applications. MF processes operate at lower pressure than UF but at a higher pressure difference driving force (PDDF) than does conventional particulate filtration (Porter 1990). Ideal membranes possess high porosity, a narrow pore size distribution, and a low binding capacity.

When separating microorganisms and cell debris from fermentation broth, a biological cake is formed. Principles of cake filtration (Porter 1990; Theodore 2015) apply to MF systems, except that the small size of yeast particles produces a cake with a relatively high resistance to flow, producing a relatively low filtration rate.

In *dead-end filtration*, feed flow is perpendicular to the membrane surface, and the thickness of the cake layer on the membrane surface increases with filtration time. Consequently, the permeation rate decreases with time as well. *Cross-flow filtration*, on the other hand, features feed flowing parallel to the membrane surface, which is designed to decrease formation of a cake by *sweeping* previously deposited solids from the membrane surface and returning them to the bulk feed stream. Cross-flow filtration is far superior to dead-end filtration for cell harvesting because the biological cake is highly compressible, which causes the accumulated layer of biomass to rapidly blind the filter surface in dead-end operation. Therefore, MF experiments are often conducted using cross-flow filtration because of the advantages that this filtration mode offers (Slater et al. 1989; Slater and Hollein 1993).

Separation principles and governing equations for MF are similar to those for RO and UF. Permeate or water flux can be expressed using a slightly modified form of Equation 15.7, ignoring the $\Delta\pi$ term as with UF and introducing a new term, R_m, the hydrodynamic resistance of the membrane:

$$J_p = \frac{K_s \Delta P}{t_m} = \frac{\Delta P}{\mu R_m} \tag{15.15}$$

where μ is the solution viscosity.

15.7 GAS PERMEATION

Gas separation processes are generally considered as relatively new and emerging technologies because they are not included in the curriculum of many traditional chemical and environmental engineering programs (Clements et al. 1986; Permea, Inc. 1986; Slater et al. 1989; Davis and Sandall 1990; Slater et al. 1990, 1992).

Gas permeation systems have gained popularity in both traditional and emerging engineering areas (Ismail et al. 2015). These systems were originally developed primarily for hydrogen recovery. There are presently numerous applications of gas permeation in industry, and other potential uses of this technology are in various stages of development. Applications today include gas recovery from waste gas streams, landfill gases, and ammonia and petrochemical products. Gas permeation membrane systems are also employed in gas generation and purification, including the production of nitrogen and enriched oxygen gases (Slater et al. 1990) and enrichment of biogas (Beil and Beyrich 2013).

Gas permeation (GP) is the term used to describe a membrane separation process in which a gaseous feed stream is fractionated into permeate and nonpermeate streams. The nonpermeating stream is typically referred to as the *nonpermeate* in gas separation terminology, although it is defined as the retentate in liquid separation. Transport separation occurs by a solution diffusion mechanism. Membrane selectivity is based on the relative permeation rates of the components through the membrane. Each gaseous component transported through the membrane has a characteristic individual permeation rate that depends on its ability to dissolve and diffuse through the membrane material. The mechanism for transport is based on solubilization and diffusion; the two describing relationships on which the transport is based are Fick's law (diffusion) and Henry's law (solubility) (Slater et al. 1990). Gas membranes generally work because individual gases differ in their solubility and diffusivity through the nonporous polymers they are made of. As noted previously, gas separation membranes separate gases from other gases. Some gas filters, which remove liquids or solids from gases, are MF membranes.

There must be a driving force for the process of permeation to occur. For gas separations, that force is the *partial pressure* of the gas on the feed side of the membrane. Because the ratio of the component fluxes determines the separation, the partial pressure of each component at each point is important in gas permeation systems. There are two ways of driving the process: employing a high partial pressure on the feed side of the membrane or providing a low partial pressure on the permeate side, the latter which may be achieved by either vacuum or inert-gas flushing.

15.8 PERVAPORATION AND ELECTRODIALYSIS

Pervaporation (PER) is a separation process in which a liquid mixture contacts a nonporous, semipermeable membrane. One component is transported through the membrane preferentially. It evaporates on the downstream side of the membrane, leaving as a vapor. The process is induced by lowering the partial pressure of the permeating component, usually by a vacuum or occasionally with an inert gas. The permeated component is then condensed or recovered as the product of interest.

Electrodialysis (ED) is a membrane separation process in which ionic species can be separated from water macrosolutes and all charged solutes. Ions are induced to move by an electric potential, and separation is facilitated by ion-exchange membranes. The membranes are highly selective, passing primarily either anions or cations. In the ED process, the feed solution containing ions enters a compartment whose walls contain either a cation exchange or an anion exchange membrane interposed in the path of a direct current generated by end electrodes. The selectivity of the membranes alternates so that those permeable to anions, such as Cl^-, are placed between membranes permeable to cations like Na^+. With saline water flowing between the membranes, the direct current passing through the water draws anions toward the anode from one compartment, while cations are drawn in a similar manner in the opposite direction toward the cathode. From this movement of ions, the water in alternate compartments is reduced in ionic concentration by increasing the salinity of the water in the other compartments. The two flows become the desalted product water and the reject brine. ED finds its principle application in the treatment of brackish water

Campione et al. (2018) recently published a comprehensive review of the fundamentals of the ED process and the development of membranes, process modeling tools, and recent applications and developments of ED-related processes and should be consulted for details of recent advances in the understanding of hydrodynamic and mass transport processes taking place in these complex membrane systems.

15.9 APPLICATIONS

Four Illustrative Examples complement the material presented in this chapter focused on membrane separation processes.

Illustrative Example 15.1

An RO unit is to demineralize 100,000 gal/d of treated tertiary effluent. The following data are available for this membrane system: the water permeability coefficient, A_w = 0.035 gal/(day-ft-psi) @ 25°C; the pressure drop across the membrane, P = 425 psi; the change in osmotic pressure, $\Delta\pi$, across the membrane = 45 psi; the lowest operating temperature will be 10°C. For 10°C, the A_{wT}/A_{w25} values is 1.58. Determine the water flux at 25°C and the corresponding membrane area required to meet production requirements as the lowest operating temperature expected for the source water.

Solution

Equation 15.7 is used to estimate the membrane water flux at 25°C. Substituting given data into Equation 15.7 yields:

$$J_p = A_w\left(\Delta P - \Delta\pi\right) \tag{15.7}$$

$$J_w\,@\,25°\text{C} = (0.035\text{ gal/(day} - \text{ft}^2 - \text{psi}))\left(425\text{ psi} - 45\text{ psi}\right) = 14.44\text{ gal/day} - \text{ft}^2$$

The area required to produce the required 100,000 gal/day at worst case feed water conditions of 10°C can be determined from the water flux calculated at 25°C and the membrane area correction factor, A_{wT}/A_{w25}. The required membrane area then is:

$$\text{Membrane area} = (100{,}000\text{ gal/day})/(14.44\text{ gal/day–ft}^2)(1.58) = (6{,}925\text{ ft}^2)(1.58)$$

$$= 10{,}942\text{ ft}^2\text{ total membrane area.}$$

Illustrative Example 15.2

An RO unit is to demineralize 200,000 gal/d of brackish water. The following data are available for this membrane system: the water permeability coefficient, A_w = 0.030 gal/(day-ft-psi) @ 25°C; the pressure drop across the membrane, P = 380 psi; the change in osmotic pressure, $\Delta\pi$, across the membrane = 45 psi; the lowest operating temperature will be 15°C. For 15°C, the A_{wT}/A_{w25} values is 1.34. The membrane area/unit volume of equipment = 2,500 ft²/ft³. Determine the water flux at 25°C and the corresponding membrane area required to meet production requirements as the lowest operating temperature expected for the source water. Also determine the space (ft³) required for the equipment.

Solution

Equation 15.7 is used to estimate the membrane water flux at 25°C. Substituting given data into Equation 15.7 yields:

$$J_p = A_w\left(\Delta P - \Delta\pi\right) \tag{15.7}$$

$$J_w\,@\,25°\text{C} = (0.030\text{ gal/(day} - \text{ft}^2 - \text{psi}))(380\text{ psi} - 45\text{ psi}) = 10.05\text{ gal/day} - \text{ft}^2$$

The area required to produce the required 200,000 gal/day at worst case feed water conditions of 15°C can be determined from the water flux calculated at 25°C and the membrane area correction factor, A_{wT}/A_{w25}. The required membrane area then is:

$$\text{Membrane area} = (200{,}000 \text{ gal/day})/(10.05 \text{ gal/day–ft}^2)(1.34) = (19{,}900 \text{ ft}^2)\ (1.34)$$

$$= 26{,}666 \text{ ft}^2 \text{ total membrane area.}$$

The total equipment space requirements is determined as follows:

$$\text{Equipment space required} = (26{,}666 \text{ ft}^2)/(2{,}500 \text{ ft}^2/\text{ft}^3) = 10.66 = 11 \text{ ft}^3$$

Illustrative Example 15.3

A low-pressure RO membrane system treating a brackish groundwater is operating at 80% recovery. The rejection for both calcium and sulfate is 95%. The feed water contains 100 mg/L of calcium and 200 mg/L of sulfate, and the K_{sp} for $CaSO_4$ is 2.5×10^{-5}. Determine the concentration of calcium and sulfate in the concentrate and determine if the solubility of $CaSO_4$ has been exceeded, posing serious scaling potential.

Solution

Equation 15.12 is used to determine the concentrate concentration for each of the ions of concern. Substituting given data yields:

$$C_c = \frac{[1-Y(1-R)]C_f}{1-Y} \tag{15.12}$$

$$C_{Ca^{2+}} = \frac{[1–0.80(1–0.95)]100 \text{ mg/L}}{1–0.80} = \frac{0.96(100 \text{ mg/L})}{0.20} = 480 \text{ mg/L}$$

$$C_{SO_4^{2-}} = \frac{[1–0.80(1–0.95)]200 \text{ mg/L}}{1–0.80} = \frac{0.96(200 \text{ mg/L})}{0.20} = 960 \text{ mg/L}$$

To check the solubility of $CaSO_4$, the individual ion concentrations must be converted to molar concentrations using their respective molecular weights, which are Ca = 40 g/gmol, SO_4 = 96 g/gmol. The respective molar concentrations are then calculated as follows:

$$\left[Ca^{2+}\right] = \frac{480 \text{ mg/L}}{40{,}000 \text{ mg/gmol}} = 0.012 \text{ gmol}$$

$$\left[SO_4^{2-}\right] = \frac{960 \text{ mg/L}}{96{,}000 \text{ mg/gmol}} = 0.010 \text{ gmol}$$

The product of these molar concentrations is then compared to the K_{sp} for $CaSO_4$

$$\left[Ca^{2+}\right] \times \left[SO_4^{2-}\right] = (0.012)(0.010) = 1.2 \times 10^{-4;} > K_{sp} = 2.5 \times 10^{-5}$$

The ion product exceeds the solubility product constant, so precipitation on the membrane is likely and an antiscalant should be used to control this potential scaling.

Illustrative Example 15.4

Using the RO membrane system described in Illustrative Example 15.1 to treat a brackish groundwater, determine the expected permeate concentration of chloride if the feed concentration is 4,500 mg/L. The solute permeability coefficient, B_s = 0.74 gal/(day-ft^2). Also determine the percent rejection of the membrane for Cl^-.

Solution

Equation 15.7 was used to determine the water flux rate through the membrane in Illustrative Example 15.1. The water flux rate from this problem was found to be 14.44 gal/day-ft^2 @ 25°C. Equation 15.9 can be used to determine the solute flux rate. Substitution into Equation 15.9 yields:

$$J_s = B_s\left(\Delta C_s\right) \tag{15.9}$$

$$J_s = B_s\left(C_f - C_p\right) \tag{15.16}$$

It can also be seen from Equation 15.8, that

$$J_s = J_p C_p \tag{15.8}$$

Substitution and rearrangement yields:

$$B_s\left(C_f - C_p\right) = J_p C_p;\ B_s C_f - B_s C_p = J_p C_p;\ B_s C_f = B_s C_p + J_p C_p \tag{15.17}$$

$$B_s C_f = \left(B_s + J_p\right)C_p \tag{15.18}$$

$$C_p = \frac{B_s C_f}{\left(B_s + J_p\right)} \tag{15.19}$$

Substituting the given values from the problem statement, keeping careful track of units yields:

$$C_p = \frac{\left[0.74\ \text{gal}/\left(\text{day}-\text{ft}^2\right)\right]\left(4{,}500\ \text{mg/L}\right)}{\left(\left[0.74\ \text{gal}/\left(\text{day}-\text{ft}^2\right)\right]+\left[14.44\ \text{gal/day}-\text{ft}^2\right]\right)} = \frac{3{,}330\ \text{mg/L}}{15.18} = 219\ \text{mg/L Cl}^-$$

The percent rejection, $\%R$, is determined from Equation 15.10 to be:

$$\%R = \left(\frac{C_f - C_p}{C_f}\right)\times 100 \tag{15.10}$$

$$\%R = \left(\frac{4{,}500\ \text{mg/L} - 219\ \text{mg/L}}{4{,}500\ \text{mg/L}}\right)\times 100 = 95.1\%$$

REFERENCES

Applegate, L. 1984. Membrane Separation Processes. *Chem. Eng.* 91(6):64–89.

Beil, M., and W. Beyrich. 2013. Chapter 15–Biogas upgrading to biomethane. In *The Biogas Handbook, Science, Production and Applications* ed. A. Wellinger, J. Murphy, and D. Baxter. Cambridge, UK: Woodhead Publishing, Limited. https://doi.org/10.1533/9780857097415.3.342

Brooks, K. 1985. Membranes Push into Separations. *Chem. Week.* 138(1):21–24.

Campione, A., L. Gurreri, M. Ciofalo, G. Micale, A. Tamburini, and A. Cipollina. 2018. Electrodialysis for Water Desalination: A Critical Assessment of Recent Developments on Process Fundamentals, Models and Applications. *Desalination* 434(2018):121–160. https://doi.org/10.1016/j.desal.2017.12.044

Clements, L., M. Otten, and P. Bhat. 1986. Laboratory Membrane Gas Separator-a New Teaching Tool. *Paper 53b presented at the AIChE Annual Meeting*, Miami Beach, FL.

Davis, R., and O. Sandall. 1990. A Membrane Gas Separation Experiments for the Undergraduate Laboratory. *Chem. Eng. Educ.* 87:10–21.

Dupont, R.R., T.N. Eisenberg, and E.J. Middlebrooks. 1982. *Reverse Osmosis in the Treatment of Drinking Water.* Water Quality Series UWRL/Q-82/05. Logan, UT: Utah Water Research Laboratory. https://digitalcommons.usu.edu/water_rep/505.

Garcia, A., B. Medina, N. Verhoek, and Moore, P. 1989. Ice Cream and Components Prepared with Ultrafiltration and Reverse Osmosis Membranes. *Biotechnol. Prog.* 5:46–50.

Hollein, H., C. Slater, R. D'Aquino, and A. Witt. 1995. Bioseparation Via Cross Flow Membrane Filtration. *Chem. Eng. Educ.* 29:86–93.

Ismail, A.F., K.C. Khulbe, and T. Matsuura. 2015. *Gas Separation Membranes: Polymeric and Inorganic.* New York: Springer.

Kaup, E.C. 1973. Design Factors in Reverse Osmosis. *Chem. Eng.* 4:46–55.

Kosikowski, F.V. 1986. Chapter 9–Membrane Separations in Food Processing. In *Membrane Separations in Biotechnology* ed. W.C. McGregor. New York: Marcel Dekker.

Maubois, J. 1980. Recent Developments of Membrane Ultrafiltration in the Dairy Industry. In *Ultrafiltration Membranes and Applications* ed. A.R. Cooper, pp. 305–318. New York: Plenum Press.

Mulder, M. 1996. *Basic Principles of Membrane Technology*, 2nd edition. Boston, MA: Kluwer Academic.

Parkinson, G. 1983. Reverse Osmosis: Trying for Wider Applications. *Chem. Eng.* 90(5):26–31.

Permea, Inc. 1986. *Prism Separators*, Bulletin PERM-6-008. St. Louis, MO: Permea Inc.

Porter, M.C. 1990. *Handbook of Industrial Membrane Technology.* Park Ridge, NJ: Noyes Publications.

Schweitzer, P. 1979. *Handbook of Separation Techniques for Chemical Engineers.* New York City: McGraw-Hill.

Slater, C.S., and H. Hollein. 1993. Educational Initiatives in Teaching Membrane Technology. *Desalination* 90:625–634.

Slater, C.S., H. Hollein, P.P. Antonechia, L.S. Mazzella, and J. Paccione. 1989. Laboratory Experiences in Membrane Separation Processes. *Int. J. Eng. Educ.* 5:369–378.

Slater, C.S., C. Vega, and M. Boegel. 1992. Experiments in Gas Permeation Membrane Processes. *Int. J. Eng. Educ.* 9:1–7.

Slater, C.S., C. Vega, and M. Boegel. 1990. Membrane Gas Separation Experiments for a Chemical Engineering Laboratory. *American Society for Engineering Education (ASEE) Annual Conference Proceedings*, pp. 648–650. Washington, DC: ASEE.

Theodore, L. 2006. *Nanotechnology: Basic Calculations for Engineers and Scientists.* Hoboken, NJ: John Wiley & Sons.

Theodore, L. 2015. *Personal Notes.* East Williston, NY.

Theodore, L. 2016. *Chemical Engineering: The Essential Reference.* New York: McGraw-Hill.

Theodore, L., and F. Ricci. 2010. *Mass Transfer Operations for the Practicing Engineer.* Hoboken, NJ: John Wiley & Sons.

Wankat, P.C. 1990. *Rate-Controlled Separation.* Boston, MA: Chapman & Hall.

16 Crystallization

16.1 INTRODUCTION

Throughout the history of the chemical industry, crystals have been produced via crystallization methods that range between something as simple as allowing vats of hot concentrated solution to cool to those as complex as continuous, carefully controlled, multistep processes that yield a crystal product of a particular size or size distribution, shape, moisture content, and purity. Today, consumer demands require crystal products meet rigid specifications that can also include specific color, odor, particle size *and* particle size distribution, and caking characteristics. Some of the developed crystallization processes have been for use in water desalination, and on a laboratory scale, the process has been used in the treatment of ammonia nitrate wastewater, paper mill bleach solutions, plating liquors, and arsenal red water. Freeze crystallization also has the potential for the removal of 1% to 10% total dissolved solids (TDS) in other aqueous waste streams. A discussion of phase diagrams and crystallization is provided by Theodore and Ricci (2010).

This chapter introduces the reader to the concepts and applications of the crystallization process to water desalination and wastewater treatment through discussion of crystallization operations, processes, equipment, describing equations, and design considerations. An application section provides six Illustrative Examples related to the general subject of crystallization.

16.2 CRYSTALLIZATION OPERATIONS

Crystallization is an important mass transfer operation that is often employed in the preparation of a pure product. In the process, a crystal usually separates out as a substance of definite composition from a solution of varying composition. Any impurities in the liquid (often referred to as the *mother liquor*) are carried in the crystalline product only to the extent that they adhere to the surface or are *occluded* (retained) within the crystals that may have grown together during or after the crystallization operation. Occlusion is a word that describes a process that "prevents the passage of" something. It is employed with reference to mother liquor retained within crystals. The word *magma* is simply "a suspension of precipitated matter in a water substance."

The separation of a solid from a solution onto a crystal occurs only if there is a state of imbalance involving a mass driving force (i.e., if a differential in chemical potential, or concentration) between the bulk of the liquid solution and the crystal interface exists. This effectively means that the solution must be supersaturated.

Although crystallization is ordinarily thought of as the deposition of a solid, crystalline phase from a liquid phase by cooling, evaporation, or both, the same principles apply to crystal formation by precipitation caused by the addition of a third substance. This other component may either react to form a precipitate or simply decrease the solubility of the precipitated material. One may also view crystallization as a phase equilibrium application. These solid crystals are usually formed from a homogenous liquid phase. If one starts with an unsaturated solution formed by dissolving some solid in liquid, more solid can be dissolved until the solution becomes *saturated.* More solid can still be dissolved in the solution causing it to become *supersaturated.* At this point, solids begin to deposit out of solution in a process often referred to as *nucleation.* Crystal growth continues until the solution reaches the previously attained saturation (equilibrium) point.

There are several different ways that crystallization can occur. The four most often encountered in practice are:

1. Cooling
2. Evaporation
3. Cooling and evaporation (also denoted adiabatic evaporation)
4. Salting out

Process 1 is most commonly employed, provided the solubility of the component being crystallized decreases with decreasing temperature.

The problems concerning, crystallization that are most frequently encountered by the practicing environmental engineer are:

1. Yield of a given product
2. Purity of the product
3. Energy requirements for cooling, evaporation, etc.
4. Shape of the individual crystals
5. Size of the crystals
6. Uniformity or size distribution of the crystals
7. Rate of production of the desired crystals
8. Caking

It should also be noted that when crystallization begins, heat is evolved (liberated). The amount of heat evolved is defined as the heat (or enthalpy) of crystallization.

16.3 THE CRYSTALLIZATION PROCESS

The crystallization process essentially involves three steps:

1. Formation of crystals
2. Crystal growth
3. Separation of the crystals from residual liquid.

A simple schematic of this overall process is provided in Figure 16.1.

There are various processes that can accomplish this phase separation while meeting the desired design requirements associated with a crystallization operation or the rate of crystallization. As with many mass transfer (as well as heat transfer) calculations, the crystallization rate is expressed in terms of an overall coefficient, which takes into account all individual resistances. This coefficient is almost always determined from experimental data or experience. Details on product yield,

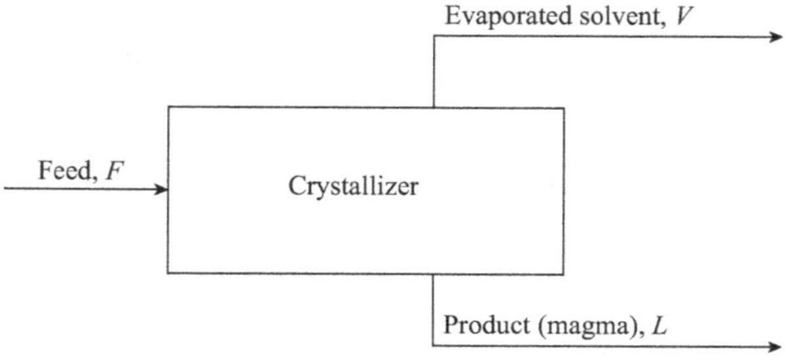

FIGURE 16.1 Generalized diagram of the crystallization process.

product purity, crystal size, or size distribution must also be specified. Each of these topics is discussed with the latter topic receiving additional treatment in the next section.

The rate of crystallization involves two distinct steps:

1. The rate of formation of new crystals, or *nucleation*
2. The rate of precipitation on crystals already present, usually referred to as *crystal growth.*

The mechanism of nucleation is essentially unknown. However, the process of crystal growth has been studied extensively, with some researchers suggesting that crystal growth occurs at erratic rates and that the solution concentration at the crystal surface is not uniform.

It is common knowledge that the solubility of small crystals is greater than that of larger crystals. Therefore, material may be *simultaneously* depositing on larger crystals while smaller crystals are dissolving when both are exposed to the same solution. As a result, it is difficult in practice to maintain uniform conditions while growing crystals. Thus, it is usually impossible to avoid locally varying conditions in the solution.

Before crystallization, it is common practice to inject small crystals, known as *seeds*, into the solution so that solids will be deposited more easily. Because the solubility of crystals of this size is less than that of submicroscopic crystals, the crystallization unit may be operated in such a manner as to retain the supersaturated solution with respect to the seeds but not supersaturated with respect to crystals of the size of the aggregations that will inevitably form under the operating conditions imposed. It should be noted that seeding is rarely used in continuous processes; it is instead used often in batch processes.

Predetermination of the yield from a crystallizer may be obtained from a material balance. The yield can usually be predicted from the solubility (Theodore and Ricci 2010) of the solid phase being precipitated. It is general practice to assume equilibrium or a saturated condition. In some applications, crystallization occurs slowly and equilibrium is not attained in a short time period. In many cases, controlled precipitation does not occur unless the aforementioned seed crystals are present. If solubilities are obtained from the equilibrium diagrams discussed previously, the prediction of the equilibrium yield can be made by applying the usual *lever arm ratio* principle (Theodore and Ricci 2010) between the phases, which are known to be present at the specified operation temperature.

The solubilities usually reported in the literature are expressed in terms of mass of "salt" per 100 units of mass of pure solvent. The prediction of a yield of a nonhydrated salt from a solution is simple because the amount of solvent present during crystallization is generally constant and the quantity remaining at the operating temperature is known (Brown and Associates 1950).

Impurities in crystallization operations are introduced primarily from the mother liquor that is not completely removed from the solid product. The extent of occlusion and the completeness of washing are important factors in the determination of the purity of a crystalline product. The agglomeration of crystals into larger crystals, which can occur during the growth of the crystals, makes washing more difficult and accordingly results in a lower purity of product. However, agitation decreases the tendency of the crystals to agglomerate.

Purification techniques have become increasingly important in recent years. Mechanisms by which the impurities discussed can be incorporated into crystalline products include adsorption of impurities on crystal surfaces, solvent entrapment in cracks, crevices and agglomerates, and inclusion of pockets of liquid. The suggested recommendation for producing high purity crystals is to maintain supersaturation at a low level so that large crystals are formed. The characteristics of an impurity can also determine whether it is positioned on the surface or in the interior of the host crystal.

Although crystallization is often thought of as solely a separation process, it is a filtration step at the end of the process that accomplishes the actual crystal separation. The residual liquid (*mixed magma underflow*) from a crystallizer typically contains approximately 30% mass solids and can be thickened further before filtration by being passed through a gravity settler. Such pre-thickening allows dewatering of the *mother liquor*, which allows either more time for the final wash step or a shorter filter cycle.

The filter cycle almost always includes a washing step at the end of the full cycle. This permits a thorough drain before the crystals are discharged. It should also be noted that the system's throughput and performance is often limited by the liquid-solid separation step. In effect, this then becomes the critical step in the overall operation (Brown and Associates 1950).

Perhaps the most troubling problem is associated with the final product. Crystalline materials have a tendency to bind together or cake on storage. Most crystalline products are required in a free-flowing form, as with sugar and table salt, or to be capable of being distributed uniformly over surfaces (e.g., fertilizers). It is also important for the crystalline product to remain in a particulate state. Caking may also require a crushing operation before the crystals can be used.

Numerous problems can arise during the operation, design, and analysis of a crystallizer. Ten of the most important problems are:

1. Crystal-size distribution (CSD)
2. Purity
3. Fouling
4. Vapor release and location
5. Scale-up
6. Stability of operation
7. Liquid-solid separation
8. Space requirements
9. Capital costs
10. Operating and maintenance costs

With reference to the crystallization process, the average size to which crystals are grown is usually dictated by the manner in which they are to be used. As noted previously, fine crystals have less opportunity for agglomeration and accordingly occlude less mother liquor; however, fine crystals have a greater surface area and offer greater difficulty of complete removal of mother liquor entrained on the crystals. In addition, producing crystalline material of a given size range (a specific upper and lower limit) is difficult to accomplish.

It should be noted that the removal rate of crystals of a given size can greatly alter a crystallizer's ability to continue the production of crystals having that same size or size distribution. This effect is particularly pronounced in a separation unit where the crystals of a desired size distribution range are preferentially removed and the undersized particles recycled to the crystallizer. This can shift the distribution to a smaller and narrower size range.

Finally, two of the goals of the design and practicing engineers who work on the CSD problem are to be able to:

1. Understand CSD so as to analyze and modify what a particular operating crystallizer can and cannot do in the way of producing acceptable crystal size and crystal size ranges.
2. Include the effects of CSD in the design of new units (i.e., to design an original process that is not yet operating) (Brown and Associates 1950).

16.4 CRYSTALLIZATION EQUIPMENT

Equipment and equipment selection procedures are usually based on maximum throughput capacities for a new crystallization process. The reason for this is to enable the equipment to perform satisfactorily under the most extreme operating conditions. Material and energy balances based on these conditions are also required before the individual equipment is selected. Most importantly, this procedure is also influenced by the manner in which crystallization occurs. As noted, the methods to accomplish this are crystallization via cooling, evaporation, cooling and evaporation (adiabatic evaporation), and a salting out process. Details on the equipment that can accomplish these tasks are provided herein.

Perhaps the simplest type of equipment for crystallization is a tank in which natural cooling is allowed to lower the temperature of a solution. A hot concentrated solution with the solute is simply poured into the tank where it cools by natural convection. However, there are several drawbacks when employing this practice. This type of crystallizer is inefficient relative to the quantity produced per unit volume or per unit time because of the slow cooling rate. It offers no control over size or size distribution of crystals, and it also favors the formation of large crystals that are subject to occlusion of the mother liquor because the crystals tend to grow together. This agglomeration phenomena can be reduced with agitation. This class of equipment is employed when precipitating small quantities of a material and where the size range of the product is not critical.

Crystallization can also be achieved by cooling a hot, concentrated solution on a surface. The cooling, in turn, results in solute solubility that favors solid-phase precipitation. This type of equipment is available in several designs. The most popular of these in previous years was the *Swenson-Walker crystallizer.* This unit contains a water-jacketed open trough with a semi-cylindrical bottom. Hot, concentrated solution enters at one end of the trough and the crystals that accumulate on the cooling surface are lifted and pushed along by a spiral stirrer to the other end of the crystallizer. Crystals are removed at the end of the crystallizer by an inclined, spiral flight conveyor that lifts them onto a draining board or a conveyor that carries the crystals to centrifuges or other drying or separation operations that may be required. This, as well as other units in this category, are usually employed for systems with a *steep solubility* slope for which a relatively large yield of crystals can be achieved with a modest drop in temperature.

The *vacuum crystallizer* accomplishes cooling via a flashing process. It is a device that cools the solution by the evaporation of a portion of the solvent. The energy for vaporization is obtained from the sensible enthalpy (heat) of the feed. The unit is usually evacuated to a low pressure by steam jet ejectors. The flashing of the feed due to the lower pressure results in both the cooling and concentration of the solution. One of the problems that can arise with this type of crystallization unit is that the operating pressure required to attain the desired crystallization is so low that the evolved vapor cannot be condensed by any cooling water available. These units work for systems of intermediate solubility slopes. It also has the advantage of vaporizing some of the solvent to bring about additional yield.

In an *evaporative crystallizer*, crystallization is caused by evaporating the solvent from the feed, which can either be a weak, unsaturated solution or a hot, concentrated mixture. These classes of crystallizers are employed when the solute has a small, low slope temperature-solubility curve. The operation of the evaporator seldom allows much flexibility as to the size, size distribution, or shape of crystals produced. Some classifying action does occur in that fine crystals are usually carried in the circulating mother liquor until they attain a size that will result in their precipitation out of the circulating stream into the solids-removal equipment (Brown and Associates 1950). These units are normally steam heated although the heating can also be accomplished by passing hot gases through the solution. If a traditional evaporator is employed, the classic equation for heat transfer may be applied (Reynolds et al. 2004):

$$Q = UA\Delta T \tag{16.1}$$

where Q is the rate of heat transfer, U is the overall heat transfer coefficient (the resistance to heat transfer), A is the available area for heat exchange, and ΔT is the temperature difference driving force across the heat transfer area. Auxiliary equipment for a crystallization process can include slurry pumps, vacuum pumps, mixers, condensers, steam traps, and entrainment separators (Theodore 2008).

16.5 DESCRIBING EQUATIONS

Overall and componential material balance calculations in the chemical process industry include transient effects that can account for process upsets, start-ups, shut-downs, and so on. The describing equations for these time-varying (unsteady-state) systems are differential. The equations usually take the form of a first-order derivative with respect to time, where time is the independent variable. However, calculations for most crystallization processes assume steady-state conditions with the ultimate or final design based on the worst-case or maximum flow conditions. This greatly simplifies the calculations because the describing equations are no longer differential but, rather algebraic.

The development of mass (and energy to follow) balances follow traditional procedures. Consider, as with other mass transfer operations, material and energy balances and any heat transfer equations that can be written for the various crystallization processes. The material balances can provide the process yield (i.e., the mass of crystals formed from a given mass of solution). The effect of evaporation or of cooling must be included if applicable.

As an example, the evaporation of solvent from a crystallizer is pictured in Figure 16.2. A solute componential and total material balance (on a mole basis) may be written:

$$F = V + L + P; \text{ total} \tag{16.2}$$

$$x_F F = V(0) + x_L L + x_C P; \text{ componential} \tag{16.3}$$

where F is the feed, V is the quantity of evaporation, L is liquid (filtrate) withdrawal, P is the crystal production, and x represents the solute content of a stream in units consistent with the flow rate. There is also an equilibrium expression relating x_L and x_C.

Once the material balance is completed, one may then proceed directly to the energy calculations, some of which can play a significant role in crystallizer design. As with material balance calculations, energy calculations for crystallizers almost always are based on steady-state conditions.

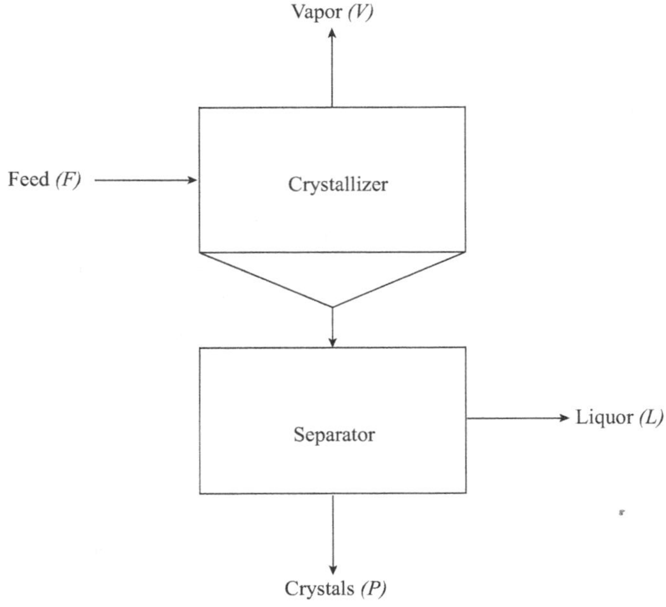

FIGURE 16.2 Line diagram of an evaporative crystallizer.

One of the principal jobs of a practicing engineer involved with crystallizers is to account for the energy that flows into and out of the unit and to determine overall energy requirements. This is accomplished by performing energy balances on the unit.

As noted previously, crystallization operations may be carried out either by cooling a solution or by evaporative concentration, or by both. The cooling of a solution may be accomplished by the removal (transfer) of heat of both the sensible heat and the heat evolved during crystallization of the product by (to) cooling water or air. If the solution is crystallized by evaporation, the required heat may be supplied by the sensible heat of solution as in vacuum crystallizers, or it may be supplied from an external source. In the former case, the heat of vaporization of the solvent from the solution may be safely assumed equal to the heat of vaporization of the pure solvent. In the latter case, the heat of solution (data for which is rarely available) may be assumed equal to the heat of solution at infinite dilution (Theodore and Ricci 2010). Obviously, for crystallization problems requiring energy balances (e.g., where evaporation occurs or when the temperature of an adiabatic crystallizer is unknown), enthalpy data must be provided or made available.

Referring once again to Figure 16.2, one may also write an energy balance

$$F\widehat{H}_F + \dot{Q} = V\widehat{H}_V + L\widehat{H}_L + P\widehat{H}_C \tag{16.4}$$

where F, V, L, P refer to the feed, vapor, liquid and product phase mass flow rates, respectively; $\dot{Q}$ is the rate of heat flow; and $\widehat{H}_F$, $\widehat{H}_V$, $\widehat{H}_L$, $\widehat{H}_C$ are the changes in enthalpy for the feed, vapor, liquid, and crystal phases, respectively. The enthalpies in this equation can be determined, provided the temperatures of the product streams are known. In addition, the operation is usually either adiabatic or may be safely assumed to be adiabatic, so that $\dot{Q} = 0$.

Several excellent examples of calculating crystallizer energy balances are available in the literature (Chopey 1994).

16.6 DESIGN CONSIDERATIONS

As with any mass transfer device, there are usually five conceptual steps to be considered with the design of equipment. These are:

1. Identification of the parameters that must be specified
2. Application of fundamentals underlying theoretical equations or concepts
3. Enumeration, explanation, and application of simplifying assumptions
4. Possible use of correction factors for "nonideal" behavior
5. Identification of other factors that must be considered for adequate equipment specification

Because design calculations are generally based on the maximum throughput capacity for the proposed crystallizer or process, these calculations are never completely accurate. It is usually necessary to apply reasonable safety factors when setting the final design for the crystallizer. Safety factors vary widely, particularly with crystallizers, and are a strong function of the accuracy of the data involved, calculational procedures, and past experience. Nonetheless, simple qualitative design procedures are provided.

Assuming a crystallizer is the mass transfer operation of choice, the class of crystallizer is normally selected first. For high feed rate (e.g., greater that 50,000 lb/hr), a continuous, as opposed to a batch process is normally selected. Overall process decisions are based on capital, operating and maintenance costs, space, location, production rate, crystal characteristics, size and size distribution of crystals, physical and chemical characteristics of the feed liquor and slurry, corrosion issues, environmental concerns, etc.

Regarding a new design, scale-up problems can be expected when the design is based on pilot-plant or laboratory units. Even after the equipment is fully constructed, significant start-up problems are the norm. Finally, it should be noted that the scaling-up of crystallization equipment is probably more difficult than that for any of the other mass transfer operation.

If a pilot or small-scale crystallizer produces a satisfactory product, then the design of the larger crystallizer must simulate a number of different conditions obtained in the smaller unit. The four most important conditions are (Mullen 1961):

1. Identical flow characteristics of liquids and solid particles
2. Identical degrees of supersaturation in all equivalent regions of the crystallizer
3. Identical initial seed sizes (if applicable) and magma densities
4. Identical contact times between growing crystals and supersaturated liquor

Marnell (2009) has provided a step-by-step procedure for the design of a single solute forced circulation crystallization unit. The nine specifications required are:

1. Production rate of crystals, P
2. Feed (liquid) mass fraction
3. Desired CSD and appropriate residence time to achieve it
4. Feed temperature
5. Design temperature and corresponding mother liquid saturation mass fractions
6. Centrifuge wash water ratio
7. Crystal bulk density, ρ_B
8. Magma density, ρ_M
9. Liquor density, ρ_L

Specifications (5) through (9) are generally drawn from experience.

As discussed previously, the design of crystallization equipment can be difficult. Knowledge of phase equilibrium and physical properties does not allow one to accurately predict the behavior of the actual process. Bench-scale and pilot-plant experiments on the actual stream are usually recommended, and some equipment vendors require a pilot-plant test before designing the equipment. Even with such testing, operational adjustments must usually be made on the actual commercial installation before the equipment will operate to yield an acceptable product.

16.7 APPLICATIONS

Six Illustrative Examples complement the material presented in this chapter.

Illustrative Example 16.1

The following crystal sizes (in mm) in a sample were recorded from a crystallizer: 22, 10, 8, 15, 13, and 18. Find the median, the arithmetic mean, the geometric mean, and the standard deviation of these crystal sizes.

Solution

The *median* is simply the medium value of a distribution, or the value above which half of the data lie and below which the other half lie. If n data points are listed in their relative numerical order, the median is the $(n + 1)/2$th value. If the number of data is even, then the numerical value of the median is the value midway between the two data nearest the middle. The median, being a positional value, is less influenced by extreme values in a distribution than the mean value. The median value for these data is 14 mm, the value midway between 13 and 15 mm.

The *arithmetic mean* of this crystal particle size distribution, $\bar{d}$, is:

$$\bar{d} = (22 + 10 + 8 + 15 + 13 + 18)/6 = (86)/6 = 14.33 \text{ mm}$$

Another measure of central tendency used in special applications is the *geometric mean*, X_G. The geometric mean is calculated using Equation (16.5)

$$X_G = \sqrt[n]{(d_1)(d_2)(d_3)\ldots(d_n)} \quad (16.5)$$

For the crystal particle size distribution given, the geometric mean of the distribution is:

$$X_G = \sqrt[6]{(22)(10)(8)(15)(13)(18)} = \sqrt[6]{6{,}177{,}600} = 13.55 \text{ mm}$$

The *standard deviation* of the crystal particle size distribution is calculated using Equation 16.6 as follows:

$$s = \sqrt{\frac{\sum(d_i - \bar{d})^2}{n-1}} = \sqrt{\frac{n\sum d_i^2 - \left(\sum d_i\right)^2}{n(n-1)}} \quad (16.6)$$

For the crystal particle size distribution given above, the standard deviation of the distribution is:

$$\sum d_i^2 = (8)^2 + (10)^2 + (13)^2 + (15)^2 + (18)^2 + (22)^2 = 1{,}366$$

$$\left(\sum d_i\right)^2 = (8 + 10 + 13 + 15 + 18 + 22)^2 = 7{,}396$$

$$s = \sqrt{\frac{6(1{,}366) - 7{,}396}{6(5)}} = \sqrt{\frac{6(1{,}366) - 7{,}396}{6(5)}} = \sqrt{\frac{800}{30}} = \sqrt{26.67} = 5.16 \text{ mm}$$

Illustrative Example 16.2

Calculate the overall efficiency of the crystallization separation unit using the data provided in Table 16.1.

Solution

The overall efficiency is obtained from the cross-product of Columns 2 and 4 for each crystal size range (CSR) and summing the results:

$$E = \sum_{i=1}^{n} (\text{Column2}_i \times \text{Column4}_i)/100; \text{ percent basis} \quad (16.7)$$

The calculated results are provided in Table 16.2. From Table 16.2, one obtains:

$$E = \frac{8{,}782}{100} = 87.82\%$$

TABLE 16.1
Crystal Size Distribution and Separation Data for Illustrative Example 16.2

CSR, mm	% in Particle Size Range	$\bar{d}$, mm	Separation Efficiency, %
0–5	2	2.5	0.6
5–10	2	7.5	7.7
10–15	1	12.5	19.2
15–20	4	17.5	38.3
20–30	6	25.0	56.2
30–50	11	40.0	79.1
50–100	13	75.0	98.9
>100	61	100+	100

Abbreviation: CSR, crystal size range.

TABLE 16.2
Crystal Separation Efficiency Calculations for Illustrative Example 16.2

CSR, mm	Column 2 × Column 4
0–5	1.2
5–10	15.4
10–15	19.2
15–20	153.2
20–30	337.2
30–50	870.1
50–100	1,287.7
>100	6100.0
Total	8782.0

Abbreviation: CSR, crystal size range.

Finally, two of the goals of the design and practicing engineers who work on the CSD problem are to be able to:

1. Understand CSD so as to analyze and modify what a particular operating crystallizer can and cannot do in the way of producing acceptable crystal size and crystal size ranges, and
2. Include the effects of CSD on the design of new units (i.e., to design an original process that is not yet operating).

Illustrative Example 16.3

As noted, crystallization is the process of forming a solid phase from a solution. It is employed heavily as a separation process in the inorganic chemistry industry, particularly where salts are recovered from aqueous media. In the production of organic chemicals, crystallization is also used to recover product, to refine intermediate chemicals, to remove undesired salts, and to recover "pure" water. The feed to a crystallization system consists of a solution from which solute is crystallized through one or more of a variety of processes. The solids are normally separated from the crystallizer liquid, washed, and discharged to downstream equipment for further processing.

Potassium nitrate is obtained from an aqueous solution of 20% potassium nitrate (KNO_3). During the process, the aqueous solution of potassium nitrate is evaporated, leaving an outlet stream with a concentration of 50% KNO_3, which then enters a crystallization unit where the outlet product is 96% KNO_3 (anhydrous crystals) and 4% water. A residual aqueous solution that contains 0.55 g of KNO_3 per gram of water also leaves the crystallization unit and is mixed with the fresh solution of KNO_3 at the evaporator inlet. Such recycling provides a means of preventing pollution by minimizing the loss of valuable raw materials. The flow diagram in Figure 16.3 depicts the process.

What is the mass flowrate, M, in kilograms/hr of the recycled material, R, when the feedstock flowrate is 5,000 kg/hr?

Solution

Using Figure 16.3, an overall material balance on KNO_3 leads to:

$$0.96\ C = 0.2\ (5{,}000)$$

This equation may be solved for the rate of product leaving the crystallizer:

$$C = 1{,}000/0.96 = 1{,}042\ \text{kg/hr}$$

An overall mass balance around the crystallizer may also be written:

$$M = R + C = R + 1{,}042$$

Because the mass concentration of the KNO_3 in R is (0.55/1.55), the equation for the KNO_3 mass balance around the crystallizer is:

$$0.5M = 0.96\ (1{,}042) + (0.55/1.55)\ R = 1{,}000 + 0.355\ R$$

These two equations may be solved simultaneously for the amount of KNO_3 leaving the evaporator in terms of the amount of recycle. Solving the second equation for M yields:

$$M = 2{,}000 + 0.71R$$

Therefore,

$$2{,}000 + 0.71R = R + 1{,}042$$

and the new mass flow rate of the recycle stream is

$$R = 3{,}301\ \text{kg/hr}$$

The reader is left with the exercise of calculating the rate of "pure" water, E, production.

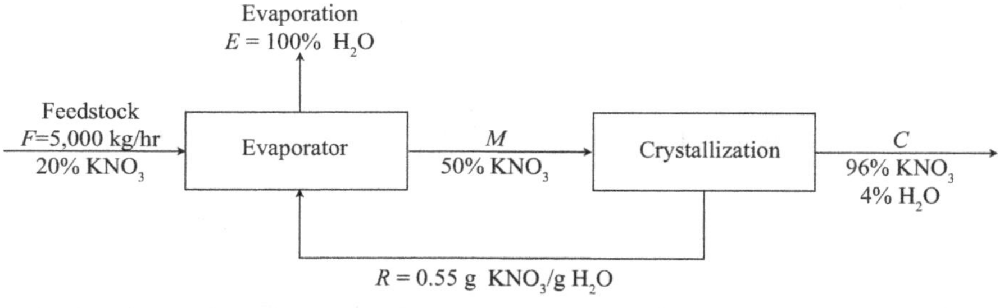

FIGURE 16.3 Process flow diagram for Illustrative Example 16.3. H_2O = water; KNO_3 = potassium nitrate.

Illustrative Example 16.4

Using the process diagram in Figure 16.4 that will produce 2,000 lb/hr of sodium nitrate ($NaNO_3$) crystals from a 90% (by mass) sodium nitrate-sodium chloride ($NaNO_3$-NaCl) mixture, write the pertinent material balance equations around the cooler and crystallizer. Write the pertinent material balance equations around the heater and the evaporator. Finally, determine the flow rates for all of the streams in the system described in Figure 16.4.

Solution

The overall material balance around the cooler and crystallizer is:

$$A+B+C-E-D=0;\ D=2{,}000\ \text{lb/hr}$$

A water balance around the unit can be written as:

$$A+0.487\,(B)-0.5506\,(E)=0$$

A NaCl balance around the unit can be written as:

$$0.1\,(C)+0.0999(B)-0.13(E)=0$$

A $NaNO_3$ balance around the unit can be written as:

$$0.9\,(C)+0.4131\,(B)-0.3194\,(E)-2{,}000=0$$

Because there are three components, water (H_2O), $NaNO_3$, and NaCl, only three independent equations may be written.

Considering the Heater and Evaporator, the overall material balance is:

$$E-G-F-B=0$$

A water balance around this unit can be written as:

$$0.5506\,(E)-F-0.487\,(B)=0$$

A NaCl balance around this unit can be written as:

$$0.13\,(E)-G-0.0999\,(B)=0$$

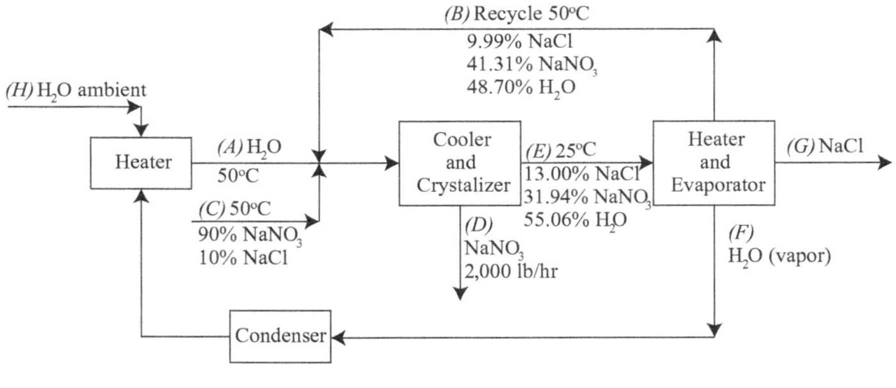

FIGURE 16.4 Process flow diagram with system data for Illustrative Example 16.4. H_2O = water; $NaNO_3$ = sodium nitrate; $NaNO_3$-NaCl = sodium nitrate-sodium chloride.

A $NaNO_3$ balance around this unit can be written as:

$$0.3194\,(E) - 0.4131\,(B) = 0$$

Once again, only three independent equations may be written.

Examining the material balance equations provided indicates that although there are eight equations, there are a total of only seven independent equations. Because there are seven unknown streams, *A, B, C, E, F, G,* and *H,* a unique solution is possible. That unique solution, in lb/hr can be shown to be:

$A = 733$ lb/hr; $B = 3{,}254$ lb/hr; $C = 2{,}222$ lb/hr; $D = 2{,}000$ lb/hr; $E = 4{,}209$ lb/hr; $F = 733$ lb/hr;

$G = 222$ lb/hr; $H = 0$ lb/hr.

Illustrative Example 16.5

Complete a material balance for the process pictured in Figure 16.5. The following information is provided:

Production rate: 3,125 kg/hr
Feed composition: 75% urea, 25% water
Magma density: 1,170 kg/m^3
Magma crystal concentration: 450 kg crystals/m^3 magma
Liquor density: 1,340 kg/m^3
Magma recirculation rate: 567,900 kg/hr
Centrifuge wash water ratio: 0.12 kg H_2O/kg urea

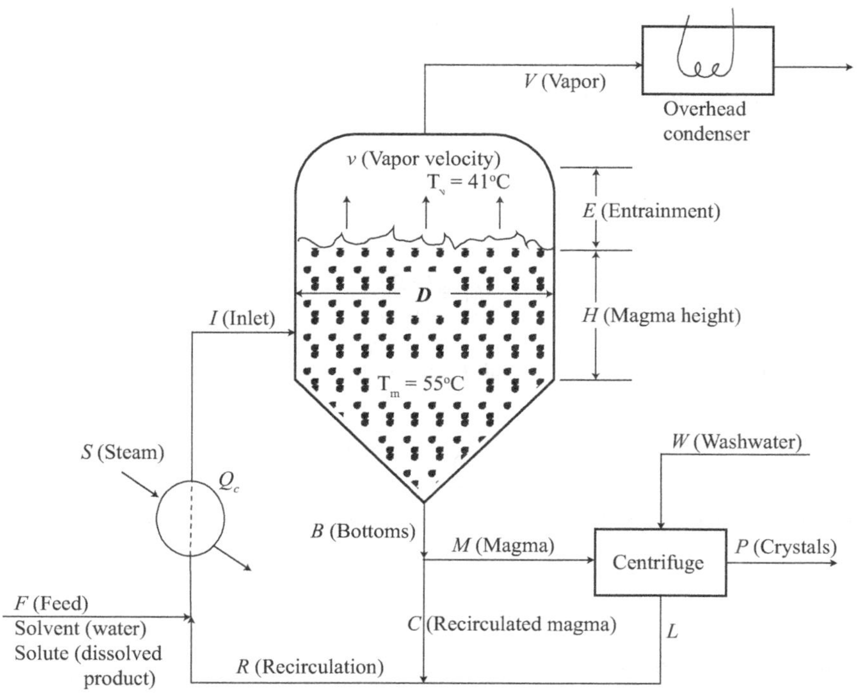

FIGURE 16.5 Forced circulation crystallizer flow diagram with system data for Illustrative Example 16.5.

Also, size the crystallizer, using the following design information:

Average residence time, τ (based on M): 3.38 h
Maximum (with safety factor) vapor velocity,

$$v = C_f\left(\frac{\rho_l - \rho_v}{\rho_v}\right)^{0.5};C_f = 0.0122 \text{ m/s} \tag{16.8}$$

Crystallizer diameter, D, is based on the maximum vapor velocity, v
Crystallizer magma cylinder height, $H = 1.15\ D$
Freeboard height (for droplet disengagement) is 2 m

Solution

First calculate the feed ratio, F:

$$F = \frac{P}{x_{F,U}} \tag{16.9}$$

where $x_{F,U}$ = mass fraction urea.

$$F = \frac{3{,}125}{0.75} = 4{,}167 \text{ kg / hr}$$

The wash rate, W, is:

$$W = (0.12)P = (0.12)(3{,}125) = 375 \text{ kg/hr}$$

The water vaporization rate, V, is given as:

$$V = (F)(x_{F,W}) + W \tag{16.10}$$

where $x_{F,W}$ = mass fraction water.

$$V = (4{,}167)(0.25) + 375 = 1{,}417 \text{ kg/hr}$$

Calculate the magma flow, M, as:

$$M = \frac{P(\rho_M)}{\rho_{C,M}} \tag{16.11}$$

where ρ_M = magma density, $\rho_{C,M}$ = crystal density in magma.

$$M = \frac{(3{,}125)(1{,}170)}{450} = 8{,}120 \text{ kg/hr}$$

The liquor flow, L, is obtained by a total mass balance around the centrifuge:

$$W + M = P + L; L = W + M - P$$

$$L = 375 + 8{,}120 - 3{,}125 = 5{,}370 \text{ kg/hr}$$

The bottoms mass flow, B, is:

$$B = M + C = 8{,}120 + 567{,}900 = 576{,}020 \text{ kg/hr}$$

The recycle mass flow, R, is:

$$R = C + L = 567,900 + 5,370 = 573,270 \text{ kg/hr}$$

Finally, the inlet mass flow to the crystallizer, I, is:

$$I = F + R = 4,167 + 573,300 = 577,467 \text{ kg/hr}$$

The crystallizer area and diameter are based on entrainment to ensure vapor droplet disengagement. The vapor velocity as given in the problem statement is:

$$v = C_f\left(\frac{\rho_l - \rho_v}{\rho_v}\right)^{0.5}; \; C_f = 0.0122 \text{ m/s} \tag{16.8}$$

For steam at atmospheric pressure, set $\rho_v = 0.0563$ kg/m^3. Therefore,

$$v = 0.0122\left(\frac{1,339.94}{0.0563}\right)^{0.5} = 1.88 \text{ m/s} = 6,768 \text{ m/hr}$$

The crystallizer area, A, is based on the water mass vaporization rate, V, the water vapor density, ρ_v, and the maximum vapor velocity, v, as follows:

$$A = \frac{V}{(\rho_v)(v)} = \frac{1,417 \text{ kg/hr}}{(0.0563 \text{ kg/m}^3)(6,768 \text{ m/hr})} = 3.72 \text{ m}^2$$

The minimum diameter, D_{min}, of the crystallizer is determined assuming a circular cross section as:

$$D_{min} = \left(\frac{4A}{\pi}\right)^{0.5} = \left(\frac{4(3.72)}{\pi}\right)^{0.5} = 2.18 \text{ m}$$

The magma occupies two sections (of volume) in the crystallizer, the cylinder of height, H, and the cone below the cylinder with a top width, D. The magma volume, MV, is therefore:

$$MV = \text{cylinder volume} + \text{cone volume} \tag{16.12}$$

The volume of a cone is equal to $\pi r^2 h/3$, where r is the radius of the top of the cone, and H is the height of the cone. Using this equation, the height of the cylinder, $H = 1.15\ D$ from the problem statement, and setting the height of the cone to the cylinder height, the following equation for MV can be written:

Total crystallizer height = Cone height + H + Freeboard height = 3.11 m + 3.11 m + 2 m
Total crystallizer height = 8.22 m

The volume of magma is controlled by the average retention within the crystallizer and can be expressed as:

$$\begin{aligned} MV &= \text{(magma volumetric flow rate)(Average retention time)} \\ MV &= M/\rho_M(\tau) = (8,120 \text{ kg/hr})/(1,170 \text{ kg/m}^3)(3.38 \text{ hr}) = 23.5 \text{ m}^3 \end{aligned} \tag{16.13}$$

This magma volume can be substituted in the previous equation to determine the required crystallizer diameter based on the required average retention time in the crystallizer as:

$$MV = 23.5 \text{ m}^3 = 1.2D^3; \; D = \sqrt[3]{19.58} = 2.7 \text{ m}$$

This diameter is larger than the minimum diameter based on maximum vapor velocity, so this should be the diameter used for the final design of the crystallizer. Other crystallizer dimensions are calculated as follows:

$$H = 1.15D = 1.15\,(2.7) = 3.11 \text{ m}$$

$$\text{Total crystallizer height} = \text{Cone height} + H + \text{Freeboard height} = 3.11 \text{ m} + 3.11 \text{ m} + 2 \text{ m}$$
$$\text{Total crystallizer height} = 8.22 \text{ m}$$

Illustrative Example 16.6

Outline the sea water desalination process using a crystallizer that was developed by Ganiaris (1975) more than 40 years ago.

Solution

In this seawater desalination process, feed seawater is pumped to a deaerator operating at a pressure of approximately 35 mm Hg absolute. The dissolved air is reduced from 30 to 40 ppm to about 0.1 ppm. The sea water then enters a filter where all suspended impurities, if any, are removed. The deaerated and filtered sea water is then pumped to a heat exchanger. This is a direct liquid-liquid contact unit, which consists of two sections with a heat, exchange medium circulating in both areas. The warm seawater enters the heat exchanger where it is cooled by direct contact with the heat exchange medium. The chilled seawater is mixed with recycle brine and flows into a crystallizer, where a portion of the inflow is converted to ice crystals.

The crystallizer is simply an agitated, horizontal vessel operating at a preset temperature and pressure. Refrigeration is provided by flashing a liquid hydrocarbon refrigerant. The ice brine slurry flows by gravity to a horizontal, two-stage, low speed centrifuge. The separated crystals are washed on a second stage screen by approximately 3% to 5% fresh water. Provision is made to permit washing of crystals on the first stage as well, but this is not normally required. The separated ice crystals, with 3% to 5% water adhering to them and a total dissolved solids of less than 500 ppm, are reslurried by a circulating stream of fresh water and pumped to a melter condenser.

The melter condenser is a vertical tower where the ice-fresh water slurry flows countercurrent to the bulk of the compressed refrigerant vapors. The bulk of the water collected in the base of the tower is circulated back to the centrifuge and the product water, produced by melting of ice, is pumped to the heat exchanger. The product water in the heat exchanger is used to chill a portion of the circulating heat exchange medium, which in turn cools the entering seawater.

The product water is then pumped to a two-stage debutanizer where residual refrigerant is removed by flashing to a suitable pressure.

The bulk of the brine separated in the centrifuge is recycled back to the freezer after combining with the seawater feed stream. The excess brine is pumped to the heat exchanger where it is used to cool a portion of the circulating heat exchange medium; namely, the butane stream.

The brine leaves the heat exchanger and is pumped to the debutanizer where residual butane is removed as with product water by flashing.

The hydrocarbon flow is described as follows: the liquid hydrocarbon refrigerant is pumped initially into the system from storage. In flash chillers the liquid hydrocarbon is flashed to the temperature corresponding to crystallizer pressure. The chilled refrigerant flows by gravity to the crystallizer. The vapor stream from the crystallizer, together with the vapor from flash chillers, are compressed and the bulk of the compressed vapor is then condensed in the melter condenser and returned to chillers. The balance is compressed to a higher pressure by an auxiliary compressor for condensation by sea water in the auxiliary condenser.

The refrigerant chosen to be used is butane. The process as developed has certain special features, which include:

1. Ice crystallization under, controlled conditions.
2. Effective ice melting.
3. Adherence of brine on ice to less than 5%.
4. Separation by a low speed centrifuge that is commercially available and needs no further development.
5. Salts in product water is less than 500 ppm.
6. In regard to energy requirements, experience indicates that about 25 to 35 kWhr of electric power are required for 1,000 gallons of water produced, with the actual consumption depending on the size of the production unit. Energy requirement is affected by variations of process schemes. One method of saving overall power is by the use of multistage freezing. Further, the yield of fresh water in each stage of a multistage unit also affects power input needed.
7. Direct contact heat exchangers are used with consequent increases in efficiency of heat transfer and savings in heat exchanger surfaces. No metallic heat transfer surfaces are involved in the process.

REFERENCES

Brown and Associates. 1950. *Unit Operations*. Hoboken, NJ: John Wiley & Sons.

Chopey, N. 1994. *Handbook of Chemical Engineering Calculations*, 2nd ed. New York: McGraw-Hill.

Ganiaris, N. 1975. U.S. Patent 3,879,956. Crystal Wash, April 29. New York. https://patentimages.storage.googleapis.com/32/44/af/c8de852d9c60fc/US3879956.pdf.

Marnell, P. 2009. *Personal Communications to L. Theodore*. Bronx, NY: Manhattan College.

Mullen, J. 1961. *Crystallization*. London, UK: Butterworth.

Reynolds, J., J. Jeris, and L. Theodore. 2004. *Handbook of Chemical and Environmental Engineering Calculations*. Hoboken, NJ: John Wiley & Sons.

Theodore, L. 2008. *Air Pollution Control Equipment Calculations*. Hoboken, NJ: John Wiley & Sons.

Theodore, L., and F. Ricci. 2010. *Mass Transfer Operations for the Practicing Engineer*. Hoboken, NJ: John Wiley & Sons.

17 Nanotechnology

17.1 INTRODUCTION

Nanotechnology. Although not to be found in any Webster's Dictionary, it is concerned with the world of invisible miniscule particles that are dominated by forces of physics and chemistry that cannot be applied at the macro- or human scale. These particles have come to be defined by some as nanomaterials, and these materials possess unusual properties not present in traditional or ordinary materials.

Regarding the word *nanotechnology*, it is derived from the words *nano* and *technology*. Nano, typically employed as a prefix, is defined as one-billionth of a quantity or term that is represented mathematically as 1×10^{-9}, or simply as 10^{-9}. Technology generally refers to "the system by which a society provides its members with those things needed or desired." The term *nanotechnology* has come to be defined as those systems or processes that provide goods or services that are obtained from matter at the nanometer level, that is, from sizes in the range of one-billionth of a meter. The new technology thus allows the engineering of matter by systems or processes that deal with atoms; or as Drexler (1986) (who some view as the godfather of this industry) put it: "entails the ability to build molecular systems with atom-by-atom precisions, yielding a variety of nanomachines." One of the major problems that remain is the development of nanomachines that can produce other nanomachines in a manner similar to what many routinely describe as mass production.

The laws of chemistry and physics work differently when particles reach the nanoscale because the powers of hydrogen bonding, quantum energy, and van der Waals' forces endow some nanomaterials with unusual properties. Carbon nanotubes, for instance, discovered in the sooty residue of vaporized carbon rods, defy standard physics. Stronger and more flexible than steel, yet measuring about 10,000 times smaller than the diameter of a human hair, these cylindrical sheets of carbon atoms are useful as coatings on computer chips and other electrical devices despite their ability to conduct heat and electricity. Nanoparticles, another manifestation of nanotechnology, are known to foster stubborn reactions because they have enormous surface area relative to their volume.

This chapter summarizes a wide range of nanotechnology process topics by addressing the early history of the technology, fundamental and basic principles, nanomaterials, current implementation in environmental technology, and environmental concerns. An application section provides four Illustrative Examples related to the general subject of nanotechnology and its environmental applications.

17.2 EARLY HISTORY

Nanoparticles arrived on scene immediately following the Big Bang some 14 billion years ago. However, it is not clear when humans first began to take advantage of nanosized materials. It is known that in the fourth-century CE Roman glassmakers were fabricating glasses containing nanosized metals. Michael Faraday published a paper in 1857 in the *Philosophical Transactions of the Royal Society*, which attempted to explain how metal particles affect the color of church windows. Gustav Mie was the first to provide an explanation of the dependence of glass color on the type and size of metal. A century later, Richard Feynman presented a lecture titled *There Is Plenty of Room at the Bottom*, where he speculated on the possibility and potential of nanosized materials (Poole and Owens 2003).

He proposed manipulating individual atoms to create new small structures having different properties. Groups at Bell Laboratories and IBM fabricated the first two-dimensional quantum cells in the early 1970s. They were made by thin-film (epitaxial) growth techniques that build a semiconductor layer one atom at a time. This work initiated the development of the zero-dimensional quantum dot, which is now one of the nanotechnologies in commercial applications. In 1996, a number of government agencies led by the National Science Foundation (NSF) commissioned a study to assess the current worldwide status of trends, research, and development in nanoscience and nanotechnology. This NSF activity provided the necessary impetus for the future for this industry to expand and flourish (Poole and Owens 2003).

17.3 FUNDAMENTALS AND BASIC PRINCIPLES

Matter is anything that has mass and can be physically observed. All matter is composed of *atoms* and *molecules*; it consists of a finite number of elements, often represented as *building blocks. Atoms* are small particles that cannot be made smaller, whereas *molecules* are groups of atoms bound together but possessing properties different from those of an atom.

The atom is composed of a small core defined as a *nucleus* that is surrounded by *electrons.* The nucleus is composed of two types of particles: *protons* and *neutrons*; however, there is significant space between the electrons and the nucleus. Protons and neutrons are themselves made up of even smaller particles, known as *quarks.* One generally depicts the charges in units of electron charge, so that the charge of an electron is written as −1 and that of a proton is written as 1+.

Matter has physical and chemical properties that are related to its size; the properties of most solids depend on the size range over which they are measured. the size range can be macroscopic, microscopic, or molecular (Figure 17.1). One may view these sizes as finite, differential, and molecular, respectively. The object of this chapter is to discuss these characteristics at the molecular, or nanometer, level. In the macro- or large-scale range ordinarily studied in traditional fields of physics such as mechanics, electricity, magnetism, and optics, the sizes of the objects under study range from millimeters to kilometers (i.e., finite sizes). The properties that one associates with these materials are averaged properties, such as density and thermal conductivity.

When familiar materials such as metals, metal oxides, ceramics, and polymers, and novel forms of carbon, such as carbon nanotubes and fullerenes (also called "buckyballs" named after R. Buckminster Fuller), are converted into or produced from infinitesimally small particle sizes (and, in the case of carbon nanotubes and buckyballs, unique structural geometries, as well), the resulting particles have an order-of-magnitude increase in available surface area. It is this remarkable surface area of particles in the nanometer range that confers on them some unique material properties, especially when compared to macroscopic particles of the same material (Shelley and Ondrey 2002).

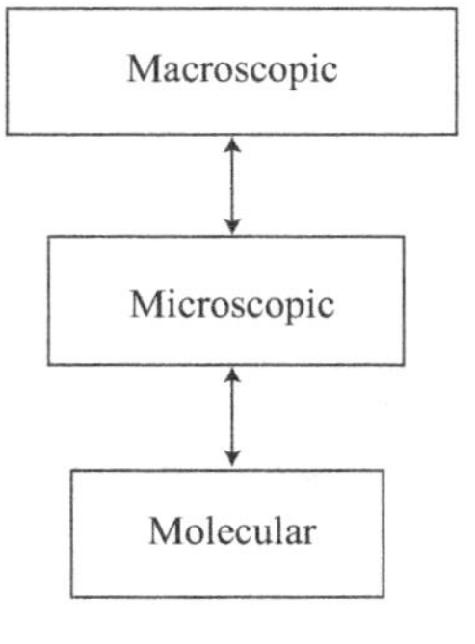

FIGURE 17.1 Traditional size ranges of matter as defined by engineering and science.

One hallmark of nanotechnology is the desire to produce and use nanometer-sized particles of various materials to explore the remarkable characteristics and performance attributes that many materials exhibit at these infinitesimally small particle sizes.

Technical individuals have traditionally conducted calculation-related studies using one of a combination of the size range approaches presented in Figure 17.1. These studies generally involve the application of a conservation law (e.g., mass, energy, and momentum). For example, if one were interested in determining changes occurring at the inlet and outlet of a system under study, the conservation law is applied on a "macroscopic" level to the entire system. The resultant equation describes the overall changes occurring to the system without regard for internal variations within the system. This approach is usually employed in a Unit Operations course. The microscopic approach is employed when detailed information concerning the behavior *within* the system is required, and this is often requested of and by technical personnel. The conservation law is then applied to a differential element within the system, which is large compared to an individual molecule but small compared to the entire system. The resultant equation is then expanded, via an integration, to describe the behavior of the entire system. This has come to be defined by some as the *transport phenomena approach*. The molecular approach involves the application of the conservation law to individual molecules. This leads to a study of statistical and quantum mechanics—both of which are beyond the scope of this text.

The unit operations and transport phenomena approaches are normally in the domain of the engineers, and the transport phenomena and molecular approaches are employed by the scientists, particularly the physicists. However, nanotechnology has disrupted the above classical approach to the describing behavior of systems. The nanotechnology field today belongs somewhere between the microscopic and molecular approaches (Figure 17.2).

Nanoparticles cannot be correctly described by applying either the microscopic or molecular method of analysis. This new, so-called, in-between field gives rise to some very unusual physics. This unusual behavior results because the (physical, chemical, and so on) properties of a substance are a strong function of its size. At microscopic or macroscopic scales, one chunk of iron (an element) has the exact same properties of another chunk of iron. At the molecular level, an atom of iron has the exact same properties of another atom of iron. However, "something happened on the way to the forum" when the size of the iron particle is in the nano range. The chemical, physical, mechanical, and electrical properties of these bulk materials are different in the nanometer range. Further, a 10-nanometer particle has different properties than a particle of different size (e.g., 20 nanometers). The same phenomenon is experienced with iron oxide or any other solid particle. What does all of this mean? It permits a new way to vary and control the properties of materials. In effect, one need only change the size of the particle rather than its composition to develop a material with new and unique properties.

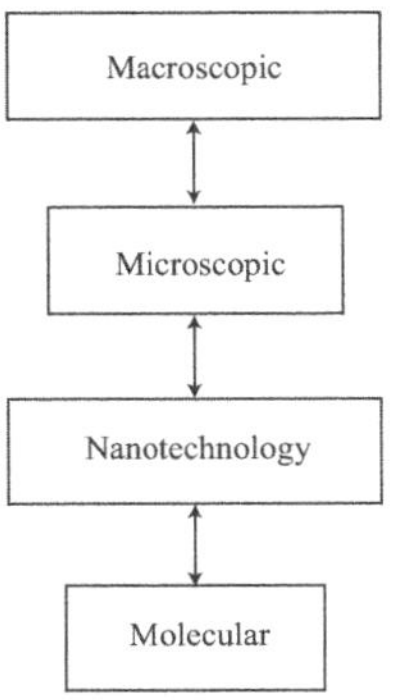

FIGURE 17.2 Nontraditional size ranges of matter as defined by nanotechnology.

One of the future challenges facing nanotechnology is to change molecules in a controlled way, creating new substances with very special properties. These changes can lead to improvement of health care, conservation of natural resources, protection of the environment, provision of everyday needs for food, clothing, and protective armaments, and for the production of clean, potable water. When familiar materials such as metals, metal oxides, ceramics, polymers, and novel forms of carbon are converted into infinitesimally small particle sizes, the resulting particles have orders of magnitude increases in available surface area. It is this remarkable surface of particles in the nanometer range that confers on them unique properties, especially when compared to macroscopic particles of the same material.

17.4 NANOMATERIALS

Nanomaterials originated from entities that are now defined as prime materials. These prime materials essentially consist of (pure) elements and compounds. The elements and compounds that have been successfully produced and deployed as nanometer-sized particles include:

1. Metals such as iron, copper, gold, aluminum, nickel, and silver
2. Oxides of metals such as iron, titanium, zirconium, aluminum, and zinc
3. Silica sols and fumed and colloidal silica
4. Clays such as talc, mica, smectite, asbestos, vermiculite, and montmorillonite
5. Carbon compounds, such as fullerenes, nanotubes, and carbon fibers

Each of these types of materials, along with the manufacturing methods used to render them into nanoscale particles, is discussed in the next section. Some of the information presented was adapted from the literature (Boxall et al. 2008).

As noted, when macro-sized metals are produced in the nanometer range, they exhibit properties not found in large particle sizes. This includes properties such as quantum effects, the ability to sinter at temperatures significantly below their standard melting points, increased catalytic activity due to higher surface area per mass, and higher (more rapid) chemical reaction rates. Additionally, when nanoscale particles of metals are incorporated into other larger structures, they often exhibit increased strength, hardness, and tensile strength when compared to structures formed from conventional *micrometer-sized* powders.

However, reducing some metals to nanoscale powder presents some problems; in particular, when reduced to sufficiently small particle sizes, many of the chemical properties of metals become more reactive and subject to oxidation, often explosively. Copper and silver powders are among the few metallic nanoparticles that are not explosive and thus can be handled in air. Many others must be stabilized with a passivation layer or handled in an inert, blanketed (neutral) environment. Moreover, in some applications, effort is required to minimize unwanted particle agglomeration.

Metals that have served as prime materials include the following:

1. Iron
2. Aluminum
3. Nickel
4. Silver
5. Gold
6. Copper

Mixed oxides include the following:

1. Iron oxides (Fe_2O_3 and Fe_2O_4)
2. Silicon dioxide (silica; SiO_2)

3. Titanium dioxide (titania; TiO_2)
4. Aluminum oxide (alumina; Al_2O_3)
5. Zirconium dioxide (zirconia; ZrO_2) and zinc oxide (ZnO)

Additional details on both metals and metal oxides are available in the literature (Theodore and Kunz 2005).

In general, there are six widely used methods for producing nanoscaled particles, on the order of 1 to 100 nm in diameter, of various metal oxides (Fink et al. 2002; Wilson et al. 2004):

1. High-temperature processes such as plasma-arc and flame-hydrolysis methods (including flame ionization)
2. Chemical vapor deposition (CVD)
3. Electrodeposition
4. Sol-gel synthesis
5. Mechanical crushing via ball milling
6. Use of naturally occurring nanomaterials

Each of these production methods are described briefly.

17.4.1 High-Temperature Processes

These high-temperature processes involve the use of a high-temperature plasma or flame ionization reactor. As an electrical potential difference is imposed across two electrodes in a gas, the gas, electrodes, or other materials ionize and vaporize if necessary and then condense as nanoparticles, either as separate structures or as surface deposits. An inert gas or vacuum is used when volatilizing electrodes.

During flame ionization, a material is sprayed into a flame to produce ions (Fink et al. 2002). Using flame hydrolysis, highly dispersed oxides can be produced via high-temperature hydrolysis of the corresponding chlorides. Flame hydrolysis produces extremely fine, mostly spherical particles with diameters in the range of 7–40 nm and high specific surface areas (in the range of 50–400 m^2/g [Wilson et al. 2004]).

In general, high-temperature flame processes for making nanoparticles are divided into two classifications—*gas-to-particle* or *droplet-to-particle* methods—depending on how the final particles are made.

In *gas-to-particle* processes, individual molecules of the product material are made by chemically reacting precursor gases or rapidly cooling a superheated vapor. Depending on the thermodynamics of the process, the molecules then assemble themselves into nanoparticles by colliding with one another or by repeatedly condensing and evaporating into molecular clusters. Gas-to-particle processes involve the use of flame, hot-wall, evaporation-condensation, plasma, laser, and sputtering-type reactors.

In *droplet-to-particle* processes, liquid atomization is used to suspend droplets of a solution or slurry in a gas at atmospheric pressure. Solvent is evaporated from the droplets, leaving behind solute crystals, which are then heated to change their morphology. Spray drying, pyrolysis, electrospray, and freeze-drying equipment are typically used in the droplet-to-particle production process.

17.4.2 Chemical Vapor Deposition (CVD)

In CVD, a starting material is vaporized and then condensed on a surface, usually under vacuum conditions. The deposit may be the original material or a new and different species formed by chemical reaction.

17.4.3 Electrodeposition

Electrodeposition uses an approach where individual species are deposited from solution, with an aim to lay down a nanoscaled surface film in a precisely controlled manner.

17.4.4 Sol-Gel Synthesis

Sol-gel processing is a wet-chemical method that allows high-purity, high-homogeneity nanoscale materials to be synthesized at lower temperatures compared to competing high-temperature methods. A significant advantage that sol-gel science affords over more conventional materials-processing routes is the mild conditions that the approach employs.

After mixing the reactants, organic or inorganic precursors undergo two chemical reactions: hydrolysis, and condensation or polymerization, typically with an acid or base as a catalyst, to form small solid particles or clusters in a liquid (either an organic or aqueous solvent). The resulting solid particles or clusters are so small (1–1000 nm) that gravitational forces are negligible and interactions are dominated by van der Waals, coulombic, and steric forces. These sols—colloidal suspensions of oxide particles—are stabilized by an electric double layer, or steric repulsion, or a combination of both. Over time, the colloidal particles link together by further condensation and a dimensional network occurs. As gelling proceeds, the viscosity of the solution increases dramatically.

The sol-gel can then be formed into three different shapes: thin film, fiber, and bulk. Thin ($\approx$100 nm) uniform and crack-free films can readily be formed on various materials by lowering, dipping, spinning, or spray-coating techniques.

17.4.5 Mechanical Crushing via Ball Milling

Progressive particle-size reduction or pulverization using a conventional ball mill is one of the primary methods for preparing nanoscaled particles of various metal oxides. High-energy ball milling is in use today, but its use is considered by some to be limited because of the potential for contamination problems. The availability of tungsten carbide components and the use of an inert atmosphere and high-vacuum processes has helped operators reduce impurities to acceptable levels for many industrial applications (Fink et al. 2002). Other common drawbacks, however, include the highly polydisperse size distribution and partially amorphous state of nanoscaled powders prepared using the pulverization method.

17.4.6 Naturally Occurring Materials

Certain naturally occurring materials, such as zeolites, can be used as found or synthesized and modified by conventional chemistry to yield nanosized materials. A zeolite is a caged molecular structure containing large voids that can admit molecules of a certain size and deny access to other, larger molecules (Wilson et al. 2004). They find application as catalysts and adsorbents, among other uses.

Depending on their application, nanomaterials may be manufactured so they are nanoscale in one dimension, such as for thin films, coatings, and quantum wells, which are used for applications such as computers and cell phones; in two dimensions, such as for nanowires and nanotubes for applications such as transistors; and in three dimensions, such as nanoparticles, fullerenes, and dendrimer, for applications such as pollution prevention/remediation, sensors, water treatment, desalination, and biomedical applications.

The ongoing challenge for the research community is to continue to devise, perfect, and scale-up viable production methodologies that can cost effectively and reliably produce the desired nanoparticles with the desired particle size, particle size distribution, purity, and uniformity in terms of both composition and structure.

17.5 CURRENT APPLICATIONS

Present-day applications of nanoparticles include chemical products that include plastics, specialty metals and powders, and computer chip and computer systems. Specific examples of nanotechnology in actual commercial use today are (Shelley and Ondrey 2002):

1. Semiconductor chips and other microelectronics applications
2. High-surface-to-volume catalysts, which promote chemical reactions more efficiently and selectively
3. Ceramics, lighter weight alloys, metal oxides, and other metallic compounds
4. Coatings, paints, plastics, fillers, and food-packaging applications
5. Polymer-composite materials, including tires, with improved mechanical properties
6. Transparent composite materials, such as sunscreens containing nanosized titanium dioxide and zinc oxide particles
7. Fuel, cells, battery electrodes, communications applications, photographic film developing, and gas sensors
8. Nanobarcoding
9. Tips for scanning probe microscopes
10. Purification of water (Sumesh et al. 2011), pharmaceuticals, and enzymes

Other applications involving water purification/desalination are certain to emerge in the future. For example, the protection of human health and ecosystems requires rapid, precise sensors capable of detecting pollutants at the molecular level. Major improvement in process control, compliance monitoring, and environmental decision making could be achieved if more accurate, less costly, more sensitive techniques were available. Examples of research in sensors include the development of nanosensors for efficient and rapid in situ biochemical detection of pollutants and specific pathogens in water. Treatment options might include the removal of very fine particulates (under 300 nm). Substances of significant concern in groundwater because of both their cancer and noncancer hazards, include heavy metals (e.g., mercury, lead, cadmium) and organic compounds (e.g., benzene, chlorinated solvents, creosote, toluene). Reducing releases to water, providing safe drinking water, and reducing quantities and exposure to hazardous wastes also are areas of interest for nanotechnology applications.

17.6 ENVIRONMENTAL CONCERNS

Any technology can have various and imposing effects on the environment and society. Nanotechnology is no exception, and the results will be determined by the extent to, which the technical community manages this technology. This is an area that has, unfortunately, been seized upon by a variety of environmental groups.

There are two thoughts regarding the environmental implications of nanotechnology: one is positive and the other is potentially negative. The positive features of this new technology are well documented. The other implication of nanotechnology has been dubbed by many in this diminutive field as, "potentially negative." The reason for this label is as simple as it is obvious. The technical community is dealing with a significant number of unforeseen effects that could have disturbingly disastrous impacts on society. Fortunately, it appears that the probability of such dire consequences actually occurring is near zero but, *not* zero. This finite, but small probability is one of the key topics that is addressed here.

Air, water, and land (solid waste) concerns with emissions from nanotechnology operations in the future, as well as companion health and hazard risks are discussed in brief. All of these issues arose with the Industrial Revolution and the development, testing, and use of the atomic bomb, the arrival of the Internet, Y2K, and so forth, and all were successfully (relatively speaking) resolved by the engineers and scientists of their period. Furthermore, to the authors' knowledge there are no documented nanomaterial-related human health problems. Statements in the literature refer only to *potential* health concerns (Papp et al. 2008).

Theodore (2006) has speculated on the need for future nano regulations. Theodore (2006) also noted that the ratio of pollutant nanoparticles (from conventional sources such as power plants) to engineered nanoparticles being released into the environment may be as high as a trillion to one (i.e., 10^{12}:1). If this is so, the environmental concerns for nanoparticles can almost certainly be dismissed.

Prior experience with materials such as polychlorinated biphenyls (PCBs) and asbestos and a variety of unintended effects of drugs such as thalidomide mean that both companies and governments have an incentive to keep a close watch on potential negative health and environmental effects of new material (National Center for Environmental Research 2003). Many environmental concerns are addressed by existing health and safety legislation. Most countries, including the United States, require a health and safety assessment for any new chemical before it can be marketed. Further, the European Commission (EC) recently introduced the world's most stringent labeling system, EC 1907/2006, and established the European Chemicals Agency in Helsinki. This agency manages the Registration, Evaluation, Authorization and Restriction of Chemical (REACH) substances systems, which is a database of information provided by manufacturers and importers on the properties of their chemical substances. Detailed analysis of various U.S. and EU laws and regulations are available in the literature (Breeggin et al. 2009; Fiorino 2010; U.S. EPA 2007, 2011).

It should be noted, however, that there are no nanoscale or nanoscale-related environmental regulations in the United States or the European Union at this time that require controls on process releases, production activities, or specific workspace safety measures. Completely new legislation and regulatory efforts may be necessary to protect the public and the environment from the potentially adverse effects of nanotechnology.

The principal U.S. agencies concerned with environmental risks are the U.S. Environmental Protection Agency (EPA) and the Occupational Safety and Health Administration (OSHA). The EPA's mission is to protect human health and the environment. One of its major goals is to ensure that all Americans are protected from significant risks to human health and the environment where they live, learn, and work (i.e., to closely monitor and control any potentially toxic or biohazardous conditions that could pose an unreasonable risk to human health or the environment). The mission of OSHA is to ensure safe and healthful working conditions for working men and women by setting and enforcing standards and by providing training, outreach, education, and assistance. Both agencies are, therefore, directly concerned with the environmental implications of nanotechnology.

It is difficult to predict what future regulations might come into play for nanomaterials. In the past, regulations have been both a moving target and confusing. What can be said is that there will be regulations, and there is a high probability that they will be contradictory and confusing. Past and current regulations provide a measure of what can be expected. Control of the production and use of nanomaterials is most likely to occur under the Clean Air Act (CAA) and the Toxic Substances Control Act (TSCA).

17.7 APPLICATIONS

Illustrative Example 17.1

Discuss future prospects for nanotechnology.

Solution

For nanotechnology's most ardent supporters, the scope of this emerging field seems to be limited only by the imaginations of those who would dream within these unprecedented dimensions. However, considerable technological and financial obstacles still need to be reconciled before nanotechnology's full promise can be realized. The future of nanotechnology in therefore not known.

Scientists, engineers, and even manufacturers can only speculate on its implications or the magnitude of its impact not only on water management but also to environmental health. The reader is referred to Brame et al. (2011), Qu et al. (2013), and Zhang et al. (2016) for a review of past and future potential applications of nanotechnology in water and wastewater treatment and water reuse. Additional applications are sure to emerge in the desalination field.

Illustrative Example 17.2

Discuss how nanotechnology can impact terrorism threats, particularly as it applies to water resources.

Solution

Regarding crime and terrorism, the techniques of nanoscience will have much to offer forensic investigations, for biological analysis, and for materials and chemical studies as well. Portable instruments with sophisticated nanosensors will be able to perform accurate, high-level analyses at contaminated water sites. These instruments should greatly improve the speed in generating reliable and defensible data and corresponding conviction rates.

Illustrative Example 17.3

Discuss environmental, health, and safety concerns associated with nanotechnology.

Solution

There are a range of potential environmental, health, and safety risks associated with nanotechnology. Questions remain about the potential toxicity of nanoscaled materials to humans in terms of all potential modes of exposure, and not only via water ingestion, to such ultrafine particles. Short- and long-term modes of exposure, such as via skin absorption, ingestion, and inhalation, among others, must be systematically studied to determine any potential for organ or tissue damage, inflammation, a triggering of autoimmune diseases, and other health-related consequences.

Illustrative Example 17.4

A spherical particle has a diameter of 100 nanometers (nm). Calculate the volume (cm^3) and surface area (cm^2) of the particle.

Solution

The volume, V, of the spherical particle is:

$$V = \pi D^3/6 = 0.524(100\text{ nm})^3 = 0.524 \times 10^6 = 5.24 \times 10^5\text{nm}^3$$

Because there are 10^7 nm/cm,

$$V = (5.24)(10^5)(10^{-7})^3 = 5.24 \times 10^{-16}\text{cm}^3$$

The surface area, A, of a sphere is given by:

$$A = \pi D^2 = 3.14\ (100\text{ nm})^2 = 3.14 \times 10^4\text{nm}^2$$

$$A = (3.14)\ (10^4)(10^{-7})^2 = 3.14 \times 10^{-10}\text{cm}^2$$

Table 17.1 provides volume and surface area data for a range of spherical particles from 1.0 nm to 1,000 μm in diameter. It is important to note the order of magnitude increase in surface area-to-volume ratio as particle diameter is decreased by an order of magnitude.

TABLE 17.1
Spherical Particle Sizes, Volumes, Surface Areas, and Surface Area-to-Volume Ratio

Particle Size, nm (μm)	Particle Volume, cm^3	Particle Surface Area, cm^2	Surface Area-to-Volume Ratio, 1/cm
1.0 (0.001)	5.24×10^{-22}	3.14×10^{-14}	6×10^{7}
10 (0.01)	5.24×10^{-19}	3.14×10^{-12}	6×10^{6}
100 (0.01)	5.24×10^{-16}	3.14×10^{-10}	6×10^{5}
1,000 (1)	5.24×10^{-13}	3.14×10^{-8}	6×10^{4}
10,000 (10)	5.24×10^{-10}	3.14×10^{-6}	6×10^{3}
100,000 (100)	5.24×10^{-7}	3.14×10^{-4}	6×10^{2}
1,000,000 (1,000)	5.24×10^{-4}	3.14×10^{-2}	6×10^{1}

Illustrative Example 17.5

Describe the general relationship between particle size and collection efficiency by various particle removal devices.

Solution

Owing to the combined action of the various collection mechanisms, the performance of particulate control devices often has unique particle size-efficiency relationships. Above 100 microns, particles are collected with very high efficiency by inertial impaction, electrostatic attraction, and even gravity settling. Efficiency remains high throughout the range of 10 to 100 μm due to the ever present inertial or electrostatic forces (depending on type f collector), both of which are approximately proportional to the square of the particle diameter. For particles less than 10 μm, the limits of inertial forces and electrostatic forces begin to become apparent, and the efficiency drops. Efficiency due to these collection mechanisms reaches negligible levels between 3 and 0.3 μm (300 nm) depending on factors such as gas velocities (inertial forces) and electrical field strengths (electrostatic attraction).

Below 0.3 μm (300 nm), Brownian motion (molecular diffusion) begins to become more pronounced. Accordingly, the overall efficiency curve begins to rise in this very small size range.

The result of these various collection mechanisms is a low collection efficiency in the particle size range of 0.1 to 10 μm. In many particle control devices, none of the collection mechanisms is highly efficient for particles in this range. These particles can be classified as "difficult-to-control" due to the inherent limitations of the collection mechanisms (Theodore 2008).

This relationship, which has been reported in a number of studies of actual sources, indicates that sources generating high concentrations of particles in the 0.1- to 1.0-μm range may pose an especially challenging control problem. The data also suggest that particles in the nanosize range (<100 nm) are captured with relative ease. However, no one can say with certainty what the behavior of particles <50 nm will be.

Illustrative Example 17.6

The EPA's *Nanotechnology White Paper* (U.S. EPA 2007) and *Nanotechnology for Site Remediation Fact Sheet* (U.S. EPA 2008) have a range of case studies highlighting potential pollution prevention opportunities for nanomaterials. Go to these industrial sector resources, and using the various pollution prevention opportunity descriptions and case studies presented in them and other resources you can find on the Internet or in your library, answer the following open-ended questions.

1. Discuss source reduction techniques that may be applied to this industrial sector.
2. Discuss recycling or reuse options that may be applied to this industrial sector.
3. Discuss how ultimate disposal options may be either enhanced or eliminated for this industrial sector.

4. Discuss how energy conservation measures may be applied to this industrial sector.
5. Discuss how health, safety, and accident prevention measures may be applied to this industrial sector.

Solution

As this is an open-ended question, solutions will vary with the case studies being examined and external sources used in the completion of the problem.

REFERENCES

Boxall, A., Q. Chaudhry, A. Jones, B. Jefferson, and C.D. Watts. 2008. *Current and Future Predicted Environmental Exposure to Engineered Nanoparticles.* Sand Hutton, UK: Central Science Laboratory.

Brame, J., Q. Li, and P.J.J. Alvarez. 2011. Nanotechnology-enabled water treatment and reuse: Emerging opportunities and challenges in developing countries. *Trends Food Sci. Technol.* 22:618–624.

Breeggin, L., R. Falkner, N. Jasoers, J. Pendergrass, and R. Porter. 2009. *Securing the Profits of Nanotechnologies: Towards Transatlantic Regulatory Cooperation.* London, UK: Chatham House.

Drexler, K.E. 1986. *Engines of Creation: The Coming Era of Nanotechnology.* New York: Anchor Books. 320 p.

Fink, U., R.E. Davenport, S.L. Bell, and Y. Ishikawa. 2002. *Nanoscale Chemicals and Materials—An Overview on Technology, Products and Applications.* Menlo Park, CA: Specialty Chemicals Update Program, SRI Consulting.

Fiorino, D.J. 2010. *Voluntary Initiatives, Regulation, and Nanotechnology Oversight: Charting a Path.* Pen 19. Washington, DC: Project on Emerging Nanotechnologies, Woodrow Wilson International Center for Scholars. (www.nanotechproject.org/process/assets/files/8347/pen-19.pdf).

National Center for Environmental Research. 2003. *Nanotechnology and the Environment: Applications and Implications*, STAR Progress Review Workshop, Office of Research and Development. Washington, DC: National Center for Environmental Research.

Papp, T., D. Schiffmann, D.G. Weiss, V. Castranova, V. Vallyathan, and Q. Rahman. 2008. Human health implications of nanomaterial exposure. *Nanotoxicology* 2:9–27.

Poole, C., and F. Owens. 2003. *Introduction to Nanotechnology.* Hoboken, NJ: Wiley.

Qu, X., P.J.J. Alvarez, and Q. Li. 2013. Applications of nanotechnology in water and wastewater treatment. *Water Res.* 47:3931–3946.

Shelley, S.A., and G. Ondrey. 2002. Nanotechnology—The Sky's the Limit. *Chem. Eng.* 12:23–27,72.

Sumesh, E., M. Bootharaju, and A. Pradeep. 2011. A Practical Silver Nanoparticle-Based Adsorbent for the Removal of Hg^{2+}. *J. Hazard. Mater.* 189:450–457.

Theodore, L. 2006. *Nanotechnology: Basic Calculations of Engineers and Scientists.* Hoboken, NJ: John Wiley & Sons.

Theodore, L., and R. Kunz. 2005. *Nanotechnology: Environmental Implications and Solutions.* Hoboken, NJ: John Wiley & Sons.

Theodore, L. 2008. Air Pollution Control Equipment Calculations, Hoboken, NJ: John Wiley & Sons.

U.S. Environmental Protection Agency (U.S. EPA). 2007. *Nanotechnology White Paper.* EPA 100/B-07/001. Washington, DC: Science Policy Council, Office of the Science Advisor, U.S. Environmental Protection Agency. (http://nepis.epa.gov/Exe/ZyPDF.cgi/60000EHU.PDF?Dockey=60000EHU.PDF).

U.S. Environmental Protection Agency (U.S. EPA). 2008. *Nanotechnology for Site Remediation Fact Sheet.* EPA 542-F-08-009. Washington, DC: Office of Solid Waste and Emergency Response, U.S. Environmental Protection Agency. (http://nepis.epa.gov/Exe/ZyPDF.cgi/P1001JIB.PDF?Dockey=P1001JIB.PDF).

U.S. Environmental Protection Agency (U.S. EPA). 2011. *EPA Needs to Manage Nanomaterial Risks More Effectively.* Report No. 12-P-0162. Washington, DC: Office of the Inspector General, U.S. Environmental Protection Agency. (www.epa.gov/sites/production/files/2015-10/documents/20121229-12-p-0162.pdf).

Wilson, M., K. Kannangara, G. Smith, M. Simmons, and B. Raguse. 2004. *Nanotechnology Basic Science and Emerging Technologies.* Boca Raton, FL: Chapman & Hall/CRC Press.

Zhang, Y., B. Wu, H. Xu, H. Liu, M. Wang, Y. He, and B. Pan. 2016. Nanomaterials-enabled water and wastewater treatment. *Nanoimpact* 3–4:22–39.

Section IV

The Future

18 New Options for Water Desalination

18.1 INTRODUCTION

The world's total water supply is enormous compared with the presently conceivable needs of man, yet there is a growing water concern. Approximately 98% of the 320 million cubic miles of water in the Earth's crust is salty and useful neither for irrigation by present techniques nor for the majority of man's other needs. Rain (the hydrological cycle) provides many times the world's annual water needs, but freshwater supplies vary widely not only over the Earth's surface but also from time to time in a given region. This accounts for what has come to be referred to as the "water problem."

In the United States, the actual "consumptive use" is only approximately 15% of the potential supply. In addition, water prices are low, and based on statistical analyses, there should be no major problem for many years in the future years. Although the United States is relatively well supplied with water, it continues to turn to sources other than rainfall and runoff. Parts of the Southwest have too little water to support rapid industrial expansion and California has embarked on projects to bring water nearly 1,000 miles to Los Angeles and the southern part of the state. Outside of the United States, expensive desalination plants have been in operation for years, including in the Virgin Islands, the Persian Gulf, the West Indies, Libya, the Union of South Africa, Israel, and many other places.

An obvious way to increase water availability is to recover fresh water through desalination from seawater or from the large underground stores of brackish water, which are available in many arid regions. An alternative is to find ways to use saline waters for agriculture and other purposes that require fresh water.

This chapter introduces the reader to the desalination process. Traditional processes are reviewed and 11 new options for desalination developed by the authors are presented. These include solar evaporation, crystallization, a greenhouse solar evaporator, a simple evaporation (still) unit, a humidification/dehumidification unit, a heat conduit evaporator, a geothermal evaporator, freeze desalination process, mangrove processes, a humidification unit, and reverse osmosis and crystallization hybrid processes. An application section provides three Illustrative Examples related to the general subject of water desalination options.

18.2 INTRODUCTION TO DESALINATION

Two definitions need to be first established. A salt is any substance that yields ions, other than hydrogen or hydroxyl ions. A salt is obtained by displacing the hydrogen of an acid by a metal; the salt of concern in this chapter is primarily sodium chloride. A solution may be described as a homogeneous mixture, constituting only one phase of two or more components, and although there is no fundamental difference between the roles of these components, it is convenient for certain purposes, to call the one present in excess the solvent and the other (i.e., the dissolved substance), the solute. For purposes of most discussions, the solute is generally assumed to be nonvolatile, whereas the solvent is volatile.

The overall desalination process is presented in Figure 18.1. Feed water (1) is introduced to the desalination unit (2) where—following the application of some form of energy, (3) it is separated into (relatively) pure water (4) and a more concentrated brine solution (5).

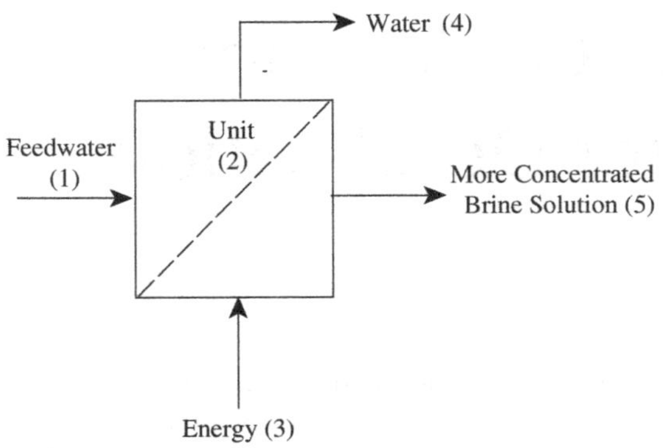

FIGURE 18.1 Simplified desalination process.

The various methods (to be discussed in the next section) for separating pure water and the aforementioned dissolved salts from seawater have two things in common:

1. They all involve the same theoretical (minimum possible) work energy—approximately 3.67 kWh is required to separate 1,000 gal of fresh water from seawater at 25°C, leaving a concentrated brine solution containing the same 1,000 gal of water (50% recovery).
2. They all involve some sort of physical separation process.

These two considerations and the accompanying mode of energy delivery serve to differentiate among the various desalination processes (i.e., the type of energy employed and its accompanying mode of delivery to the solution plus the subsequent separation technique employed are the key factors). Each are briefly described.

Skipka and Theodore (2014) have reviewed the various forms of energy that are available to do useful work (Theodore et al. 2007). They include:

1. Fossil fuels—coal, oil, natural gas, shale oil, and tar sands
2. Solar
3. Nuclear
4. Geothermal
5. Hydroelectric
6. Wind
7. Hydrogen
8. Biomass

For purposes of desalination, the three major sources of energy are the fossil fuels, solar, and nuclear, although geothermal does receive consideration later. However, because most of the aforementioned water problems exists in arid regions, it would seem that solar energy would be the most attractive option.

Any of the traditional unit operations (Theodore 2016; Theodore et al. 2017) are potential candidates for providing the separation technology. These can include:

1. Evaporation
2. Reverse osmosis
3. Crystallization
4. Electrodialysis
5. Solvent extraction

6. Thermal diffusion
7. Polarization
8. Ion exchange, and
9. Hydration

The energy requirement is proportional to the quantity of water recovered, in separation processes in which *water* crosses a phase boundary, and the actual work requirements are effectively independent of the salt concentration of the water processed, whether it is seawater or slightly brackish water. In processes in which the *salt* crosses the phase boundary, the energy requirement increases roughly in proportion to the salt concentration of the feed. Therefore, processes of the second type are generally less expensive to operate with brackish water than with seawater.

Of all the separation processes, the only ones that appear to be economically feasible are evaporation, reverse osmosis, and crystallization. At the present, the economic feasibility of all other processes, irrespective of the energy employed, is considered doubtful with the exception of solvent extraction (Theodore 2016; Theodore et al. 2017) and hydrate formation, which may have some possibility in the future.

18.3 TRADITIONAL SEAWATER DESALINATION PROCESSES

As noted, three prime candidate desalination processes have emerged and have been employed in the past. Each are briefly reviewed in the subsections to follow.

18.3.1 Evaporation Processes

The oldest and best developed process for saline water conversion is the evaporation (or what some define as distillation) method. Evaporation remains the major method today for commercial production of fresh water from seawater.

In principle, traditional evaporation is the simplest method. Seawater is boiled in an evaporator (Theodore 2011, 2016; Theodore et al. 2017) by passing hot steam through a steam chest where the steam condenses on the inside of the tubes of the chest and is usually returned to the boiler (Theodore 2011). The vapors rising from the seawater feed are cooled in the condenser and, thus, converted into pure liquid water, which is collected in a storage vessel. The resulting concentrated brine solution is continuously or intermittently withdrawn from the evaporator.

The most advanced commercial evaporation processes are the following:

1. Multiple-effect evaporation
2. Flash evaporation, and
3. Vapor compression evaporation

In multiple-effect units, the vapors from the first evaporator condense in the second effect where their heat of condensation is employed to boil the seawater in the second effect (chest), etc. In the flash evaporation process, seawater is first heated in tubes and then made to evaporate in chambers where the pressure is lower than in the heating tubes. Although multiple-effect and flash evaporation units use an external supply of heated steam as the primary heat source, vapor compression evaporation literally uses its own steam as a heat source after it has been compressed. Here, seawater is preheated in a tubular heat exchanger by the outgoing streams of concentrated brine and fresh water and boiled in the tubes of the still. The (steam) vapors are compressed and fed back to the still to condense outside the tubes, thus providing the heat necessary for the boiling process.

The two major drawbacks to the evaporation processes are corrosion and scale formation (Flynn et al. 2019).

18.3.2 Reverse Osmosis

Electrodialysis was the membrane separation process employed for desalination a century ago. Here, the ions forming the salt are separated from the saltwater by electric forces and concentrated in different compartments. As noted previously, the higher the salinity of the raw water, the more electric power is needed for this process. Hence, this process was applied primarily in the treatment of moderately brackish waters containing 1,000 to 2,000 ppm of dissolved salts.

In recent years, reverse osmosis (RO) (Theodore and Ricci 2010) has displaced electrodialysis as the membrane separation desalination process of choice, leaving the latter as the choice for medical kidney applications. An RO system consists of an intake, a pretreatment component, a high-pressure pump, a membrane apparatus, remineralization, and pH adjustment components, as well as a disinfection step. Generally, a pressure of approximately 1.7 to 6.9 MPa is required to overcome the osmotic pressure of salt water (the pressure needed to stop the flux of solvent through the membrane).

This advanced separation technique may be used whenever low-molecular-weight solutes such as inorganic salts or small organic molecules (e.g., glucose) are to be separated from a solvent (usually water). In normal (as opposed to reverse) osmosis, water flows from a less concentrated salt solution to a more concentrated salt solution as a result of an innate concentration driving force (thermodynamically referred to as the chemical potential). As a result of the migration of water, an "osmotic pressure" is created on the side of the membrane to which the water flows. In reverse (as opposed to normal) osmosis, the membrane is permeable to the solvent or water and relatively impermeable to the solute or salt. To make water pass through an RO membrane in the desired direction (i.e., away from a concentrated salt solution), a pressure must be applied that is *higher* than the osmotic pressure.

The membranes used for RO processes are characterized by a high degree of semipermeability, high water fluxes, good mechanical strength, chemical stability, and relatively low operating and capital costs. Early RO membranes were composed of cellulose acetate, but restrictions on process stream pressure, temperature, and organic solute rejection spurred the development of noncellulosic and composite materials (membrane "sandwiches"). These membranes may be configured into certain geometries for system operation, including plate and frame, tubular, spiral wound (composite), and hollow fiber.

Obviously, the heart of the membrane process is the membrane itself. As noted, a membrane is an ultrathin semipermeable barrier separating two fluids that permits the transport of certain species through the barrier from one fluid to the other. The membrane is selective because it permits the transport of certain species while rejecting others. The term *semipermeable* is frequently used to describe this selective action. New and improved membranes will probably involve nanomaterials (Theodore and Kunz 2005).

18.3.3 Crystallization Processes

Crystallization is an important mass transfer operation that is often employed in the preparation of a pure product. In the process, a crystal usually separates out as a substance of definite composition from a solution of varying composition. Any impurities in the liquid (often referred to as the *mother liquor*) are carried in the crystalline product only to the extent that they adhere to the surface or are *occluded* (retained) within the crystals that may have grown together during or after the crystallization operation (Theodore and Ricci 2010, Theodore 2016).

The separation of a solid from a solution onto a crystal occurs only if there is a state of imbalance involving a mass driving force; namely, a decrease in chemical potential (or concentration) between the bulk of the liquid solution and the crystal interface exists. This effectively means that the solution must be supersaturated.

Although crystallization is ordinarily thought of as the deposition of a solid crystalline phase from a liquid phase by cooling, evaporation, or both, the same principles apply to crystal formation by precipitation caused by the addition of a third substance. This other component may either react to form a precipitate or simply decrease the solubility of the precipitated material. It should also be noted that when crystallization begins, heat is *evolved* (liberated). The amount of heat evolved is defined as the heat (or enthalpy) of crystallization.

There are several different ways that crystallization can occur. The four most often encountered in practice are (Theodore and Ricci 2010, Theodore 2016):

1. Cooling
2. Evaporation
3. Cooling and evaporation (also denoted adiabatic evaporation), and
4. A salting out process

Process (1) is most commonly employed, provided the solubility of the component being crystallized decreases with decreasing temperature.

The problems concerning crystallization that are most frequently encountered by the practicing engineer are (Theodore and Ricci 2010, Theodore 2016):

1. Yield of a given product
2. Purity of the product
3. Energy requirements for cooling, evaporation, etc.
4. Shape of the individual crystals
5. Size of the crystals
6. Uniformity or size distribution of the crystals
7. Rate of production of the desired crystals
8. Caking

18.4 NEW OPTIONS FOR WATER DESALINATION

Eleven new processes are proposed by GADUTH (GAniaris, DUpont, and THeodore) Consulting Engineers that may be employed for desalination purposes. These include one that is uses solar evaporation, one that uses crystallization, one that uses a greenhouse solar system, one that is specific for domestic applications, two that involve a humidification/dehumidification unit, one involving a heat conduit evaporator, one involving a geothermal evaporator, one involving freeze desalination, one involving mangrove processes, and one that is a hybrid RO and crystallization process. New systems involving RO apparently have already been studied and exploited, and all that appears to remain is developing improved membranes, perhaps involving nanomaterials (Theodore and Kunz 2005). The aforementioned new options are briefly discussed.

18.4.1 The GADUTH Solar Evaporation Process (Ganiaris et al. 2018a)

Refer to Figure 18.2. Preheated seawater (1) is fed to an evaporator (2) and heated by solar energy (3). The resulting steam produced (4) is fed to a condenser (5) where the overhead vapors (4) are condensed using heated seawater (6) from the heat exchanger to produce pure water (7). The heated seawater (8) from the evaporator is then directed to a heat exchange where the concentrated brine (9) releases some of its thermal energy to the cool seawater feed stream (10) and is discharged from the process. The preheated seawater feed from the heat exchanger gains additional heat when it is directed to the condenser (5) where the overhead vapors from the evaporator are condensed. This warm preheated seawater (1) is then fed to the evaporator (2), completing the cycle.

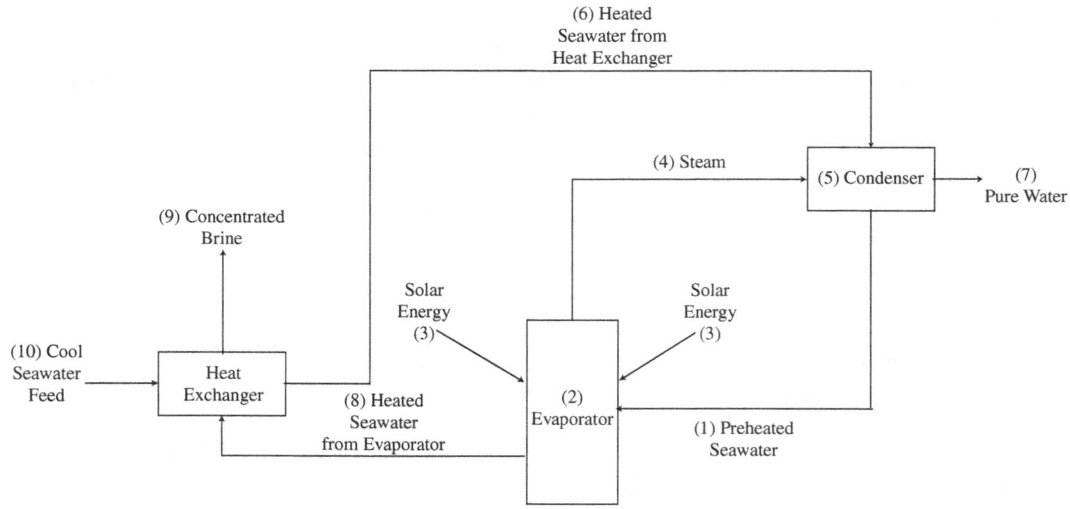

FIGURE 18.2 The GADUTH solar evaporation process.

An enthalpy balance (Theodore et al. 2007, Theodore 2016, Flynn et al. 2018) around the process indicates that the energy provided by the sun is transferred to the concentrated brine solution and pure water discharge streams. In effect,

$$H_{solar} + H_{feed} = H_{conc.\,brine} + H_{pure\,water} \tag{18.1}$$

or

$$H_3 + H_{10} = H_9 + H_7 \tag{18.2}$$

Additional energy conservation measures could be included in the process subject to economic considerations.

18.4.2 The GAniaris Crystallization Process (GACP)

In the GACP Process (Ganiaris 1975), following the freezing/refrigeration of seawater feed by any means, the slurry of ice crystals and brine is forced through a conduit against a counter flow of butane or freon as a wash fluid. The wash fluid emerges from the conduit through perforated or porous wall portions to enter a brine tank in which brine separates from the wash fluid and is removed. Washed crystals emerge from the conduit into a melt tank containing wash fluid. The crystals are melted, separated from the wash fluid, and withdrawn as fresh water. The wash fluid is pumped from the brine tank to the melt tank to cause its counterflow through the ice crystals. Thus, the freeze desalination of seawater is accomplished using the refrigeration potential released during the regassification of liquefied natural gas. Seawater feed is cooled in an ice melter and feed cooler to near freezing. The feed is mixed with cold brine, and the feed is passed to a crystallizer wherein ice crystals are formed by direct contact with a refrigerant that has been cooled by the evaporation of the liquid natural gas. Ice crystals are separated from slurry led from the crystallizer and the ice crystals are fed into the ice melter and feed cooler in which a refrigerant in a closed cycle condenses to melt ice and evaporates to cool feed. Fresh water is separated from condensed refrigerant in the ice melter and feed cooler. Refrigerant is stripped from the fresh water and brine passing from the system. In the crystallizer, gaseous refrigerant emerging from the feed is condensed by liquid refrigerant spraying into a tray suspended above the feed.

18.4.3 The GADUTH Greenhouse Solar Evaporator

In this process (Ganiaris et al. 2018b) (see Figure 18.3a), solar radiation (1) passes through the glass in a greenhouse (2) and heats seawater in a large pan (3) to a temperature that exceeds that of the surrounding air. The steam vapor from the heated seawater (4) condenses on the cooler inside of the glass (5) and flows by gravity into freshwater collectors (6). Concentrated brine is continuously removed from the pan (7). The glass roof of the house is constructed in a manner that allows the fresh water to condense on the inside and seawater feed (8) to flow downward outside (or inside) the glass at

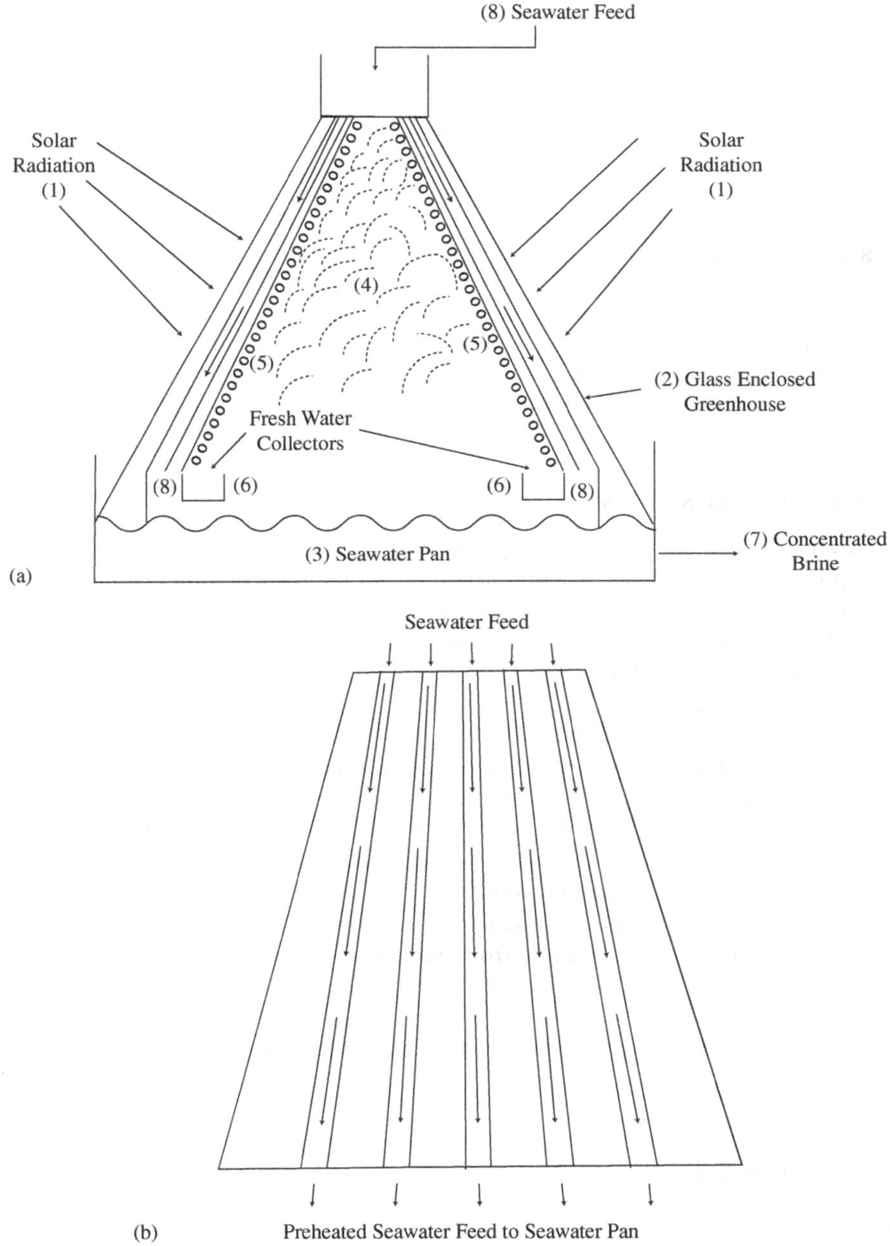

FIGURE 18.3 The GADUTH greenhouse solar evaporator. (a) Front view and (b) Side view.

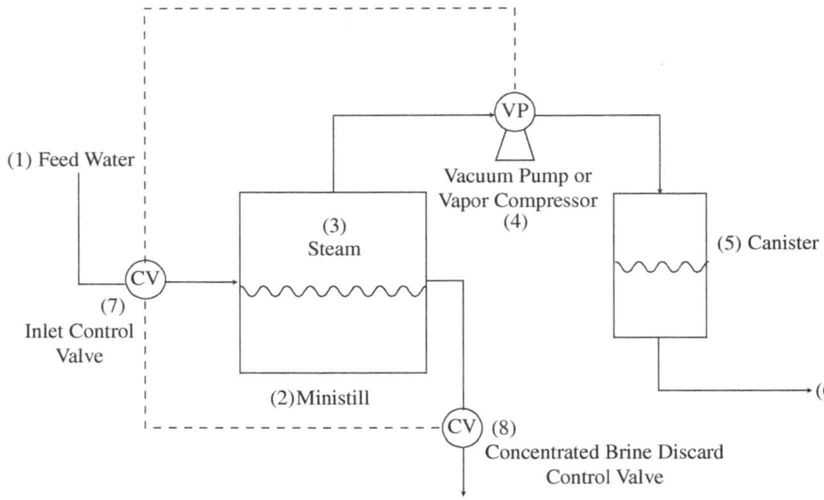

FIGURE 18.4 The Theodore simple still.

spaced intervals. In this manner, a certain percentage of the glass receives unobstructed sunrays, and the feed water is heated in the remaining glass during its approach to the basin pan in the greenhouse. A side view of the greenhouse design is provided in Figure 18.3b.

18.4.4 The Theodore Simple Still

This fourth water desalination process proposed by GADUTH is one that would be suitable for domestic applications (Ganiaris et al. 2018c). The main requirement is that relatively inexpensive energy be available for domestic households. The only equipment of a unit operations (Theodore and Ricci 2010, Theodore 2016) nature would be an efficient vacuum pump or vapor compressor (or the equivalent). As indicated in Figure 18.4, feed water (1) is introduced to a ministill (2) where vapors in the form of steam (3) are suctioned off by a vacuum pump or vapor compressor (or the equivalent) (4). When a vacuum pump is employed, the steam is fed to a canister (5) where condensation occurs due to radiation losses in the line and clean water is available (6). When the vacuum pump (4) is activated, a proportional controller opens the valve (7) to allow the feed water to enter the ministill. This in turn proportionally opens a valve (8) allowing part of the more concentrated brine solution to leave the still. The control valves can be set during operation for 50% of the feed water to be evaporated and the remaining to be discarded as a 50% more concentrated salt solution. A slightly modified process is employed when a vapor compressor replaces the vacuum pump. Here, the vapor compressor takes in water vapor from the evaporator and compresses it to both a higher pressure and temperature (Theodore 2016). This vapor passes through the coils in the evaporator and heats the seawater while the vapor is condensing into fresh water. This heat exchanger can salvage heat from both the discharged concentrated brine and fresh water for reuse in the evaporator, subject to economic considerations.

18.4.5 The GADUTH Dewdrop Process

The evaporation of water into air for the purpose of increasing the air humidity is known as *humidification*. Closely allied to this is the evaporation of water into air for the purpose of cooling the water. *Dehumidification* consists of condensing water from air to decrease the air humidity. All of these processes are of considerable industrial importance and involve the contacting of air and water accompanied by heat transfer. The cooling of water in a cooling tower is accomplished by bringing the water into contact with unsaturated air under such conditions that the air is humidified and the water is brought approximately to the wet-bulb temperature (Theodore and Ricci 2010). When air at the wet-bulb temperature enters the tower, it receives sensible heat from the hot water and its temperature is raised so that it no longer is saturated. Water then evaporates continuously into the air as it travels upward in the tower. Generally, the temperature of the water sent to a cooling tower rarely exceeds 120°F and is usually less. The lowest possible temperature to which water can be cooled in a cooling tower is the wet-bulb temperature of the air. The difference between the outlet water temperature from the tower and the wet-bulb temperature is defined as the *approach temperature*. When water is present in air, it is possible to extract the water by cooling the air-water mixture below the mixture's dew point temperature. If the temperature is reduced further, water is forced out of the air-water mixture as condensation (dew) is formed (i.e., dehumidification). The emphasis in recent years has been to employ *mechanical-draft* equipment that include spray columns and spray chambers, as well as open surface spray ponds and cooling ponds. Other methods of dehumidification also include adsorption and absorption processes (Theodore and Ricci 2010).

The GADUTH Dewdrop Process (Theodore 2015) uses a humidification followed by dehumidification approach where relatively dry air (1) is employed as a carrier gas to evaporate water from a feed of seawater (2). This evaporation process decreases the temperature of both the air and feed while saturating the air (3) with water from the feed. This is the humidification part of the process. The saturated air is then cooled in a condenser (4), a dehumidification process, where water in the carrier air condenses as dew to produce pure water (5) via recycled cooled concentrated brine solution (6). A flow diagram of the process is provided in Figure 18.5. Various forms of energy recovery can be included in the GADUTH Dewdrop process, subject to economic considerations.

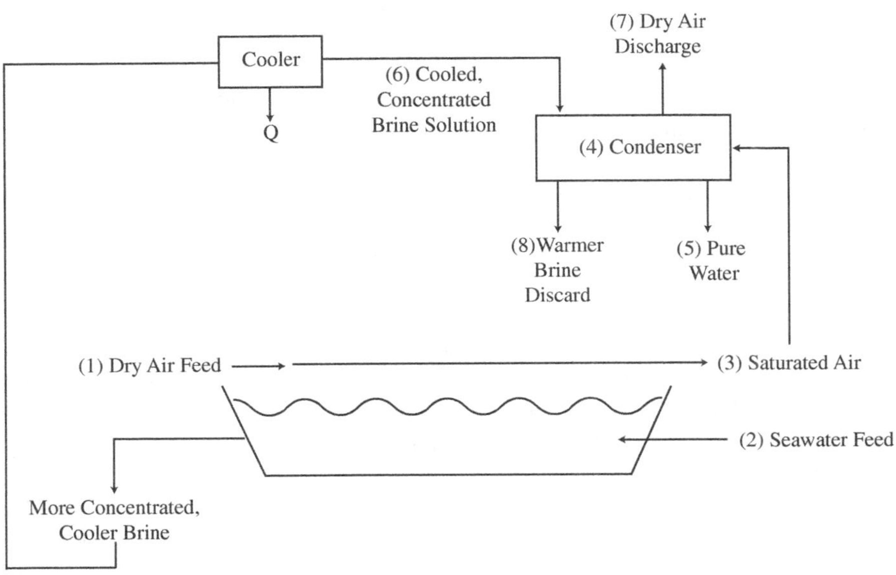

FIGURE 18.5 The GADUTH dewdrop process.

18.4.6 The Heat Conduit Evaporator (HCE) Process

The HCE process (Theodore 2017) involves both geothermal energy and the utilization of a heat pipe. Both geothermal energy and heat pipes are briefly described before proceeding to a description of the HCE processes.

Geothermal energy generally refers to the utilization of energy from the stored heat in the Earth's crust. Geothermal energy (also referred to by some as *geothermics*) is virtually limitless and is based on the fact that the Earth is hotter the deeper one drills below the surface. The Earth's core lies almost 4,000 miles beneath the Earth's surface. The double-layered core is made up of very hot molten iron surrounding a solid iron center. Estimates of the temperature of the core range from 5,000 to 11,000°F! Surrounding the Earth's core is the mantle, which is partly rock and partly magma. The mantle is about 1,800 miles thick. The outermost layer of the Earth is the insulating crust; it is not one continuous sheet of rock but is broken into pieces called *plates*.

The magma can reach the surface and form volcanoes, but most remains underground where it can underlie regions as large as spacious mountain ranges. The magma can take from 1,000 to 1,000,000 years to cool as its heat is transferred to surrounding rocks by conduction. In areas where there is underground water, the magma can fill rock fractures and porous rocks. The water becomes heated and can circulate back to the surface to create hot springs and mud pots, or it can become trapped underground, forming deep geothermal reservoirs. The Earth's geothermal resources are theoretically more than adequate to supply humanity's energy needs, but only a very small fraction may be profitably exploited at this time. Unfortunately, drilling and exploration for deep resources is very expensive.

Seismically active hotspots are not the only places where geothermal energy can be found. There is a steady supply of milder heat, useful for direct heating purposes, virtually in any location on Earth at depths of anywhere from 10 to a few hundred feet below the surface. In addition, there is a vast amount of heat energy available from dry rock formations deeper below the surface (4 to 10 km). Some of the emerging technologies may be able to capture this energy on a much larger scale than some of the conventional technologies.

The areas with the highest underground temperatures are in regions with active or geologically young volcanoes. These "hot spots" occur at plate boundaries or at places where the crust is thin enough to allow the heat to pass. The Pacific Rim, often called the "Ring of Fire" for its many volcanoes, has many hot spots in the United States, including those in Alaska, California, and Oregon. In addition, Nevada has hundreds of hot spots, covering much of the northern part of the state. A similar "ring" extends through southern Europe and across the middle of Asia connecting with some of the South Sea Islands.

The bulk of thermal energy stored on Earth is located within its core. Some of this energy is transmitted from the interior by means of conduction through the Earth. The average temperature difference driving force is approximately 10°C/km. The average rate of flow of heat to the surface has been estimated to be 0.063 W/m^2. Because the Earth's surface is 510×10^{12} m^2, the total heat flow amounts to 32×10^{12} W. Only 1% of this total can be attributed to heat transfer by convection (Harder 1982). The heat flow rate to the surface by conduction is approximately 30 to 40 terawatts (TW). This energy is replenished by radioactive decay at a rate of approximately 30 TW. This nuclear process has been going on since the Earth's formation and will effectively continue for many millennia. Harder (1982) has provided approximate temperature data at various depths. This information is provided in Table 18.1. Understandably, future development potential is highest for hot water/hot temperature resources.

A heat pipe (Kern 1950, Farber 2017) may be circular, square, or rectangular and is a relatively new heat-transfer device that combines the principles of both thermal conductivity *and* phase transition to efficiently manage the transfer of heat between two interfaces (Figure 18.6). When heat is applied to the heat pipe, the working liquid in the wick (chosen for compatibility

TABLE 18.1
Earth Temperatures at Various Depths

Depth	Temperature, F°
Surface	50
1,000 ft (305 m)	65
1 mile (1.61 km)	129
1.4 miles (2.25 km)	161
2.05 miles (3.29 km)	212
10 miles (16.1 km)	842

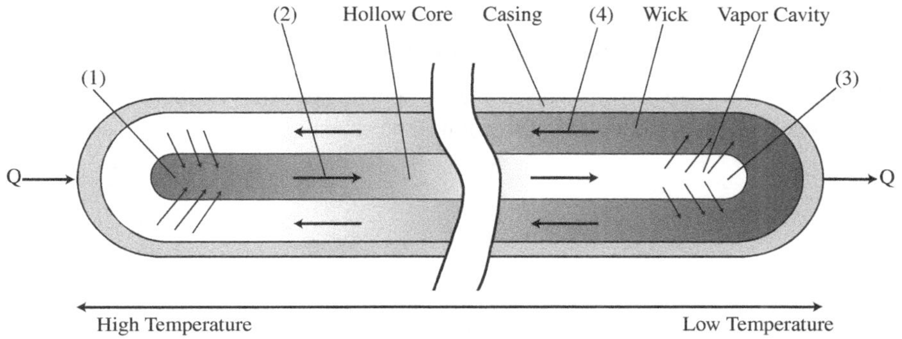

FIGURE 18.6 Heat pipe principles.

with both the heat pipe material of construction and the temperatures of the hot and cold surfaces) heats and evaporates (high temperature applications almost always use liquid metals such as sodium and potassium). As the evaporating fluid fills the heat pipe's hollow center, it diffuses axially along its length. Condensation of the downstream vapor occurs wherever the temperature is even slightly below that of the evaporation area. As it condenses, the vapor gives up the heat it acquired during evaporation and establishes a reverse vapor flow pattern in the pipe. The condensed liquid flows, through capillary action within the wick, back to the heat source, completing the heat transfer cycle. Thus, the heat pipe thermal cycle operates in the following manner (refer to Figure 18.6).

1. Working fluid evaporates to vapor, absorbing thermal energy.
2. Vapor migrates along the cavity to the lower temperature end.
3. Vapor condenses back to the fluid releasing thermal energy in the cooler end and is absorbed by the wick.
4. The working fluid flows back to the higher temperature end.

The unit process itself consists of three operating sections: the evaporator section, the adiabatic section, and the condenser section.

Heat pipes have been employed within computer systems to transfer heat from central processing units (CPUs) to heat sinks so the heat can be rejected to the environment. Heat pipes are also used in heating, ventilation, and air-conditioning (HVAC) systems to recover heat by transfer from an exit stream to a fresh air stream in the winter and to cool an entering air stream in the summer.

For example, a heat pipe buried deep in the ground transports energy from below the Earth's surface to the ground where the accumulation of ice and snow can be reduced or eliminated. Thus, heat pipes are capable of transferring heat from a high temperature source to a cooler surface or location. Additional information is available in the literature (Kern 1950). Calculational details regarding heat pipes are provided by Hagen (1999). Information on the pipe thermal resistances, the heat source, the cooling (sink) source, evaporator and condenser pipes, and for the wick on both the evaporator and condenser sections are also provided.

The HCE process (Theodore 2017) combines the principles associated with both geothermal energy and heat pipes. This appears to be a promising prospect for those locations on Earth where volcanic (or the equivalent) activity can provide temperatures in excess of 212°F at reasonable depths below the Earth's upper surface. The discharge of heat from the top section of conduit can be transferred to seawater that will produce steam that can be recovered as clean water. The first (original) diagram developed by the authors is provided in Figure 18.7.

18.4.7 The Geothermal Evaporator (GEO) Process

The GEO process (Theodore 2018) involves a very simple extension of the aforementioned HCE process. Refer, once again, to Figure 18.7. Once a well has been dug, seawater is fed directly into the

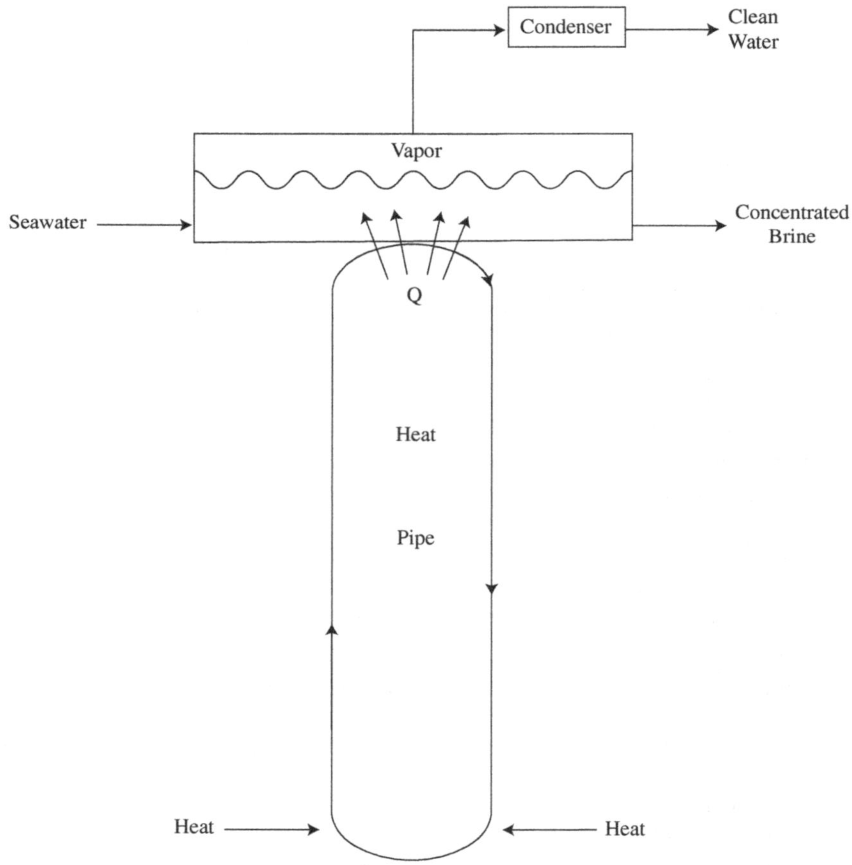

FIGURE 18.7 The heat conduit evaporator process.

Earth's cavity (without the presence of a heat pipe) where steam is generated at greater depths due to the high underground temperatures. Additional horizontal cavities at various depths can be included to increase the contact area for transferring heat from the hot underground sources of the downward flowing seawater. The steam output can be recovered by any of the means described previously. The hot concentrated brine solution that remains at the bottom of the well is withdrawn through a separate conduit attached to the bottom of the well that discharges at ground level. The energy content of this hot brine can be used to either ensure that the rising steam does not condense or heat the seawater downflow or both.

18.4.8 The GADUTH Freeze Desalination Process (FDP)

Two key operation in every FDP are crystallization and ice separation. In a direct contact process, a refrigerant is mixed with seawater in a crystallization zone with the subsequent formation of ice crystals. The crystals are then separated. This concept was developed and demonstrated 50 years ago; however, the economics did not favor this process at the time. The two main problems with the FDP were:

1. The cost of refrigeration, an expensive form of energy for ice crystal formation, and
2. The separation and washing of ice crystals using centrifuges is technically feasible but again, very expensive, and a "wash column" is extremely difficult to scale up to very large sizes.

GADUTH has proposed a process that uses the refrigeration potential from liquefied natural gas (LNG), thus eliminating the cost of refrigeration (Ganiaris 2017). The LNG has to be re-gasified at the point of utilization. The process would also replace the separation step or "wash column" with a refrigerant tank. The process is shown in Figure 18.8.

Early details of the process follow. In 1965, a refrigerant desalination process development project was awarded to Struthers Scientific and was directed by one of the authors (Ganiaris et al. 1969). The primary objective of the contract was to optimize the three main unit operations of the process: crystallization, ice separation, and ice melting and to establish the economic

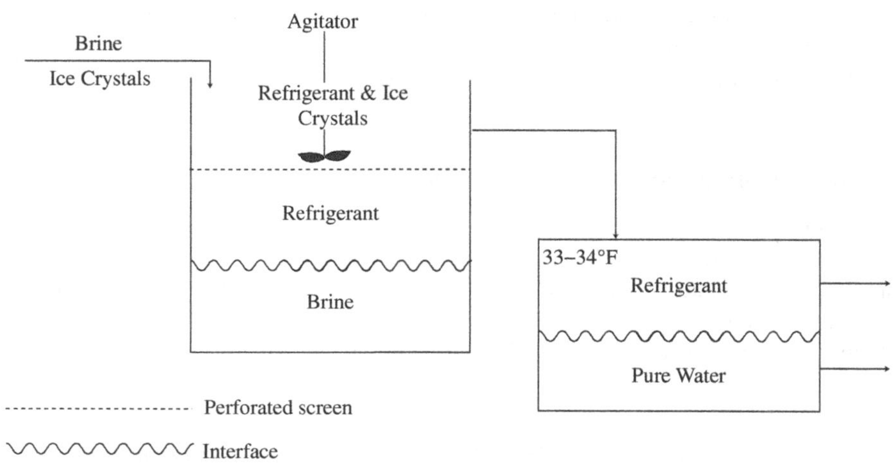

FIGURE 18.8 The GADUTH freeze desalination process.

potential of the process on the basis of data obtained at a 15,000 gal/d pilot plant. The following objectives were achieved at the conclusion of the project.

1. The sparged, mechanically agitated, crystallizer appeared to be the optimum concept at this stage of development.
2. The cylindrical, gravity-drained, wash column appeared to be the least expensive ice separation device.
3. Ice melting in a slurry-type melter-condenser column appeared to be the most economical approach.
4. Normal butane was the most desirable refrigerant for the freezing process. It is a low-cost hydrocarbon, commercially available, without any undesirable odor and does not form hydrates.
5. The unit cost of producing water in a plant in the range of 1.0 to 10.0 MGD appeared to be on the order of $0.45 to $1.00 per 1,000 gallons.

Further engineering and economic studies are currently being investigated by GADUTH to more completely determine the potential of this refrigerant FDP.

18.4.9 The GADUTH Mangrove Process

Mangrove is a common name applied to several kinds of flowering plants that are members of three different families. From a desalination perspective, mangroves are a group of trees and shrubs that grow in shallow coastal saline and brackish waters that are usually muddy. They can survive in a salinity of 9%. Mangroves protect shorelines from storms and prevent erosion by stabilizing sediments and are excellent at adsorbing carbon dioxide (CO_2) from the atmosphere and storing carbon (as leaves, roots, and branches die) at the sea floor. Typically, they produce tangled masses of arching roots that are exposed to the air during low tide. These vertical branches, called *pneumalophores*, act as aerating organs. Mangroves thus have the ability to grow where no other trees can, thereby making significant contributions that benefit coastal ecologies.

Perhaps the key property of mangroves is that they absorb saline water and either reject salt with their roots or excrete salt through glands of their leaves. There is little in the technical literature on this unique separation process and the mechanism has not been investigated to any great extent for the purpose of accomplishing desalination. The GADUTH engineers are currently investigating the commercialization of this potential desalination process (Ganiaris 2018a).

18.4.10 The Theodore HUMidification Process (THUMP)

THUMP (Flynn 2018) involves three steps: heating the brackish/seawater by any of a host of means, contacting the heated water with air by any of a host of means that results in a hotter humidified air, and condensing the saturated air by any of a host of means that separates the resulting condensed water from a less saturated air. The user also has the option of recycling the hot discharge air from the cooler. Figure 18.9 shows a schematic of THUMP. Technical details follow.

18.4.11 Reverse Osmosis and Crystallization Hybrid (ROACH) Desalination Process

RO has been commercialized successfully for the desalination of both seawater (35,000 ppm) and brackish water. However, the process has certain limitations:

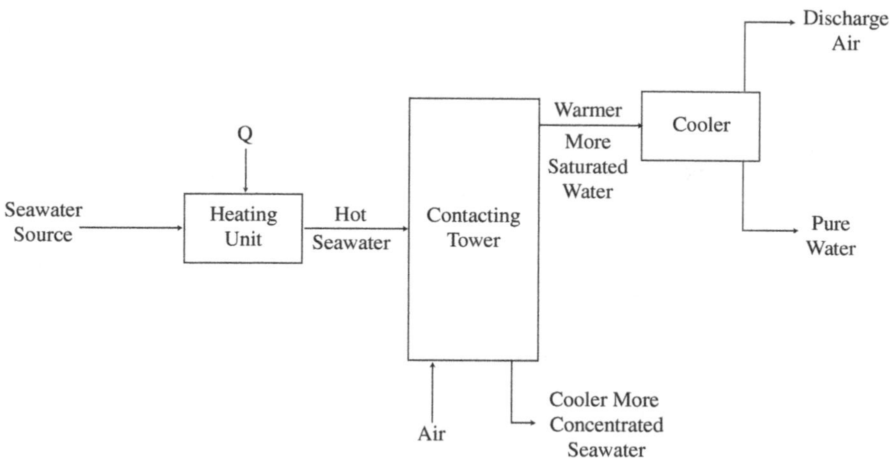

FIGURE 18.9 The Theodore humidification process.

1. The cost of equipment and operation is substantially higher for seawater than brackish water.
2. RO is not adaptable to large plants because the filtration area (membrane) is proportional to capacity and the pumps required to deliver high pressure (1,000 psi) are limited in size; thus, there are no economies of scale.
3. The RO system becomes extremely costly for high salinity (Red Sea 40,000 to 45,000 ppm) feed or other wastewaters from mining operations, whereas the cost of freeze concentration (crystallization) does not increase proportionately with the higher salinity feedwater.

The FDP also has certain limitations:

1. The ice crystals must be separated from the brine and washed to produce potable water (<500 ppm).
2. These unit operations (wash column or centrifuge) are costly and difficult to scale up.

If the end product is not potable water (<500 ppm) and freeze-crystallization is used to reduce the salinity from 35,000 ppm to 2,000 to 3,000 ppm, the overall cost would be substantially lower. A hybrid process using freeze crystallization for pretreatment of the feed (seawater) followed by a reverse osmosis system (ROACH I, Figure 18.10) could overcome the limitations of

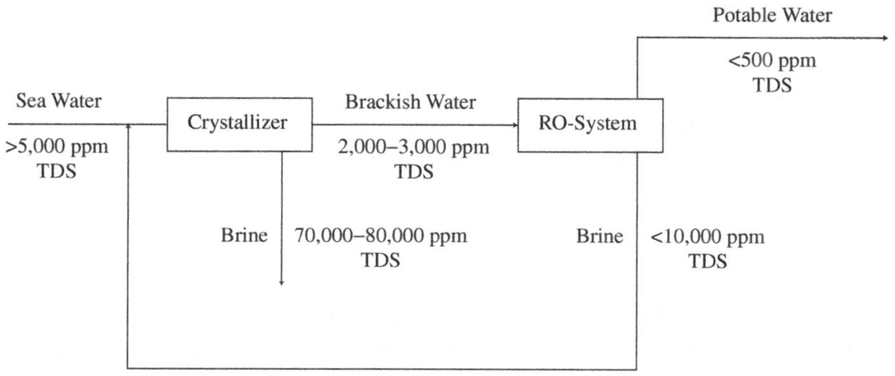

FIGURE 18.10 The reverse osmosis and crystallization hybrid I process.

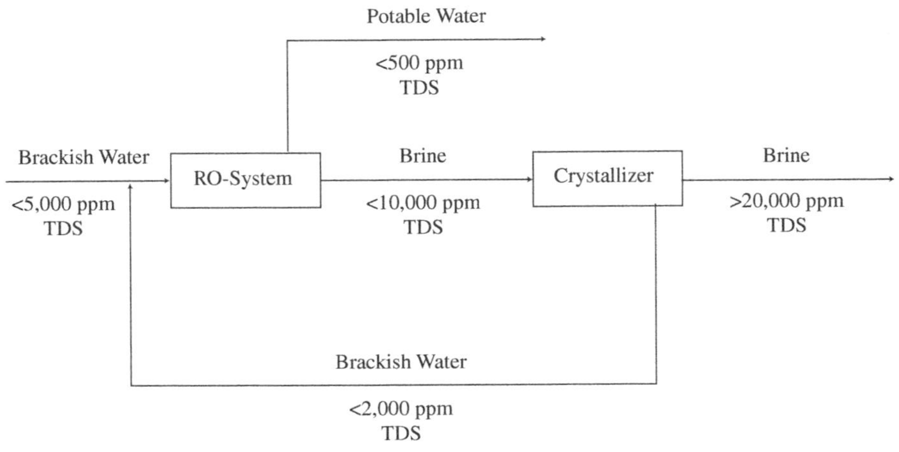

FIGURE 18.11 The reverse osmosis and crystallization hybrid II process. RO = reverse osmosis; TDS = total dissolved solid.

both processes (Ganiaris 2018b). If the feed is brackish water (<5,000 ppm), the process can be reversed (ROACH II, Figure 18.11) with freeze crystallization being used for posttreatment of RO system effluent. Both ROACH flow diagrams contain some data and information from initial studies conducted by GADUTH.

18.5 FUTURE PROSPECTS

The proposed GADUTH processes offer the technical community the opportunity to commercialize these ideas. The sequence in the development of a process from an idea to a plant will vary with the magnitude and nature of each of the GADUTH projects. Each process should be studied so that any unique factors are analyzed. As the projects develop, information should become more complete for calculating the return on investment, which is the key to commercialization.

The following suggested sequence pertains to a new plant and is subject to modification to suit each specific application, that is, the logical evolution of a process may proceed via the following stages:

1. The idea
2. Process development
3. Preliminary design
4. Estimating
5. Final design

It is important to evaluate the idea at an early stage so that the probable success can be determined and if further work is justified. This step can include a search of the literature for pertinent authoritative references.

18.6 CONCLUSIONS

Seawater conversion is one of the most difficult problems ever proposed for chemical engineering process analysis. This is mainly because of process complexities and the unusual economic requirements. However, the authors believe that the market for producing potable water from seawater is not only near unlimited but also astronomical from a financial perspective. Domestic clean water ministills in arid regions short on pure water supplies may soon be as common as filtered water systems in many domestic homes in the United States.

18.7 APPLICATIONS

Three Illustrative Examples complement the material presented in this chapter.

Illustrative Example 18.1

A small evaporator is fed 1,000 kg/min of se water containing 0.005 mass fraction salt. The overhead flow rate, *D*, of pure water is 400 kg/min. Calculate the rate of flow, *W*, and the salt concentration, x_w, of the brine wash.

Solution

The calculations are based on a feed rate of 1,000 kg/min. An overall flow balance can be described as:

$$1{,}000 \text{ kg/min} = D + W = 400 \text{ kg/min} + W$$

$$W = 600 \text{ kg/min}$$

The material balance for the salt yields the following:

$$(1{,}000 \text{ kg/min})(0.005) = (400 \text{ kg/min})(0.0) + W(x_w)$$

$$5 \text{ kg/min} = 0 + 600 \text{ kg/min}(x_w)$$

$$x_w = 5/600 = 0.0083 = 0.83\% = 8{,}300 \text{ mg/L}$$

Illustrative Example 18.2

A tank contains 1,000 kg of seawater. It is desired to crystallize 400 kg of ice, *D*, from this seawater. Calculate the resulting brine volume, *W*, and concentration, x_w.

Solution

Assuming a sea water salt concentration of 0.5% and a pure water salt concentration of 0.0%, this problem can be solved in a manner identical to that of Illustrative Example 18.1. Here the volumes are substituted for volumetric flow rate in Illustrative Example 18.1 to yield:

$$1{,}000 \text{ kg} = D + W = 400 \text{ kg} + W$$

$$W = 600 \text{ kg}$$

The material balance for the salt yields the following:

$$(1{,}000 \text{ kg})(0.005) = (400 \text{ kg})\ (0.0) + W(x_w)$$

$$5 \text{ kg} = 0 + 600 \text{ kg}\ (x_w)$$

$$x_w = 5/600 = 0.0083 = 0.83\% = 8{,}300 \text{ mg/L}$$

Illustrative Example 18.3

Plans are underway to construct and operate a commercial desalination plant. The company is still undecided as to whether employ an RO unit or an evaporator (EV) to accomplish the separation. The RO unit is less expensive to purchase and operate than an EV unit. However, the projected income from the EV unit is higher because it can handle a larger quantity and different seawater concentrations. Based on economic and financial data provided, select the process that will yield the higher annual profit. Calculations should be based on an interest rate, *i*, of 12% and a lifetime, *n*, of 12 years for both processes.

Solution

Calculate the capital recovery factor (CRF) to generate annual costs for the capital and installation costs of the units using Equation 18.7.

$$\text{CRF} = \frac{i}{1-(1+i)^{-n}} = \frac{0.12}{1-(1.12)^{-12}} = \frac{0.12}{1-0.257} = 0.1614 \qquad (18.3)$$

Using this CRF, the annualized cost for capital and installation for the two units are:
For RO, Annualized Cost for Capital + Installation is

$$(\text{Capital} + \text{Installation})(CRF) = (\$2,625,000 + \$1,575,000)(0.1614) = \$677,880/\text{yr}$$

For EV, Annualized Cost for Capital + Installation is

$$(\$2,975,000 + \$1,700,000)(0.1614) = \$754,545/\text{yr}$$

Table 18.2 summarizes the total annual costs for the RO and EV units, and provides the annual profit based on the total annual costs minus the annual income. Based on results shown in

Table 18.3, the EV unit provides a higher annual income, and therefore should be the unit of choice for the company's new desalination plant.

TABLE 18.2
Cost and Income Data for Desalination Options for Illustrative Example 18.3

Cost and Income	RO Unit	EV Unit
Capital ($)	2,625,000	2,975,000
Installation ($)	1,575,000	1,700,000
Operation ($/yr)	400,000	550,000
Maintenance ($/yr)	650,000	775,000
Income ($/yr)	2,000,000	2,500,000

Abbreviations: EV = evaporator; RO = reverse osmosis.

TABLE 18.3
Comparison of Total Annual Cost and Net Annual Income Data for Desalination Options for Illustrative Example 18.3

Cost and Income	RO Unit	EV Unit
Total Installed Annual Cost ($)	677,880	754,545
Operation ($/yr)	400,000	550,000
Maintenance ($/yr)	650,000	775,000
Total Annual Cost ($/yr)	1,727,880	2,079,545
Income ($/yr)	2,000,000	2,500,000
Net Annual Profit ($/yr)	27	420,455

Abbreviations: EV = evaporator; RO = reverse osmosis.

REFERENCES

Farber, P. 2017. *Personal Notes*. Chicago, IL.

Flynn, A.M. 2018. Personal Notes. Floral Park, NY.

Flynn, A.M., T. Akoshige, and L. Theodore. 2019. *Kern's Process Heat Transfer*, 2nd edition. Scrivener-Wiley, Salem, MA.

Ganiaris, N. et al. 1969. *Secondary Refrigerant Freezing Desalting Process: Operations of a 15,000 GPD Pilot Plant*. Report No. 416, Department of the Interior, Washington, DC.

Ganiaris, N. 1975. Adapted from U.S. Patent 3,879,956, Crystal Wash, April 29, 1975.

Ganiaris, N. 2017. *The GADUTH Freeze Desalination Process*. New York.

Ganiaris, N. 2018a. *The GADUTH Mangrove Process*. New York.

Ganiaris, N. 2018b. *The Reverse Osmosis and Crystallization Hybrid Process*. New York.

Ganiaris, N., R. Dupont, and L. Theodore. 2018a. *The GADUTH Desalination Process*. East Williston, NY.

Ganiaris, N., R. Dupont, and L. Theodore. 2018b. *The GADUTH Greenhouse Solar Evaporator*. East Williston, NY.

Ganiaris, N., R. Dupont, and L. Theodore. 2018c. *The Theodore Simple Still*. East Williston, NY.

Harder, E. 1982. Adapted from *Fundamentals of Energy Production*. John Wiley & Sons, Hoboken, NJ.

Hagen, K. 1999. *Heat Transfer with Applications*. Prentice-Hall, Upper Saddle River, NJ.

Kern, D. 1950. *Process Heat Transfer*. McGraw-Hill, New York City.

Skipka, K., and L. Theodore. 2014. *Energy Resources: Availability, Management, and Environmental Impacts*. CRC Press/Taylor & Francis Group, Boca Raton, FL.

Theodore, L. 2011. *Heat Transfer Applications for the Practicing Engineer*. John Wiley & Sons, Hoboken, NJ.

Theodore, L. 2015. *The Dewdrop Process*. East Williston, NY.

Theodore, L. 2016. *Chemical Engineering: The Essential Reference*. McGraw-Hill, New York.

Theodore, L. 2017. *The Heat Conduit Evaporator*. East Williston, NY.

Theodore, L. 2018. *The Geothermal Evaporator*. East Williston, NY.

Theodore, L., R. Dupont, and K. Ganesan. 2017. *Unit Operations in Environmental Engineering*. Scrivener-Wiley, Salem, MA.

Theodore, L., and R. Kunz. 2005. *Nanotechnology: Environmental Implications and Solutions*. John Wiley & Sons, Hoboken, NJ.

Theodore, L., and F. Ricci. 2010. *Mass Transfer Operations for the Practicing Engineer*. John Wiley & Sons, Hoboken, NJ.

Theodore, L., F. Ricci, and T. VanVliet. 2007. *Thermodynamics for the Practicing Engineer*. John Wiley & Sons, Hoboken, NJ.

19 Terrorism Considerations

19.1 INTRODUCTION

Terrorism! Well, what about terrorism? Webster (1971) defines terrorism as "the act of terrorizing, use of force or threats to demoralize, intimidate, and subjugate, especially, such use as a political weapon or policy." It also refers to terror as "a person, or thing causing intense fear, the quality of causing such fear." Can the term be applied in the context of water, and in particular drinking water? The answer is *yes*.

Using the Global Terrorism Database, which includes more than 170,000 terrorism incidents from 1970 to 2016, the Wilson Center (Veilleux and Dinar 2018) developed a method to codify water-related terrorism across the globe and found 675 water-related incidents in 71 countries conducted by 124 known terrorist organizations, resulting in approximately 3,400 dead or wounded people. The most common target of water-related terrorism was found to be water infrastructure: the pipes, dams, weirs, levees, and treatment plants associated with water storage, treatment, and delivery. Terrorists target infrastructure to inconvenience government authorities, influence populations, and cripple corporations. Although these water-related terrorist incidents are not distributed evenly across the world (i.e., more than 200 events have occurred in South Asia, primarily Pakistan, Afghanistan, and India), these events are not uncommon in South America (>60 in Columbia, >40 in Peru) and have been attempted in the United States in the recent past (Veilleux and Dinar 2018). In 2016, the Justice Department revealed that a group affiliated with Iran's Islamic Revolutionary Guard had hacked into the control system of a small dam in Rye Brook, New York, north of New York City (Thompson 2016). Although the attack was not successful because during the attack the sluice gate was offline for maintenance, this event did underscore the vulnerability of critical infrastructure systems even in developed nations. Risks of such cyberattacks on the dams along the Columbia or Missouri rivers, for example, are significantly higher because of the large populations downstream of these structures.

In 2003, President George W. Bush issued Homeland Security Presidential Directive 7 (HPSD-7), which affirmed the Environmental Protection Agency (EPA) as the lead federal agency for coordinating the protection of the nation's critical water infrastructure. Under this directive, EPA is responsible for developing and providing tools and training on improving security to roughly 52,000 community water systems and 16,000 municipal wastewater treatment facilities.

The EPA established a Water Security Division within the Office of Ground Water and Drinking Water in 2003. This division works with drinking water and wastewater utilities, states, tribes, and other stakeholders to improve the security of these utilities and improve their ability to respond to security threats and breaches. Among its responsibilities and activities, the Water Security Division provides security and antiterrorism-related technical assistance and training to the water sector.

Security-related activities undertaken by the EPA and the water sector have fallen into five general categories, including: (1) establishing an information center for drinking water alerts or incidents, (2) developing vulnerability assessment tools, (3) identifying actions to minimize vulnerabilities, (4) revising emergency operations plans, and (5) supporting research on biological and chemical contaminants considered to be potential weapons of mass destruction.

This chapter addresses terrorism issues relevant to the water sector through the following topics: the need for Emergency Response Planning, assessing the risk, the planning process, training of personnel, emergency communication, emergency response, and the emergency recovery process. An application section completes the chapter with six Illustrative Examples related to the general subject of water security and water utility threat assessment.

19.2 THE NEED FOR EMERGENCY RESPONSE PLANNING

In response to the terrorist events of September 11, 2001, across the United States, the 107th Congress passed the Public Health Security and Bioterrorism Preparedness and Response Act of 2002 (P.L. 107–188, the Bioterrorism Act of 2002) to address a wide range of security issues. Title IV of the Bioterrorism Act amended the Safe Drinking Water Act (SDWA) to address threats to drinking water security. One key provision of the act requires that all Public Water Systems (PWSs) serving populations of more than 3,300 persons conduct assessments of their vulnerabilities to terrorist attack or other intentional acts and to defend against adversarial actions that might substantially disrupt the ability of a system to provide a safe and reliable supply of drinking water. In addition to the vulnerabilities assessment (VA), the act requires a PWS to certify and submit a copy of the VA to the EPA Administrator, prepare or revise an emergency response plan (ERP) based on the results of the vulnerability assessment, and certify that an ERP has been completed or updated within 6 months of completing the assessment. The law also required the EPA to conduct research on preventing and responding to terrorist or other attacks.

Improving the security of the nation's drinking water and wastewater infrastructure has become a top priority. Significant actions are underway in response to guidance from the U.S. EPA (2003, 2004a, 2004b) and others (Ross and Associates 2009) to assess and reduce vulnerabilities to potential terrorist attacks; to plan for and practice emergencies and incidents; and to develop new security technologies to detect and monitor contaminants and prevent security breaches.

On October 23, 2018, America's Water Infrastructure Act (AWIA) was signed into law that updates requirements of the Bioterrorism Act of 2002. The law requires community (drinking) water systems serving more than 3,300 people to develop or update risk assessments and ERPs. The law expands on components that the risk assessments and ERPs must address, and establishes new deadlines by which water systems must certify to the EPA completion of the risk assessment and ERP.

Emergencies have occurred in the past and will continue to occur in the future. The frequency of intense storms and natural disasters is increasing, and unsuccessful acts of terrorism involving drinking water (Rye Brook, NY) and wastewater systems (Mesa, AZ [WEF 2013]) have occurred in the United States in the past. Both natural and man-made disasters are likely to continue to occur in the future, and planning for these types of emergencies is a must. A few of the many commonsense reasons to plan ahead for emergencies are as follows (Krikorian 1982, WEF 2013):

1. Emergencies will happen; it is only a question of when.
2. When emergencies do occur, minimization of loss and the protection of people, property, and the environment can be achieved through the proper implementation of an appropriate emergency response plan.
3. Minimizing the losses caused by an emergency requires planned procedures, understood responsibility, designated authority, accepted accountability, and trained, experienced people. A fully implemented plan can do this.
4. If an emergency occurs, it may be too late to plan. Lack of preplanning can turn an emergency into a disaster.

A particularly timely reason to plan ahead is to ease the "chemophobia" or fear of chemicals, which is so prevalent in society today. So much of the recent attention to emergency planning and newly promulgated laws are a reaction to the tragedy at Bhopal. Either a total lack of information or misinformation is the probable cause of "chemophobia." Fire is hazardous, and yet it is used regularly at home. Most adults have understood the hazard associated with fire since the time of the caveman. By the same token hazardous chemicals, necessary and useful in our technological society, are not something to be afraid of. Chemicals need to be carefully used and their hazards understood by the general public. An emergency plan that is well designed, understood by the individuals responsible for action, and understood by the public can ease the concern over emergencies and reduce chemophobia. People will react during an emergency; how they react can be somewhat controlled through education. The likely behavior during an emergency when ignorance is pervasive is panic.

An emergency plan can minimize loss by helping to ensure the proper response in an emergency. "Accidents become crises when subsequent events and the actions of people and organizations with a stake in the outcome combine in unpredictable ways to threaten the social res involved" (Beranek et al. 1987). The wrong response can turn an accident into a disaster as easily as no response. One example is a chemical fire that is doused with water, causing the fire to emit toxic fumes; the same fire would be better left to burn itself out. Another example is the evacuation of people from a building into the path of a toxic vapor cloud; they might well be safer staying indoors with closed windows. Still another example is the members of a rescue team becoming victims because they were not wearing proper breathing protection. The proper response to an emergency requires an understanding of the hazards. A plan can provide the right people with the information needed to respond properly during an emergency.

Other than these mentioned commonsense reasons to plan, there are legal reasons. Recognizing the need for better preparation to deal with chemical emergencies, Congress enacted the Superfund Amendments and Reauthorization Act (SARA) of 1986. One part of SARA is a free-standing act called Title III, the Emergency Planning and Community Right-to-Know Act of 1986. This act requires federal, state, and local governments to work together with industry in developing emergency plans and "community right-to-know" reporting on hazardous chemicals. These requirements build on the EPA's Chemical Emergency Preparedness Program and numerous state and local programs that are aimed at helping communities deal with potential chemical emergencies. As indicated, the Bioterrorism Act of 2002 and the recent AWIA of 2018 both extend legal requirements for water utilities to complete risk assessments and emergency response plans.

Most larger industries have long had emergency plan designed for on-site personnel. The protection of people, property, and thus, profits has made emergency plans and prevention methods common in industry. On-site emergency plans are often a requirement by insurance companies. Expansion of these existing industry plans to include all significant hazards and all people in the community is a way to minimize the effort required for additional emergency planning.

19.3 UTILITY RISK ASSESSMENT

The VAs required by the SDWA must include, but are not be limited to, a review of pipes and constructed conveyances, physical barriers, water collection, pretreatment, treatment, storage and distribution facilities, electronic, computer or other automated systems that are used by the public water system, the use, storage, or handling of various chemicals, and the operation and maintenance of such systems.

Based on the results of the VA, the community water system should prepare or revise, where necessary, an emergency response plan. The emergency response plan must include, but is not limited to, the following:

- Plans, procedures, and identification of equipment that can be implemented or used in the event of a terrorist or other intentional attack on the PWS
- Actions, procedures, and identification of equipment which can obviate or significantly lessen the impact of terrorist attacks or other intentional actions on the public health and the safety and supply of drinking water provided to communities and individuals.

VAs help water systems evaluate susceptibility to potential threats and identify corrective actions that can reduce or mitigate the risk of serious consequences from vandalism, insider sabotage, terrorist attack, etc. Such an assessment for a water system takes into account the vulnerability of the entire water supply system from raw water supply (either groundwater or surface water), to raw water transmission, water treatment, and final water distribution. It also considers risks posed to the surrounding community related to attacks on or releases from the water system. An effective VA serves as a guide to the water utility by providing a prioritized plan for security upgrades, modifications of operational procedures, or policy changes to mitigate the risks and vulnerabilities to the utility's critical assets identified in the VA. The VA also provides a framework for developing risk reduction options and associated costs and ranking risk reduction efforts based on a cost/benefit analysis. Water systems should review their VAs periodically to account for changing threats or upgrades to the system to ensure that security objectives are continually being met. Preferably, a VA is "performance-based," meaning that it evaluates the risk to the water system based on the effectiveness (performance) of existing and planned measures to counteract adverse actions.

Common elements of a VA are (U.S. EPA 2002):

1. Characterization of the water system, including its mission and objectives
2. Identification and prioritization of adverse consequences to avoid
3. Determination of critical assets that might be subject to malevolent acts that could result in undesired consequences
4. Assessment of the likelihood (qualitative probability) of such malevolent acts from adversaries
5. Evaluation of existing countermeasures
6. Analysis of current risk and development of a prioritized plan for risk reduction

The nature and extent of the VA process will vary based on the complexity of the design and operation of the water system as well as the system size, potential affected population, source water, system infrastructure, sensitivity of the affected area in terms of sociopolitical characteristics, etc. Security and safety evaluations may also vary based on knowledge and types of threats, available security technologies, and applicable local, state, and federal regulations. A summary of each of these six basic elements is provided below.

19.3.1 Characterization of Water System

Characterization of a water system should include the mission and objectives of the facility, including the highest priority services provided by the utility and critical facilities, customers, processes and assets to achieve these objectives. Priority services provided by a facility may include water provision to the general public, the government, the military, industrial or critical care facilities, retail operations, and firefighting. In identifying critical facilities and assets, the following should be considered: critical customers, dependence on other infrastructures (e.g., electricity, transportation, other water utilities), contractual obligations, single points of failure (e.g., critical aqueducts, transmission systems, aquifers etc.), chemical hazards and other aspects of the utility's operations, or availability of other utility capabilities that may increase or decrease the criticality of specific facilities, processes, and assets (U.S. EPA 2002).

19.3.2 Identification and Prioritization of Consequences to Avoid

Specific consequences to be identified should be of significant concern in that they could substantially disrupt the ability of the system to provide a safe and reliable supply of drinking water or otherwise present significant public health concerns to the surrounding community. Characteristics of an event that should be considered when identifying "significant" consequences include:

1. Magnitude of service disruption
2. Economic impact (such as replacement and installation costs for damaged critical assets or loss of revenue due to service outage)

3. Number of illnesses or deaths resulting from an event
4. Impact on public confidence in the water supply
5. Chronic problems arising from specific events

19.3.3 Determination of Critical Assets at Risk

Consider the operation of critical facilities, assets, and/or processes and assess what an adversary could do to disrupt these operations. Such acts may include physical damage to or destruction of critical assets, contamination of water, intentional release of stored chemicals, and interruption of electricity or other infrastructure interdependencies. Critical assets that must be evaluated include pipes and other constructed conveyances; physical barriers; water collection, pretreatment and treatment facilities; storage and distribution facilities; electronic, computer, or other automated systems that are used by a PWS; the use, storage, or handling of various chemicals; and the operation and maintenance processes and procedures for all of these systems.

19.3.4 Assessment of the Likelihood of Intentionally Disruptive Acts

The objective of this step of the assessment is to move beyond what is merely possible and determine the likelihood of a particular attack scenario or sequences of disruptive acts that might be launched against a facility, water system, or ancillary support facilities. The threats and mode of attack considered during a vulnerability assessment will dictate, to a great extent, the risk reduction measures that should be designed to counter the threat(s). Some vulnerability assessment methodologies refer to this as a "Design Basis Threat" (DBT) where the threat serves as the basis for the design of countermeasures, as well as the benchmark against which vulnerabilities are assessed. An early DBT-based guideline document was published by the American Society of Civil Engineers (ASCE), the American Water Works Association (AWWA), and the Water Environment Federation (ASCE and AWWA 2006) titled the *Draft American National Standard for Trial Use, Guidelines for the Physical Security of Water Utilities.* A final version of this document was published by ASCE (2010) under the title *Guidelines for the Physical Security of Water Utilities and of Wastewater/Stormwater Utilities* providing national American National Standards Institute (ANSI)/ASCE/Environmental & Water Resources Institute (EWRI) final standards for the assessment of vulnerabilities and selection of mitigation measures for the protection of water supply and wastewater treatment facilities. These guideline documents direct water utilities to consider threats from vandals, criminals, saboteurs and insiders, all with different objectives and motivations for their actions, and identify baseline security measures necessary to protect raw water facilities, wells and pumping stations, treatment plants, finished water storage facilities, distribution and collection systems, and water system support facilities.

It should be noted that there is no single DBT or threat profile for all water systems in the United States. Differences in geographic location, size of the utility, previous attacks in the local area, and many other factors will influence the threat(s) that water systems should consider in their assessments. Water systems should also refer to the EPA's "Baseline Threat Information for Vulnerability Assessments of Community Water Systems" to help assess the most likely threats to their system. This document is available to community water systems serving populations greater than 3,300 people. Water systems may also want to review their incident reports to better understand past breaches of security. Two example computer-based approaches to VAs are the Risk Assessment Methodology for Water Utilities (RAM-W™) developed by Sandia National Laboratory (AWWARF and SNL 2002), and the Vulnerability Self-Assessment Tool (VSAT™) currently maintained and upgraded by the U.S. EPA (2017).

The VSAT™ Tool (https://vsat.epa.gov/vsat/) is a qualitative risk assessment tool designed to support water and wastewater utility assessments of vulnerabilities to both man-made and natural hazards. VSAT™ is a web-based application that has been redesigned for use on mobile devices and incorporates another EPA-developed tool, Water Health and Economic Analysis Tool (WHEAT), that assists utilities in estimating utility financial and regional economic consequences, as well as public health consequences from loss of system assets, release of stored hazardous gas, or intentional contamination of a drinking water distribution system (U.S. EPA 2017).

It should also be noted that there are a wide range of hazard identification and risk assessment methodologies that have been used for many years in the chemical and process industry that may be helpful in developing qualitative risk assessments in the water sector. These techniques include fault and event trees, hazard and operability studies (HAZOPs) and qualitative risk assessments as described by Theodore and Dupont (2012).

19.3.5 Evaluation of Existing Countermeasures

Depending on countermeasures already in place, some critical assets may already be sufficiently protected. This step will aid in identification of the areas of greatest concern and help to focus priorities for further risk reduction in water facilities. A facility should identify the capabilities it currently employs for detection, delay, and response to security incidents. Current intrusion detection systems, water quality monitoring, operational alarms, guard post orders, and employee security awareness programs should be identified. Delay mechanisms, including locks and key control, fencing, structure integrity of critical assets, and vehicle access checkpoints should be cataloged along with existing policies and procedures for evaluation and response to intrusion and system malfunction alarms, adverse water quality indicators, and cyber system intrusions. Above all, it is important to determine the performance characteristics of all of these systems and policies as poorly operated and maintained security technologies provide little or no protection.

Particular attention should be paid to cyber protection system features in place for the SCADA and business-related computer information systems. Cyber security protection assessment should include firewalls, modem access, Internet and other external connections, including wireless data and voice communications, and cyber security policies and protocols. It is important to identify whether vendors have access rights and/or "backdoors" to conduct system diagnostics remotely to assess potential insider vulnerability via this route.

Finally, all security policies and procedures and compliance records for them should be reviewed. Policies and procedures that should be reviewed include personnel security, physical security, key and access badge control, control of system configuration and operational data, scheduling and control of chemical and other vendor deliveries, and security training and security exercise records

19.3.6 Analysis of Current Risk and Prioritization for Risk Reduction

Information gathered on threat, critical assets, water utility operations, consequences, and existing countermeasures should be analyzed to determine the current level of risk. The utility should then determine whether current risks are acceptable or risk reduction measures should be pursued. Recommended actions should measurably reduce risks by reducing vulnerabilities or consequences through improved deterrence, delay, detection, and/or response capabilities or by improving operational policies or procedures. Selection of specific risk reduction actions should be completed prior to considering the cost of the recommended action(s). Utilities should carefully consider both short- and long-term solutions. An analysis of the cost of short- and long-term risk reduction actions may impact which actions the utility chooses to reach its security goals.

Utilities may also want to consider security improvements in light of other planned or needed improvements. Security and general infrastructure may provide significant multiple benefits. For example, improved treatment processes or system redundancies can both reduce vulnerabilities and enhance day-to-day operation.

Generally, strategies for reducing vulnerabilities fall into three broad categories: sound business practices, system upgrades, and security upgrades. Sound business practices affect policies, procedures, and training to improve the overall security-related culture at a facility. System upgrades include changes in operations, equipment, processes, or infrastructure itself that make the system fundamentally safer. Security upgrades improve capabilities for detection, delay, or response.

An action plan defines the specific actions that should be taken to respond to events where high priority vulnerabilities have been compromised. Even if a VA did not identify any vulnerabilities, contingency planning should be considered for the possibility of at least the following high consequence events:

- Contamination of the drinking water
- Structural damage/physical attack
- SCADA, computer, or cyber attack
- Intentional hazardous chemical release (e.g., release of chlorine or ammonia from storage facilities)

Example action plans for the four intentional events listed here are provided by the U.S. EPA (2004a) and are included in an appendix to this chapter.

19.4 THE PLANNING PROCESS

Emergency planning for water utilities is required by the Bioterrorism Act of 2002 but is less prescriptive than similar planning requirements of SARA's Community-Right-to-Know Act. The Bioterrorism Act required all utilities to complete a VA and then respond to findings from the VA through specific action plans that define specific response actions to identified vulnerabilities (EPA 2004a). These action plans are completed by the local water utilities as part of the creation or revision of existing emergency response plans (ERPs). Preparation and updates of ERPs are required by Section 1433(b) of the Bioterrorism Act to include the following planning partners to the extent possible: Local Emergency Planning Committees (LEPCs) established under the Community Right-to-Know Act such as local law enforcement departments, fire departments, health departments, local environmental agencies, hospitals, broadcast and print media, community groups, and nearby utilities. Other partners could include state and federal agencies, and laboratories. Once the relevant ERP partners are identified, the ERP creation/revision process is to take place for the development of plans that have the following eight core elements (EPA 2004a): system-specific information; community water system (CWS) roles and responsibilities; communication procedures; personnel safety; identification of alternate water sources; replacement equipment and chemical supplies; property protection; and water sampling and monitoring. Each of these core elements are discussed briefly.

19.4.1 System-Specific Information

During a major event, basic technical information about a water system should be readily available. This information would include distribution system maps, detailed "as-built" drawings, site plans, source water locations, system/SCADA/ process control/communication operations manuals, staffing rosters, and chemical handling and storage facility locations and release impact analyses. This technical information should be clearly documented and readily accessible so it can be distributed quickly to those who may be involved in responding to a major event. This information should have been reviewed while conducting the VA, so the ERP should only need to identify it and its location. The level of technical documentation should reflect the complexity of a given CWS.

19.4.2 CWS Roles and Responsibilities

An emergency response lead (ER lead) and alternate ER lead should be identified in this section of the ERP. The ER lead is the main point of contact and decision maker during a major event. This person has responsibility for evaluating incoming information, managing resources and staff, and deciding on appropriate response actions. This person also has the responsibility of coordinating emergency response efforts with first responders. The ER lead should be heavily involved in forming the planning partnerships described previously.

An alternate ER lead is identified who would step in should the ER lead be unavailable or incapacitated during an event. The ER lead and the alternate ER lead need to be reachable 24 hours a day, 7 days a week. A clear chain of command or command structure should also be established so that personnel and staff know their individual roles and responsibilities.

The incident command system (ICS) is a model for command, control, and coordination of an emergency response and provides a means to coordinate the efforts of first responders as they work toward the common goal of stabilizing a major event and protecting life, property, and the environment. ICS uses a well-defined command structure to specify roles and responsibilities in responding to a major event. In ICS, the main contact person and decision maker is the incident commander. At the CWS level, the ER lead has the role of incident commander, unless the incident is of such significance that local, state, or federal officials take over the command. First responders may use ICS when responding to a major event, and a CWS may be required to abide by this command structure where other first responders take over the role and responsibilities of incident commander in the latter stages of such an event. The ER lead should address roles, responsibilities, and the command structure for both minor and major events when forming planning partnerships.

19.4.3 Communication Procedures

Appropriate and timely communication is essential during an emergency. The ERP should identify clear communication channels for CWS staff and personnel, external non-CWS entities (local and state first responders, law enforcement, regulatory agencies), and the public/media. As part of the ERP, internal and external notification lists should be maintained that contain information on all appropriate entities to be contacted, including their names, titles, mailing addresses, e-mail addresses, and all applicable land line and cellular phone numbers. These lists should be updated as necessary. Provisions need to be made for an efficient and fail-safe forms of communication to be available during conditions when the use of normal means may not be possible. Communication procedures with the public and media may already be part of a CWS's day-to-day operations, but these procedures need special attention during a major event to provide the public and media with timely, accurate, and complete information.

A list of critical users should be developed as part of the ERP. Some of these users should be given priority notification because of their public health mission and because they may serve customers considered "sensitive sub-populations" (e.g., senior residential housing, child care centers, medical facilities). Commercial and industrial customers that incorporate water into their product (e.g., bottling and canning companies) should also be considered for priority notification. More discussion of emergency communication issues is provided.

19.4.4 Personnel Safety

Protecting the health and safety of CWS personnel as well as the surrounding community is a key priority during an emergency when personnel may be at risk of harm, injury, or even death. The ERP should provide direction to personnel on how to safely implement a variety of response actions including evacuation planning, identifying evacuation routes and exits, locations of assembly areas and accounting for personnel, shelter in place procedures, safety training, procedures for using and maintaining emergency response equipment, and proper first aid and emergency medical treatment for employees and others who are onsite at CWSs.

19.4.5 Alternate Water Sources

A CWS should consider the amount of water needed to address short-term (hours to days) versus long-term (weeks to months) outages. Alternate water supplies available during both types of outages should be identified as part of an ERP. To do this, a comprehensive understanding of the current water supply, water distribution system, and water system demand requirements is required. The location and capabilities of other regional CWSs, including available excess capacity and ease of connection to the affected distribution system should be identified. In addition, interconnection agreements with partner utilities that may be in place should be reviewed, and potential issues that could arise if multiple CWSs are affected should be identified and discussed with planning partners.

Short-term outages might be due to electrical power outages or contamination. As part of an ERP, the potential effects of a power outage should be considered. If a utility could is without power due to a major event, and it may take several days for power to be restored. The ERP should include contingencies for back-up power generation and alternative power sources.

If contamination has occurred, a public health notification such as "boil water," "do not drink," or "do not use," may be issued by the drinking water primacy agency. If a "boil water" notice is issued, no alternative water source is needed. If a "do not drink" order is issued, then the suspect water can still be used for activities other than direct ingestion of the water. In this situation, it will only be necessary to provide an alternate drinking water supply for consumption and related activities such as food preparation.

A "do not use" order is much more restrictive, and sufficient alternate water sources will be required to supply water for consumption, hygiene, and emergency needs. A "do not use" notice may also have implications with respect to water used for firefighting. Although a prohibition on use of water for firefighting is likely to occur only if the water is contaminated with certain substances, an alternate source of firefighting water, such as a pond, river, or stream, may be also be necessary.

As part of an ERP, agencies or private companies that could provide water supplies (bottled or bulk) in the event of a major event should be identified, and mutual aid agreements with surrounding communities, industries, contractors and related utilities should be established as appropriate. This alternative source list should be maintained and continually updated to include accurate information on points of contacts. Additional equipment may be available from local businesses such as dairies, well drillers, irrigation supply firms, or distributors that may have tank trucks that can be made suitable for carrying water, chlorinators or generators that can be used for emergency disinfection, and pipe that can be used to extend water supply lines. Other water utilities in the area may have spare parts (such as valves, pumps, and pipe) available for use in an emergency, and Federal Emergency Management Agency (FEMA), U.S. Army Corps of Engineers (USACE), and the U.S. Forest Service that may be able to provide firefighting equipment

Plans could also be prepared for water conservation measures to be implemented if a major event causes a reduction in service or a "do not use" notice is issued. For example, if a major event causes a reduction in service, water use could be limited by advising customers not to do laundry, run the dishwater, or water the garden and to limit the duration of showers. To plan for a "do not use" notice, consumers could be advised to maintain an emergency supply of water in their homes.

If a CWS requires extensive cleaning, or if portions of the system have been destroyed, a long-term alternate water supply will be needed. The following are examples of possible long-term water supply options:

- Connection of the water distribution system to an existing municipal or private water supply (assumes existing water treatment plant and distribution system is intact and useable)
- Connection of the water distribution system to a new uncontaminated groundwater or surface water source (assumes existing water treatment plant and distribution system is intact and useable)
- Development of oversized community storage facilities to compensate for loss of existing system capacity.

19.4.6 Replacement Equipment and Chemical Supplies

An ERP should identify equipment that can significantly lessen the impact of a major event on public health and protect the safety and supply of drinking water. An updated inventory should be maintained of current equipment (e.g., pumps), repair parts, chemical supplies for normal maintenance and operations, and information on mutual aid agreements.

Based on findings of the VA, how and where to find the equipment, repair parts, and chemicals needed to respond adequately to a particular vulnerability should be identified. Mutual aid agreements with other CWSs to address any deficiencies should be established. These agreements should identify the equipment, parts, and chemicals available to members under the agreement.

19.4.7 Property Protection

Protecting CWS facilities, equipment, and vital records is essential to restoring operations once a major event has occurred. The ERP should identify measures and procedures that are aimed at securing and protecting the CWS following a major event. Items that should be considered include:

- "Lock down" procedures
- Access control procedures
- Establishing a security perimeter following a major event
- Evidence protection measures for law enforcement (should the major event also be declared a crime scene)
- Securing buildings against forced entry, and
- Other property protection procedures and measures.

19.4.8 Water Sampling and Monitoring

Water sampling and monitoring should be an integral part of an ERP and not an afterthought. During the stage of forming partnerships, the State Drinking Water Primacy Agency should be consulted with on issues of water sampling and monitoring that may arise during and after a major event. Some water sampling and monitoring issues to consider include:

- Identifying proper sampling procedures for different types of contaminants
- Obtaining sample containers
- Determining the quantity of required samples
- Identifying who is responsible for taking samples
- Identifying who is responsible for transporting samples (in time-sensitive situations)
- Confirming laboratory capabilities and certifications, and
- Interpreting monitoring or laboratory results

A more detailed discussion of water sampling and monitoring issues can be found in the U.S. EPA's *Water Security Handbook* (2006), which outlines the need for planning for site characterization and sampling and analytical laboratory protocol development and planning both before and in response to a potential terrorist incident.

19.5 TRAINING OF PERSONNEL

Once an ERP is completed or updated, training is an essential element in ensuring appropriate and timely repose to emergency events. Training is important for the emergency teams to ensure that their roles are clearly understood and that accidents can be reacted to safely and properly without delay. Emergency teams include the police, firefighters, medical people, and volunteers who will be required to take action during an emergency. These people must be knowledgeable about the potential hazards. For example, specific antidotes for different types of medical problems must be known by medical personnel. The whole emergency team must also be taught the use of personal protective equipment.

Local government officials also need training because these officials have the power to order an evacuation. They must be aware of under what circumstances such action is necessary. The timing of an evacuation may be critical; this must be understood by these people prior to the emergency itself. Local officials also control the use of city equipment and, therefore, must be knowledgeable as to what is needed for an appropriate response to a given emergency.

Media personnel must also be involved in the training program. The public receives information through the media, and it is vitally important that the information the public receives is accurate. If the information is incorrect or distorted, an emergency can easily cause panic. For this reason, it is important that the media be knowledgeable about the potential hazards and the details of specific emergency responses.

Refresher training should be conducted on a regular basis. It is important for emergency procedures to be performed as planned, and this requires training on a regular basis so that people understand and remember how to react. The best plan on paper is likely to fail if the persons involved are reading it for the first time as the emergency is occurring. People must be trained *before* an emergency happens.

Training can include briefing sessions, classroom sessions, or mock exercises. Training should include testing of the ERP. Drills and exercises that challenge the information in the ERP should be conducted at least annually. There are many sources (state, federal, and industry specific) that describe what should be included in emergency training (U.S. EPA 2011). These typically include the following four types of training:

- Orientation sessions: Orientation sessions work well for basic instruction and explaining ERP procedures.
- Table-top workshops: Table-top workshops involve developing scenarios that describe potential problems and provides certain information necessary to address the problems. The idea is to present staff and emergency response officials with a fabricated event, have them verbally respond to a series of questions, and then evaluate whether the responses match what is written in the ERP. The Tabletop Exercise Tool for Drinking Water and Wastewater Utilities (TTX Tool) from the U.S. EPA provides users with the resources to plan, conduct, and evaluate tabletop exercises that focus on water sector-related incidents and challenges. The 2018 version of the TTX Tool contains 12 all-hazards scenarios (e.g., natural disasters, man-made incidents) related to emergency preparedness and response and can be downloaded at the following URL: https://www.epa.gov/waterresiliencetraining/develop-and-conduct-water-resilience-tabletop-exercise-water-utilities.
- Functional exercises: The functional exercise is considered the most-effective training tool, next to a real emergency, because a team of simulators is trained to develop a realistic major event. By using a series of prescripted messages, the simulation team sends information in to personnel

assigned to carry out the ERP procedures. Both the simulators and personnel responding to the simulation are focused on carrying out the procedures to test the validity of the ERP.

- Full-scale drills: These are the most costly and time-consuming training programs but can be extremely effective. In a full-scale drill, emergency response personnel and equipment are mobilized to a scene, an emergency scenario is presented, and they respond as directed by the ERP.

The bottom line is that time, resources, and personnel need to be dedicated to accomplishing the training. The training should be used to identify lessons learned to enhance future response and recovery efforts and to update plans as necessary to continually improve their effectiveness.

19.6 ERP ACTIVATION

The ERP may be activated when a utility receives a "threat warning" about a potential threat and then proceeds through a standard threat decision process to verify that a potential threat is real. A threat warning is an occurrence or discovery that indicates a threat of a malevolent act and triggers an evaluation of the threat. These warnings should be evaluated in the context of typical CWS activity and previous experience in order to avoid false alarms. Threat warnings can take a variety of forms that can include a physical security breach, a witness account, notification by the perpetrator, notification by law enforcement, notification by news media, unusual water quality monitoring results, consumer complaints, or a public health notification. Once a threat warning is received, a threat decision/evaluation process should be initiated. This threat decision process involves three stages (Threat possible? Threat credible? Threat confirmed?) discussed and reviewed in the example action plans summarized in the appendix to this chapter.

19.6.1 Stage 1—Threat Possible?

The first stage in evaluating a potential threat is to determine if the threat as reported is actually possible, that is, could something as reported actually have happened? Evidence of a threat is stronger from unusual water quality monitoring results, consumer complaints, or reported emergency departments and health clinic visits, but findings of the VA should be reviewed to determine whether a specific threat has been identified as possible. If the threat is possible, immediate operational response actions might be implemented as suggested in the example action plans in the appendix, and the next stage of threat evaluation should be carried out.

19.6.2 Stage 2—Threat Credible?

There must be corroborating information regarding the threat for it to be considered credible. If for example, threat warning information sources are highly credible, such as from hospitals and from unusual monitoring results, the threat can be considered credible and additional portions of an ERP might be activated, such as initiating internal and external notifications, conducting additional water sampling and analysis, or issuing public health advisories. At this stage, the occurrence of a major event has not been confirmed, but response preparations are underway should the threat actually lead to a major event, and Stage 3 is undertaken.

19.6.3 Stage 3—Threat Confirmed?

Confirmation implies that definitive evidence has been collected to establish that an incident has occurred. Confirmation of an incident may be obvious, such as when structural damage occurs at a CWS. In such a case, Stages 1 and 2 would not be necessary before an ERP is fully implemented. Upon confirmation of the incident, specific action plans within an ERP that address specific major events should be implemented *immediately* as suggested in the example action plans in the appendix.

19.7 EMERGENCY COMMUNICATION

Notifying the public of an emergency is a task which must be accomplished with caution. People will react in different ways. Many will simply not know what to do, others will not take the warning seriously, and others will panic. Proper training in each community can help minimize any panic and condition the public to make the right response (Dupont 2013). Consumers may be instructed to boil water, limit their water uses to activities that do not involve consumption, or not come in contact with the water at all. A good public communications strategy will help get the message out effectively and outline who needs to do what and when.

The communication plan or strategy should be a set of general guidelines for a spokesperson to follow to craft clear and concise messages for the public and to also deal with the media. The plan should be targeted to reach multiple audiences that make up a CWS's stakeholders. These stakeholders will include customers (both residential and business), local health professionals, and others. Creating draft press releases and public water restriction notices in advance of an emergency as part of the ERP will expedite the communication process when timing may be critical. The key is that the message should be clear, accurate, and easily understood by the stakeholders.

The CWS spokesperson should be someone who is knowledgeable and credible, has good communication skills, and, if possible, is not a key person needed for implementing response actions during a major event. In communicating with the media, the lead spokesperson may be someone external to the CWS if another organization has taken over the role of lead agency or incident commander (e.g., a representative from the health department or the State Drinking Water Primacy Agency). Both field and office staff should defer questions to the designated spokesperson during an incident.

There are no easy prescriptions for successful risk communication. However, those who have studied risk communication generally agree on seven cardinal rules (U.S. EPA 1988). These personalized rules apply equally well to the public and private sectors. Although many of these rules may seem obvious, they are continually and consistently violated in practice. Thus, a useful way to consider these rules is to focus on why they are frequently not followed (Theodore and Dupont 2012).

19.7.1 Accept and Involve the Public as a Legitimate Partner

A basic tenet of risk communication in democracy is that people and communities have a right to participate in decisions that affect their lives, their property, and the things they value. Demonstrate respect for the public and underscore the sincerity of effort involving the community early, before important decisions have to be made that may affect their health and/or well-being. Involve all parties that have an interest or stake in the issue under consideration. The goal is to produce an informed public that is involved, interested, reasonable, thoughtful, solution-oriented, and collaborative. It should not be to diffuse public concerns or replace action.

19.7.2 Plan Carefully and Evaluate Communication Efforts

Emergency communication will be successful only if carefully planned. Train staff, including technical staff, in communication skills and pretest messages. Carefully evaluate communication efforts and learn from mistakes. Carefully consider that there are many publics, each with their own interests, needs, concerns, priorities, preferences, and organizations.

19.7.3 Listen to the Public's Specific Concerns

If the public is not listened they cannot be expected to listen. Communication is a two-way activity. Do not make assumptions about what people know, think or want done about risks. Take the time to find out what people are thinking. People in a community are often more concerned about such issues as trust, credibility, competence, control, voluntariness, fairness, caring, and compassion than about mortality statistics and the details of a quantitative risk assessment.

19.7.4 Be Honest, Frank, and Open

In emergency communication, trust and credibility are the most valuable assets. Once lost, they are almost impossible to regain completely. State credentials; but do not ask or expect to be trusted by the public. If an answer to a question from the public is not known or is uncertain, say so. Get back to the public with answers. Admit mistakes. Disclose risk information as soon as possible (emphasizing any reservations about reliability). Do not minimize or exaggerate the level of risk. Speculate only with great caution. If in doubt, lean toward sharing more information and not less, or people may think something is being hidden from them. Discuss data uncertainties, strengths, and weaknesses, including the ones identified by other credible sources. Identify worst-case estimates as such, and cite ranges of a risk estimate when appropriate.

19.7.5 Coordinate and Collaborate with Other Credible Sources

Allies can be effective in helping communicate emergency information. Few things make emergency communication more difficult than conflicts or public disagreements with other credible sources. As part of the ERP, coordination of interorganizational and intra-organizational communications should be a priority. Joint communications should be issued with other trustworthy sources (e.g., credible university scientists and/or professors, physicians, or trusted local officials).

19.7.6 Meet the Needs of the Media

The media are a prime transmitter of information during emergencies. Be open and accessible to reporters. Respect their deadlines. Provide risk information tailored to the needs of each type of media (e.g., graphics and other visual aids for television, web sites, e-mails, blogs, etc.). Try to establish long-term relationships of trust with specific editors and reporters.

19.7.7 Speak Clearly and with Compassion

Technical language and jargon are useful as professional shorthand but is a barrier to successful communication with the public. Use simple, nontechnical language. Be sensitive to local norms, such as speech, dress, and predominant language(s) of impacted minority communities that may require the use of translators in both written and oral communication. Use vivid, concrete images that communicate on a personal level. Use examples and anecdotes that make technical data come alive. Avoid distant, abstract, unfeeling language about deaths, injuries, and illnesses. Acknowledge and respond (both in words and with action) to emotions that people express that can include anxiety, fear, anger, outrage, and helplessness. Never let efforts to inform people prevent acknowledging and saying that any illness, injury, or death is a tragedy. Always try to include a discussion of actions that are under way or can be taken. Promise only what can be done, and be sure to do what is promised.

19.8 THE EMERGENCY RECOVERY PROCESS

Once a threat is confirmed, the ERP and specific action plans have been implemented to manage the risk and reduce loss of life and property, the remediation and recovery process begins (Dupont 2013). EPA (2006) suggests there are nine steps in this overall remediation and recovery process. Each of these steps is briefly discussed. It should be noted that although Step 9 focuses on communication with the public post recovery to restore public confidence in the CWS, continual communication with the public during the remediation and recovery process is essential, using the principles discussed in the previous section, if there is any hope of restoring consumer confidence in the remediated system.

19.8.1 Long-Term Alternative Water Supply

If remediation and recovery actions are going to take some time, safe water must be provided to the public in the meantime. An alternate safe water supply should be identified during the planning process before a contamination incident happens. If a CWS, working with local agencies, cannot provide a long-term alternate water supply, then help should be requested from state and federal emergency planning agencies.

19.8.2 System Characterization and Feasibility Study

After the contamination has been contained, "unified command" will most likely need more information to choose the right remediation method, such as the identity of the contaminant involved and how much of the water system is affected. Command should also identify possible remediation options and find out how effective and feasible those options are. Doing a System Characterization and Feasibility Study should provide this information. Because no single agency is likely to have all the resources necessary to carry out remediation and recovery, it is likely that several agencies will be involved in the recovery effort. Under unified command, the utility should coordinate with their drinking water primacy agency, public health agencies and other agencies involved in protecting public health and drinking water.

19.8.3 Risk Assessment

It is important to find out how unsafe the incident site is to workers and the public. Unified command should do a quick risk assessment as part of the System Characterization and Feasibility Study to determine required worker safety levels, area access restrictions, and for deciding what response actions to take and remediation goals to establish. Any post-containment risks associated with the remedial action, together with necessary personal protection and risk reduction measures, should be communicated by the incident commander or unified command to the CWS, public health agencies, drinking water primacy agency and the public.

19.8.4 Remediation and Rehabilitation Alternatives

Once a System Characterization and Feasibility Study and associated risk assessment has been carried out, unified command can begin to evaluate and compare possible remedial actions, remediation technologies and rehabilitation methods to restore the water system to normal. Remediation can also be done in stages, such as emergency actions to reduce dangerous levels of a contaminant, followed by long-term remediation to remove remaining low levels of the contaminant to allow resumption of normal operation of the CWS.

19.8.5 Select Remediation Alternative

A remedial alternative should be selected to protect human health and the environment, comply with all applicable regulations (such as the SDWA), and be feasible, effective, and affordable.

19.8.6 Design Remedial Alternative

After remedial actions are selected, the engineering design, planning, and documentation of the remediation begins. Personnel from the affected water utility and other community technical support staff should be involved in providing technical assistance to the engineering design firm to help prevent unforeseen impacts on the remaining unaffected water system and associated infrastructure.

19.8.7 Implement Remedial Alternative

Once the remedial design has been approved, the remediation and rehabilitation of the contaminated parts of the water system can be carried out. Contractors may assist in these procedures, but the expertise of CWS personnel in the normal operation of the water system is essential during this step to avoid unforeseen impacts on the water system infrastructure.

19.8.8 Post-Remediation Monitoring

After remediation is completed, both water quality and the water system infrastructure should be monitored to ensure that the remedial alternative was effective. If remedial action was not effective, additional remedial action or infrastructure replacement should be carried out to remedy the remaining contamination problem and the water should be tested again.

19.8.9 Communication with Public to Restore Confidence in CWS

During all stages of remediation, the CWS and unified command should keep the community informed about the progress of the remediation process, who is in charge, how it could affect human health, what is being done to restore safe water, and when things are expected to return to normal. Otherwise, the public may continue to use unsafe water, hoard water, or act in other ways that could cause unforeseen problems and even interfere with remediation.

Once a remedial action is complete and its performance has been validated through post-remediation monitoring, the system should be ready to resume normal operation. The overall success of the emergency response will hinge on the level of preparation the CWS has achieved prior to an incident, and their effectiveness in communicating with the public during the course of the recovery effort.

19.9 APPLICATIONS

The following Illustrative Examples complement the material presented in this chapter.

Illustrative Example 19.1

Describe the difference between an ERP and an action plan as they related to CWS risk management and planning.

Solution

U.S. EPA (2004) defines an ERP as a comprehensive planning document that contains the following eight core elements:

1. System-specific information;
2. CWS roles and responsibilities;
3. Communication procedures: who, what, and when;
4. Personnel safety;
5. Identification of alternate water sources;
6. Replacement equipment and chemical supplies;
7. Property protection;
8. Water sampling and monitoring.

Action plans are the specific emergency actions that are to be taken for specific major events that a CWS is particularly vulnerable to. A VA is used to define these specific major events that the CWS must address in its ERP through these action plans. The action plans are specific actions to be taken for specific major events (i.e., contamination of the drinking water supply, physical attack, cyberattack, or intentional hazardous chemical release) and keys in on specific special notification requirements and unique response steps or recovery actions necessary to address these high-risk events. Example action plans are provided as an appendix to this chapter.

Illustrative Example 19.2

One of the required elements of an ERP is the identification of both short- and long-term alternative water sources in the case of a CWS contamination event. List some alternative water sources that could provide necessary water supplies during short-term events.

Solution

Possible short-term alternate water supply options include (but are not limited to) the following:

- Bottled water provided by outside sources
- Bottled water provided by local retailers
- Bulk water provided by certified water haulers
- Bulk water transported or provided by military assets (i.e., National Guard or USACE)
- Bulk water provided by neighboring water utilities by truck or via pipeline
- Bulk water from hospitals, universities, and local industry that maintain backup water supplies for consumption
- Interconnections with nearby public water systems
- Water treated by the plant and hauled to distribution centers (i.e., in the case of water distribution system contamination)
- Water pumped from surface water sources, treated at the plant or nearby plants, and hauled to distribution centers
- Water for firefighting from Federal agencies such as the USACE and FEMA
- Water from unaffected wells owned by local citizens and businesses.

Illustrative Example 19.3

Communication with the public prior to and during an incident is critical to manage the public's response to a crisis and their expectations regarding their health and safety. There are many recent cases in the literature of poorly conceived communication methods during a potential drinking water health emergency (delayed, then door-to-door notification of elevated fluoride levels in drinking water due to a dosing pump failure with no written notification left when people were not home during the notification effort, Salt Lake City area 2019). What are the guidelines established by the EPA regarding a decision-making timeline for the threat evaluation process?

Solution

The threat evaluation process involves determining if a threat is possible, determining if it is credible, and then confirming the threat with definitive evidence so that proper emergency action can be taken. The EPA (2006) recommends that the possibility of a threat, once a threat warning is received, should be made rapidly, within an hour, so that the threat's credibility can be evaluated as quickly as possible. Once a threat has been deemed possible, the EPA (2006) recommends that the credibility of the threat be determined within 2–8 hours, based on site information, the

TABLE 19.1
Examples of Public Notices Regarding Drinking Water Contamination Events

Type of Notice	Appropriate Use	Impact on Public
Boil Water Before Use	Use if boiling makes water safe to drink and boiling does not create other health problems, particularly through routes of exposure other than drinking (e.g., inhalation or skin contact with water).	Least burdensome. Commercial facilities using large volumes of water for drinking or food preparation are most affected.
Do Not Drink	Use if boiling is not an option to remove contaminant, or if inhalation or skin contact do not pose risks.	More burdensome. An alternative water supply for drinking and food preparation will be required.
Do Not Use	Use if the contaminant is unknown, or if health risks exist for the public from inhalation or dermal exposure to the contaminant.	Greatest burden. An alternate water supply for all used, including fire fighting and flushing toilets will be needed.

Source: U.S. Environmental Protection Agency (U.S. EPA), *A Water Security Handbook: Planning for and Responding to Drinking Water Contamination Threats and Incidents*, Office of Ground Water and Drinking Water, Water Security Division, Washington, DC, https://www.epa.gov/sites/production/files/2015-06/documents/watersecurity_water_security_ handbook_rptb_1.pdf, 2006.

nature of the threat, the circumstances and other reliable information as public health and safety may be at risk. Once a threat is deemed credible, a utility should work with the drinking water primacy agency, public health agencies and law enforcement agencies to confirm the threat, and initial implementation of the ERP should begin at this stage. To confirm a credible threat, definitive lab results or a preponderance of evidence are required indicating that the threat is plausible and believable. Once a threat is confirmed, it becomes a contamination incident and the ERP should be fully activated.

Illustrative Example 19.4

There are generally three levels of public health warnings related to drinking water contamination and water use restrictions. What are these three levels of public health warning, when should they be used, and what is their relative burden on the public when they are implemented?

Solution

The three levels of public health warnings related to drinking water contamination are: Boil Water Before Use, Do Not Drink, and Do Not Use. The appropriate use of each warning level depends on the contaminant's properties and their potential effects on the public. A summary of these three notice types and their impact on the public are provided in Table 19.1.

REFERENCES

American Society of Chemical Engineers (ASCE). 2010. *Guidelines for the Physical Security of Water Utilities and of Wastewater/Stormwater Utilities.* Reston, VA: ASCE. 126 pp.

American Society of Chemical Engineers (ASCE) and American Water Works Association (AWWA). 2006. Draft American National Standard for Trial Use, Guidelines for the Physical Security of Water Utilities. Reston, VA: ASCE. https://www.waterboards.ca.gov/drinking_water/certlic/drinkingwater/documents/security/WISE-Phase3WaterUtilityGuidelines.pdf.

American Water Works Association (AWWA) Research Foundation and Sandia National Laboratories (SNL). 2002. *Risk Assessment Methodology for Water Utilities (RAM-W™)*, 2nd edition. Denver, CO: AWWA Research Foundation.

Beranek, W., J.P. McCullough, S.H. Pine, and R.L. Soulen. 1987. Getting involved in community-right-to-know. *Chem. Eng. News* 65(43):62–65.

Dupont, R.R. 2013. *Emergency Planning, Response, and Recovery*. Chapter 5. Case Studies. Special Publication, Water Environment Federation, Manual of Practice. Alexandria, VA: Water Environment Federation. 312 pp.

Krikorian, M. 1982. *Disaster and Emergency Planning*. Loganville, GA: Institute Press.

Ross & Associates Environmental Consulting, Ltd. 2009. *All-Hazard Consequence Management Planning for the Water Sector*. Report prepared for the Association of State Drinking Water Administrators. Arlington, VA: ASDWA. https://asdwasecurity.files.wordpress.com/2014/03/all-hazard-cmp_final.pdf

Theodore, L. and R.R. Dupont. 2012. *Environmental Health and Hazard Risk Assessment, Principles and Calculations*. Boca Raton, FL: Taylor & Francis Group. 612 p.

Thompson, M. 2016. Iranian Cyber Attack on New York Dam Shows Future of War. *Time. Com*. http://time.com/4270728/iran-cyber-attack-dam-fbi/ (accessed January 27, 2018).

U.S. Environmental Protection Agency (U.S. EPA). 1988. *Seven Cardinal Rules of Risk Communication*. Washington, DC: Office of Policy Analysis.

U.S. Environmental Protection Agency (U.S. EPA). 2002. *Vulnerability Assessment Factsheet*. Washington, DC: Office of Water. https://nepis.epa.gov/Exe/ZyPDF.cgi/P1004AYS.PDF?Dockey=P1004AYS.PDF.

U.S. Environmental Protection Agency (U.S. EPA). 2003. *Large Water System Emergency Response Plan Outline: Guidance to Assist Community Water Systems in Complying with the Public Health Security and Bioterrorism Preparedness and Response Act of 2002*. Washington, DC: Office of Water. https://www.epa.gov/sites/production/files/2015-03/documents/erp-long-outline.pdf.

U.S. Environmental Protection Agency (U.S. EPA). 2004a. Emergency Response Plan Guidance for Small and Medium Community Water Systems to Comply with the Public Health Security and Bioterrorism Preparedness and Response Act of 2002. Washington, DC: Office of Water. https://www.epa.gov/sites/production/files/2015-04/documents/2004_04_27_watersecurity_pubs_small_medium_erp_guidance040704.pdf.

U.S. Environmental Protection Agency (U.S. EPA). 2004b. *Response Protocol Toolbox: Planning for and Responding to Drinking Water Contamination Threats and Incidents, Interim Final – August 2004. Response Guidelines*. Washington, DC: Office of Ground Water and Drinking Water, Water Security Division. https://www.epa.gov/sites/production/files/2015-06/documents/2004_11_24_rptb_response_guidelines.pdf.

U.S. Environmental Protection Agency (U.S. EPA). 2006. *A Water Security Handbook: Planning for and Responding to Drinking Water Contamination Threats and Incidents*. Washington, DC: Office of Ground Water and Drinking Water, Water Security Division. https://www.epa.gov/sites/production/files/2015-06/documents/watersecurity_water_security_ handbook_rptb_1.pdf.

U.S. Environmental Protection Agency (U.S. EPA). 2011. *How to Develop a Multi-Year Training and Exercise (T&E) Plan: A Tool for the Water Sector*. EPA 816-K11-003. Washington, DC: Office of Water. https://www.epa.gov/sites/production/files/2015-05/documents/how_to_develop_ a_multi-year_training_and_exercise_plan_a_tool_for_the_water_sector.pdf.

U.S. Environmental Protection Agency (U.S. EPA). 2017. *VSAT Web User Guide*. Washington, DC: Office of Ground Water and Drinking Water, Water Security Division. https://vsat.epa.gov/vsat/.

Veilleux, J., and S. Dinar. 2018. New Global Analysis Finds Water-Related Terrorism Is on the Rise. *NewSecurityBeat, the Blog of Environmental Change and Security Program*. Washington, DC: Wilson Center. https://www.newsecuritybeat.org/2018/05/global-analysis-finds-water-related-terrorism-rise/ (accessed January 27, 2018).

Water Environment Federation (WEF). 2013. *Emergency Planning, Response, and Recovery. WEF Special Publication*. Alexandria, VA: Water Environment Federation. 283 pp.

Webster. 1971. *New World Dictionary*, Second College Edition. Upper Saddle River, NJ: Prentice Hall.

APPENDIX

EXAMPLE ACTION PLANS (U.S. EPA 2004A)

*Water System Contamination**

Threat Warning Stage

Threat Warning Received	Special actions and notifications to be taken: Notify ER Lead or Alternate ER Lead Record and document all information pertaining to the threat warning Do not disturb site if the threat warning could be a possible crime scene Return to normal operations if no further action is required (i.e., the threat warning can be explained) Begin the "Threat Decision Process" if the threat warning cannot be explained

Threat Decision Process Stage

Is the Threat Possible? (Stage 1)	Special actions and notifications to be taken: Notify local law enforcement Notify State Drinking Water Primacy Agency Evaluate threat warning and make decisions in consultation with State Drinking Water Primacy Agency and local law enforcement Initiate basic precautionary measures: 1. Alert staff and personnel about threat warning 2. Prepare additional notification lists if the situation escalates to the "Is the Threat Credible?" stage

If the threat is not possible, then return to normal operations. Otherwise, proceed to "Is the Threat Credible" stage.

Is the Threat Credible? (Stage 2)	Special actions and notifications to be taken: Activate notification and personnel safety portions of ERP Evaluate whether the threat is credible in consultation with assisting agencies Visually inspect physical evidence and determine whether there is a change in normal system operating parameters (i.e., chlorine residuals, turbidity, odor, color, pH, etc.) Conduct actions and testing as recommended by monitoring and sampling experts

If the threat is not credible, then return to normal operations. Otherwise, proceed to "Has the Threat been Confirmed" stage.

Has the Incident Been Confirmed? (Stage 3)	Special actions and notifications to be taken: Initiate full ERP activation Follow State Incident Command System Isolate portion of system or backflush Shut down system if obvious or confirmed contamination warrants Issue public notice and issue follow-up media press releases Continue sampling and water monitoring Assess need to remediate storage tanks, filters, sediment basins, solids handling, etc.

*This is a simplified Action Plan example that includes the threat decision process to determine if the major event is just a threat or actual event. You should develop this "Action Plan" specific to the needs of your CWS and surrounding community.

*Structural Damage/Physical Attack to Water System or Facility(ies)**

Threat Warning Stage

Threat Warning Received	Special actions and notifications to be taken: Notify ER Lead or Alternate ER Lead Record and document all information pertaining to the threat warning Do not disturb site if the threat warning could be a possible crime scene Return to normal operations if no further action is required (i.e., the threat warning can be explained) Begin the "Threat Decision Process" if the threat warning cannot be explained

Threat Decision Process Stage

Is the Threat Possible? (Stage 1)	Special actions and notifications to be taken: Notify local law enforcement Notify State Drinking Water Primacy Agency Evaluate threat warning and make decisions in consultation with State Drinking Water Primacy Agency and local law enforcement Initiate basic precautionary measures: 1. Alert staff and personnel about threat warning 2. Heighten security at critical facilities 3. Prepare additional notification lists if the situation escalates to the "Is the Threat Credible?" stage

If the threat is not possible, then return to normal operations. Otherwise, proceed to "Is the Threat Credible" stage.

Is the Threat Credible? (Stage 2)	Special actions and notifications to be taken: Activate notification and personnel safety portions of ERP Physically secure water system facilities Evaluate whether the threat is credible in consultation with assisting agencies

If the threat is not credible, then return to normal operations. Otherwise, proceed to "Has the Threat been Confirmed" stage.

Has the Incident Been Confirmed? (Stage 3)	Special actions and notifications to be taken: Initiate full ERP activation Follow State Incident Command System Deploy damage assessment team Isolate damaged facility from rest of water system Coordinate alternative water supply, as needed, or consider alternate (interim) treatment schemes Issue public notice and issue follow-up media press releases Repair damaged facilities Assess need for additional protection/security measures

*This is a simplified Action Plan example that includes the threat decision process to determine if the major event is just a threat or actual event. You should develop this "Action Plan" specific to the needs of your CWS and surrounding community.

*Cyber Attack on SCADA or Operational Computer System**

Threat Warning Stage

Threat Warning Received	Special actions and notifications to be taken: Notify ER Lead or Alternate ER Lead Record and document all information pertaining to the threat warning Do not disturb site if the threat warning could be a possible crime scene Return to normal operations if no further action is required (i.e., the threat warning can be explained) Begin the "Threat Decision Process" if the threat warning cannot be explained

Threat Decision Process Stage

Is the Threat Possible? (Stage 1)	Special actions and notifications to be taken: Notify local law enforcement Notify State Drinking Water Primacy Agency Evaluate threat warning and make decisions in consultation with State Drinking Water Primacy Agency and local law enforcement Initiate basic precautionary measures: 1. Alert staff and personnel about threat warning 2. Temporarily shut down SCADA system and go to manual operation using established protocol 3. Prepare additional notification lists if the situation escalates to the "Is the Threat Credible?" stage

If the threat is not possible, then return to normal operations. Otherwise, proceed to "Is the Threat Credible" stage.

Is the Threat Credible? (Stage 2)	Special actions and notifications to be taken: Activate notification and personnel safety portions of ERP Continue manual operation using established protocol Consider whether to isolate source water Consider whether to shut down system and provide alternate water Evaluate whether the threat is credible in consultation with assisting agencies Conduct actions/testing recommended by monitoring and sampling experts

If the threat is not credible, then return to normal operations. Otherwise, proceed to "Has the Threat been Confirmed" stage.

Has the Incident Been Confirmed? (Stage 3)	Special actions and notifications to be taken: Initiate full ERP activation Follow State Incident Command System Continue manual operation, source water isolation, or system shut down and alternate water supply, as appropriate Issue public notice and issue follow-up media press releases Make image copy of all system logs to preserve evidence With law enforcement assistance, check for implanted backdoors and other malicious code before restarting SCADA system Install safeguards before restarting SCADA system Bring SCADA system up and monitor system Assess/implement additional precautions for SCADA system

*This is a simplified Action Plan example that includes the threat decision process to determine if the major event is just a threat or actual event. You should develop this "Action Plan" specific to the needs of your CWS and surrounding community.

*Hazardous Chemical Release from Water System Facility(ies)**

Threat Warning Stage

Threat Warning Received	Special actions and notifications to be taken: Notify ER Lead or Alternate ER Lead Record and document all information pertaining to the threat warning Do not disturb site if the threat warning could be a possible crime scene Return to normal operations if no further action is required (i.e., the threat warning can be explained) Begin the "Threat Decision Process" if the threat warning cannot be explained

Threat Decision Process Stage

Is the Threat Possible? (Stage 1)	Special actions and notifications to be taken: Notify local law enforcement Notify State Drinking Water Primacy Agency Evaluate threat warning and make decisions in consultation with State Drinking Water Primacy Agency and local law enforcement Initiate basic precautionary measures: 1. Alert staff and personnel about threat warning 2. Post full-time operations personnel at the chemical treatment areas of the facility 3. Verify that monitoring, leak detection, and personal protection equipment are fully operational 4. Prepare additional notification lists if the situation escalates to the "Is the Threat Credible?" stage

If the threat is not possible, then return to normal operations. Otherwise, proceed to "Is the Threat Credible" stage.

Is the Threat Credible? (Stage 2)	Special actions and notifications to be taken: Activate notification and personnel safety portions of ERP Physically secure water system facilities and specifically the chemical treatment areas of the facility Identify potentially hazardous chemical(s) to appropriate assisting agencies Based on Risk Management program and ERP, evaluate potential extent of public evacuation or shelter in place order Evaluate whether the threat is credible in consultation with assisting agencies

If the threat is not credible, then return to normal operations. Otherwise, proceed to "Has the Threat Been Confirmed" stage.

Has the Incident Been Confirmed? (Stage 3)	Special actions and notifications to be taken: Initiate full ERP activation Follow State Incident Command System Determine extent/concentration of chemical release and deploy damage assessment team Turn off chemical treatment equipment and isolate chemical treatment areas from rest of water system Depending on extent and concentration of release, issue evacuation or shelter in place order per Risk Management Program and ERP Coordinate alternative water supply, as needed, or consider alternate (interim) treatment schemes Issue public notice and issue follow-up media press releases Repair damaged facilities Assess need for additional protection/security measures

*This is a simplified Action Plan example that includes the threat decision process to determine if the major event is just a threat or actual event. You should develop this "Action Plan" specific to the needs of your CWS and surrounding community.

20 The Pollution Prevention Approach

20.1 INTRODUCTION

Applying pollution prevention strategies—the environmental management option of the future—will not eliminate all wastes from all production processes. Rather, pollution prevention strategies offer a cost-effective means of minimizing the generation of waste while facilitating compliance with local, state, and federal environmental and health and safety regulations. Source reduction is the first step in the traditional hierarchy of integrated waste management options and represents the most preferable of the available waste generation and pollution prevention steps. The next step in such a hierarchy is the responsible recycling of any wastes that cannot be reduced at the source. When recycling is conducted in an environmentally sound manner, it shares many of the same advantages of source reduction, such as conserving energy and other resources, reducing the reliance on raw materials, and reducing the need for end-of-pipe treatment or containment of wastes. Wastes that cannot be "feasibly" recycled should be treated in accordance with environmental standards that are designed to reduce both the hazard and volume of waste streams. Finally, any residues remaining from the treatment of waste should be disposed of safely to minimize their potential for release into and impact on the environment. Although this integrated waste management approach includes treatment and ultimate disposal, the concepts of pollution prevention that are highlighted in this chapter are focused on the first two waste management options (i.e., source reduction and closed-looped recycling), particularly as they apply to water resource management.

This chapter provides an overview of pollution prevention concepts through a discussion of the shifting waste management paradigm, regulations, the Environmental Protection Agency's (EPA) pollution prevention strategy, the pollution prevention hierarchy, pollution prevention opportunity assessments, pollution prevention incentives, deterrents to pollution prevention, and options for water recycling and reuse. An applications section provides six Illustrative Examples related to pollution prevention principles and approaches.

20.2 THE SHIFTING WASTE MANAGEMENT PARADIGM

Most environmental protection efforts have traditionally emphasized control of pollution from wastes after they have been generated. Although it may, in some instances, be effective in protecting the environment, this method of waste management has certain disadvantages. Specifically, this type of pollution control does not always solve the problem of pollution; rather, it alters the problem often by transferring pollution from one medium to another or expending significant energy and materials for the treatment of the generated waste, resulting in no net environmental benefit. This has caused interest to shift from focusing on how to deal with the amounts of waste generated to reducing the amounts of waste produced. The underlying principle of this new direction, defined as waste reduction or *pollution prevention*, is based on limiting the amount of waste produced up front, rather than developing extensive treatment processes downstream, to ensure that the waste poses no threat to human health or to the environment.

Pollution prevention and material and energy use reduction clearly have now become a top priority for both industries and regulators. As the commitment to pollution prevention increases, the problems created by the generation of wastes will begin to lessen. For industry, this means that pollution prevention, and material and energy use reduction must be integral to a company's overall operational strategy. Industry must also view pollution prevention as a continuous process of searching for new areas to assess, investigating new methods of pollution prevention and material and energy use reduction, and analyzing and implementing these methods.

20.3 REGULATIONS

The concept of pollution prevention was first defined as "waste minimization" and was conceptually introduced when the U.S. Congress specifically stated in the Hazardous and Solid Waste Amendments (1984) to the Resource Conservation and Recovery Act (RCRA):

> The Congress hereby declares it to be the national policy of the United States that wherever feasible, the generation of hazardous waste is to be *reduced* or *eliminated* as expeditiously as possible. Waste that is nevertheless generated should be treated, stored. or disposed of so as to minimize the present and future threat to human health and the environment.

In its 1986 report to Congress, the Environmental Protection Agency (U.S. EPA 1986) detailed the concept of waste minimization as:

> The *reduction*, to the extent feasible, of hazard u. waste that is *generated* or subsequently treated, stored or disposed of. It includes any *source reduction* or *recycling* activity undertaken by a generator that results in either (1) the reduction of total volume or quantity of hazardous waste or (2) the reduction of toxicity of hazardous waste, or both, so long as such reduction is consistent with the goal of minimizing present and future threats to human health and the environment.

However, the most important piece of legislation enacted to date is the Pollution Prevention Act. The Pollution Prevention Act of 1990 signed into law in November 1990, established pollution prevention as a "national objective." The act notes that:

> There are significant opportunities for industry to reduce or prevent pollution at the source through coat-effective changes in production, operation, and raw material use. The opportunities for source reduction are often not realized because existing regulations, and the industrial resources they require for compliance, focus upon treatment and disposal, rather than source reduction. Source reduction is fundamentally different and more desirable than waste management and pollution control.
>
> The Act establishes the pollution prevention hierarchy as·[a] national policy, declaring that pollution should be prevented or reduced at the source wherever feasible, while pollution that cannot be prevented should be recycled in an environmentally safe manner. In the absence of feasible prevention or recycling opportunities, pollution should be treated; disposal or other releases into the environment should be used as a last resort.
>
> Source reduction is defined to mean any practice which reduces the amount of any hazardous substance, pollutant, or contaminant entering any waste stream or otherwise released into the environment (including fugitive emissions) prior to recycling, treatment or disposal; and which reduces the hazards to public health and the environment associated with the release of such substances, pollutants, or contaminants.

The Pollution Prevention Act also formalized the establishment of the EPA's Office of Pollution Prevention, independent of the single-media programs, to carry out the functions required by the act and to develop and implement a strategy to promote source reduction. Among other provisions, the law directs the EPA to:

1. Facilitate the adoption of source reduction techniques by businesses and by other federal agencies.
2. Establish standard methods of measurement for source reduction.
3. Review regulations to determine their effect on source reduction.
4. Investigate opportunities to use federal procurement to encourage source reduction.

5. Develop improved methods for providing public access to data collected under federal environmental statutes.
6. Develop a training program on source reduction opportunities, devise model source reduction auditing procedures, establish a source reduction clearinghouse, and establish an annual awards program.

Under the act, facilities required to report releases to the EPA for the Toxic Release Inventory (TRI) must now also provide information on pollution prevention and recycling for each facility and for each toxic chemical. The information includes: (a) the quantities of each toxic chemical entering the waste stream and the percentage change from the previous year, (b) the quantities recycled and the percentage change from the previous year, (c) source reduction practices, and (d) changes in production from the previous year. Finally, the Act requires the EPA to report to Congress within 18 months (and biennially afterward) on the actions needed to implement the strategy to promote source reduction.

States have been at the forefront of the pollution prevention movement, providing a direct link to industry, local governments, and consumers. Through grants to states, the EPA has enhanced the capabilities of states to demonstrate innovative and results-oriented programs and has assisted states in implementing a waste prevention approach. Local governments can also play a significant role in promoting pollution prevention in the industrial, consumer, transportation, agricultural, and public sectors of the community. Many local governments have already taken the lead in putting successful recycling programs into place. A variety of tools are available to promote prevention. Local governments (including cities, counties, sewer and water agencies, planning departments, and other special districts) can provide:

1. Educational programs to raise awareness in business and the community of the need to reduce waste and pollution and conserve resources.
2. Technical assistance programs that provide on-site help to companies and organizations in reducing pollution at the source.
3. Regulatory programs that promote prevention through mechanisms such as codes, licenses, and permits.
4. Procurement policies regarding government purchase of recycled products, reusable products, and products designed to be recycled.

Many local governments have passed resolutions and ordinances relating to water reduction, energy conservation, automobile use, procurement policies, and so on. Such resolutions and ordinances can be useful steps in signaling a public commitment to operate under environmentally sound principles. They also help to define goal and targets and to delineate the specific responsibilities of different local agencies.

20.4 THE EPA'S POLLUTION PREVENTION STRATEGY

The EPA's Pollution Prevention Strategy, released in February 1991, was developed by the agency in consultation with all program and regional offices. The strategy provides guidance on incorporating pollution prevention into the EPA's ongoing environmental protection efforts and includes a plan for achieving substantial voluntary reductions of targeted high-risk industrial chemicals. The strategy is aimed at maximizing private sector initiatives while challenging industry to achieve ambitious prevention goals. A major component of the initial strategy was the *Industrial Toxics Project*, known more generally as the *33/50 Program*. The overall goals of the 33/50 Program were to reduce national pollution releases and off-site transfers of priority toxic chemicals and to encourage pollution prevention. The specific program goal was to voluntarily reduce releases and transfers of 17 toxic chemicals by 33% by the end of 1992 and 50% by the end of 1995. Both goals were reached 1 year ahead of schedule (U.S. EPA 1999).

TABLE 20.1
The 33/50 Program Target Chemicals and Estimated Releases in 1988

Chemical	Total Releases (Million lb)
Benzene	33.1
Cadmium	2.0
Carbon Tetrachloride	5.0
Chloroform	26.9
Chromium	56.9
Cyanide	13.8
Dichloromethane	153.4
Lead	58.7
Mercury	0.3
Methyl Ethyl Ketone (MEK)	159.1
Methyl Isobutyl Ketone (MIBK)	43.7
Nickel	19.4
Tetrachloroethylene (PCE)	37.5
Toluene	344.6
1,1,1-Trichloroethane (TCA)	190.5
Trichloroethylene (TCE)	55.4
Xylene	201.6

The 17 pollutants identified as targets of the industrial toxics project presented both significant risks to human health and the environment and significant opportunities to reduce such risks through prevention. The list (Table 20.1) was drawn from recommendations submitted by program offices, taking into account such criteria as health and ecological risk, potential for multiple exposures or cross-media contamination, technical or economic opportunities for pollution prevention, and limitations for their effective treatment using conventional technologies.

All of the targeted chemicals are included on the EPA's TRI; thus, reductions in their releases can be measured in each year's TRI reports. Despite company participation being strictly voluntary, almost 1,300 parent companies, operating more than 6,000 facilities in the United States, participated in the 33/50 Program. Companies examined their own industrial processes and identified and implemented cost-effective pollution prevention practices for toxic chemicals. Information on a variety of "success stories" is available from the EPA (1994) and Khanna and Damon (1997). These success stories detail not only technical innovations, but new managerial and cost-accounting approaches that led to statistically significant reductions in hazardous materials releases by participating companies. Khanna and Damon (1997) indicated that although there were statistically significant negative impacts on short-term returns on investments, long-term profitability of participating firms was projected to be positive and statistically significant.

The EPA's pollution prevention strategy also provides guidance on incorporating pollution prevention into the agency's existing program, emphasizing the need for continued strong regulatory and enforcement programs. At the same time, the strategy favors flexible, cost-effective approaches that involve market-based incentives where practical. For example, the strategy calls for the use of "regulatory clusters," through which the EPA categorizes the rules it intends to propose over the next several years for certain chemicals and their sources. The clusters are intended to foster improved cross-media evaluation of the cumulative impact of standards, provide more certainty for industry, and encourage early investment in pollution prevention activities.

The strategy outlines several short-term measures that address various institutional barriers within the agency's own organization that limit its ability to develop effective prevention strategies.

Such measures include designating special assistants for pollution prevention in each assistant administrator's office, developing incentives and awards to encourage agency staff to engage in pollution prevention efforts, incorporating prevention into the comprehensive four-year strategic plans by each program office, and providing pollution prevention training to agency staff.

The 33/50 Program for the manufacturing sector represented the first focus of a comprehensive agency strategy. After the completion of this project in 1995, the EPA staff of the program focused on dissemination of results from the program in the form of company profiles that summarize successful programs at industrial facilities; a generic waste reduction manual, the Industry Pollution Prevention Guide (U.S. EPA 2001); numerous industry-specific guides to Pollution Prevention; supplementary information in reports, handbooks, reference manuals, bibliographic reports, and videos; the development of seven manufacturing technology centers across the nation for research and training purposes; state programs for grants and awards; and several hotlines for pollution prevention support. One recent initiative is the *Design for Environment (DfE) Program*, which was a cooperative, voluntary effort by the EPA, industry, professional organizations, state and local governments, other federal agencies (including the Small Business Administration), and the public aimed at developing specific pollution prevention information in a number of small business-dominated industries including printing (screen printing and lithography), dry cleaning, automotive refinishing, and electronics/printed wiring boards. The program also developed guidance on the creation and use of environmental management systems in small manufacturers to more carefully monitor and management the use of hazardous materials (U.S. EPA 2002). Specifically, the DfE Program used its industry projects to bring together comparative information on the environmental and human health risks, exposures, performance, and costs of alternative products and technologies so that businesses can make more informed environmental management decisions, realizing that many businesses often do not have the resources or technical expertise to develop this kind of technical information on their own.

Another important goal of the EPA's Pollution Prevention Program is to ensure that pollution prevention training and education are available to government, industry, academic institutions, and the general public. Training and education are needed to help institutionalize prevention as the strategy of choice in all environmental decision-making activities. The U.S. EPA created the *National Pollution Prevention Center for Higher Education (NPPC)* in 1991 to collect, develop, and disseminate educational materials on pollution prevention; the University of Michigan was selected as the site. The NPPC, now the Center for Sustainable Systems (CSS) as of 2001, represents a collaborative effort between business and industry, government, nonprofit organizations, and academia and has as its mission to promote sustainable development by educating students, faculty, and professionals about pollution prevention, industrial ecology, and sustainability; create educational materials; provide tools and strategies for addressing relevant environmental problems; and establish a national network of pollution prevention educators. In addition to developing educational materials and conducting research, the NPPC also offers an internship program, professional education and training, and conferences. The original NPPC web site address is: http://www.umich.edu/~nppcpub/resources/, and the site for the CSS is http://css.snre.umich.edu.

The *Pollution Prevention Information Clearinghouse (PPIC)* is a free information service dedicated to reducing and eliminating industrial pollutants through education and public awareness. The PPIC is operated by the EPA's Office of Pollution Prevention and Toxics and can be reached online at: http://www.epa.gov/p2/pollution-prevention-resources#ppic. The PPIC serves as a repository for hard copy reference materials and provides online information retrieval and ordering system. In addition, the PPIC provides online interactive access to a wide range of pollution prevention topics that include general pollution prevention information, enforcement and compliance issues, environmentally preferable purchasing, pollution prevention incentives for states, design for the environment, environmental accounting, hospitals for a healthy environment, voluntary standards network, education and training, environmental labeling, and information on persistent and bioaccumulative chemicals. The PPIC provides a national calendar of conferences and workshops relating to pollution prevention, as well as a PPIC technical assistance hotline (202/566-0799) to answer pollution prevention questions, access information in the PPIC, and assist in document searches and ordering.

20.5 WASTE MANAGEMENT HIERARCHY

In the current working definition used by the EPA, source reduction and recycling are considered to be viable pollution prevention techniques, preceding treatment and disposal in the waste management hierarchy. In its original *Pollution Prevention Policy Statement* published in the January 26, 1989, *Federal Register*, the EPA encouraged organizations, facilities, and individuals to fully use source reduction and recycling practices and procedures to reduce risk to public health, safety, and the environment.

Figure 20.1 depicts the EPA hierarchy of preferred approaches to sustainable materials management. As one proceeds upward through the hierarchy, more and more waste is prevented, recycled, or treated, thereby reducing the amount of waste generated or released to the environment. The highest priority waste management technique in the hierarchy is source reduction, also known as *waste reduction* or *pollution prevention*. The following provides a brief description of each of these waste management techniques.

1. *Source reduction*, consisting of technologies to reduce the volume of wastes initially generated, is the primary approach. Some techniques are applied to the production process before the point of generation. Methods that eliminate or reduce the amount of waste generated by a particular process, either through procedure modifications or through material substitution, are employed. Some techniques are applied at the consumer level and can include reuse or product or process substitution. Substitution of reusable shopping bags or multiuse straws for single use products is an example of source reduction techniques applied at the consumer level.
2. *Recycling*, the secondary approach for sustainable materials management, attempts to recover a usable material from a waste stream. These methods can take place within a process or at the end of the process, either on site or off site, and can be employed off site at the consumer level.

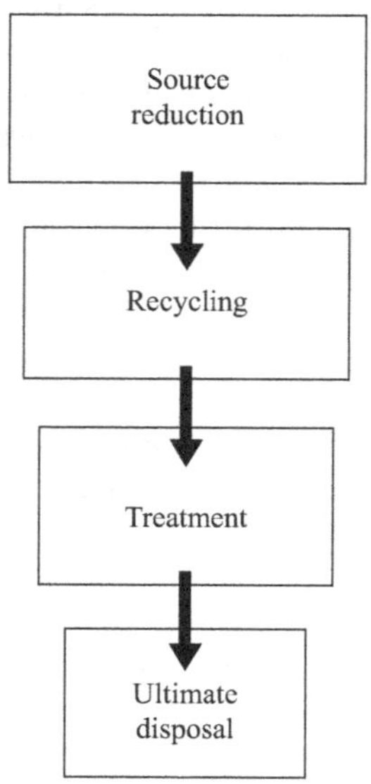

FIGURE 20.1 The Environmental Protection Agency's (EPA) sustainable materials waste management hierarchy.

3. For the wastes remaining after all possible source reduction and recycling techniques have been employed, the next approach is the use of physical, biological, and chemical *treatment methods*, including incineration. This results in a reduction of the toxicity and volume of waste requiring ultimate disposal.
4. The last approach for managing wastes is *ultimate disposal*, consisting of landfilling, landfarming, and deep-well injection.

Many individuals, particularly those employed by the EPA and other (regulatory) environmental organizations, have interpreted the Pollution Prevention Act and the EPA's subsequent policy statement to mean that pollution prevention refers only to source reduction and, as such, is the preferred method of sustainable materials management. Recycling, although possessing environmental advantages over other management techniques, is relegated to a secondary position in the pollution prevention hierarchy.

20.5.1 Source Reduction

Once opportunities for reduction have been identified by a pollution prevention assessment (see Section 20.6), source reduction techniques should be implemented first. As described previously, source reduction involves the reduction of pollutant wastes at their source, usually within a process, and is the most desirable option in the pollution prevention hierarchy. By avoiding the generation of wastes, source reduction eliminates the problems associated with the handling and disposal of wastes. A wide variety of facilities can adopt procedures to minimize the quantity of waste generated. Many source reduction options involve a change in procedural or organizational activities, rather than a change in technology. For this reason, these options tend to affect the managerial aspect of production and usually do not demand large capital and time investments. This makes implementation of many source reduction options affordable to companies of any size. Figure 20.2 depicts the source reduction scheme for manufacturers. As depicted, source reduction can be broken down into options involving product substitution or source control.

Additional source reduction options can be effectively used by consumers. These options also include product substitution or source control options but with a slightly different focus. Every time a new product is used instead of a recycled one, some form of natural resource is depleted forever.

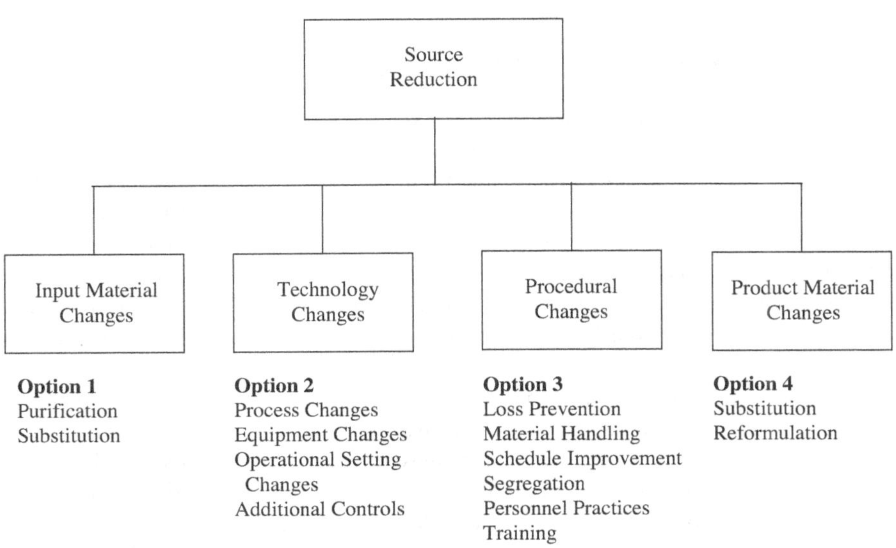

FIGURE 20.2 Source reduction options for manufacturers.

Source reduction of plastics can be achieved in a number of ways in the home: Avoid the use of disposable items, try to buy concentrates or other products that require less packaging, and so on. Use reusable water bottles rather than single-use containers. Single-use plastic bag and plastic straw bans have recently been passed in a number of jurisdictions across the United States, and many durable, multiuse options are now available for these and many other (i.e., K-cup coffee holders) plastic-based consumer goods. Repairing rather than throwing away electronic equipment or garments, or donating used equipment and clothing to charities are additional ways to reduce the generation of waste at the consumer level.

20.5.2 Recycling and Reuse

Recycling or reuse can take two forms: pre-consumer and postconsumer applications. Pre-consumer recycling involves raw materials, products, and by-products that have not reached a consumer for an intended end use but are typically reused within an original process. Postconsumer recycled materials are those that have served their intended end use by a business, consumer, or institutional source and have been separated from municipal solid waste for the purpose of recycling. Regarding pre-consumer recycling, recycling techniques allow waste materials to be used for a beneficial purpose. A material is recycled if it is used, reused, or reclaimed. Recycling through use and/or reuse involves returning waste material either to the original or to another process as a substitute for input of virgin material. Recycling through reclamation is the processing of a waste for recovery of a valuable material or for regeneration. Recycling of wastes can provide a very cost-effective waste management alternative. This option can help eliminate waste disposal costs, reduce raw material costs, and provide income from salable waste.

Because recycling is the second-most preferred option in the pollution prevention hierarchy, it should be considered only when all source reduction options have been investigated and implemented. Reducing the amount of waste generated at the source will often be more cost-effective than recycling because waste primarily is lost raw material or product, which requires time and money to recover. Because recycling reduces the amount of waste requiring disposal and reuses materials that carry with them the embedded energy, water, and material use required to manufacture and transport them, it can provide significant reduction in the consumption of natural resources. A national recycling goal of 35% of municipal solid waste was set by the EPA in 2000, and as of 2015, the national recycling rate has exceeded 34%.

20.6 POLLUTION PREVENTION OPPORTUNITY ASSESSMENTS

A *pollution prevention opportunity assessment (PPOA)* is defined as a systematic, planned procedure with the objective of identifying ways to reduce or eliminate waste, preferably at the source. Generally, the assessment is preceded by careful planning and organization to set overall pollution prevention goals. Next, the actual assessment procedure begins with a thorough review of a plant's operations and waste streams, and the selection of specific areas of the plant to assess. Once an area is selected as a possible "minimization" area, various options with the potential for reducing waste generation can be developed and screened. The technical and economic feasibility of each option is evaluated, and finally, the most cost effective and technically feasible options are implemented.

Initiating a successful program must begin with a secure commitment from top management, allocation of adequate funding and technical expertise, appropriate organization, and a good understanding of the goals of the assessment and planning required to make it effective. To be successful, pollution prevention must become an integral part of a company's operations. The PPOA offers opportunities to increase production efficiency, reduce operating costs, reduce potential liability, and improve the environment, while also improving regulatory compliance. Much of the material presented in this chapter has been drawn from the EPA's original "waste minimization opportunity

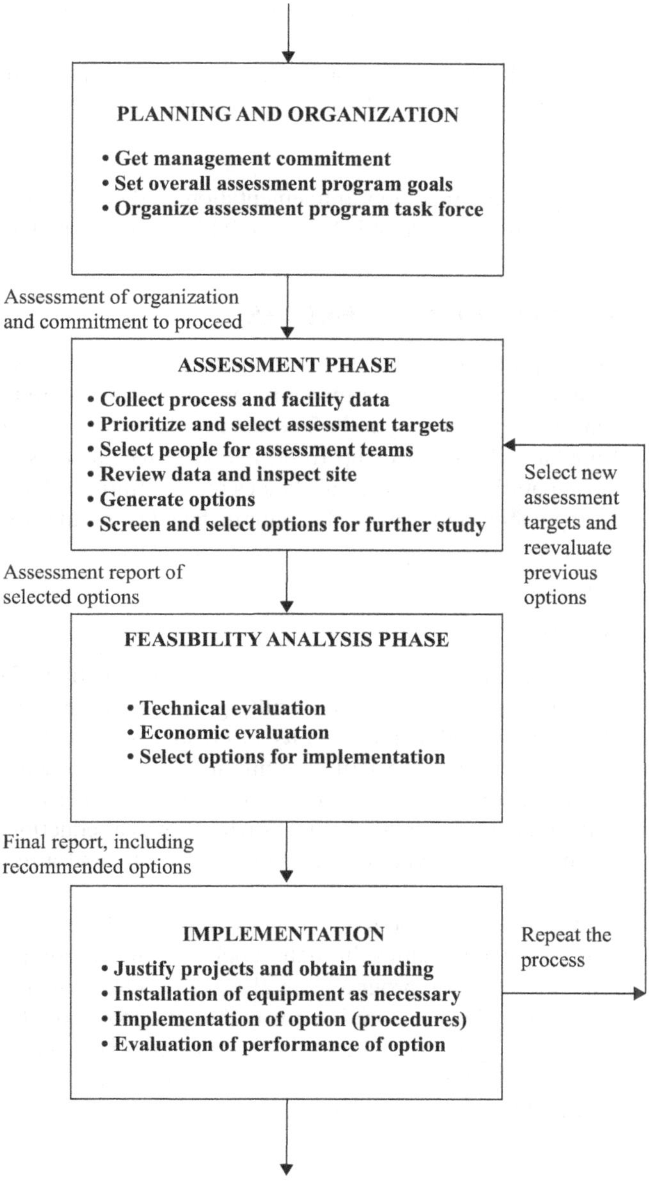

FIGURE 20.3 Pollution prevention opportunity assessment procedure.

assessment" procedure (U.S. EPA 1988). This procedure has been modified and adapted to apply to what is defined as a pollution prevention opportunity assessment.

Figure 20.3 depicts the pollution prevention opportunity assessment procedure. As shown, the assessment procedure can be divided into four phases which include the following:

1. *Planning and organization phase*, during which management buy-in and commitment are obtained, overall assessment goals are identified, and an assessment team is assembled.
2. *Assessment phase*, consisting of the collection of data and identification and screening of potential pollution prevention options.

3. *Feasibility analysis phase*, consisting of the technical and economic evaluation of each option.
4. *Implementation phase*, where recommended options are put into place, their performance is monitored, and feedback is provided for the next round of pollution prevention assessment activities at a facility.

Additional details regarding the organization, implementation, and evaluation of pollution prevention opportunity assessments are available in U.S. EPA (1988) and Dupont et al. (2016).

20.7 POLLUTION PREVENTION INCENTIVES

As described previously, the problems associated with the handling of waste streams have become a major concern for industry. Overall costs for managing waste, coupled with increasing regulatory and economic consequences, are escalating daily. There are real incentives for avoiding the production of hazardous and nonhazardous wastes up front, and they will increase as more regulations are passed and treatment capabilities for waste substances remain limited. These incentives usually result in monetary encouragements and consist of economic benefits, regulatory compliance, reduction in liability, and enhanced public image.

20.7.1 Economics Benefits

Reducing the amount of waste produced initially will result in a reduction of the costs associated with the handling of that waste. The obvious costs resulting from the treatment, transportation, and disposal of wastes will all be lowered as the quantity of created waste is lowered. Current trends predict that the cost of disposing of waste will continue to increase. In addition, income can be derived through the sale, reuse, or recycling of certain wastes. A number of not-so-obvious gains that are associated with the reduction of waste include such items as a reduction in health and safety costs, reduction in insurance costs, reduction in raw material costs, and a reduction in reporting, manifesting, and permitting costs.

Regulatory agencies can influence companies to investigate pollution prevention techniques through offers of reduced fines and penalties through Supplemental Environmental Projects (SEPs; U.S. EPA 2015). A pollution prevention program can also be used to bring a firm into compliance with regulations. Permit fees may be structured in a way that promotes pollution prevention activities. For instance, instead of basing a permit fee on the size of a company, an agency might base the fee on the volume and/or toxicity of waste substances produced. Firms actively pursuing pollution prevention would be rewarded by paying a lower fee.

20.7.2 Regulatory Compliance

Federal and state laws require all firms classified as hazardous waste generators or small-quantity generators (SQGs) to implement a pollution prevention program to reduce the quantity of waste to the extent that it is economically feasible. SQGs are facilities that generate more than 10 kg/month of waste but less than 1,000 kg/month of hazardous waste. A facility generating more than 1,000 kg/month is classified as a large quantity generator. Large quantity generators are required to sign the following statement on all manifests to certify their pollution prevention efforts:

> Unless I am a small-quantity generator who has been exempted by statute or regulation from the duty to make a waste minimization certification under Section 3002(b) of RCRA, I also certify that I have a program in place to reduce the volume and toxicity of waste generated to the degree I have determined to be economically practicable, and I have selected the method of treatment, storage, or disposal currently available to me which minimizes the present and future threat to human health and the environment.

Firms permitted as hazardous waste generators are generally required under RCRA to report sampling data on a regular basis. When hazardous wastes are minimized, the frequency of sampling and data reporting is also minimized, thereby saving the firm a considerable amount of personnel time money.

In addition, land disposal restrictions and treatment standards provide indirect incentives for reducing (hazardous) waste. Under these regulations, only wastes meeting specified treatment standards are allowed to be disposed of in landfills. The level of treatment standards are based on pollution prevention technologies, and therefore, those facilities already employing these reduction techniques will be best suited to meet the standards. With these reduction methods already in place, a company can save significantly on costs associated with waste management.

20.7.3 Reduction in Liability

Both short- and long-term liabilities can be reduced with pollution prevention programs. Short-term liabilities associated with releases to the environment resulting in noncompliance with permits can be reduced through the overall reduction of waste generation rates. Short-term liabilities connected with personnel exposure and workplace safety will also be lessened. Long-term liabilities resulting from on-site or off-site disposal of wastes can be reduced as well. Disposing of wastes at a permitted disposal facility does not end a firm's liability for that waste. If a disposal facility is shown to be releasing contaminants into the environment, not only are the owners/operators of that facility liable and responsible for carrying out remedial actions to clean up the contamination, the generators whose wastes were disposed of in that facility are also liable. These generators can be ordered to finance both (a) the investigation of the extent of contamination and (b) the full cost of remedial action to address the contamination. Reduction or elimination of hazardous waste through active pollution prevention programs goes a long way to reducing long-term liabilities associated with long-term hazardous waste management.

20.7.4 Enhanced Public Image

The public has placed a great deal of emphasis on environmental issues. Recent election campaigns at all levels of government have made the environment a top priority platform issue. Because of this level of concern and awareness, it is becoming increasingly important for companies to share information with the public. Under the Superfund Amendments and Reauthorization Act (SARA), the Emergency Planning and Community Right-to-Know Act (EPCRA) includes mandatory reporting of releases to the environment and optional reporting of pollution prevention activities. Implementation of a pollution prevention program provides a good community relations baseline for improving a firm's public image as a good corporate citizen.

20.8 DETERRENTS TO POLLUTION PREVENTION

The previous section presented the advantages of developing and implementing a pollution prevention program. This section briefly reviews some of the deterrents or impediments to pollution prevention. A "dirty dozen" list is provided, followed by a short description (and some comments) on each of these deterrents (Theodore and McGuinn 1992).

1. Management apathy
2. Lack of financial commitment
3. Production concerns
4. Research, development, and design concerns
5. Failure to monitor program success
6. Middle-management decisions

7. Information exchange within organization
8. Confusion regarding regulations
9. Confusion about economic advantages
10. Bureaucratic resistance to change
11. Lack of awareness of pollution prevention advantages
12. Failure to apply multimedia approach

20.8.1 Management Apathy

It is not uncommon for upper-level management in most companies, particularly large ones, to take an indifferent attitude toward pollution prevention. Because administrators are often not technically oriented, most have difficulty realizing the potentially enormous benefits that can be gained from a pollution prevention program.

20.8.2 Lack of Financial Commitment

It is no secret that many considerations, including resources, individuals, money, economic incentives, and so forth, are critical to the success of a pollution prevention program. These "resources" are often just not available to invest in pollution prevention activities within small companies or within large companies that may be experiencing a financial downturn.

20.8.3 Production Concerns

The classic song of the production supervisor and/or plant operator is, "I'm meeting deadlines and making money, so don't rock the boat." Sending a member of the pollution prevention team out to a plant is somewhat akin (at times) to letting a bull loose in a china closet.

20.8.4 Research, Development, and Design Concerns

The somewhat simplistic mentality described in Section 20.8.3 can be extended to apply to new projects at the research, development, or design stages. A very common misconception is that any pollution prevention activity is going to delay a (new) project and cost money.

20.8.5 Failure to Monitor Program Success

Pollution prevention programs are often instituted with high hopes of success, but later abandoned because of a failure to monitor the program properly. Responsible individuals must continuously monitor and record both the success and failures of the program to ensure successful approaches are adopted, and unsuccessful ones are improved upon.

20.8.6 Middle-Management Decisions

In most companies, it is usually middle management that is directly responsible to the top company administrator(s) for earnings. Unfortunately, it is often this group that views this type of program as an added expense that will eat into profits and so are inclined not to consider or implement them when proposed.

20.8.7 Information Exchange within Organization

As indicated in Section 20.8.1, management often has a poor understanding of the advantage of pollution prevention. In addition, the economic, environmental, liability, social, and other advantages that various organizations within a company ascribe to a program of this nature will differ widely. If successes and importance of pollution prevention approaches are not communicated within and throughout an organization, pollution prevention opportunities may go largely unrealized.

20.8.8 Confusion Regarding Regulations

It may be hard to believe, but many companies, particularly small ones, are not fully aware of the applicable regulations regarding their process or operation. Any permit review process is almost certainly doomed to failure if the responsible individual(s) or group is not cognizant of the pertinent regulations. Without a full understanding of relevant regulations, the significant benefits of avoiding some or all of these regulations through pollution prevention programs will go largely ignored.

20.8.9 Confusion about Economic Advantages

Most companies are simply not aware of the true costs associated with generating and treating wastes. However, treatment or disposal costs have increased dramatically in recent years, and liability concerns continue to mount. As indicated, many companies do not realize that pollution prevention opportunities, in many instances, result in increased profits. This situation is prevalent today because, for many companies, particularly large ones, it is difficult to evaluate the economic advantages of pollution prevention. The accounting system is often so complex and not properly integrated within the various divisions or organizations of a company that it is nearly impossible to quantify the true benefits arising from a pollution prevention program.

20.8.10 Bureaucratic Resistance to Change

It is natural, particularly in large corporations, to resist any change to an existing process or method of operation. This reluctance to adapt to a changing environmental, regulatory, liability, etc., climate is commonplace in industry.

20.8.11 Lack of Awareness of Pollution Prevention Advantages

Middle-level managers and upper-level administrators are often unaware of both the waste management problems at their facility and the associated true cost of waste treatment. Thus, it is understandable why these individuals are not aware of the overall advantages of a pollution prevention program.

20.8.12 Failure to Apply Multimedia Approach

Unfortunately, an overall approach that examines a system or process from a multimedia point of view is rarely found in industry today. There is much work to be done in this area, and many environmental improvements can be expected in the years to come once systems thinking is applied within a pollution prevention and waste minimization context.

As has been noted, many companies and individuals are unaware of the advantages of pollution prevention. Until this situation is changed, progress in the environmental arena will, as it has done in the past, come slowly.

20.9 WATER RECYCLING AND REUSE

Water recycling is a critical element for managing water resources. Through water conservation and water recycling society can meet environmental needs and still support sustainable development and a viable economy. Water recycling is reusing treated wastewater for beneficial purposes such as agricultural and landscape irrigation, industrial processes, toilet flushing, and replenishing a groundwater basin (referred to as *groundwater recharge*). Water is sometimes recycled and reused onsite; for example, when an industrial facility recycles water used for cooling processes. A common type of recycled water is water that has been reclaimed from municipal wastewater, or sewage. The term *water recycling* is generally used synonymously with water reclamation and water reuse.

Through the natural water cycle, the Earth has recycled and reused water for millions of years. Water recycling, though, generally refers to projects that use technology to speed up these natural processes. Water recycling is often characterized as "unplanned" or "planned." A common example of unplanned water recycling occurs when cities draw their water supplies from rivers, such as the Colorado River and the Mississippi River, that receive wastewater discharges upstream from other cities. Water from these rivers has been reused, treated, and piped into the water supply a number of times before the last downstream user withdraws the water. On the other hand, planned water recycling projects are those that are developed with the goal of beneficially reusing a recycled water supply.

Recycled water can satisfy most water demands as long as it is adequately treated to ensure the water quality is appropriate for the end use. Table 20.2 summarizes the suggested uses of water treated to different levels of treatment typical of municipal wastewater treatment plants. In uses where there is a greater chance of human exposure to the water, more treatment is required. As for any water source that is not properly treated, health problems could arise from drinking or being exposed to recycled water if it contains disease-causing organisms or other contaminants. The EPA regulates many aspects of wastewater treatment and drinking water quality, and most states have established criteria or guidelines for the beneficial use of recycled water.

Recycled water is most commonly used for non-potable (not for drinking) purposes, such as agriculture, landscape, public parks, and golf course irrigation. Other nonpotable applications include cooling water for power plants and oil refineries, industrial process water for such facilities as paper mills and carpet dyers, toilet flushing, dust control, construction activities, concrete mixing, and artificial lakes.

Although most water recycling projects have been developed to meet non-potable water demands, a number of projects use recycled wastewater *indirectly* for potable purposes. These projects include recharging groundwater aquifers and augmenting surface water reservoirs with

TABLE 20.2
Suggested Water Recycling and Uses as a Function of Different Levels of Municipal Wastewater Treatment

Levels of Treatment	Suggested Reuse Options
Primary Treatment	• No use recommended at this treatment level
Biological Secondary Treatment with Disinfection	• Surface irrigation of orchards and vineyards • Nonfood crop irrigation • Restricted landscape impoundments • Groundwater recharge of non-potable aquifer • Wetlands, wildlife habitat, stream augmentation • Industrial cooling
Tertiary/Advanced Treatment (Chemical Coagulation, Filtration, Disinfection)	• Indirect potable reuse • Groundwater recharge of potable aquifer • Surface water reservoir augmentation

Source: U.S. Environmental Protection Agency (U.S. EPA), *Water Recycling and Reuse: The Environmental Benefits*, Water Division Region IX, San Francisco, CA, 1998.

recycled water. In groundwater recharge projects, recycled water can be spread or injected into groundwater aquifers to augment groundwater supplies, and to prevent salt water intrusion in coastal areas.

In addition to providing a dependable, locally controlled water supply, water recycling provides tremendous environmental benefits. By providing an additional source of water, water recycling can help decrease the diversion of water from sensitive ecosystems. Other benefits include decreasing wastewater discharges and reducing and preventing pollution. Recycled water can also be used to create or enhance wetlands and riparian habitats.

When pollutant discharges to oceans, rivers, and other water bodies are curtailed, the pollutant loadings to these bodies are decreased. Moreover, in some cases, substances that can be pollutants when discharged to a body of water can be beneficially reused for irrigation. For example, recycled water contains higher levels of nutrients, such as nitrogen, than potable water. Application of recycled water for agricultural and landscape irrigation can provide an additional source of nutrients to the users of this reclaimed water, can save communities the high cost of nutrient removal, can reduce the production of nutrient rich biosolids, and lessen the need to apply synthetic fertilizers throughout the community.

The U.S. EPA has produced a variety of resources for the public and commercial sectors highlighting the benefits of water use efficiency, recycling, and reuse for pollution prevention and resource conservation. The EPA's *Preventing Pollution through Efficient Water Use* (U.S. EPA 1990) is directed toward individuals and communities, summarizing pollution prevention, energy conservation, source water protection, and economic and habitat benefits to efficient use of water. A series of documents highlighting the environmental benefits of *Lean Manufacturing* has been published by the EPA for the manufacturing and commercial sectors. These documents include a general analysis of environmental benefits from Lean Manufacturing (U.S. EPA 2003), a general guide to Lean Manufacturing advanced production techniques (i.e., Value Stream Mapping, 6S Techniques) and their pollution prevention benefits (U.S. EPA 2007), and a specific water-related *Lean & Water Toolkit* (U.S. EPA 2011) manual. These latter documents are attempts to address some of the deterrents to pollution prevention efforts highlighted previously, and to focus material use efficiency activities on water use and reuse opportunities in manufacturing and commercial settings.

20.10 APPLICATIONS

Six Illustrative Examples complement the material presented in this chapter on the pollution prevention concepts.

Illustrative Example 20.1

Describe how environmental management efforts have changed from pollution control to pollution prevention.

Solution

Traditionally, environmental management efforts have emphasized control of pollution after it has been generated. Although this approach may, under many circumstances, be effective in protecting human health and the environment, this method of waste management has certain disadvantages. Specifically, this type of pollution control does not always solve the problem of pollution; rather, it often transfers pollution from one medium to another, resulting in no net environmental benefit. In addition, waste management after it is generated requires investment in pollution control equipment and expenditures of materials, energy, and personnel that would not be required if the waste is not generated. The underlying principle of pollution prevention is based on limiting

TABLE 20.3
Evolution of Waste Management Approaches in the United States

Prior to 1945	No Control
1945–1960	Little control
1960–1970	Some control
1970–1975	Greater control, U.S. EPA established
1975–1980	More sophisticated control
1980–1985	Beginnings of waste reduction management
1985–1990	Waste reduction management
1990–1995	Formal pollution prevention programs (Pollution Prevention Act of 1990)
1995–2015	Widespread acceptance of pollution prevention
>2015	Green manufacturing, sustainable development, Design for the Environment …

the amount of waste produced "up front," rather than developing extensive treatment processes "downstream" to ensure that the waste poses no threat to human health or the environment.

Because of the growing appreciation of the benefits of waste avoidance, the overall approach to waste management in the United States has begun to shift from pollution control-driven activities to pollution prevention activities as suggested in Table 20.3.

Illustrative Example 20.2

Describe the list of 126 Priority Pollutant chemicals that need to be addressed to control releases to surface water.

Solution

The Clean Water Act includes National Pollutant Discharge Elimination System (NPDES) permits, effluent guidelines, and zero discharge goals designed to control releases to surface waters of 126 Priority Pollutants. This Priority Pollutant list (U.S. EPA 2014) contains chemicals of concern because of their toxicity or hazards for which the EPA has published analytical test methods. The list includes volatile organic substances such as benzene, chloroform, and vinyl chloride; acid compounds such as phenol and its derivatives; pesticides such as chlordane, DDT, and toxaphene; heavy metals such as lead and mercury; polychlorinated biphenyls (PCBs); and other organic and inorganic compounds.

Illustrative Example 20.3

Discuss some of the problems associated with the traditional type of economic analysis of pollution prevention projects.

Solution

The main problem with the traditional type of economic analysis is that it is difficult and, in some cases, impossible to quantify some of the not so obvious economic merits of a pollution prevention program. The following list is a summary of some of the economic aspects of pollution prevention projects that should be captured but are difficult or practically impossible to quantify accurately during an economic analysis of project merits.

1. Decreased long-term liabilities
2. Regulatory compliance costs
3. Regulatory recordkeeping costs

4. Costs associated with dealing with the EPA
5. Costs associated with dealing with state and local regulatory bodies
6. Elimination or reduction of fines and penalties
7. Potential tax benefits
8. Improved customer relations
9. Stockholder support (corporate image)
10. Improved public image
11. Reduced technical support costs
12. Potential reductions in insurance costs and claims
13. Effects on improved borrowing power
14. Improved mental and physical well-being of employees
15. Reduced health-maintenance costs
16. Improved employee morale
17. Other process benefits
18. Avoidance of rising costs of waste treatment and/or disposal
19. Reduced training costs
20. Reduced emergency response planning costs

Many proposed pollution prevention programs have been squelched in their early stages because a comprehensive economic analysis was not performed. Until the effects described are included, the true merits of a pollution prevention program may be clouded by incorrect and/or incomplete economic data. Can something be done by industry to remedy this problem? One approach is to use a modified version of the standard Delphi panel. To estimate these other economic benefits of pollution prevention, several knowledgeable individuals within and perhaps outside the organization are asked to independently provide estimates, with explanatory details, of these economic benefits. Each individual in the panel is then allowed to independently review all responses. The cycle is then repeated until the group's responses approach convergence (Dupont et al. 2016).

More details on quantifying the economic merits of a pollution prevention project are available in the literature (U.S. EPA 1993; U.S. EPA 1995; Theodore and Theodore 2010; Theodore 2014).

Illustrative Example 20.4

Explain material balances as they apply to water pollution prevention studies.

Solution

The conservation law for mass can be applied to any process or system. The general form of this law is given in Equation 20.1:

$$\text{mass in} - \text{mass out} + \text{mass generated} = \text{mass accumulated} \tag{20.1}$$

This equation may be applied to the total mass involved or to a particular species, on either a mole or mass basis. The conservation law for mass can be applied to steady-state or unsteady-state processes and to batch or continuous systems. To isolate a system for study, it is separated from the surroundings by a boundary or envelope. This boundary may be real (e.g., the walls of a vessel) or imaginary. Mass crossing the boundary and entering the system is part of the mass-in term, whereas that leaving the system is part of the mass-out term. The equation may be written for any compound whose quantity is not changed by chemical reaction or any chemical element, regardless of whether it has participated in a chemical reaction. It may be written for one piece of equipment, around several pieces of equipment, or around an entire process. It may be used to calculate an unknown quantity directly, to check the validity of experimental data, to check the validity of experimental data, or to express one or more of the independent relationships among the unknown quantities in a particular problem.

A steady-state process is one in which there is no change in conditions (temperature, pressure, etc.) or rates of flow with time at any point in the system. The accumulation term is then zero. If there is no chemical reaction, the generation term is also zero. All other processes are classified as unsteady-state. In a batch process, the container holds the product or products. In a continuous process, reactants are fed in an unending flow to a piece of equipment or to several pieces in series, and products are continuously removed from one or more points in the system. A continuous process may or may not be steady-state. As indicated previously, Equation 20.1 may be applied to the total mass of each stream (referred to as an *overall* or *total material balance*) or to the individual components of the streams (referred to as a *componential* or *component material balance*). Often the primary task in preparing a material balance is to develop the quantitative relationships among the streams. For more information on the conservation of mass, refer to the literature (Theodore 2014).

The conservation law for mass finds its major application in performing pollution prevention assessments and water use efficiency evaluations. Once a pollution prevention or water use efficiency assessment is initiated, and specific areas of a plant or process are selected, careful review of a plant's operation and water use and waste streams should lead to the selection of a number of options with the potential to minimize water use and waste generation. These options are screened using a material balance approach, and the most promising option(s) is (are) selected for implementation based on the cost effectiveness of these water use and waste reduction alternatives.

Illustrative Example 20.5

A large, deep cavern (formed from a salt dome) located north of Houston, Texas, has been proposed as an ultimate disposal site for both industrial and municipal wastewater sludge. Preliminary geological studies indicate that there is little chance that the wastewater sludge will penetrate the cavern walls and contaminate adjacent soil and aquifers. A risk assessment was also conducted during the preliminary study and the results indicate that there was a greater than 99% probability that no material would "meander" beyond the cavern walls during the next 25 years.

The company preparing the permit application for the Texas Water Pollution Board has provided the following data and information:

Approximate total volume of cavern = 0.78 mi^3
Approximate volume of cavern available for sludge disposal = 75% of total volume
Proposed maximum wastewater sludge feed rate to cavern = 20,000 lb/d
Feed rate schedule = 6 days/wk
Average bulk density of wastewater sludge = 60 lb/ft^3

Based on the above data, estimate the minimum time to fill the cavern with this wastewater sludge.

Solution

The volume of the cavern, V, in cubic miles available for the wastewater sludge is:

$$V = (0.75)\left(0.78 \text{ mi}^3\right) = 0.585 \text{ mi}^3$$

This volume is converted to cubic feet as:

$$V = \left(0.585 \text{ mi}^3\right)\left(5{,}280 \text{ ft/mi}\right)^3 = 8.61 \times 10^{10} \text{ft}^3$$

The daily volume of sludge disposed of within the cavern in ft^3/d, q, is:

$$q = \left(20{,}000 \text{ lb/d}\right)/\left(60 \text{ lb/ft}^3\right) = 333 \text{ ft}^3\text{/d}$$

The daily volume can now be converted to ft^3/year as:

$$q = \left(333\ \text{ft}^3/\text{day}\right)\ \left(6\text{d/wk}\right)\ \left(52\ \text{wk/yr}\right) = 103{,}896 = 104{,}000\ \text{ft}^3/\text{yr}$$

The time to fill the cavern is therefore:

$$t = V/q\ = \left(8.61 \times 10^{10}\text{ft}^3\right) / \left(104{,}000\ \text{ft}^3/\text{yr}\right) = 829{,}000\ \text{yr!}$$

The proposed operation could extend well beyond the 25 years upon which the risk assessment was based. Although this cavern system has adequate capacity for many centuries of waste sludge disposal, pollution prevention principles would suggest that alternative methods for sludge handling (i.e., source reduction or recycling) should be considered before the ultimate disposal option is selected. For wastewater sludges, recycling of this material for its nutrient and soil conditioning value could be a viable alternative to ultimate disposal as long as no toxic or hazardous constituents from the industrial sludge source are found in the final material. The decision whether to grant the permit is somewhat subjective because there is a finite, though extremely low probability that the cavern walls will be penetrated. Another, but more detailed and exhaustive, risk analysis study should be considered, both for cavern disposal as well as a sludge recycling option.

Note:

This sludge disposal method, deep-well injection, is an ultimate disposal method that transfers wastes far underground and away from freshwater sources. This disposal process has been used for many years by the petroleum industry. It is also used to dispose of salt water in oil fields. When the method first came into use in the oil fields, the injected brine would often eventually contaminate groundwater and freshwater sands because the sites were poorly chosen. The process has since been improved, and laws such as the Safe Drinking Water Act of 1974 ensure that sites for potential wells are better surveyed and installed to protect drinking water aquifers from contamination.

Many factors are considered in the selection of a deep-well injection site. For example, the rock formation surrounding the disposal zone must be rigid but permeable enough to absorb the waste, and the site must be far enough from drinking water sources to prevent contamination. Once a site is selected it must be tested by drilling a pilot well. The performance data from the pilot well, besides testing permeability and water quality, also aids in the design of the final well and in determining the proper injection rate.

Finally, the type of waste injected into the well is a determinant in how deep the injection will be made. The more toxic the waste, the deeper the disposal zone must usually be.

Illustrative Example 20.6

The Institute for Local Self-Reliance (1986) published a compendium of case studies highlighting the cost savings, resource conservation, and waste reduction benefits from implementing pollution prevention techniques in various industries. Select one of these case studies, describe the pollution prevention options implemented, and summarize the cost, water and waste savings that were realized.

Solution

This is an open-ended question. An example of a response for a Defluorinated Phosphate manufacturer Case Study 20 is provided. Many other water reduction–related examples can be found in this compendium.

Problem Description: Fluorides were being released to air and water from the manufacture of defluorinated phosphate. The plant combines phosphate ore with soda ash and phosphoric acid and processes the mixture through a kiln mixer to remove the fluorides and some sulfur compounds present in phosphate rock. The fluorides exit from the mixer as gases but are routed to a venturi scrubber before being discharged to the atmosphere. Therefore, the scrubber wastewater contains fluorides.

Pollution Prevention Technique(s) Used: A closed-loop process water-recycling system was installed to remove inorganic fluorides from the scrubber discharge stream, replacing the single-pass system.

Benefit(s): Approximately 280,000 gallons of water were saved per day. Combined annual savings of \$1.8 million in water purchases and waste disposal costs alone were achieved without compromising human health or environmental quality. Total annual savings were \$2 million, with an investment payback period of 1 year.!

REFERENCES

Dupont, R.R., K. Ganesan, and L. Theodore. 2016. *Pollution Prevention, Industrial Ecology, Green Science and Engineering, and Sustainability.* Boca Raton, FL: CRC Press/Taylor & Francis Group.

Institute for Local Self-Reliance. 1986. *Proven Profits from Pollution Prevention: Case Studies in Resource Conservation and Waste Reduction.* Washington, DC: Institute for Local Self Reliance. (http://ilsr.org/wp-content/uploads/1986/02/Proven-Profits-from-Pollution-Prevention.pdf).

Khanna, M., and A.L. Damon. 1997. EPA's Voluntary 33/50 Program: Impact on Toxic Releases and Economic Performance of Firms. *Journal of Environmental Economics and Management.* 37:1–25.

Theodore, L. 2014. *Chemical Engineering: The Essential Reference.* New York: McGraw-Hill.

Theodore, L., and Y.C. McGuinn. 1992. *Pollution Prevention.* New York: Van Nostrand Reinhold.

Theodore, M.K., and L. Theodore. 2010. *Introduction to Environmental Management*, 2nd edition. Boca Raton, FL: CRC Press/Taylor & Francis Group.

U.S. Environmental Protection Agency (U.S. EPA). 1986. *Report to Congress. Minimization of Hazardous Waste Appendices.* Washington, DC: Office of Solid Waste and Emergency Response. https://nepis.epa.gov/Exe/ZyPURL.cgi?Dockey=400011DV.TXT.

U.S. Environmental Protection Agency (U.S. EPA). 1988. *Waste Minimization Opportunity Assessment Manual.* Cincinnati, OH: Office of Research and Development, Hazardous Waste Engineering Research Laboratory.

U.S. Environmental Protection Agency (U.S. EPA). 1990. *Preventing Pollution through Efficient Water Use.* OW WH-556, OPPE PM-222. Washington, DC: Office of Water, Office of Pollution Prevention.

U.S. Environmental Protection Agency (U.S. EPA). 1993. *A Primer for Financial Analysis of Pollution Prevention Projects.* Washington, DC: Office of Research and Development.

U.S. Environmental Protection Agency (U.S. EPA). 1994. *33/50 Program Company Profiles: Reduction Highlights.* Washington, DC: Office of Pollution Prevention and Toxics.

U.S. Environmental Protection Agency (U.S. EPA). 1995. *An Introduction to Environmental Accounting As A Business Management Tool: Key Concepts and Terms.* Washington, DC: Office of Pollution Prevention and Toxics.

U.S. Environmental Protection Agency (U.S. EPA). 1998. *Water Recycling and Reuse: The Environmental Benefits.* San Francisco, CA: Water Division Region IX.

U.S. Environmental Protection Agency (U.S. EPA). 1999. *33/50 Program. The Final Record.* Washington, DC: Office of Pollution Prevention and Toxics.

U.S. Environmental Protection Agency (U.S. EPA). 2001. *Guide to Industrial Assessments for Pollution Prevention and Energy Efficiency.* Cincinnati, OH: Office of Research and Development, National Risk Management Research Laboratory.

U.S. Environmental Protection Agency (U.S. EPA). 2002. *Design for the Environment Projects.* Washington, DC: Office of Prevention, Pesticides and Toxic Substances.

U.S. Environmental Protection Agency (U.S. EPA). 2003. *Lean Manufacturing and the Environment: Research on Advanced Manufacturing Systems and the Environment and Recommendations for Leveraging Better Environmental Performance.* Washington, DC: Office of Solid Waste and Emergency Response. http://www.epa.gov/lean/environment/pdf/leanreport.pdf.

U.S. Environmental Protection Agency (U.S. EPA). 2007. *The Lean and Environment Toolkit.* Washington, DC: Office of Policy. http://www.epa.gov/lean/environment/toolkits/ environment/resources/LeanEnviro Toolkit.pdf.

U.S. Environmental Protection Agency (U.S. EPA). 2011. *Lean & Water Toolkit. Achieving Process Excellence Through Water Efficiency.* Washington, DC: Office of Policy. http://www.epa.gov/lean/environment/toolkits/water/resources/lean-water-toolkit.pdf.

U.S. Environmental Protection Agency (U.S. EPA). 2014. *Priority Pollutant List.* Washington, DC: Office of Water. https://www.epa.gov/sites/production/files/2015-09/documents/priority-pollutant-list-epa.pdf.

U.S. Environmental Protection Agency (U.S. EPA). 2015. *U.S. Environmental Protection Agency Supplemental Environmental Projects Policy 2015 Update.* Memo dated March 10, 2015, from Cythia Giles, Assistant Administrator to Regional Administrators. Washington, DC: Office of Enforcement and Compliance Assurance. https://www.epa.gov/sites/production/files/2015-04/documents/sepupdatedpolicy15.pdf.

21 Sustainability

21.1 INTRODUCTION

The term *sustainability* has many different meanings to different people. To sustain is defined as to "support without collapse" (Theodore and Theodore 2010, Shipka and Theodore 2014). Discussion of how sustainability should be defined was initiated by the Brundtland Commission. This group was assigned a mission to create a "global agenda for change" by the General Assembly of the United Nations in 1984. They defined *sustainable* very broadly (Theodore and Theodore 2010): Humanity has the ability to make development sustainable—to ensure that it meets the needs of the present without compromising the ability of future generations to meet their own needs. In a very real sense, sustainability involves simultaneous progress in four major areas: human, economic, technological, and environmental. The United Nations (1987) defined sustainable development as development that meets the need of the present without compromising the ability of future generations to meet their own needs. Sustainability requires conservation of resources, minimizing depletion of nonrenewable resources, and using sustainable practices for managing renewable resources. There can be no product development or economic activity of any kind without available resources. Except for solar energy, the supply of resources is finite. Efficient designs conserve resources while also reducing impacts caused by material extraction and related activities. Depletion of nonrenewable resources and overuse of otherwise renewable resources limits their availability to future generations.

Water is truly at the core of sustainable development. It is inextricably linked to climate change, agriculture, food security, health, equality, gender, and education, and there is already international agreement that water and sanitation are essential to the achievement of many sustainable development goals.

Another principal element of sustainability is the maintenance of ecosystem structure and function. Because the health of human populations is connected to the health of the natural world, the issue of ecosystem health is a fundamental concern to sustainable development. Thus, sustainability requires that the health of all diverse species as well as their interrelated ecological functions be maintained. As only one species in a complex web of ecological interactions, humans cannot separate their survivability from that of the total system.

This chapter provides a general overview of sustainability concepts that can be applied to water resource management. Chapter topics include a historical perspective on sustainability issues, the topic of resource limitations, sustainable development considerations, ways to benchmark sustainability, resources to support moves toward sustainability, and future trends in sustainable development and assessment. An application section provides five Illustrative Examples exploring sustainability concepts.

As a side note, to add perspective to this issue, sustainability in many ways conflicts with conventional changes expected in the socioeconomic landscape, politics, human evolution, and inevitable basic principles of entropy as well as other thermodynamic principles (Theodore et al. 2008).

21.2 HISTORICAL PERSPECTIVE

To develop an understanding of why sustainability is a topic of urgency today, one should understand the history behind it. As agriculture developed, social structure supporting agriculture grew as well. Social stratification became increasingly widespread as humanity proceeded from agriculture to industry. Eventually, a new class-based society led to differences in standards of living between the rich and the poor. As population grew, technical development spiraled up as well. The increase in demand for goods and more powerful machines led to increased extraction of natural resources at the expense of the environment. Environmental effects built up slowly, gaining momentum as the problem worsened. Due to the populations' uncertainty and limited understanding when a problem is finally identified, it is often so bad that even an immediate response may not be able to solve it. Examples of such lag and momentum have been exhibited by damage to the ozone layer and global warming (Bishop 2000).

Activity in the sustainability area was born with the World Commission on Environment and Development (WCED). It was formally known as the Brundtland Commission (named after its Chair Gro Harlem Brundtland) and was convened by the United Nations in 1984. The commission was created to address growing concerns "about the accelerating deterioration of the human environment and natural resources and the consequences of that deterioration for economic and social development." In establishing the commission, the UN General Assembly recognized that environmental problems were global in nature. It was determined that it was in the common interest of all nations to establish policies for sustainable development (United Nations 1987).

Later, the United Nations Conference on Environment and Development, also known as the Earth Summit, was held in Rio de Janeiro in June 1992. A total of 178 governments participated, with 118 sending their heads of state or government (Schneider 1992). Some 2,400 representatives of nongovernmental organizations (NGOs) attended, with 17,000 people at the parallel NGO Forum, who had the so-called Consultative Status, also present. One of the issues addressed, which deals with carbon dioxide related global warming, was development of alternative sources of energy to replace the use of fossil fuels which are linked to global climate change. An important achievement of the Summit was an agreement on a Climate Change Convention, which in turn led to the Kyoto Protocol. The Earth Summit resulted in the following documents: Rio Declaration on Environment and Development; Agenda 21; Convention on Biological Diversity; Forest Principles; and, Framework Convention on Climate Change (Schneider 1992). The trends in *Sustainable Development Report*, published by the United Nations Department of Economic and Social Affairs, highlighted key developments and recent trends in the areas of energy for sustainable development, industrial development, atmosphere/air pollution, and a host of other related topics. The 2018 edition of this *Sustainable Development Goals Report* (United Nations 2018) is available online at https://unstats.un.org/sdgs/files/report/2018/TheSustainableDevelopmentGoalsReport2018-EN.pdf.

21.3 RESOURCE LIMITATIONS

Most have defined the Earth as consisting of four parts:

1. Atmosphere
2. Lithosphere
3. Hydrosphere
4. Barysphere

The atmosphere is the gaseous envelope that surrounds the solid body of the planet. The lithosphere is the solid rocky crust of the earth, extending to a depth of perhaps 40 km (25 miles). The hydrosphere is the layer of water, in the form of the oceans, which covers approximately 70% of the surface of the earth. The barysphere, sometimes called the *centrosphere*, is below the lithosphere. It is the heavy interior of the Earth constituting more than 99.6% of the Earth's mass. From a sustainability prospective, the two major resources available to humans are the lithosphere, and to a lesser degree the hydrosphere. These two resources are finite and for all intents and purpose are nonrenewable. Both are briefly discussed.

The rocks of the lithosphere primarily consist of 11 elements, which together account for about 99.5% of its mass. The most abundant is oxygen (about 46.60% of the total), followed by silicon (about 27.72%), aluminum (8.13%), iron (5.0%), calcium (3.63%), sodium (2.83%), potassium (2.59%), magnesium (2.09%), and titanium, hydrogen, and phosphorus (totaling less than 1%). In addition, 11 other elements are present in trace amounts of 0.1% to 0.02%. These elements, in order of abundance, are carbon, manganese, sulfur, barium, chlorine, chromium, fluorine, zirconium, nickel, strontium, and vanadium. The elements are present in the lithosphere almost entirely in the form of compounds rather than in their free state. The most common compounds of the Earth's crust are silicates and aluminosilcates of various metals. In addition, the surface of the earth is largely covered with sedimentary rocks and soil.

The hydrosphere consists chiefly of the oceans but technically includes all water surfaces in the world, including inland seas, lakes, rivers, and groundwater.

Traditionally, humans have viewed Earth's resources as a source of economic wealth—water, minerals, food, forests, and land on which to place buildings and other structures. These were looked upon as assets to be exploited, not necessarily as precious attributes to be used sustainably and preserved insofar as possible. The loss of these resources would be catastrophic. For example, the loss of Earth's food productivity would certainly adversely affect sustainability and, in the worst case, could lead to massive starvation of human populations. Although a number of human activities have adversely affected food productivity, these effects have been largely masked by remarkable advances in agriculture, including increased use of industrial-scale fertilizer, development of highly productive hybrid crops, and widespread irrigation.

In addition to food, humans obtain shelter, health, security, mobility, and other necessities through activities involving resources that are carried out by individuals, businesses, and government entities. By their very nature, these use resources (renewable and nonrenewable) and all tend to produce wastes.

The "energy resource" is a sustainability topic within itself. Consider fossil fuels. One of the greatest challenges facing humanity during the twenty-first century will surely be that of providing everyone on the planet access to safe, clean, and sustainable energy supplies. The use of energy has been central to the functioning and development of human societies throughout history. However, in recent years fossil fuel energy usage and pricing has run amuck. World petroleum resources currently appear to be abundant with the recent accelerated expansion of oil and gas extraction from alternative oil supplies (i.e., tar sands, shale) and the use of new extraction techniques (i.e., fracking). Despite very recent painfully high prices for petroleum (the price of crude oil exceeded $100/barrel in 2014), the price of crude oil at the time of the preparation of this manuscript was below $67/barrel due to this expansion of natural gas and crude oil supplies. Furthermore, the International Energy Agency projected that more than 80% of the world energy demand will continue to be met by fossil fuels through 2030. Fossil fuels will be a major energy source into the near future, and therefore, there is an immediate need to increase the present efficiency of fossil fuel usage to reduce its impact on human health and the environment (i.e., attempt to make this fuel more sustainable. This can include:

1. Increasing the mileage efficiency of transportation sources
2. Improving the energy efficiency of new power plants
3. Developing "green buildings" and sustainable communities

As noted previously, natural resources were initially abundant relative to needs. In the earlier years of the Industrial Revolution, production was limited by technology and labor. However, population is in surplus and technology has reduced the need for human labor. Increasingly, production is becoming limited by the Earth's natural environment that includes the availability of natural resources. The demand for most resources has increased at a near exponential rate. The emergence of newly developing economies, particularly those in the highly populated countries of China and India, has further increased the demand for resources. Humans need to realize that reduced material demand, particularly those from nonrenewable sources, is essential to sustainability. There are some elaborate changes in place to reduce material demand and the potential exists for much greater reductions. Naturally, wherever possible, materials should come from renewable sources and materials should be reused, and then recycled insofar as possible.

21.4 SUSTAINABLE DEVELOPMENT CONSIDERATIONS

Water is at the core of sustainable development. Water resources, and the range of services they provide, underpin economic growth, poverty reduction and environmental sustainability. From food and energy security to human and environmental health, water has been shown to contribute to improvements in social well-being, affecting the livelihoods of billions. Progress toward the achievement of most sustainable development goals requires significant improvement of water management across the globe.

Sustainable development demands change. Consumption of energy, natural resources, and products must eliminate waste. The manufacturing industry must develop green products that can meet sustainability requirements. Life cycle analysis, design for environment, and toxic use reduction are elements that support sustainability. Sustainable manufacturing, for example, extends the responsibility of industry into material selection, facility and process design, marketing, cost accounting, and waste disposal. Extending the life of a manufactured product is likely to minimize waste generation. Design engineers must consider many aspects of their products including their durability, reliability, re-manufacturability, and adaptability.

Designing a product that can withstand wear, stress, and degradation extends its useful life. This, in many cases, reduces the cost and impact on the environment. Reliability is the ability of a product or system to perform its function for the length of an expected period under the intended conditions of use. Reducing the number of components in a system and simplifying the design can enhance its reliability. Screening out potentially unreliable parts and replacing them with more reliable parts helps to increase overall system reliability. Increases in reliability reduce failures and enhance a system's sustainability.

Adaptable designs rely on interchangeable parts. With adaptable designs, products are designed to allow consumers to upgrade components as needed to maintain state-of-the-art performance. In remanufacturing, used worn products are restored to "like-new" condition. Thus, remanufacturing minimizes the utilization of virgin materials and the generation of waste. Products that are expensive but not subject to rapid change are the best candidates for remanufacturing. Design continuity between models in the same product line increases interchangeable parts. The parts must be designed for easy disassembly to encourage remanufacturing.

Design of products that emphasizes efficient use of energy and materials, and reuse and recycling of manufactured materials reduces waste and supports sustainability. By effective recycling, material life can be extended. Materials can be recycled through open-loop or closed-loop pathways. Postconsumer material is recycled in an open loop one or more times before disposal. However, in a closed-loop pathway, such as with solvents or water, materials within a process are recovered and used as substitutes for virgin material. Minimizing the use of virgin materials supports sustainability. Thus, resource conservation can reduce waste and directly lower environmental impact. Manufacturing a less material-intensive product not only saves materials and energy but will also be lighter, thus reducing energy and costs related to product transportation and use.

Process modifications and alterations specifically focused on replacing toxic materials with more benign ones minimize the health risk and the environmental impact of a process and/or product and improves the safety of employees during manufacturing. Process redesign may also yield "zero discharge" by completely eliminating waste production. Thus, sustainability can be accomplished through many different approaches. Evaluating these options upfront will aid in developing truly sustainable processes and products and is much more desirable than implementing control measures after unacceptable waste releases occur.

Current design practices for suitability projects usually fall into the category of state-of-the-art and pure empiricism. Past experience with similar applications is commonly used as the sole basis for the design procedure. In designing a new process, historical records are consulted for similar applications and old designs are heavily relied upon. By contrast, the engineering profession in general, and the chemical engineering profession in particular, has developed well-defined procedures for the design, construction, and operation of chemical plants. These techniques, tested and refined for better than a half-century, are routinely used by today's engineers. These same procedures should be used in the design of sustainable "facilities" (Reed 2001, Metzger and Salmoned 2004).

Finally, responsible businesses can begin moving toward sustainability by taking seven "simple" steps:

1. Foster a company culture of sustainability.
2. Initiate voluntary performance improvements.
3. Apply eco-efficiency (material, water and energy conservation, toxic use reduction, recycling, etc.) concepts.
4. Grasp opportunities for sustainable business growth.
5. Invest in creativity, innovation, and technology for the future.
6. Reward employee commitment and action.
7. Encourage thinking "outside the box."

21.5 BENCHMARKING SUSTAINABILITY

A variety of sustainability indices have been published that mostly measure a companies' corporate responsibility and environmental performance (Cobbetal 2007). Starting in 2001, the American Institute of Chemical Engineers (AIChE) decided to strike out on a new strategic direction and a number of new initiatives were begun. These new areas included biotechnology, materials technology, and sustainable development. The AIChE ultimately formed the Institute for Sustainability (IfS) in 2004 to promote the societal, economic, and environmental benefits of sustainable and green engineering. IfS serves the needs, and influences the efforts of professionals in industry, academia, and government. Scientists and engineers working with IfS have defined sustainability as the "path of continuous improvement, wherein the products and services required by society are delivered with progressively lesser negative impacts upon the Earth."

IfS established an industry group, Center for Sustainable Technology Practices (CSTP), to address practical issues of sustainability implementation with member companies, including BASF, Dow, Cytec, Honeywell, DuPont, Air Products, FMC, and Shell. One area of focus for CSTP is the development of a Sustainability Roadmap, which is designed to improve decision making relative to sustainability, and is expressed as a Sustainability Index (SI).

The AIChE SI is composed of seven critical elements:

1. Strategic commitment to sustainability
2. Safety performance
3. Environmental performance
4. Social responsibility
5. Product stewardship
6. Innovation
7. Value-chain management

Details are provided in Table 21.1.

A related sustainability metric recently developed specifically for physical infrastructure is Envision? (https://sustainableinfrastructure.org/envision/), a joint collaboration between the Zofnass Program for Sustainable Infrastructure at the Harvard University Graduate School of Design and the Institute for Sustainable Infrastructure. Envision? is an online infrastructure project rating system that assigns credit to sustainability related approaches in project planning, design, construction, and operation across five categories of major project impact that include Quality of Life, Leadership, Resource Allocation, Natural World, and Climate and Resilience. Table 21.2 expands

TABLE 21.1
Examples of Indicator Areas in AIChE's SI

Indicator Area	Components
Strategic Commitment to Sustainability	Stated commitment
	Presence and extent of sustainability goals
Safety Performance	Process safety
	Employee safety
Environmental Performance	Resource use
	Waste and emissions (including greenhouse gases)
	Compliance history
Social Responsibility	Community investment
	Stakeholder partnership and engagement
Product Stewardship	Product safety and environmental assurance process
	System in place for compliance with emerging regulations (e.g., research)
Innovation	R&D in place to address societal needs (e.g., Millennium Development Goals)
	Integration of sustainability concepts and tools in R&D new products related to sustainability
Value Chain Management	Environmental management systems
	Supplier standards and management process

Abbreviations: AIChE = American Institute of Chemical Engineers; R&D = research and development; SI = Sustainability Index.

TABLE 21.2
Examples of Sustainability Criteria in ISI's Envision™ Sustainability Metric

Sustainability Category	Example Sustainability Criteria
Quality of Life	Improve Community Quality of Life
	Enhance Public Health and Safety
	Improve Community Mobility and Access
	Advance Equity and Social Justice
Leadership	Foster Collaboration and Teamwork
	Provide for Stakeholder Involvement
	Plan for End of Life
Resource Allocation	Use Recycled Materials
	Reduce Operational Energy Consumption
	Preserve Water Resources
	Reduce Operational Water Consumption
Natural World	Preserve Sites of High Ecological Value
	Preserve Undeveloped Land
	Manage Stormwater
	Protect Surface and Groundwater Quality
	Enhance Functional Habitats
Climate and Resilience	Reduce Greenhouse Gas Emissions
	Reduce Air Pollutant Emissions
	Assess Climate Change Vulnerability
	Improve Infrastructure Integration

Abbreviation: ISI = Institute for Sustainable Infrastructure.

on these project impact category characteristics that contribute to the expected sustainability of civil works projects and highlight the human, economic, technological, and environmental dimensions of sustainable systems. Envison™ is a free, voluntary guide to enhancing the sustainability of infrastructure investments and can be used with optional third-party verification to pursue ISI recognition of project sustainability achievements (ISI 2019) (Table 21.2).

21.6 RESOURCES FOR SUSTAINABILITY

Some key website, software, and document resources for sustainability include:

1. National Institute of Standards and Technology's (NIST) Building for Environmental and Economic Sustainability (BEES) Lifecycle Tool. https://www.nist.gov/services-resources/software/bees.
2. U.S. Environmental Protection Agency's Tool for the Reduction and Assessment of Chemical and Other Environmental Impacts (TRACI) Tool. https://www.epa.gov/chemical-research/tool-reduction-and-assessment-chemicals-and-other-environmental-impacts-traci.
3. *The Ecology of Commerce: A Declaration of Sustainability.* Paul Hawken, Harper Business, 1994.
4. *Industrial Ecology: An Introduction.* University of Michigan's National Pollution Prevention Center for Higher Education. http://www.umich.edu/~nppcpub/resources/compendia/ind.ecol.html.
5. *Industrial Ecology and "Getting the Prices Right."* Resources for the Future https://www.resourcesmag.org/archives/industrial-ecology-and-quotgetting-the-prices-rightquot/.
6. *Journal of Industrial Ecology.* Yale School of Forestry & Environmental Studies. https://onlinelibrary.wiley.com/journal/15309290.
7. *Mid-Course Correction: Toward a Sustainable Enterprise: The Interface Model.* Ray Anderson. Peregrinzilla Press, 1999.
8. *Natural Capitalism: Creating the Next Industrial Revolution.* Paul Hawken, Amory Lovins, and L. Hunter Lovins. Rocky Mountain Institute, 1999. www.naturalcapitalism.org.
9. *The Next Bottom Line: Making Sustainable Development Tangible.* World Resources Institute. www.wri.org.
10. *The NEXT Industrial Revolution.* The Atlantic Monthly. October 1998. www.theatlantic.com/issues/98oct/industry.htm.
11. *Envision™ sustainable infrastructure rating system.* Institute for Sustainable Infrastructure. www.sustainableinfrastructure.org.

21.7 FUTURE TRENDS

The fact is there is enough water available to meet the world's growing needs but not without dramatically changing the way water is used, managed, and shared. The global water crisis is one of governance, much more than of resource availability, and this is where the bulk of the action is required to achieve a water secure world.

Over the next 50 years, projections suggest that the world's population could increase by 50%. Global economic activity is expected to increase by 500%. Concurrently, global energy consumption and manufacturing activity are likely to rise to three times current levels. These trends could have serious social, economic, and environmental consequences unless a way can be found to use fewer resources in a more efficient way. The task ahead is to help shape a sustainable future in a cost-effective manner, recognizing that economic and environmental considerations, supported by innovative science and technology, can work together and promote societal benefits. However, unless humans embrace sustainability, they will ultimately deplete Earth's resources and damage its environment to an extent that conditions for human existence on the planet will be seriously compromised and even human survival may become impossible.

As stated previously, sustainable development is feasible but requires a significant change both in consumption patterns toward environmentally more benign products and in investment patterns. It will require a shift in the way economic progress is pursued. Environmental concerns must be properly integrated into rearrangement policies and the environment must be viewed as an integral part of human well-being.

Finally, some very pivotal steps that must be taken in the near future must include implementing greatly improved technologies for harnessing fossil and nuclear fuels to ensure that their use, if continued, creates much lower environmental and societal impact; developing and deploying renewable energy sources on a much wider scale; and making major improvements in the efficiency of energy conversion, distribution and use (Reed 2001, Metzger and Salmoned 2004).

21.8 APPLICATIONS

These Illustrative Examples complement the material presented in this chapter related to the concepts and practices of sustainability.

Illustrative Example 21.1

Discuss the connection between sustainability and ecosystem structure and function.

Solution

A principal element of sustainability is the maintenance of ecosystem structure and function. Because the health of human populations is connected to the health of the natural world, the issue of ecosystem health is a fundamental concern to sustainable development. Sustainability requires that the health of all diverse species as well as their interrelated ecological functions be maintained. As only one species in a complex web of ecological interactions, humans cannot separate our survivability from that of the total system.

Illustrative Example 21.2

Describe the role of water in a sustainable world in the future.

Solution

Unsustainable development approaches have affected the quality and availability of water resources throughout the world, compromising their capacity to generate social and economic benefits. Demand for fresh water is growing, and a balance between demand and finite high-quality water supplies must be restored or the world will face an increasingly severe global water deficit.

In a sustainable world, water and related resources would be managed in support of human well-being and ecosystem integrity in a robust economy. Sufficient and safe water would be made available to meet every person's basic needs, with reliable and affordable water supply and sanitation services supported by equitably extended and efficiently managed infrastructure. Water resources management, infrastructure, and sanitary and potable water service delivery would be sustainably financed. Water would be appropriately valued in all its forms, with wastewater treated as a resource that supplies energy, nutrients, and fresh water for reuse. Human development would coexist in harmony with the natural water cycle and the ecosystems that support it, with measures in place that reduce water supply vulnerability and improve infrastructure resilience to water-related disasters. Integrated approaches to water resources development, management, and equitable availability and use would be the norm. Water resources would be managed in a participatory way that draws on the full potential of all people as professionals and citizens, guided by a number of able and knowledgeable organizations, within a just and transparent institutional framework.

Illustrative Example 21.3

A nickel electroplating line uses a dip-rinse tank to remove excess plating metals from parts. Currently, a single tank is used that requires R gal/hr of fresh rinse water to clean F parts/hr (see Figure 21.1). Assume the cleaning is governed by the following equilibrium relation:

$$\lambda = \frac{f_i}{r_i} = \frac{\text{ounces of metal residue / part}}{\text{ounces of metal residue / gal bath}} \quad (21.1)$$

Calculate the reduction in rinse, water flowrate (a sustainability metric) if a two-stage countercurrent rinse tank is used (as compared to the single-stage unit), and 99% of the residue must be removed. Assume the drag-out volume is negligible.

Solution

Using the flow diagram provided in Figure 21.1, a material balance for the residue may be written as:

$$f_{in}F + r_{in}R = f_1F + r_1R \quad (21.2)$$

This equation may be rearranged in terms of λ. For the single-stage operation, set $i = 1$:

$$R / F = (f_{in} - f_1) / r_1 = \lambda (f_{in} - f_1)/f_1 \quad (21.3)$$

The fraction of residue removed, x, is:

$$x = (f_{in} - f_1)/f_{in} \quad (21.4)$$

Now R/F may be expressed in terms of λ and x as:

$$R / F = \lambda\, [x / (1 - x)] \quad (21.5)$$

The flow diagram for a two-stage (Stage 1, Stage 2) countercurrent operation is provided in Figure 21.2. Material balances on the residue for each stage can be written as follows:

$$\text{Stage 1: } f_{in}F + r_2R = f_1F + r_1R \quad (21.6)$$

$$\text{Stage 2: } f_1F + r_{in}R = f_2F + r_2R \quad (21.7)$$

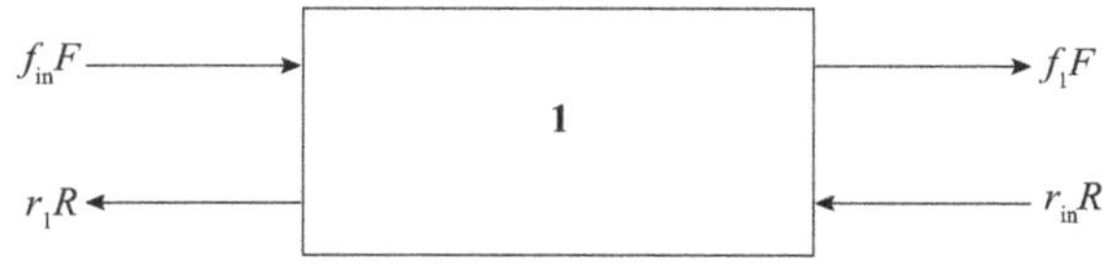

FIGURE 21.1 Flow diagram for one-stage dip-rinse tank for Illustrative Example 21.3.

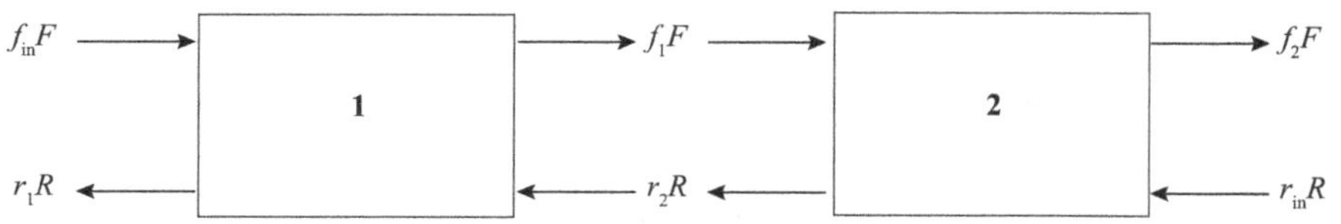

FIGURE 21.2 Flow diagram for two-stage dip-rinse operation for Illustrative Example 21.3.

Each of these equations may be solved for R/F in terms of λ and f_i using the defining equation for λ as:

$$\text{Stage 1: } R/F = \lambda\left(f_{in} - f_1\right)/\left(f_1 - f_2\right) \tag{21.8}$$

$$\text{Stage 2: } R/F = \lambda\left(f_1 - f_2\right)/f_2 \tag{21.9}$$

If x is substituted into the mass balance expressions in the form of Equation 21.4, the following equations can be written:

$$\text{Stage 1:} \frac{R}{F} = \frac{f_{in} - f_1}{f_1 - f_{in}\left(1 - x\right)} \tag{21.10}$$

$$\text{Stage 2:} \frac{R}{F} = \frac{f_1 - f_{in}\left(1 - x\right)}{f_{in}\left(1 - x\right)} \tag{21.11}$$

The right-hand sides (RHS) of the Stage 1 and Stage 2 equations may be set equal to each other to yield:

$$\frac{f_{in} - f_1}{f_1 - f_{in}\left(1 - x\right)} = \frac{f_1 - f_{in}\left(1 - x\right)}{f_{in}\left(1 - x\right)} \tag{21.12}$$

This equation can be rearranged to obtain a quadratic equation in f_1 as:

$$f_1^2 - \left[\left(1 - x\right)f_{in}\right]f_1 - \left[x\left(1 - x\right)f_{in}^2\right] = 0 \tag{21.13}$$

Solving for f_1 yields:

$$f_1 = (f_{in}\left(1 - x\right) \pm \left\{\left[-f_{in}\left(1 - x\right)\right]^2 - 4\left(1\right)x\left(1 - x\right)f_{in}^2\right\}^{0.5} \Big/ 2 \tag{21.14}$$

This can be rewritten as:

$$f_1 = \left(f_{in}/2\right)\{1 - x \pm \left[\left(3x + 1\right)\left(1 - x\right)\right]^{0.5} \tag{21.15}$$

The following expression results if (1–x) is factored out:

$$f_1 = \left(\frac{f_{in}\left(1 - x\right)}{2}\right)\left[1 \pm \left(\frac{3x + 1}{1 - x}\right)^{0.5}\right] \tag{21.16}$$

Only the "+" term is physically reasonable; therefore,

$$f_1 = \left(\frac{f_{in}\left(1 - x\right)}{2}\right)\left[1 + \left(\frac{3x + 1}{1 - x}\right)^{0.5}\right] \tag{21.17}$$

The term f_1 may be substituted into the Stage 2 mass balance equation as:

$$\text{Stage 2:} \frac{R}{F} = \lambda\left\{\left(\frac{f_{in}\left(1 - x\right)}{2}\right)\left[1 + \left(\frac{3x + 1}{1 - x}\right)^{0.5}\right] - f_2\right\}/f_2 \tag{21.18}$$

However, f_2 can be expressed in terms of x and f_{in}:

$$f_2 = f_{in}(1-x) \tag{21.19}$$

This can then be substituted into Equation 21.18 to yield:

$$\text{Stage 2:} \frac{R}{F} = \frac{\lambda\left\{\left(\frac{f_{in}(1-x)}{2}\right)\left[1+\left(\frac{3x+1}{1-x}\right)^{0.5}\right]-f_{in}(1-x)\right\}}{f_{in}(1-x)} \tag{21.20}$$

which reduces to:

$$\text{Stage 2:} \frac{R}{F} = \left(\frac{\lambda}{2}\right)\left[\left(\frac{3x+1}{1-x}\right)^{0.5}-1\right] \tag{21.21}$$

The rinse water requirements for both a single- and a two-stage countercurrent unit becomes:

Single Stage:

$$R = \lambda\left(\frac{0.99}{1-0.99}\right)F = 99.0\lambda F \tag{21.22}$$

Two Stage:

$$R = \left(\frac{\lambda}{2}\right)\left[\left(\frac{3(0.99)+1}{1-0.99}\right)^{0.5}-1\right]F = 9.46\ \lambda F \tag{21.23}$$

The rinse water flowrate reduction is therefore:

$$\frac{99.0\lambda F - 9.46\lambda F}{(99.0\lambda F)} = 0.904 = 90.4\%\ \text{reduction}$$

Illustrative Example 21.4

Refer to Illustrative Example 21.3. Has the total metal content of the exit rinse water been altered?

Solution

For fixed residue removal, the total mass of metals in the rinse water will be same; with reduced water duty, the metals concentration increases. A smaller volume now needs to be processed (or disposed of).

Illustrative Example 21.5

Refer to Illustrative Example 21.4. Discuss implications for further wastewater treatment/reuse.

Solution

With reduced water duty metals concentrations do increase. If the metals concentrations are sufficiently high, the rinse water can economically and sustainably be returned to the plating bath for reuse or can be recovered.

REFERENCES

Bishop, R.L. 2000. *Pollution Prevention.* Prospect Heights, IL: Waveland Press.

Cobbetal, C. 2007. *Benchmarking Sustainability.* New York: CEP.

Institute for Sustainable Infrastructure (ISI). 2019. *Envision Version 3.* Washington, DC: Institute for Sustainable Infrastructure. www.sustainableinfrastructure.org.

Metzger, B., and D. Salmoned. 2004. Managing for Sustainability. *EM Magazine* 6:26.

Reed, D. 2001. *Stalking the Elusive Business Case for Corporate Sustainability.* Washington, DC: World Resources Institute.

Schneider, K. 1992. White House Snubs U.S. Envoy's Plea to sign Rio Treaty. *New York Times*, June 5, 1992.

Skipka, K. and L. Theodore. 2014. *Energy Resources: Availability, Management, and Environmental Issues.* Boca Raton, FL: CRC Press/Taylor & Francis Group.

Theodore, M. K., and L. Theodore. 2010. *Introduction to Environmental Management.* Boca Raton, FL: CRC Press/Taylor & Francis Group.

Theodore, L., F. Ricci, and T. VanVliet. 2008. *Thermodynamics for the Practicing Engineer.* Hoboken, NJ: John Wiley & Sons.

United Nations. 1987. *Report of the World Commission on Environment and Development. General Assembly Resolution 42/187,* December 11. New York: United Nations.

United Nations. 2018. *The Sustainable Development Goals Report 2018.* Department of Economic and Social Affairs. New York: United Nations. https://unstats.un.org/sdgs/files/report/2018/TheSustainableDevelopmentGoalsReport2018-EN.pdf.

22 The Role of Optimization

22.1 INTRODUCTION

This chapter is concerned with the role of optimization in water management studies. It attempts to present mathematical optimization procedures that are available to the practicing engineer—particularly as they apply to the environment. The presentation includes both a traditional and a linear programming approach to solving water management-related problems.

Investing in all aspects of water resource management, services, and infrastructure (development, operation, and maintenance) can generate significant social and economic benefits. The authors believe that appropriating capital toward drinking water supply, water management, and sanitation is highly cost effective from a health perspective. Investments in disaster and antiterrorism preparedness, improved water quality, and the reduction/elimination of wastewater and stormwater management requirements are also highly cost effective.

Traditional mathematical assessments of water resources are often inadequate for addressing the optimization of water utilization and demands. Optimization studies are necessary to make informed investment and management decisions, facilitate cross-sector decision-making, and address compromises and trade-offs among the multitudes of stakeholder groups. Increased financial investment and support for water "development" is essential. When this support is efficiently managed and optimized, these investments can generate social, economic, financial, and other benefits that greatly outweigh the costs of this process. In addition to adequately financing the development, operation, and maintenance of infrastructure, increased funding is also needed to develop institutional capacity and ensure well-functioning governance structures. Greater financial support is equally essential to enhance the knowledge base and provide reliable and objective information about the state of water resources. Their use can help water managers, engineers, and administrators inform decision-makers about policy/process options and their associated potential impacts. Finally, one's ability to understand optimization and any associated risks is the key to solving some of the more complex water and water-related problems in the future.

As noted, this chapter attempts to provide a description of elementary optimization processes that can be applied to water resource management. Hopefully, it will allow water resource management administrators and technical personnel to manage water more efficiently in the future. Chapter topics included are an introduction to the optimization process, the history of optimization, the scope of optimization, general analytical formulation of optimization, traditional optimization applications, and linear programming applications. An application section provides six Illustrative Examples on traditional applications, and three Illustrative Examples of linear programming applied to water management–related problems.

22.2 INTRODUCTION TO THE OPTIMIZATION PROCESS

The optimization process has been described by Aris as "getting the best you can out of a given situation" (Aris 1964). Problems amenable to solutions by mathematical optimization techniques generally have one or more independent variables whose values yield one or more viable solutions. In addition to the optimization definition presented by Aris, one might also offer the following

Contributing Author: Ann Marie Flynn

generic definition for engineers: "Optimization is concerned with determining the "best" solution to a given problem" (Theodore 2011). Alternately, a dictionary would offer the following definition: "to make the most of... develop or realize to the utmost extent... often the most efficient or optimum use of." Mathematical optimization techniques are also used for guiding the problem solver to the choice of variables that maximizes a "goodness" measure (e.g., profit) or that minimizes a "badness" measure (e.g., cost).

As noted in its most elementary and basic form, one may say that optimization is concerned with the determination of the "best" solution to a given problem. This process is required in the solution of many general problems in engineering and applied science in the maximization (or minimization) of a given function(s), in the selection of a control variable to facilitate the realization of a desired condition, in the scheduling of a series of operations or events to control completion dates of a given project, or in the development of optimal layouts of organizational units within a given design space, etc.

One of the more important areas for the application of mathematical optimization techniques is in engineering design. Once a particular subject or process scheme has been selected for study, it is common practice to optimize the process from a capital cost and operation and maintenance (O&M) standpoint. In chemical engineering design (some of the author's area of expertise) optimization applications include:

1. Generation of best functional representation (e.g., curve fitting)
2. Design of optimal control systems
3. Determining the optimal height (or length) of a mass (pollutant) transfer control (or recovery) unit
4. Determining the optimal diameter of a unit
5. Finding the best equipment material of construction
6. Generating operating schedules
7. Selecting operating conditions

The process of optimization in water management and environmental practice is required in the solution of many problems and often involves the maximization or minimization of a mathematical function. As one might suppose, many of the applications involve economic considerations. Overall, optimization is viewed by many as a tool in decision-making. It often aids in the selection of values that allow the practicing engineer to better solve a problem.

There are many optimization procedures available, most of them too detailed for meaningful application in a text of this nature. These sophisticated optimization techniques, some of which are routinely used in the design of conventional chemical and petrochemical plants, invariably involve computer calculations. Although the use of these techniques in the majority of water management and environmental applications was not warranted in the past, more and more real-world problems are requiring the use of such optimization techniques.

In almost every situation, optimum conditions can be reduced to a consideration of costs or profits. Thus, an optimum design could be based on conditions giving the least cost per unit of time, or the maximum profit per unit of production. When one design variable is changed, it is often found that some costs increase and others decrease. Under these conditions, the total cost may go through a minimum at one value of a design variable. This value would be considered as optimum.

Optimum design can take many forms. One example is in fluid flow, where piping and pipe fittings can amount to a significant portion of the total investment costs for a plant (Theodore 2007, 2014). It is necessary, therefore, to choose pipe sizes, which yield a minimum total cost for pumping and fixed charges. For any given set of flow conditions, the use of an increased pipe diameter will increase fixed costs for the piping system, while decreasing pump or blower operating costs. Therefore, an optimum economic pipe diameter must exist. The value of this optimum diameter can be determined by combining the principles of fluid mechanics with cost considerations. The optimum economic pipe diameter is found at the point at which the sum of pump or blower operating costs and fixed costs is a minimum.

A second example is in heat exchanger design, where the objective is to obtain a specified duty with water as the cooling medium. The temperature of the cooling water affects the log mean temperature-difference (LMTD) driving force for heat transfer. The cooling water can either be circulated at a high rate with a small change in water temperature, or at a low rate with a large change in water temperature (Flynn et al. 2019; Theodore 2010). An increased water flow rate will cause a reduction in the heat transfer area, which will result in a decrease in the capital investment (and fixed costs). On the other hand, water costs increase if more water is used. An economic balance between conditions of high water rate/low surface area and low water rate/high surface area is based on the fact that an optimum flow rate of cooling water exists. It occurs at the point of minimum total cost for cooling water and equipment fixed costs. Furthermore, increased fluid velocities result in larger heat-transfer coefficients and consequently, less heat-transfer area and heat exchanger costs for a given rate of heat transfer. On the other hand, the increased fluid velocities cause an increase in pressure drop and greater pumping costs. The optimum economic design occurs at the conditions where the total cost is a minimum for a given duty. The basic problem, therefore, is to minimize the sum of the system capital costs and variable annual costs for the operation of the heat exchanger.

A final example involves mass transfer, where the design of a distillation unit is based on a required separation for a given feed, at a known composition, temperature, and flow rate (Ricci and Theodore 2010; Theodore 2014). The engineer must determine the size of the distillation column and reflux ratio necessary to meet the specifications. As the reflux ratio is increased, the number of theoretical stages required for a given separation decreases. An increase in reflux ratio also results in lower fixed costs for the distillation column. However, optimization is necessary because an increased reflux results in increased costs for the reboiler heat supply and condenser coolant.

Difficulty in predicting the future is one of the problems that an optimization proposal involving capital investment must address. A statistical approach is being employed by some engineers responsible for process-plant design in an attempt to evaluate the risks that occur. In the past, uncertainties confronting investors were resolved by intuition or "professional judgment." In estimating a projected return, the design engineer may already have used a life expectancy and a minimum acceptable return rate based on such judgments. Strict quantitative interpretation of the results may not be justified in most instances because basic assumptions of statistical theory are violated. The bottom line is that optimization produces a substantially clearer picture of the effect of errors on profitability than is possible by a nonanalytical approach.

22.3 THE HISTORY OF OPTIMIZATION

This brief overview of the history of optimization traces the evolution of mathematical ideas and concepts—beginning in prehistory. Mathematics is nearly as old as humanity itself; evidence of a sense of geometry and interest in geometric patterns has been found in the designs of prehistoric pottery, textiles, and cave paintings. Primitive counting systems were almost certainly based on using the fingers of both hands, as evidenced by the predominance of the number 10 as the base for many number systems employed today.

The subject of mathematics encompasses the study of relationships among quantities, magnitudes, and properties of logical operations by which unknown quantities and magnitudes may be deduced. At one point in history, mathematics was regarded as a science involving only geometry and algebra. Toward the middle of the nineteenth century, mathematics came to be regarded as the science of relations that draws necessary conclusions. This latter view encompasses mathematical or symbolic logic—the science of using symbols to provide a logical deduction and reference based on definitions, axioms, postulates, and rules for transforming primitive elements into more complex relations and theorems. Enter optimization.

Interestingly, the mathematics of the late nineteenth and twentieth centuries is characterized by an interest in unifying elements across numerous fields of mathematical endeavor, especially in logic. For example, group theory has proven a particularly effective unifier. The unified set theory, intuitive geometry, modular arithmetic and measurement, and the development of number systems contributed to increased accuracy, precision, and error study. It also included studies of linear algebra, modern algebra, vectors, matrices, logic, truth tables, the nature of proofs, Venn and Euler diagrams, relations, functions, probability and statistics, linear programming, and computer programming.

As to the origin of optimization, it depends on who provides the response as there are many aspects of optimization of interest to the practitioner. Some claim it was Thomas Edison when he developed a long-lasting, high-quality light bulb in the 1870s. His success was primarily the result of an extensive trial-and-error search for the optimum filament material. Some now refer to this trial-and-error process as the *Edisonian* approach. One of the authors of this book refers to it as the perturbation approach; on occasion, he has modestly termed it the *Theodore Approach*.

The ability to perform complicated calculation efficiently has not only reduced the cost and time required to perform each iteration but also provides a better understanding of how this method works. The success of this approach still relies heavily on the limitations of the user's intuition and experience because one cannot often predict or even comprehend the effects of changing numerous variables in a complex system. Despite this barrier, computational calculations have expanded and improved the classical methods of optimization.

With the promise of reducing design time and cost while improving product quality, design optimization held tremendous potential. Starting with a suboptimal design, a numerical optimization algorithm could be used to iteratively adjust a set of preselected design parameters to achieve a set of design targets. This new class of optimization technology enabled broader, more comprehensive, faster searches for innovative designs than was possible using previous generations of tools. Moreover, it required no expertise in optimization theory, so it was easier to use for nonexperts and experts alike. By leveraging an engineer's potential to discover new design concepts, this new class of optimization technology overcomes the limits of human intuition and extends the designer's professional capability to achieve breakthrough designs and accelerated innovation.

As noted previously, the current level of computational horsepower allows optimization to more easily select the best element from a set of available alternatives. In the simplest case, an optimization problem consists of maximizing (or minimizing) an objective function by systematically changing input values from an allowed set of variables. Mathematically speaking, optimization allows one to find the best available value(s) of an objective function, given a defined domain (or a set of constraints). This also includes optimization of an array of objective functions with an array of domains. This process is discussed later in this chapter.

The increased capability of computational power has also had a tremendous impact on engineering design and data processing. The ability of computers to handle large quantities of data and perform the mathematical operations described permits the analysis of many more applications and engineering variables than could have ever been imagined only 20 years ago. Scientific calculations previously estimated in lifetimes of computational time are currently generated in nanoseconds.

22.4 THE SCOPE OF OPTIMIZATION

Optimization is mathematical in nature, and typically involves the maximization or minimization of a function that represents the "performance" of a "system." This is carried out by quantifying the values of the input variables that produce an optimal value for the function.

Perhaps the most important tool employed in optimization by engineers and scientists is *linear programming*. Linear programming consists of a linear, multivariable function, which is to be optimized (maximized or minimized), subject to a particular number of constraints. The constraints are normally expressed in linear form. Integer linear programming refers to optimization problems in which at least some of the variables assume integer values. The reader should note that the terms *linear* programming and *nonlinear* programming are similar from an application perspective.

One of the major responsibilities in optimization is to construct an accurate objective function. For example, in an environmental problem, the objective function might be the profit associated with the operation of a landfill that treats two categories of waste: "inert" and "non-inert" waste. Essentially, the optimization problem is to maximize profit (from handling the two waste streams) by employing an operating schedule subject to the landfill's capacity and environmental regulatory requirements. Rephrased, the objective function should represent a mathematical description of the combination of two wastes, which will produce maximum profit subject to the constraints connected with the landfill's processing capabilities and the regulation requirements imposed on each type of waste.

The objective function might be simple to identify and quantify, or may be very illusive as a result of the presence of conflicting or dimensionally incompatible objectives. One might be asked to optimize the profit for the aforementioned landfill that not only minimizes air and water pollution but is also *aesthetically pleasing*. Thus, it may not always be possible to quantify an objective function using strictly mathematical terms.

Optimization problems can sometimes be divided into parts, where it is possible to attain the optimum for the original problem by optimizing the constituent parts. This process is very powerful because it allows one to solve a series of smaller, easier problems rather than one large one. One of the best-known techniques to attack these problems is *dynamic programming*. This approach is characterized by a process that is performed in stages. One such example is a manufacturing process. Rather than solving the problem as a whole, dynamic programming optimizes one stage at a time to produce an optimal set of decisions for the entire process. Although dynamic programming has application in some systems or processes, it is not reviewed in this chapter. If the assumption of linearity cannot be made, a collection of techniques called *nonlinear programming* is used. This is addressed later in this chapter.

Finally, most engineering optimization applications involve economics: maximizing profit, minimizing cost, or both. Environmental applications can involve the purification of water, minimizing toxic emissions, maximizing energy use efficiency and conservation, minimizing health and safety risks, etc. The reader should not lose sight, however, that most real-world industrial applications involving optimization often require simple solutions. The following is a comment from a retired process engineer after 30 years in industry (Altomare 2016), "Generally, the consumer goods industry where I worked does not rely on sophisticated models. Basically, optimization was conducted to determine how to use minimum resources to minimize costs, subject to constraints. In the determination of consumer preference, we typically designed experiments to examine the broad range of attributes that led to maximum 'product liking' or decision to purchase. Occasionally, we modeled processes and used regression models. In that case, we could search or compute the desired maximum. The last set of techniques used is what the business world calls *Operations Research.* For example, linear programming would be used in the production of America's favorite barbecue item—the hot dog. The objective function was to produce a combination of meats that maximized flavor, met or exceeded FDA requirements for protein, fat, and water content, while minimizing overall production costs."

22.5 GENERAL ANALYTICAL FORMULATION OF THE OPTIMUM

When more than one independent variable is involved in determining the objective function (e.g., profitability), the objective function y that depends on a system of $x_1, x_2, x_3, \ldots, x_n$ independent variables is written as follows,

$$y(\mathit{profit}) = \mathrm{f}(x_1, x_2, x_3, \ldots, x_n) \tag{22.1}$$

In other words, one needs to specify x_1, x_2, x_3,...,x_n so that a maximum (or minimum) value of y can be determined. Differentiation will then result in a series of partial derivatives,

$$\frac{\partial y}{\partial x_1}, \frac{\partial y}{\partial x_2}, \text{ etc.} \tag{22.2}$$

Setting these partial derivatives equal to zero results in the same number of simultaneous equations as number of variables involved. The maximum occurs at a point where the first derivatives equal zero, accounting for restrictions on the second derivative, and continuity of the function throughout the range of independent variables. These conditions can, on occasion, restrict the applicability of this approach (Happel 1955). For example, for the case of a two-variable function where $y = f(x_1, x_2)$, points $x_1 = a$ and $x_2 = b$ exist for a maximum value of $y = f(x_1, x_2)$ if the following conditions hold:

$$\frac{\partial y}{\partial x_1} = \frac{\partial y}{\partial x_2} = 0 \tag{22.3}$$

with,

$$\frac{\partial^2 y}{\partial x_1^2} < 0; \frac{\partial^2 y}{\partial x_2^2} < 0 \tag{22.4}$$

and,

$$\left(\frac{\partial^2 y}{\partial x_1^2}\right)\left(\frac{\partial^2 y}{\partial x_2^2}\right) > \left(\frac{\partial^2 y}{\partial x_1 \partial x_2}\right)^2 \tag{22.5}$$

In contrast, if $\partial^2 y / \partial x_1^2$ and $\partial^2 y / \partial x_2^2$ are positive, and Equation 22.5 holds, there will be a minimum. However, if,

$$\left(\frac{\partial^2 y}{\partial x_1^2}\right)\left(\frac{\partial^2 y}{\partial x_2^2}\right) < \left(\frac{\partial^2 y}{\partial x_1 \partial x_2}\right)^2 \tag{22.6}$$

the point concerned will be a "saddle" point. The case where,

$$\left(\frac{\partial^2 y}{\partial x_1^2}\right)\left(\frac{\partial^2 y}{\partial x_2^2}\right) = \left(\frac{\partial^2 y}{\partial x_1 \partial x_2}\right)^2 \tag{22.7}$$

is open and the value may be a maximum, a minimum, or neither.

The following general comments relate the preceding expressions to both economics and engineering practice.

1. For a process with a fixed output that involves a number of pieces of equipment (e.g., an absorber, a heat exchanger, a preheater, a stripper, a cooler, etc.), there are a number of design variables which affect the cost of several of the operations (Theodore 2007; Ricci and Theodore 2010; Flynn et al. 2019). Assume that there are two major independent variables, x_1 and x_2. For a given value of x_2, the total costs are tabulated for various values of x_1, and the minimum total cost, C, is noted. This procedure is repeated for other values of x_2 and the combination of values of x_1 and x_2 is selected, which gives the minimum of the minima. If the values of x_1 and x_2 can be varied by differential increments, the same result

can be obtained by first setting the partial derivatives $\partial C/\partial x_1$ and $\partial C/\partial x_2$ equal to zero, and solving simultaneously for the optimum values of x_1 and x_2 (as discussed previously). Similar procedures could be used where there are more than two design variables. Careful study of this process might show that certain design variables affect costs of several but not all parts of the process. This allows the problem to be simplified.

2. In the important problem of estimating the optimum production of a plant, similar methods may be followed. In this case, maximum annual profit/minimum annual loss are the criteria. However, the output can be varied, whereas in the cases considered before, the output was fixed (Aris 1964). Hence, the minimum sum of variable costs could be used as the objective function. For example, in the case of operating at a loss, the plant should not be shut down until the out-of-pocket income would fall below the out-of-pocket expenses, excluding fixed costs. If continuity of supply is stipulated in a contract, the plant may be operated at a loss on an out-of-pocket basis.
3. In the case of a decision regarding a proposed expenditure that is not essential to the success of a process, the decision is often based on whether or not the *payoff* time is attractive. The payoff time is defined as that period in which the actual savings because of the proposed change would equal the original investment. The novice sometimes credits more toward savings than actually would be realized. For example, the cost of 1,000 lb/hr of steam at a water treatment facility might be \$0.35/hr. If the steam lines were insulated, the required steam load on the plant might be reduced by 1,000 lb/hr. At first inspection, the insulation is credited with a saving of \$0.35/hr. However, a more in-depth analysis of the steam costs show that the \$0.35/1,000 lb is broken down into \$0.10 for fuel and \$0.25 for fixed costs. Because the fixed costs would not change by insulating the steam lines, the true savings is only the value of fuel saved.

22.6 APPLICATIONS

In many industrial applications, the "factor" being optimized is a function of a single variable, which greatly simplifies the problem; these are illustrated in examples in the first part of this section. However, the last example, Illustrative Example 22.6, involves an application where the factor (the energy requirement) is a function of two variables. Six Illustrative Examples involving traditional optimization applications, along with three Illustrative Examples addressing linear programming applications complement this chapter's presentation on optimization techniques applied to water resource management.

Illustrative Example 22.1

The RDLT Corporation is considering three approaches to optimizing an existing, evaporative water purification unit. The three modifications are focused on energy conservation (Theodore 2010; Dupont et al. 2017):

1. The installation of a double pipe heat exchanger
2. The installation of a shell-and-tube heat exchanger (Dupont et al. 2017)
3. Process modification of the existing water purification unit

Based on the data provided, you have been asked to use the "system of criteria" method to determine which option is most attractive. RDLT Corporation has determined that annualized cost is the most important criterion, employing a weighting factor of 100 (out of a maximum of 100). Other significant criteria include regulations (weight of 80), potential recovery/reuse of water (weight of 70), and reliability (weight of 50). Options A, B, and C have also been assigned effectiveness factors. Options A is given a rating of 80 for the annualized cost, 60 for regulations, 40 for recovery/reuse, and 20 for reliability. The corresponding effectiveness factors for B and C are 60, 30, 40, 20 and 30, 80, 50, 80, respectively.

The overall rating (OR) for each option is defined as follows:

$$OR_j = \sum_{1}^{n} (\text{Criteria})_i (\text{Weight})_i \tag{22.8}$$

Solution

In solving this example, first generate the OR for Option A:

$$OR_A = (100)(80) + (80)(60) + (70)(40) + (50)(20) = 16{,}000$$

Generate the OR for Option B:

$$OR_B = (100)(60) + (80)(30) + (70)(40) + (50)(20) = 12{,}200$$

Generate the OR for Option C:

$$OR_C = (100)(30) + (80)(80) + (70)(50) + (50)(80) = 16{,}900$$

The results show that Option C produces the highest OR with a score of 16,900, and may be viewed as the most optimal of the three plans with respect to energy conservation for the water purification unit.

Illustrative Example 22.2

The annual operating cost of an outdated water purification process is \$75,000. Under a proposed plan, the installation of a new pumping system will require an initial cost of \$150,000 and an annual operating cost of \$15,000/yr for the first 5 years. Determine the annualized cost for the new processing system by assuming the system has only 5 years (n) of operational life. The interest rate (i) is 7%. The capital recovery factor (CRF) or annual payment of a capital investment can be calculated as follows (Theodore 2014):

$$\text{CRF} = \left(\frac{A}{P}\right)_{i,n} = \frac{i(1+1)^n}{\left[(1+1)^n - 1\right]} \tag{22.9}$$

where A is the annual cost and P is the present worth.

Compare the costs for both the outdated and proposed operations.

Solution

The annualized cost for the new operation is determined based on the following input data: Capital cost = \$150,000, Interest, i = 7%, payment period, n = 5 yr.

For i = 0.07 and n = 5, the CRF is,

$$\text{CRF} = \left(\frac{A}{P}\right)_{i,n} = \frac{0.07(1+0.07)^5}{\left[(1+0.07)^5 - 1\right]} = 0.2439$$

The total annualized cost for the new operation is then,

$$\begin{aligned} \text{Annualized Cost} &= \text{Capital Cost} + \text{Operational Cost} \\ &= (0.2439)(\$150{,}000) + (\$15{,}000) = \$51{,}585 \end{aligned}$$

Because this cost is lower than the annual cost of \$75,000 for the old treatment process, the proposed plan should be instituted.

Illustrative Example 22.3

A process emits 50,000 ft³/d of water containing dissolved solids at a loading of 2.0 gr/ft³. Reverse osmosis (RO) is employed for dissolved solids recovery because the dissolved solids recovered from the unit are worth \$0.03/lb. Experimental data have shown that the collection efficiency, E, is related to the RO process system pressure drop, ΔP, by the formula,

$$E = \frac{\Delta P}{\Delta P + 15.0} \tag{22.10}$$

where, ΔP is the pressure drop across the RO membrane, lb/ft².

1. If the pump is 55% efficient (overall) and electric power costs \$0.18/kW.hr, at what collection efficiency is the cost of power equal to the recovered value of the material (RV)? What is the pressure drop in 22.14 inches of water (in H_2O) at this condition?
2. What pressure drop will maximize profits?

RV may be expressed in terms of the water volumetric flowrate, q, the chemical concentration of dissolved solids, c, the dissolved solids value (DV), and the fractional collection efficiency, E, as (Theodore 2007; Ricci and Theodore, 2010):

$$RV = (q)(c)(DV)(E) \tag{22.11}$$

Solution 1

Substitution yields,

$$RV = \left(\frac{50{,}000\ \text{ft}^3}{\text{d}}\right)\left(\frac{2.0\ \text{gr}}{\text{ft}^3}\right)\left(\frac{1\ \text{lb}}{7{,}000\ \text{gr}}\right)\left(\frac{\$0.03}{\text{lb}}\right)(E) = \$0.43\ E/\text{d}$$

The RV can be expressed in terms of pressure drop, by substituting Equation 22.10 for E as,

$$RV = \frac{(\$0.43/\text{d})(\Delta P)}{\Delta P + 15.0} \tag{22.12}$$

The cost of power (CP) in terms of ΔP, q, the cost of electricity (CE), and the fan fractional efficiency, E_f, can be written as:

$$CP = \frac{(q)(\Delta P)(CE)}{E_f} \tag{22.13}$$

Substitution and unit conversion yields,

$$CP = \left(\frac{50{,}000\ \text{ft}^3}{\text{d}}\right)\left(\frac{\Delta P\ \text{lb}_f}{\text{ft}^2}\right)\left(\frac{\$0.18}{\text{kW}\cdot\text{hr}}\right)\left(\frac{1\ \text{min}\cdot\text{kW}}{44{,}200\ \text{ft}\cdot\text{lb}_f}\right)\left(\frac{1}{0.55}\right)\left(\frac{1\ \text{hr}}{60\ \text{min}}\right)$$
$$= \$0.006/\text{d}\Delta P \tag{22.14}$$

The pressure drop at which the cost of power is equal to the value of the recovered material is found by equating RV to CP, that is,

$$RV = CP;\ \frac{(\$0.43/\text{d})(\Delta P)}{\Delta P + 15.0} = \$0.006/\text{d}\Delta P$$

Solving yields,

$$\frac{(\$0.43/\text{d})}{\$0.006/\text{d}} = \Delta P + 15.0;\ \Delta P = 71.67 - 15 = 56.67\frac{\text{lb}_\text{f}}{\text{ft}^2} = 10.9 \text{ in } H_2O$$

The dissolved solids collection efficiency corresponding to the above calculated ΔP is,

$$E = \frac{\Delta P}{\Delta P + 15.0} = \frac{56.67}{56.67 + 15.0} = 0.79 = 79\%$$

It should be noted that operating below this efficiency (or the corresponding pressure drop) will produce a profit; operating above this value leads to a loss.

Solution 2

Set the first derivative of the profit (*RV–CP*) with respect to ΔP equal to zero. The describing equation for the profit, *P*, is given by,

$$P = RV - CP \tag{22.15}$$

Substitution of Equations 22.12 and 22.14 into Equation 22.15, and replacing ΔP with the variable, *x*, yields,

$$P = \frac{(0.43)x}{x + 15.0} - 0.006x \tag{22.16}$$

To proceed analytically, *dP*/*dx* is calculated and set it equal to zero,

$$\frac{dP}{dx} = \frac{0.43}{x + 15.0} - \frac{(0.43)x}{(x + 15.0)^2} - 0.006 = 0 \tag{22.17}$$

Solving for *x* results in the following:

$$0.43 - \frac{(0.43)x}{(x + 15)} - 0.006(x + 15.0) = 0; \ldots x^2 + 143.33x - 1{,}300 = 0$$

$$x = 8.55 \text{ lb/ft}^2$$

The corresponding maximum profit is then,

$$P = \frac{(0.43)(8.55)}{8.55 + 15.0} - 0.006(8.55) = \$0.105/\text{d}$$

This analytical result compares favorably with results produced by either a trial-and-error or graphical solution (Prochaska and Theodore 2018; Theodore & Behan 2018).

Illustrative Example 22.4

A stream of 100,000 gal/d of seawater is to be desalinated. You have been asked to find the optimal process to accomplish this task. A reputable vendor has provided information on the cost of three process units, as well as installation, operation, and maintenance costs that are summarized in Table 22.1. Determine which process you would select to minimize cost on an annualized basis.

Solution

From the cost information in Table 22.1, the equipment, installation, and operating costs are converted to total costs by multiplying each by the seawater flow, 100,000 gal/d (Theodore 2010).

TABLE 22.1
Desalination System Cost Data

	Evaporation (EV)	Reverse Osmosis (RO)	Crystallization (CR)
Equipment cost	\$3.10/(gal/d)	\$1.90/(gal/d)	\$2.50/(gal/d)
Installation cost	\$0.80/(gal/d)	\$1.40/(gal/d)	\$1.00/(gal/d)
Operating cost	\$0.06/(gal/d-yr)	\$0.06(gal/d-yr)	\$0.095/(gal/d-yr)
Maintenance cost	\$14,000/yr	\$28,000/yr	\$9,000/yr
Lifetime of equipment	20 yr	15 yr	20 yr

Note: Costs are based on comparable performance. Interest is 10%, and there is zero salvage value.

For example, the total costs for the evaporation process, *EV*, are calculated in the following manner:

$$\text{Equipment Cost} = 100{,}000\ \text{gal/d}\ (\$3.10/\text{gal/d}) = \$310{,}000$$

$$\text{Installation Cost} = 100{,}000\ \text{gal/d}(\$0.80/\text{gal/d}) = \$80{,}000$$

$$\text{Operating Cost} = 100{,}000\ \text{gal/d}(\$0.06/\text{gal/d}) = \$6{,}000$$

A summary of results is provided in Table 22.2.

Note that the operating cost is on an annualized basis. The equipment cost and the installation cost are capital costs and must then be converted to an annual basis using the CRF Equation 22.9.

$$\text{CRF} = \frac{(0.1)(1+0.1)^{20}}{\left[(1+0.1)^{20}-1\right]} = 0.1175$$

The annual cost for the equipment and the installation is given by the product of the CRF and the total costs of each as shown for the evaporation unit:

$$\text{Annual Equipment Cost} = (\$310{,}000)(0.1175) = \$36{,}425$$

$$\text{Annual Installation Cost} = (\$80{,}000)(0.1175) = \$9{,}400$$

The calculation for the RO and CR desalinators are performed in the same manner. The three desalinators can then be compared after the annual costs are added. The tabulated results are

TABLE 22.2
Desalination System Cost Calculations

	Evaporation (EV)	Reverse Osmosis (RO)	Crystallization (CR)
Equipment cost	\$310,000	\$190,000	\$250,000
Installation cost	\$80,000	\$140,000	\$100,000
CRF	0.11746	0.13147	0.11746
Annual equipment cost	\$36,425	\$22,325	\$29,375
Annual installation cost	\$9,400	\$16,450	\$11,750
Annual operating cost	\$6,000	\$6,000	\$9,500
Annual maintenance cost	\$14,000	\$28,000	\$9,000
Total annual cost	\$65,825	\$72,775	\$59,625

provided in Table 22.2. According to the analysis, the RO process is the least economically attractive process because the annual cost is the highest. Crystallization is determined to be the optimal process for this application since the total annual cost is lowest.

Illustrative Example 22.5

Consider the two-stage, reversible, adiabatic compression of an instrumentation gas at a wastewater treatment facility from an initial pressure, P_1, to a final pressure, P_3. If the gas enters at a temperature, T_1, and is cooled to a temperature, T_2, between stages, the total work, E, is given by Happel (1955) as

$$E = NRT_1\left(\frac{k}{k-1}\right)\left[\left(\frac{P_2}{P_1}\right)^{\frac{k-1}{k}} + \left(\frac{P_3}{P_2}\right)^{\frac{k-1}{k}} - 2\right] \tag{22.18}$$

where E is the total work, ft lbf; N is the number of moles of gas, lb mol; R is the molar gas constant, 1545.35 ft lbf/lb mol-R; T_1 is the inlet absolute temperature, R; and $k = 1.4$, the ratio of heat capacity at constant pressure to heat capacity at constant volume for the compressed gas.

Determine the intermediate stage pressure, P_2, if E is to be minimized. The value of the initial and final pressures are $P_1 = 1$ atm and $P_3 = 10$ atm, respectively.

Solution

If E in Equation 22.35 is to be optimized, then the derivative with respect to P_2 must be zero, that is,

$$\frac{dE}{dP_2} = \left(\frac{k}{k-1}\right)\left[\left(\frac{k-1}{k}\right)P_1^{\left(\frac{1-k}{k}\right)}P_2^{\left(-\frac{1}{k}\right)} - \left(\frac{k-1}{k}\right)P_3^{\left(\frac{k-1}{k}\right)}P_2^{\left(\frac{1-2k}{k}\right)}\right] = 0 \tag{22.19}$$

where NRT has been conveniently set equal to unity. Solving for P_2 yields,

$$P_2^{2\left(\frac{k-1}{k}\right)} = P_1^{\left(\frac{k-1}{k}\right)}P_3^{\left(\frac{k-1}{k}\right)} \tag{22.20}$$

Rearranging yields,

$$P_2 = (P_1P_3)^{1/2} = \sqrt{10} = 3.16 \text{ atm}$$

This result compares favorably with the answers produced if the example were solved employing trial-and-error or graphical methods.

The conclusion indicates that when it is extremely difficult to develop one analytical function for differentiation, graphical methods or trial-and-error solutions are equally valid. However, each application should be analyzed on the basis of the existing circumstances. For example, if numerous repeated calculations are necessary, the extra time required to develop an analytical solution may be justified.

The graphical method has one distinct advantage over the analytical method. The shape of the curve provides information on the importance of operating at or very close to optimum conditions. If the maximum or minimum occurs at a point where the curve is flat with only gradual change in slope, there will be considerable flexibility in the choice of final conditions. On the other hand, if the maximum or minimum is not flat, it may be essential to control operations at or near the exact optimum conditions.

Illustrative Example 22.6

Happel (1955) addressed the problem of minimizing the work required in the previous example employing a three-stage compressor, operating between 1 atm (P_1) and 10 atm (P_4). The describing equation is (Happel 1955):

$$E = NRT_1\left(\frac{k}{k-1}\right)\left[\left(\frac{P_2}{P_1}\right)^{\frac{k-1}{k}} + \left(\frac{P_3}{P_2}\right)^{\frac{k-1}{k}} + \left(\frac{P_4}{P_3}\right)^{\frac{k-1}{k}} - 3\right] \tag{22.21}$$

Using this expression, calculate the intermediate (interstage) pressures, P_2 and P_3.

Solution

If the quantity E is to be a minimum, $\partial E/\partial P_2$ and $\partial E/\partial P_3$ must both be zero.

$$\frac{dE}{dP_2} = \left[P_1^{\left(\frac{1-k}{k}\right)} P_2^{\left(-\frac{1}{k}\right)} - P_3^{\left(\frac{k-1}{k}\right)} P_2^{\left(\frac{1-2k}{k}\right)}\right] = 0 \tag{22.22}$$

$$\frac{dE}{dP_2} = \left[P_2^{\left(\frac{1-k}{k}\right)} P_3^{\left(-\frac{1}{k}\right)} - P_4^{\left(\frac{k-1}{k}\right)} P_3^{\left(\frac{1-2k}{k}\right)}\right] = 0 \tag{22.23}$$

Solving Equations 22.40 and 22.41 simultaneously, ultimately leads to,

$$P_2^2 = P_1P_3 \text{ and } P_3^2 = P_2P_4 \tag{22.24}$$

so that,

$$P_2 = \left(P_1^2P_4\right)^{1/3} = [(1^2)(10)]^{1/3} = 2.15 \text{ atm}$$

$$P_3 = \left(P_4^2P_1\right)^{1/3} = [(10^2)(1)]^{1/3} = 4.64 \text{ atm}$$

Although several different methods may be available to find a solution to an optimization problem, it is necessary to select the one best suited for the existing application as shown in this Illustrative Example.

Illustrative Example 22.7

In the environmental field, optimization and linear programming are pushing engineers and industry to address the unknowns and uncertainties associated with water use and wastewater management. There is increased recognition and appreciation for developing long-term water management plans that include solutions to problems such as water scarcity, water quality, climate change, stakeholder expectations, and government regulations.

Linear programming has the capability to impact the way in which both society and industry use and impact water resources. For instance, many industries, such as oil refining, need large volumes of low-quality water primarily for cooling purposes, whereas others such as the electronic industry, need smaller amounts of ultra-pure water. Differences in the nature of water use can therefore influence an industry's reliance on a high quality or large quantity water supply.

The following three illustrative problems involve linear programming to complement the discussion on optimization for problems that reflect multiple and more complex optimization objectives.

You are required to obtain the maximum value of *P*, subject to the following conditions,

Objective function:

$$P = 3x_1 + 5x_2 \tag{22.25}$$

Constraints:

$$x_2 \le 4 \tag{22.26}$$

$$x_1 \le 3 \tag{22.27}$$

$$x_2 \le -3x_1 + 6 \tag{22.28}$$

$$x_2 \le -0.5x_1 + 3 \tag{22.29}$$

Solution

The graphical solution is provided in Figure 22.1.

Constraints (1) and (2) indicate the range of values for x_1 and x_2 that will maximize *P* and lie within the rectangle provided in Figure 22.1a. Constraints (3) and (4) require that the solution be within the area ABCD shown in Figures 22.1c. When the objective function is superimposed on Figure 22.1d and e results. The maximum, *P*, is located at x_1 = 1.2 and x_2 = 2.4, for which *P* = 15.6. Readily available software, such as Microsoft Excel, provides the same solution.

Illustrative Example 22.8

The AMF Water Purification Company has two RO desalination plants on its grounds that treat brackish water of different compositions from Lake 1 and Lake 2. Lake 1 and Lake 2 are located 20 and 50 miles, respectively, from the AMF plant. The processed water from each of the two lakes provide different quantities and various levels of water purity. This includes *Ultrapure* (UP) water used in semiconductor and pharmaceutical industries; *Pure* (PU) water used in hospitals; and *Semipure* (SP) water used primarily in areas such as household work. However, production of UP, PU, and SP is limited. Production information is provided in Table 22.3. The profit, *P*, of processing water from Lake 1 and Lake 2 are \$2.00/ft^3 and \$1.60/ft^3, respectively. Determine the water processing ratio of the two lakes that maximizes profit, *P*.

Solution

Set up the following expressions,

$$q_1 = \text{ft}^3 \text{ of Lake 1 water/d} \tag{22.30}$$

$$q_2 = \text{ft}^3 \text{ of Lake 2 water/d} \tag{22.31}$$

The objective function for the daily profit, *P*, is,

$$P = 2.0\,q_1 + 1.6\,q_2 \tag{22.32}$$

The constraints are,

$$0.08\,q_1 + 0.11\,q_2 \le 1{,}500 \tag{22.33}$$

$$0.29\,q_1 + 5.4\,q_2 \le 5{,}500 \tag{22.34}$$

$$0.63\,q_1 + 3.5\,q_2 \le 11{,}000 \tag{22.35}$$

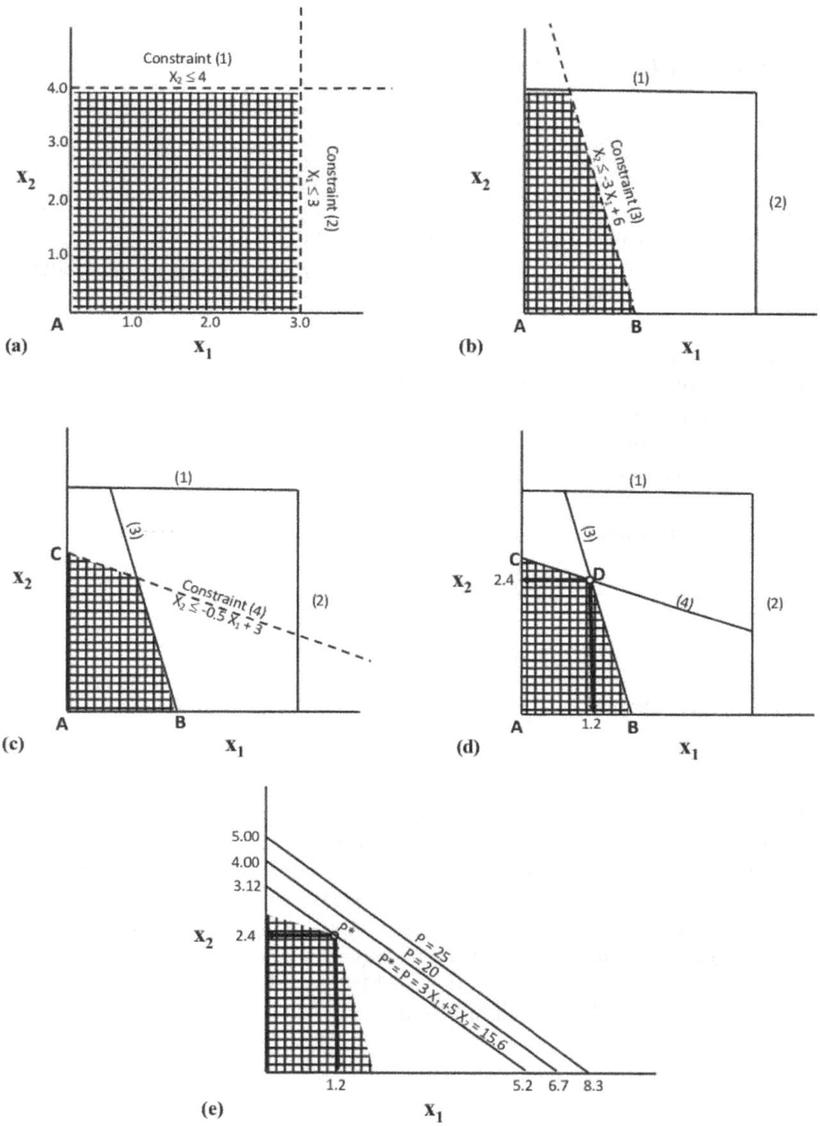

FIGURE 22.1 Graphical solution for Illustrative Example 22.7.

TABLE 22.3
Desalination Plant Information for Illustrative Example 22.8

Water Quality	Water Source (% Distribution by Volume of Processed Water) Lake 1	Lake 2	Maximum Production (ft^3/d)
Ultrapure	8	11	1,500
Pure	29	54	5,500
Semipure	63	35	11,000

	A	B	C	D	E	F	G	H
1								
2	**LP Illustrative Example 22.8**							
3								
4		Profit						
5	Max P=	$ 4						
6								
7	Variables	q1	q2					
8	Objective Coefficients	2	1.6					
9	Values	1	1	<--Optimal values appear here after solving				
10								
11	Coefficients	q1	q2	Total		Limit		
12	Constraint Equation 1	0.08	0.11	0.19	<=	1500		
13	Constraint Equation 2	0.29	0.54	0.83	<=	5500		
14	Constraint Equation 3	0.63	0.35	0.98	<=	11000		
15	Constraint Equation 4	1	0	1	>=	0		
16	Constraint Equation 5	0	1	1	>=	0		

FIGURE 22.2 Excel worksheet to solve the linear programming problem of Illustrative Example 22.8.

with,

$$q_1 \geq 0 \tag{22.36}$$

$$q_2 \geq 0 \tag{22.37}$$

The problem can be set up in Microsoft Excel and solved using Solver function. Figure 22.2 shows the original spreadsheet with the profit function calculated as the combination of objective coefficients and values for q_1 and q_2 according to Equation 22.32, and the five constraint equations given above. Initial values of 1 for both variables is shown in Figure 22.2. Figure 22.3 shows the Solver window used to generate an optimal solution for this problem, and Figure 22.4 shows the final optimal solution to this problem under the specified constraints. As indicated in Figure 22.4, 16,820 ft^3/day of water from Lake 1 and 1,152 ft^3/day from Lake 2 should be processed to result in a maximum total daily profit of $35,484.

Illustrative Example 22.9

The drinking water supply for the town of Floral Park in Nassau County, New York, is provided by two groundwater wells, W_1 and W_2. The water from W_1 contains 25 mg of 1, 4-Dioxane, 60 mg of Hexavalent Chromium, and 68 mg of Dibromochloromethane in each MG of water. The second well, W_2, contains 75 mg of 1, 4-Dioxane, 60 mg of Hexavalent Chromium, and 34 mg Dibromochloromethane in each MG. The owner of the wells generates a profit of $30/MG and $40/MG from W_1 and W_2, respectively, for each MG purchased by the town. Due to stringent local Nassau County water regulations, no more than 450 mg/day of 1, 4-Dioxane, 480 mg/day of Hexavalent Chromium, and 476 mg/day of Dibromochloromethane are allowed in the water sold by the supplier. How many gallons of W_1 and W_2 should be sold daily by the provider to maximize profits? In addition, calculate the corresponding maximum daily profit.

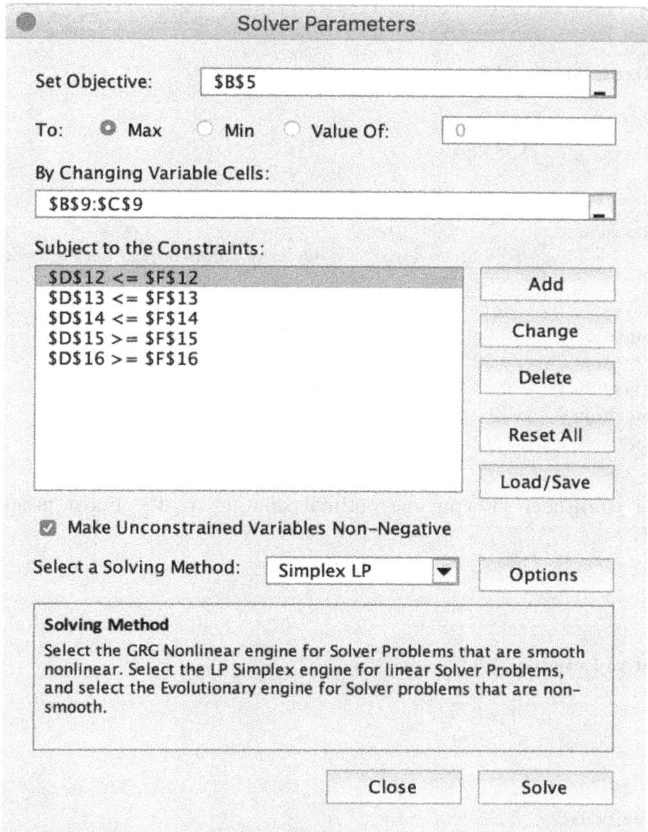

FIGURE 22.3 Excel Solver Parameters window used to solve the linear programming problem of Illustrative Example 22.8.

Solution

The linear programming format for this water purification process employs the units of million gallons/day (MGD) for x_1 and x_2.

The objective function for the daily profit, P, is set up as follows,

$$P = 30\,x_1 + 40\,x_2 \tag{22.38}$$

The constraints are,

$$25\,x_1 + 75\,x_2 \le 450 \tag{22.39}$$

$$60\,x_1 + 60\,x_2 \le 480 \tag{22.40}$$

$$68\,x_1 + 34\,x_2 \le 476 \tag{22.41}$$

	A	B	C	D	E	F	G	H	I
1									
2	**LP Illustrative Example 22.8**								
3									
4		Profit							
5	Max P=	$35,484							
6									
7	Variables	q1	q2						
8	Objective Coefficients	2	1.6						
9	Values	16,820	1,152	<--Optimal values appear here after solving					
10									
11	Coefficients	q1	q2	Total		Limit			
12	Constraint Equation 1	0.08	0.11	1472.4	<=	1500			
13	Constraint Equation 2	0.29	0.54	5500	<=	5500			
14	Constraint Equation 3	0.63	0.35	11000	<=	11000			
15	Constraint Equation 4	1	0	16820	>=	0			
16	Constraint Equation 5	0	1	1152.1	>=	0			

FIGURE 22.4 Excel worksheet showing the optimal solution for the linear programming problem of Illustrative Example 22.8.

	A	B	C	D	E	F	G	H	I
1									
2	**LP Illustrative Example 22.9**								
3									
4		Profit							
5	Max P=	$ 70							
6									
7	Variables	x1	x2						
8	Objective Coefficients	30	40						
9	Values	1	1	<--Optimal values appear here after solving					
10									
11	Coefficients	x1	x2	Total		Limit			
12	Constraint Equation 1	25	75	100	<=	450			
13	Constraint Equation 2	60	60	120	<=	480			
14	Constraint Equation 3	68	34	102	<=	476			
15	Constraint Equation 4	1	0	1	>=	0			
16	Constraint Equation 5	0	1	1	>=	0			

FIGURE 22.5 Excel worksheet to solve the linear programming problem of Illustrative Example 22.9.

with,

$$x_1 \geq 0 \tag{22.42}$$

$$x_2 \geq 0 \tag{22.43}$$

The problem can be set up in Microsoft Excel and solved using Solver function. Figure 22.5 shows the original spreadsheet with the profit function calculated as the combination of objective coefficients and values for x_1 and x_2 according to Equation 22.38, and the five constraint equations given. Initial values of 1 for both variables is shown in Figure 22.5. Figure 22.6 shows the Solver window used to generate an optimal solution for this problem, and Figure 22.7 shows the final optimal solution to this problem under the specified constraints. As indicated in Figure 22.7, 3 MGD of water from Well 1 and 5 MGD from Well 2 should be processed to result in a maximum total profit of $290/day. This result also ensures that Floral Park's water is in compliance with regulatory requirements for all specified pollutants.

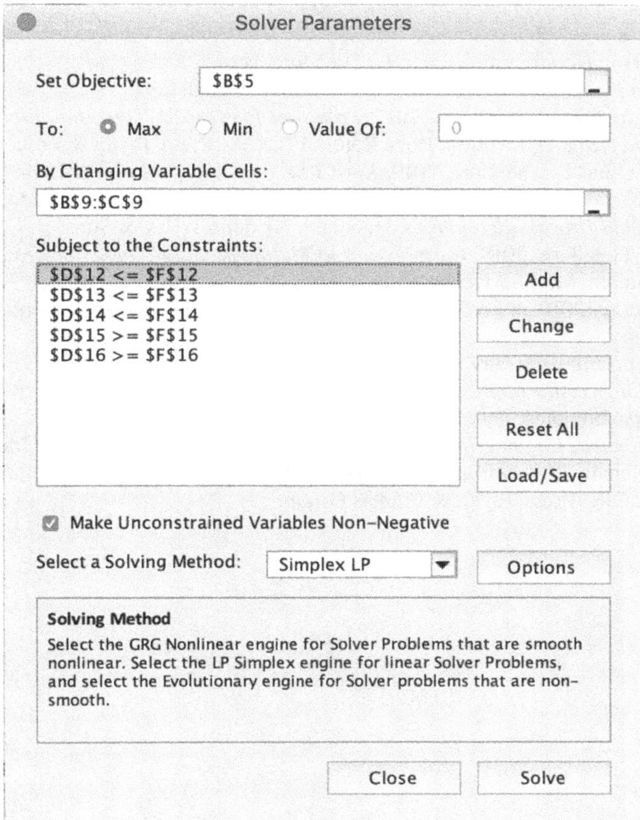

FIGURE 22.6 Excel Solver Parameters window used to solve the linear programming problem of Illustrative Example 22.9.

	A	B	C	D	E	F	G	H	I
1									
2	**LP Illustrative Example 22.9**								
3									
4		Profit							
5	Max P=	$ 290							
6									
7	Variables	x1	x2						
8	Objective Coefficients	30	40						
9	Values	3	5	<--Optimal values appear here after solving					
10									
11	Coefficients	x1	x2	Total		Limit			
12	Constraint Equation 1	25	75	450	<=	450			
13	Constraint Equation 2	60	60	480	<=	480			
14	Constraint Equation 3	68	34	374	<=	476			
15	Constraint Equation 4	1	0	3	>=	0			
16	Constraint Equation 5	0	1	5	>=	0			

FIGURE 22.7 Excel worksheet showing the optimal solution for the linear programming problem of Illustrative Example 22.9.

REFERENCES

Altomare, R. 2016. Personal communication to L. Theodore. Bronx, NY.

Aris, R. 1964. *Discreet Dynamic Programming*. Boston, MA: Blaisdell.

Dupont, R., K. Ganesan, and L. Theodore. 2017. *Pollution Prevention: Sustainability, Industrial Ecology, and Green Engineering*, 2nd edition. Boca Raton, FL: CRC Press/Taylor & Francis Group.

Flynn, A., T. Akashige, and L. Theodore. 2019. *Kern's Second Edition Process Heat Transfer*. Salem, MA: Scrivener Wiley.

Happel, J. 1955. *Chemical Process Economics*. Hoboken, NJ: John Wiley & Sons.

Prochaska, C., and L. Theodore. 2018. *Introduction to Mathematical Methods of Environmental Engineers and Scientists*. Salem, MA: Scrivener-Wiley.

Ricci, F., and L. Theodore. 2010. *Mass Transfer Operations for the Practicing Engineer*. Hoboken, NJ: John Wiley & Sons.

Theodore, L. 2007. *Fluid Flow for the Practicing Chemical Engineer*. Hoboken, NJ: John Wiley & Sons.

Theodore, L. 2010. *Heat Transfer Applications for the Practicing Engineer*. Hoboken, NJ: John Wiley & Sons.

Theodore, L. 2011. Personal notes. East Williston, NY.

Theodore, L. 2014. *Chemical Engineering: The Essential Reference*. New York: McGraw-Hill.

Theodore, L., and K. Behan. 2018. *Introduction to Optimization for Environmental and Chemical Engineers*. Boca Raton, FL: CRC Press/Taylor & Francis Group.

23 Ethical Considerations

23.1 INTRODUCTION

The primary responsibility of an engineering professional is to protect public health and safety. However, engineering professionals also have a responsibility to their employers or clients, to their families, to themselves, to the environment, and for civil and environmental engineers, to "strive to comply with the principles of sustainable development in the performance of their professional duties" (American Society of Civil Engineers [ASCE] 2008). Meeting all of these responsibilities will challenge the practicing engineer to draw upon a system of ethical values and ethical professional practice.

Well, what about ethics? Ethics means "doing the right thing" as opposed to "what you have the right to do." But doing the right thing is not always obvious or easy. In fact, ethical decisions are often difficult and may involve a certain amount of self-sacrifice. Doing the right thing for a practicing engineer can be especially challenging. Furthermore, the corporate and government world has confused this concept by developing ethics programs that emphasize only what you have the right to do. An organization, for example, may have a list—often called a Code of Conduct—of what an employee can and cannot get away with. Employees are required to sign an acknowledgment that they have read and understood the list. The company unfortunately calls this "ethics training" (Tabak 2002).

One difficulty in some situations is recognizing when a question of ethics is involved. Frequently, in the area of water resource management, a breach of ethics involves a practice that endangers public health and safety or covers up a violation of a rule or regulation. Occasionally, however, a breach may involve a case of the exact opposite. This might seem an unlikely scenario. How can someone be too honest, too caring, or too professional?

One example of this might be lying to save a life. Suppose you are standing on a street and a woman runs past you chased by two men. She screams, "They are trying to attack me!" as she dashes into the entry of a building around a corner. The men ask you, "Which way did she go?" What do you tell them? Clearly, the right thing is to lie. In this case, the value of caring overrides the value of honesty. This situation is exaggerated to illustrate that sometimes it is appropriate to violate certain values to protect public health and safety. In doing the right thing ideally one should not have to make snap decisions and should take the time to investigate all of the facts, for instance, in this case whether or not the woman was a thief and the men were police.

Sometimes one must decide how much to sacrifice to ensure public health and safety. In establishing environmental regulations, the regulating agency must decide how safe and how stringent to make the regulations. For example, in the case of drinking water regulations, one standard may result in 10 cancer cases per one million people. But why is the standard not established for 1 or 0 cases per one million people? Who should decide this acceptable level of risk?

The role of ethics in water resource management can be complex and confusing in the context of the engineer's role in providing safe drinking water to the public. This chapter provides a discussion of ethics with Illustrative Examples specifically focused on drinking water production and safe drinking water supply. Topics covered include the ethics of water access, the do's and don'ts, integrity, moral issues, guardianship, engineering and environmental ethics, and future trends in professional ethics. An applications section provides three Illustrative Examples related to applied ethics that have been primarily drawn from the work of Wilcox and Theodore (1998).

23.2 THE ETHICS OF WATER ACCESS

As noted, ethics is defined as moral principles that govern a person or group's behavior or the conduct of an activity (Oxford University Press 2019). Matters of ethics cover every aspect of life, from a personal view of right and wrong to a country's treatment of and responsibility to its people. Because water is a basic necessity for life, then access to clean water for everyone becomes an important and timely ethical issue. In fact, in 2010, the United Nations formally declared that access to safe, drinkable water and sanitation was a basic human right (United Nations 2010).

Ethical issues often arise whenever a basic human right such as access to water is threatened. Governmental oversight and public policy are among the tools that can ensure public health and safety when it comes to access to clean water and sanitation. However, limited funding and competing priorities can affect access and water supply even for governments with the best resources, infrastructure, and intentions. Climate change, and the increasingly frequent extreme weather events that come with it, threatens to exacerbate already-tenuous accessibility and public health issues in many poverty-stricken and developing areas around the world. These threats create a range of new challenges even where water access is reliable and infrastructure is currently well developed.

In the absence of regulation and oversight, even where water scarcity is not a current problem, water pollution in its many forms threatens the health and livelihood of just about every human and living creature on the planet. The detrimental effects of water pollution are well-documented and the sources are numerous (i.e., toxic runoff seeping into water supplies, illegal discharge of manufacturing waste into rivers and streams, wastewater from hydraulic fracturing contaminating groundwater, agricultural runoff causing algal blooms and dead zones in marine environments, etc.). There are many causes of water pollution, and the solutions for each will be quite different, whether necessitating new technological implementation, the mobilization of individuals or communities to speak out, or political involvement and policy interventions for change to be achieved.

Water scarcity is a growing concern throughout the world as a result of ever-increasing demands and dwindling supplies of usable water resources. Water scarcity can be due to the natural state of the climate such as in permanent desert and arid regions but can be further exacerbated by droughts or water reallocation. Sometimes droughts are temporary but can last for years. This has caused great suffering in regions of Africa, Asia, and the Middle East, as well as areas in the western United States, especially recently in Southern California. These dry seasons ultimately can lead to the suffering and death of millions, and not just through lack of water, but also as a result of famine and fires as was recently seen in Northern and Central California in 2018.

23.3 DO'S AND DON'TS

The need to make ethical decisions does not start the day one enters his or her first job. Encounters with ethical concepts begin early in life. Early on, before most have any active memory, parents almost certainly encourage their children to tell the truth. The conflict between choosing the ethical act of telling the truth and accepting the punishment may follow or choosing the unethical path of lying and going unpunished introduces virtually everyone to the struggle for ethical behavior that lasts a lifetime. In school, it is not long before students learn that it is unethical to copy homework and to cheat on tests. The unethical nature of stealing from a classmate is another area that surely received some attention early in life. As one grows up, ethical situations often become more complex. Conflicts develop. Situations arise in which one may have to decide if it is more ethical to hide an unpleasant truth that will hurt someone or stay loyal to the truth one was taught to value so highly.

As one moves into professional life, an individual learns a long list of ethical do's and don'ts. Some examples include:

1. Never falsify data.
2. Never offer or accept a bribe.

3. Judge colleagues, employees, employers, clients, and others on their professional merits and not on any other basis such as race, gender, age, religion, national origin, or disability.
4. Never draw misleading conclusions from the data you have collected.
5. Avoid conflicts of interest.
6. Accept work only in those areas in which you are properly trained.
7. Obey all laws, rules, and regulations.
8. Never break confidentiality.
9. Always do the best possible work of which you are capable.
10. Statements to the public must always be the truthful and objective.
11. Health and safety of the public must always be the highest priority.
12. Always act to protect the environment.

Ethics becomes a dilemma for the professional when confronted with difficult choices or apparent conflicts between ethical imperatives. A modern term for the outcome of such dilemmas is *situational ethics*, which refers to the temptation to bend the rules when a situation seems to have no satisfactory or satisfying solution. Examples of difficult choices are:

- A choice between keeping one's job and income that is needed to care for the children who depend on you and refusing to certify substandard work as ordered by your employer (job versus family)
- A choice between whether or not to get involved when you become aware of unethical behavior on the part of one's colleagues or employer even though not personally touched by it (job versus conscience)

Conflicts between two ethical imperative situations could include:

- The need to observe the confidentiality of a client after learning of the client's unethical behavior (protection of the client versus unethical behavior)
- The ethical imperative to faithfully serve an employer or client versus the ethical imperative to protect the public (protection of the employer versus the public). At what point does a professional stop working within the system and report a violation of regulations to authorities?

The second conflict of ethical imperatives stated above introduces issues that a water resources management professional is likely to their encounter. Most professionals belong to one or more organizations that commonly adopt a code of ethics for their members. Each code is designed provide members with an ethical foundation from which to work.

23.4 INTEGRITY

Scenarios are, for the most part, designed to reflect ambiguity in work situations. The ethicist hopes to get his or her hands dirty, dealing with the bottom-line motives of survival, competitiveness, and profitability as well as the mixed motives of self-interest, respect for the rights of others, and altruism. Obtaining an ethical solution to a difficult moral problem or dilemma is based on much more than choosing the correct ethical framework with its normative frame of reference. One must also be ready to examine fundamental assumptions and the values to which the assumptions give rise. Carter (1996) made this point in a discussion of "integrity."

- *Honesty in relation to integrity.* Carter explores integrity in relation to the value society places on honesty. On this subject, one of the best-known and most popular ethics books of the last four decades is Bok's (1978) *Lying: Moral Choice in Public and Private Life*. Without taking away from the merits of lying, Carter notes: "Plainly, one cannot have integrity without being honest (although, as we shall see, the matter gets complicated), but one can certainly be honest and yet have little integrity." Honesty is far easier to practice than the tough work of figuring out what it takes to have integrity in a situation. Integrity requires a high degree of moral reflectiveness. Honesty may result in harm to another person. Furthermore, "if forthrightness is not preceded by discernment, it may result in the expression of an incorrect moral judgment." The racist may be transparently honest, Carter declares, but he certainly lacks integrity because his beliefs, deeply held as they might be, are wrong. He has not engaged in the hard work of examining his fundamental assumptions, values, and beliefs.

- *Personal integrity without public responsibility?* It would appear that one cannot have integrity without responsibility since any consideration of integrity addresses the effects of our conduct on other people. In our work life and our community life, we have public responsibilities for our clients and fellow citizens.

23.5 MORAL ISSUES

The conflict of interest between Chief Seattle (and Native Americans in general) and President Franklin Pierce (and the European-American expansion) provides a perfect example of how ethics and the resulting codes of behavior they engender can differ drastically from culture to culture, from religion to religion, and even from person to person. This enigma, too, is noted again and again by Seattle (Martin and Schinzinger 1989):

> I do not know. Our ways are different from your ways.... But perhaps it is because the red man is a savage and does not understand.... The air is precious to the red man, for all things share the same breath... the white man does not seem to notice the air he breathes.... I am a savage and do not understand any other way. I have seen a thousand rotting buffaloes on the prairie, left by the white man who shot them from a passing train. I am a savage and I do not understand how the smoking iron horse can be more important than the buffalo we kill only to stay alive.

Chief Seattle sarcastically uses the European word *savage* and all its connotations throughout his address. When one finishes reading the work, it becomes obvious which viewpoint (President Pierce's or his own) Chief Seattle feels is the savage one. What his culture holds dearest (the wilderness), the whites see as untamed, dangerous, and savage. What the whites hold in highest regard (utilization of the Earth and technological advancement), the Native Americans see as irreverent of all other living things. Each culture maintains a distinct and conflicting standard for the welfare of the world. Opposing viewpoints and moralities such as these are prevalent throughout the world and have never ceased to present a challenge to international, national, state, community, and interpersonal peace (Burke et al. 2000).

It is generally accepted, however, that any historical ethic can be found to focus on one of four different underlying moral concepts:

1. *Utilitarianism* focuses on good consequences for all.
2. *Duties ethics* focus on one's duties.
3. *Rights ethics* focus on human rights.
4. *Virtue ethics* focus on virtuous behavior.

Utilitarians hold that the most basic reason why actions are morally right is that they lead to the greatest good for the greatest number. "Good and bad consequences are the only relevant considerations, and, hence all moral principles reduce to one: 'We ought to maximize utility'" (Martin and Schinzinger 1989).

Duties ethicists concentrate on an action itself rather than the consequences of that action. To these ethicists there are certain principles of duty such as "Do not deceive" and "Protect innocent life" that should be fulfilled, even if the most good does not result. The list and hierarchy of duties differs from culture to culture, religion to religion. For Judeo-Christians, the Ten Commandments provide an ordered list of duties imposed by their religion (Martin and Schinzinger 1989).

Often considered to be linked with duties ethics, *rights ethics* also assesses the act itself rather than its consequences. Rights ethicists emphasize the rights of the people affected by an act rather than the duty of the person(s) performing the act. For example, because a person has a right to life, murder is morally wrong. Rights ethicists propose that duties actually stem from a corresponding right. Because each person has a right to life, it is everyone's duty to not kill. It is because of this link and their common emphasis on the actions themselves that rights ethics and duty ethics are often grouped under the common heading *deontological ethics* (Barbour 1993).

The display of virtuous behavior is the central principle governing *virtue ethics*. An action would be wrong if it expressed or developed vices, for example, bad character traits. Virtue ethicists, therefore, focus on becoming a morally good person.

To display the different ways that these moral theories view the same situation, one can explore their approach to the following scenario that Martin and Schinzinger (1989) present:

> On a midnight shift, a botched solution of sodium cyanide, a reactant in organic synthesis, is temporarily stored in drums for reprocessing. Two weeks later, the day shift foreperson cannot find the drums. Roy, the plant manager, finds out that the batch has been illegally dumped into the sanitary sewer. He severely disciplines the night shift foreperson. Upon making discreet inquiries, he finds out that no apparent harm has resulted from the dumping. Should Roy inform government authorities, as is required by law in this kind of situation?

If a representative of each of the four different theories on ethics just mentioned were presented with this dilemma, their decision-making process would focus on different principles.

The utilitarian Roy would assess the consequences of his options. If he told the government, his company might suffer immediately under any fines administered, and later (perhaps more seriously) because of exposure of the incident by the media. If he chose not to inform authorities, he risks heavier fines (and perhaps even worse press) in the event that someone discovers the cover-up. Consequences are the utilitarian Roy's only consideration in his decision-making process.

The duties-ethicist Roy would weigh his duties and his decision would probably be more clear-cut than his utilitarian counterpart. He is obliged foremost by his duty to obey the law and must inform the government.

The rights-ethicist mind-set would lead Roy to the same course of action as the duties ethicist but, not necessarily because he has a duty to obey the law, but because the people in the community have the right to informed consent. Even though Roy's inquiries informed him that no harm resulted from the spill, he knows that the public around the plant has the right to be informed of how the plant is operating.

Vices and virtues would be weighed by the virtue-ethicist Roy. The course of his thought process would be determined by his own subjective definition of what things are virtuous, what things would make him a morally good person. Most likely, he would consider both honesty and obeying the law virtuous, and withholding information from the government and the public as virtueless and would, therefore, tell the authorities (Burke et al. 2000).

23.6 GUARDIANSHIP

There are two important questions that must be kept in mind when an ethical problem is being assessed:

1. *Who are the guardians of the system?* This question addresses the issue of who, among engineering or science professionals, is responsible for the ethical standards in the organization. If professionals point the finger at senior management, the legal department, the

Environmental Protection Agency (EPA), or the Department of Justice (DOJ), they have indeed misunderstood the nature of a professional calling. The first line of defense is the willingness of professionals themselves to maintain and enhance the integrity of the engineering or scientific profession through their own personal adherence to the highest standards of conduct and to assume responsibility for commitment to these standards within the companies where they work. Moreover, ethics is a positive task and not a list of dos and don'ts. To achieve excellence in one's work presumes a commitment to the client's contract, public safety, and environmental integrity, among several factors that are all too often thought of as "management" issues. They are, in reality, the ethical standards of the work itself. Thus, the ethical engineer or scientist is the one who identifies with the profession and all that is involved in the work assigned or contracted.

2. *Who gives support to the guardians of the system?* This second issue goes to the heart of the assessment problem but also has an impact on the first issue. Unless the organization backs those who assume positive responsibility for the ethical tenor of the group, very little will change. Why would someone risk ostracism or retaliation by confronting a person engaging in unethical behavior or illegal behavior if there is no institutional support for the one assuming responsibility?

Effective guardianship is facilitated if:

1. There are clear-cut standards of behavior and high expectations of the membership.
2. The standards are brought to the attention of the members through a well-developed training program.
3. The standards are taken seriously by the senior leadership team of the firm. They must demonstrate that seriousness by taking an active role in the training, without, at the same time, creating a chilly climate stifling discussion and participation in the training. The ethics program must be seen not as frosting on the cake or as a value added to forestall legal problems through better compliance. The CEO needs to demonstrate a commitment to the values and principles that drive the business. Ethics training is no add-on. Ethics is what drives the organization: trust, integrity, fidelity to the client, to the public, and to the environment.
4. It is evident that the leadership "walks the talk" in all aspects of its decision making and actions.
5. There are mechanisms in place to address the concerns of the members, mechanisms such as an ombudsperson, a hotline, etc.
6. Those who adversely affect the integrity of the business are effectively and fairly disciplined.

Another way of addressing the question of who supports the guardians is to emphasize the importance of organizational or corporate culture. A positive response to the six points just raised has a great impact on the culture of the organization. Unless there is what is sometimes called a "thick" culture, wherein respect for and adherence to guardianship and the tenets of integrity and trust are palpable, individually ethical persons can do very little to raise the moral climate of the organization. An organization is more than the sum total of the individuals who constitute the membership. The attitudes conveyed, values expressed, and ways of doing business in an organization profoundly affect the perceptions of the members therein and set the tone of the company. Having a positive impact on culture is a great challenge that is not easily achieved. Culture is so subtle that one often does not even realize or understand its dimensions until a significantly different culture is experienced.

An organization will not have effective guardianship of the system unless there is a concerted attempt to create, enhance, or reinforce a culture where values and ethics are clear and fully supported. There is little doubt, however, that the twin issues of guardianship and culture are much more difficult to address than the institutionalization of the ethics program itself.

23.7 ENGINEERING AND ENVIRONMENTAL ETHICS

In the ethical theories presented here, established hierarchies of duties, rights, virtues, and desired consequences exist so that situations where no single course of action satisfies all of the maxims can still be resolved. The entry of environmentalism into the realm of ethics raises questions concerning where it falls in this hierarchy (Theodore and Theodore 2009). Much debate continues over these questions of how much weight the natural environment should be given in ethical dilemmas, particularly in those where ecological protection seems to oppose economic profitability and technological advances.

Those wrapped up in this technology, economy, or ecology debate can generally be divided into three groups:

1. Environmental extremists.
2. Technologists to whom ecology is acceptable provided it does not inhibit technological or economic growth.
3. Those who feel technology should be checked with ecological responsibility.

Each is briefly discussed.

After his year-and-a-half of simple living on the shores of Walden Pond, Henry David Thoreau rejected the pursuit of technology and industrialization. Although most would agree with his vision of nature as being inspirational, few would choose his way of life. Even so, the movement rejecting technological advances in favor of simple, sustainable, and self-sufficient living is being embraced by more and more people who see technology as nothing but a threat to the purity and balance of nature. Often called *environmental extremists* by other groups, they even disregard "environmental" technologies that attempt to correct pollution and irresponsibilities, past and present. They see all technology as manipulative and uncontrollable and choose to separate themselves from it. To them, the preservation of the environment is the top priority.

On the other extreme are the pure technologists. They view the natural world as a thing to be subdued and manipulated in the interest of technological and economic progress. This is not to say one will not find technologists wandering in a national park admiring the scenery. They do not necessarily deny the beauty of the natural environment, but they see themselves as separate from it. They believe that technology is the key to freedom, liberation, and a higher standard of living. It is viewed, therefore, as inherently good. They see the environmental extremists as unreasonable and hold that even the undeniably negative side effects of certain technologies are best handled by technological solutions. The technologists place environmental responsibility at the bottom of their ethical hierarchy.

Somewhere in the middle of the road is the third group. Although they reap the benefits of technology, they are concerned much more deeply than the technologists with the environmental costs associated with industrialization. It is in this group that most environmental engineers find themselves. They are unlike the environmental extremists because, as engineers, they inherently study and design technological systems and have faith in the ability of such systems to have a positive effect on the state of the environment. They also differ from the technologists because they scrutinize the effects of technologies on human health and the environment much more closely and critically than the technologists do. Although they may see a brief, dilute leak of a barely toxic chemical as an unacceptable side effect of the production of a consumer product, the technologists may have to observe destruction the magnitude caused by Chernobyl before they consider rethinking a technology they view as economically and socially beneficial. In general, this group sees the good in technology but stresses that it cannot be reaped if technological growth goes on unchecked at the expense of human health or environmental quality.

The ethical behavior of engineers is more important today than at any time in the history of the profession (Theodore and Theodore 2009). The engineers' ability to direct and control the technologies they master has never been greater. In the wrong hands, the scientific advances and technologies of today's engineers could become the worst form of corruption, manipulation, and exploitation. Engineers, however, are bound by a code of ethics that carry certain obligations associated with the profession. Some of these obligations include:

- Support one's professional society
- Guard privileged information
- Accept responsibility for one's actions
- Employ proper use of authority
- Maintain one's expertise in a state-of-the-art world
- Build and maintain public confidence
- Avoid improper gift exchange
- Practice conservation of resources and pollution prevention
- Avoid conflict of interest
- Apply equal opportunity employment
- Practice health, safety, and accident prevention
- Maintain honesty in dealing with employers and clients

here are many codes of ethics that have appeared in the literature. The preamble for one of these codes is (Martin and Schinzinger 1989):

> Engineers, in general, in the pursuit of their profession, affect the quality of life for all people in our society. Therefore, an Engineer, in humility and with the need for Divine guidance, shall participate in none but honest enterprises. When needed, skill and knowledge shall be given without reservation for the public good. In the performance of duty and in fidelity to the profession, Engineers shall give utmost.

23.8 FUTURE TRENDS

Although the environmental movement has grown and matured in recent years, its development is far from stagnant. To the contrary, change in individual behavior, corporate policy, and governmental regulations are occurring at a dizzying pace.

Because of the Federal Sentencing Guidelines, the Defense Industry Initiative, as well as a move from compliance to a values-based approach in the marketplace, corporations have inaugurated company-wide ethics programs, hotlines, and senior line positions responsible for ethics training and development. The Sentencing Guidelines allow for mitigation of penalties if a company has taken the initiative in developing ethics training programs and codes of conduct.

In the near future, these same guidelines will apply to infractions of environmental law. As a result, the corporate community will undoubtedly welcome ethics integration in engineering and science programs generally, but more so in those that emphasize environmental issues. Newly hired employees, particularly those in the environmental arena who have a strong background in ethics education, will allay fears concerning integrity and responsibility. Particular attention will be given to the role of public policy in the environmental arena as well as in the formation of an environmental ethic (Theodore and Theodore 2009).

Regulations instituted by federal, state, and local agencies have over the years become more and more stringent. The deadlines and fines associated with these regulations encourage corporate and industrial compliance of companies (the letter of the law), but it is the personal conviction of the corporate individuals that lies in the spirit of the law and the heart of a true ecological ethic.

To bolster this conviction of the heart, there must be the emergence of a new *dominant social paradigm* (Barbour 1993). This is defined as "the collection of norms, beliefs, values, habits, and survival rules that provide a framework of reference for members of a society. It is a mental image of social reality that guides behavior and expectations" (Barbour 1993). The general trend in personal ethics is steadily "greener" and is being achieved at a sustainable pace with realistic goals.

It has been suggested (Gleick 1987) that "[t]he flap of one butterfly's wings can drastically affect the weather." Although this statement sounds much like one conceptualized by a romantic ecologist, it is actually part of a mathematical theory explored by the contemporary mathematician Gleick (1987) in his book *Chaos, Making a New Science*. The "butterfly" theory illustrates that the concept of interdependence, as Chief Seattle professed it, is emerging as more than just a purely environmental one. This embracing of the connectedness of all things joins the new respect for simplified living and the emphasis on global justice, renewable resources, and sustainable development (as opposed to unchecked technological advancement) as the new, emerging social paradigm. The concept of environmentalism is now *widely* held; its future is it becoming *deeply* held.

Finally, one should note that an ethical analysis generally produces no absolute answers. This can be quite disconcerting because engineers expect valid, correct, and useful answers to problems. When one studies statics, for example, one knows what the rules are and all agree on the right answer. In ethics, however, the best an individual can do is argue that some answers are better than others, and, of course, these answers are always open to disagreement.

Engineers, as professionals, have a special responsibility to the public, and this responsibility is often expressed in terms of professional ethics. Engineers invariably face situations in which values become variables in the decision-making process. Indeed, the ethical aspects of a decision often prove more difficult than the technical ones. Recognizing this, the engineering profession has strongly encouraged engineering schools to introduce more professional ethics into engineering curricula. Professional ethics has rightly become an integral part of engineering education.

Engineering ethics is tricky enough when it concerns only how engineers relate to each other and the public; it becomes trickier still when one also considers how engineers ought to react to the nonhuman environment (Theodore and Theodore 2009). That is, when one asks "what is the engineers' environmental ethic?"

Environmental ethics, to an even greater degree than ordinary ethics, is a subject without definition and without consensus. And yet, every person on this planet makes everyday decisions that relate to environmental ethics. Questions as simple as "What should I eat?" or "How should I move from place to place?" all raise environmental and ethical issues.

Environmental ethics is especially important for engineers because so much of their work affects the environment. How should the engineer balance the human gains of development against environmental damage? When should the engineer maintain client confidentiality in the face of potential environmental problems?

Finally, now that a regulatory framework and general guidelines are in place, it is up to the individual to follow these procedures and strive to provide a safer, cleaner environment for the future of the planet. In any ethical dilemma, it will always be up to individuals, such as the engineers and scientists working on the ground, whom best understand the shifting dynamic between the changing climate, the local environment, and the specific work they do to speak up if and when they see something that goes against their knowledge and belief.

As indicated previously, ethics is the branch of philosophy dealing with the rules of right conduct. When defining environmental ethics, right conduct with respect to the environment must be evaluated. The American Consulting Engineers Council Professional and Ethical Conduct Guidelines state that "consulting engineers shall hold paramount the safety, health and welfare of the public in the performance of their professional duties. Consulting engineers shall at all times recognize that their primary obligation is to protect the safety, health, property and welfare of the public. If their professional judgement is overruled under circumstances where the safety, health, property or welfare of the public are endangered, they shall notify their client and such other authority as may be appropriate. Consulting engineers shall approve only engineering work which, to the best of their knowledge and belief, is safe for public health, property and welfare and in conformity with accepted standards."

This statement of ethics can be extended to all individuals and to all aspects of engineering and technology including water resources management. Whether one's particular area of responsibility is with the government, industry, or public utility, the actions of an engineer, attorney, politician, or scientist should promote the general health and well-being of the community and the surrounding environment. If every individual makes even a small contribution to the reduction of pollution and the protection of the environment, the sum of the contributions will be great.

23.9 APPLICATIONS

Two of the three Illustrative Examples have been drawn from the work of Wilcox and Theodore (1998) keying primarily on clean water issues. Each of the last two applications are presented in case-study format, containing both a Fact Pattern and Questions for Discussion.

Illustrative Example 23.1

Provide a historical overview of trends in corporate policies related to their environmental responsibility and stewardship.

Solution

In the early 1990s, the defense industry led the initiation of the Federal Sentencing Guidelines for Organizations (FSGO), which arose from an increase in white-collar crime during the previous decade. The FSGO allowed for increased oversight into corporations' fraudulent and other unethical behaviors. The fact that senior management could now be prosecuted for the behavior of its employees led to widespread implementation of company ethics training and ethics programs. Compliance programs covered a host of topics including antitrust, equal opportunity, and environmental compliance. In today's corporate culture, 30 years after these initial efforts began, ethics and compliance programs and employee training have become commonplace in most companies due to the legacy of FSGO and other legislation. Companies no longer treat ethics and compliance as an extra drain on resource allocation because proper handling of environmental issues and serving as positive influences in their communities have helped many companies raise their visibility among consumers, which in turn can improve brand loyalty and sales. Walmart, for example, annually publishes a corporate Global Compliance and Ethics Report (https://corporate.walmart.com/global-responsibility/global-compliance-program-report-on-fiscal-year-2018), and it has a stated commitment to enhance the sustainability of its operations and global supply chains through goals of zero waste, operations with 100% renewable energy, and selling products that sustain resources and the environment (https://corporate.walmart.com/global-responsibility/sustainability/).

Illustrative Example 23.2

Fact Pattern. Laura is an engineer working in a chemical plant. She has recently received a job offer from another company, which she accepts because she knows that the new job could be a big step in her career.

Laura is responsible for one of the production lines in the plant she will soon be leaving. She has always been a reliable worker and an effective manager. However, having handed in her letter of resignation, she has been less attentive to her work over the past couple of weeks. She figures that there is no need to worry about this job anymore; she has to concentrate on her future.

On Laura's next-to-last day of work at the plant, Harry a coworker on the same production line, finds out that there is a problem with the purity of a wastewater discharge stream. The level of contaminants is a little higher than acceptable. Harry decides to consult with Laura.

He says, "The effluent stream coming out of the process exceeds the required water quality limit. I think you should investigate it so we can solve this problem."

Laura replies, "I would love to help you, Harry, but tomorrow is my last day here. I don't want to start dealing with this problem because it could take a while to solve. Let my replacement worry about it."

Harry answers, "Laura, if we let this problem go, we'll continue to have a discharge that doesn't meet regulations. The problem could also get *worse*. You are the expert here, so you could easily fix this mess."

"Harry, you're a friend of mine. Please don't ask me to get involved in this problem. It's not my concern anymore. I just want to relax during my last 2 days at work," pleads Laura. "It's not like the plant will blow up. Wait for 2 days. You can pretend that you didn't notice anything until then."

Reluctantly, Harry agrees. "I know you're really looking forward to your new job. It's just that I'll feel guilty knowing that something is wrong, and I'm not doing anything about it. But I guess I can wait for 2 days."

"Harry, don't worry. Take it easy for a couple of days. Just think of it as a minor delay," replies Laura.

Questions for Discussion

1. What are the facts in this case?
2. Do you think Laura should stay focused on her current job?
3. Should Laura handle the problem?
4. Do you think it's okay for Harry to ignore the problem for the next 2 days?
5. Should Harry consult someone else now that Laura has refused to deal with the problem?
6. What are the regulatory requirements to report a permit violation and who will be held responsible if the violation is responsible for harm to human health or the environment?

Illustrative Example 23.3

Fact Pattern. Joe Murray, an environmental engineer at the Pristine Water Company, is responsible for the production of potable drinking water the company distributes to its municipal and industrial customers. Joe has been reviewing results of routine water quality monitoring conducted within the plant for process monitoring purposes but that are not required to be reported to the local regulatory agency. Joe notices that there is a disturbing trend in pH that, although not currently violating permit requirements, has potentially significant implications for water quality within the distribution system for the mobilization of lead and for the production of undesirable disinfection by-products during treated water chlorination.

Joe decides to go to his manager and proposes that expensive source water and distribution system monitoring be carried out, first to attempt to determine if changes in source water quality could be the cause of these process changes and, secondly, to ensure that undesirable conditions within the distribution system are not occurring. His manager, concerned about quarterly profits and control of operating costs, feels that if no permit violations are occurring then no actions are warranted so no more money or personnel time are wasted.

Ethically, Joe feels strongly that this is wrong and feels this situation needs special attention before conditions within the distribution system become unstable and their customers are subjected to unnecessary risk if the plant's response to water quality changes are delayed. Joe's only alternative is to go above his manager and try to convince upper management to expend the necessary funds to conduct the required monitoring and perhaps process modifications to ensure the quality of the produced water and the health and safety of their customers. However, doing this could anger his manager and could also lead to Joe's dismissal. But it might also make him appear to be confident and conscientious, could prevent a major water crisis such as happened in Flint, Michigan, in 2014, and could also serve to publicize his abilities as an engineer.

Questions for Discussion

1. What are the facts in this case?
2. What issues are involved?
3. What are the choices Joe must make?
4. What could be the consequences of each decision he could make?
5. What were the outcomes of the Flint, Michigan, water crisis, and what can be learned for this real-life event that pertains to Joe's problem?

REFERENCES

American Society of Civil Engineers (ASCE). 2008. *Ethics Guidelines for Professional Conduct for Civil Engineers*. ASCE, Reston, VA.

Barbour, I. 1993. *Ethics in the Age of Technology*. Harper, San Francisco, CA.

Bok, S. 1978. *Lying: Moral Choice in Public and Private Life*. Pantheon, New York.

Burke, G., B. Singh, and L. Theodore. 2000. *Handbook of Environmental Management and Technology*, 2nd edition. John Wiley & Sons, Hoboken, NJ.

Carter, S. L. 1996. The insufficiency of honesty. *Atlantic Monthly*. February, 74–76.

Gleick, J. 1987. Chaos, Making a New Science. Viking Books, New York.

Martin, M., and R. Schinzinger. 1989. *Ethics in Engineering*. McGraw-Hill, New York.

Oxford University Press. 2019. *Definition of ethics in English*. https://en.oxforddictionaries.com/definition/ethics (accessed March 13, 2019).

Tabak, H. 2002. Ethics corner: Doing the right thing. *Environmental Management*, 29(1): 7–8.

Theodore, M. K., and L. Theodore. 2009. *Introduction to Environmental Management*. CRC Press/Taylor & Francis Group, Boca Raton, FL.

United Nations. 2010. *Resolution A/EWS/64/292: The Human Right to Water and Sanitation*. United Nations General Assembly, New York. https://undocs.org/E/A/RES/64/292.

Wilcox, J., and L. Theodore. 1998, *Engineering and Environmental Ethics: A Case Study Approach*. John Wiley & Sons, Hoboken, NJ.

24 Future U.S. Water Security

24.1 INTRODUCTION

As indicated in Chapters 9 and 10, the world's freshwater resources are under increasing pressure. Growth in population, increased economic activity, improved standards of living, and global conflicts are issues that continue to exacerbate water availability around the globe. It is estimated that more than 1 billion people live without access to safe water (World Health Organization [WHO] 2017). This number more than doubles to over 2.3 billion people when basic sanitation services are considered. And as indicated previously, despite availability of renewable freshwater resources, equitable and affordable supply and access to these resources remain a significant challenge, particularly to minority and low-income communities, even in the most developed countries in the world.

Securing a safe and sustainable water supply to industry and all citizens relies then on not only available renewable freshwater supplies and investment in secure and sustainable infrastructure for the collection, treatment, and reliable distribution of safe and secure water but also, the will to make those investments and ensure equitable rights to this water to all sectors of the population.

This chapter focuses on reviewing challenges to a secure water future for the United States and considers a number of options available, which when used together will improve the outlook for water availability and security into the future. Topics covered include a reconsideration of water's role in sustainable development, the role of water conservation and pollution prevention in water security, federal initiatives for the protection of water infrastructure from malicious attacks and growing natural disasters, and the concept of "effective utility management" as applied to future water security. An applications section provides three Illustrative Examples related to future water security in the United States.

24.2 WATER AND SUSTAINABLE DEVELOPMENT

Water is truly at the core of sustainable development as water is inextricably linked to climate change, agriculture, food security, health, equality, gender, and education. The critical role of water to the sustainability of human populations was clearly stated by government leaders and nongovernmental organizations from across the globe as early as 1992 through the Dublin Statement produced from the International Conference on Water and the Environment (ICWE) in Dublin, Ireland, in January that year. Four principles were established from the Dublin conference that were later adopted at the United Nations Conference on Environment and Development (UNCED) in Rio de Janeiro in June 1992. These four principles were stated as follow:

1. **Fresh water is a finite and vulnerable resource, essential to sustain life, development and the environment**. Because water sustains life, effective management of water resources demands a holistic approach, linking social and economic development with protection of natural ecosystems. Effective management links land and water uses across the whole of a catchment area or groundwater aquifer.
2. **Water development and management should be based on a participatory approach, involving users, planners and policy makers at all levels**. The participatory approach involves raising awareness of the importance of water among policy makers and the general public. It means that decisions are taken at the lowest appropriate level, with full public consultation and involvement of users in the planning and implementation of water projects.

3. **Women play a central part in the provision, management, and safeguarding of water.** This pivotal role of women as providers and users of water and guardians of the living environment has seldom been reflected in institutional arrangements for the development and management of water resources. Acceptance and implementation of this principle requires positive policies to address women's specific needs and to equip and empower women to participate at all levels in water resources programs, including decision making and implementation, in ways defined by them. (This third principle is of importance for developing countries where women's rights are weak, but their responsibilities are large. It is appropriate in developed countries as well, especially when the representation of women in engineering and water resource management positions still lags behind that of their male colleagues.)
4. **Water has an economic value in all its competing uses and should be recognized as an economic good.** It is vital to recognize first the basic right of all human beings to have access to clean water and sanitation at an affordable price. Past failure to recognize the economic value of water has led to wasteful and environmentally damaging uses of the resource. Managing water as an economic good is an important way of achieving efficient and equitable use and of encouraging conservation and protection of water resources.

These principles appear as relevant today as they were in 1992 and relate to the concepts of integrated watershed management and permitting that is beginning to be implemented by the U.S. Environmental Protection Agency (EPA), the call for inclusive and equitable water resource management, and the growing interest in the improved valuing of water and treatment of water for reuse and recovery.

Water resources and the range of services they provide underpin economic growth, poverty reduction, and environmental sustainability. As indicated in Chapter 21, Envison™ is a free, voluntary guide to enhancing the sustainability of infrastructure investments and can be used to incorporate sustainability concepts into infrastructure projects (ISI 2019). Of the range of sustainability concepts embodied within the guide, the following are specifically focused on water sustainability and security (Table 21.2: Preserve Water Resources, Reduce Operational Water Consumption, Manage Stormwater, and Protect Surface and Groundwater Quality.)

In a sustainable United States, water and related resources would be managed in support of human well-being and ecosystem integrity in a robust economy. Sufficient and safe water would be made available to meet every person's basic needs with reliable and affordable water supply and sanitation services supported by equitably extended and efficiently managed infrastructure. Water resources management, infrastructure, and sanitary and potable water service delivery would be sustainably financed. Water would be appropriately valued in all its forms, with wastewater treated as a resource that supplies energy, nutrients, and fresh water for reuse. Human development would coexist in harmony with the natural water cycle and the ecosystems that support it, with measures in place that reduce water supply vulnerability and improve infrastructure resilience to water-related disasters. Integrated approaches to water resources development, management, and equitable availability and use would be the norm. Water resources would be managed in a participatory way that draws on the full potential of all people as professionals and citizens, guided by a number of able and knowledgeable organizations within a just and transparent institutional framework.

24.3 WATER CONSERVATION AND POLLUTION PREVENTION

Chapter 20 highlighted the role that pollution prevention and water conservation plays in water resource management and sustainable development. Recycled water can satisfy most water demands as long as it is adequately treated to ensure the water quality is appropriate for the end use, and the concept of "fit for purpose," or the variable treatment of wastewater streams to meet the demands of specific reuse needs, rather than treating an entire waste flow to the strictest standards, is gaining interest. In addition to providing a dependable, locally controlled water supply, water recycling provides tremendous environmental benefits such as decreasing the diversion of water from sensitive ecosystems and using the recycled water to create or enhance wetlands and riparian habitats. Finally, the application of recycled water for agricultural and landscape irrigation rather than discharging it to surface water can provide an additional source of nutrients to the users of this reclaimed water, can save communities the high cost of nutrient removal, can reduce the production of nutrient rich biosolids, and can lessen the need to apply synthetic fertilizers throughout the community.

The U.S. EPA has produced a variety of resources for the public and commercial sectors highlighting the benefits of water conservation, water use efficiency, recycling, and reuse for pollution prevention and resource conservation. The EPA's *Preventing Pollution through Efficient Water Use* (U.S. EPA 1990) is directed toward individuals and communities, summarizing pollution prevention, energy conservation, source water protection, and economic and habitat benefits to efficient use of water. The EPA's *Cleaner Water Through Conservation* (U.S. EPA 1995) focuses on the affect of excessive water use (mostly related to outside landscape and agricultural water use) on water quality and provides residential and industrial/commercial best practices for water conservation and water use efficiency improvements.

24.4 FEDERAL INITIATIVES FOR WATER INFRASTRUCTURE RESILIENCY AND SUSTAINABILITY

In addition to concerns regarding sustainable water supplies and water availability, the federal government also considers water and wastewater treatment utilities as points of vulnerability to terrorist attacks within our infrastructure system. As indicated in Chapter 19, the U.S. EPA established a Water Security Division within the Office of Ground Water and Drinking Water in 2003 to protect water and wastewater utilities against cyberattacks and intentional acts of water system contamination and disruption of water treatment and/or distribution.

Security-related activities undertaken by EPA and the water sector in response to perceived terrorist threats have fallen into five general categories, including:(i) establishing an information center for drinking water alerts or incidents, (ii) developing vulnerability assessment tools, (iii) identifying actions to minimize vulnerabilities, (iv) revising emergency operations plans, and (v) supporting research on biological and chemical contaminants considered to be potential weapons of mass destruction. The requirements for utilities to conduct vulnerability analyses, risk assessments, response planning, emergency communication, and post-event recovery are summarized in Chapter 19.

The EPA and the Department of Homeland Security have also developed guidance for water utilities for the planning, response, and recovery following natural disasters to improve water system resilience (https://www.epa.gov/waterresilience) in response to the America's Water Infrastructure Act of 2018. The *Water and Wastewater Systems Sector-Specific Plan* (U.S. Department of Homeland Security and U.S. EPA 2015) outlines processes and activities related to the protection and increased resilience of drinking water and wastewater utilities. The objectives of the plan are to enable water utilities, in response to natural disasters and aging infrastructure to reduce the magnitude and/or duration of disruptive events by sustaining protection of public health and the environment, recognizing and reducing risk, maintaining a resilient infrastructure, and increasing communication and public confidence in our water systems. Much of the guidance for response to terrorist acts is applicable to water resilience action (i.e., risk and vulnerability assessment, cyber infrastructure integrity, emergency response planning, utility recovery action, communication with emergency responders and the public, etc.). Regarding resilient infrastructure, guidance focuses on continuity of operation and emergency preparedness, response, and recovery planning to maintain public health and community confidence; mutual aid agreements among utilities and states to provide support during response and recovery; and recognition and strengthening of interdependent relationships with other infrastructure sectors (i.e., healthcare, public health, emergency services, chemical, and energy sectors) for ensuring support during response and recovery efforts.

Sustainable water infrastructure support from the U.S. EPA (https://www.epa.gov/sustainable-water-infrastructure) and the U.S. Department of Energy (https://www.energy.gov/articles/doe-launches-water-security-grand-challenge) focus on building sustainable water infrastructure with energy and water efficiency in mind (e.g., source water protection for pollution prevention, green infrastructure for stormwater management, decentralized waste treatment systems for rural and low density development, increased resource recovery from municipal, and energy sector wastewater), and sustainably managing water utilities through effective utility management, effective asset management, affordably pricing water services, and sustainably financing water system operations and improvements.

The U.S. EPA and 10 collaborating organizations have produced the *Effective Utility Management* (U.S. EPA et al. 2017) primer, a self-assessment guide for water and wastewater utilities. It provides guidance to water and wastewater utility managers to enable them to make organization-wide performance improvements toward the goal of creating a sustainable water utility. Of particular interest here are the "Ten Attributes of an Effectively Managed Utility" that the document identifies. These attributes of a sustainably managed water utility are stated as follows (U.S. EPA et al. 2017):

1. Product Quality. A sustainable utility produces "fit for purpose" water and other recovered resources (e.g., energy, nutrients, biosolids) that meet or exceed full compliance with regulatory and reliability requirements and are consistent with customer, public health, ecological, and economic needs. Products include treated drinking water, treated wastewater effluent, recycled water, stormwater discharge, and recovered resources.
2. Customer Satisfaction. A sustainable utility provides reliable, responsive, and affordable services in line with explicit, customer-derived service levels; provides tailored customer service and outreach to traditional residential, commercial, and industrial customers; and understands and exercises as appropriate the opportunities presented by emergent customer groups (e.g., high strength waste producers, power companies).
3. Stakeholder Understanding and Support. A sustainable utility engenders understanding and support from stakeholders including customers, oversight bodies, community and watershed interests, and regulatory bodies for service levels, rate structures, operating budgets, capital improvement programs, and risk management decisions; actively promotes an appreciation of the true value of water and water services, and water's role in the social, economic, public and environmental health of the community; engages in partnerships; involves stakeholders in the decisions that will affect them; understands what it takes to operate as a "good neighbor"; and positions the utility as a critical asset (anchor institution) to the community.
4. Financial Viability. A sustainable utility understands and plans for the full life-cycle cost of utility operations and value of water resources; establishes and maintains an effective balance between long-term debt, asset values, operations and maintenance expenditures, and operating revenues; establishes predictable rates consistent with community expectations and acceptability adequate to recover costs, provide for reserves, maintain support from bond rating agencies, plan and invest for future needs, and taking into account affordability and the needs of disadvantaged households; implements sound strategies for collecting customer payments; and understands the opportunities available to diversify revenues and raise capital through adoption of new business models, including revenues from resource recovery.
5. Operational Optimization. A sustainable utility ensures ongoing, timely, cost-effective, reliable, and sustainable performance improvements in all facets of its operations in service to public health and environmental protection; makes effective use of data from automated and smart systems, and learns from performance monitoring; minimizes resource use, loss, and impacts from day-to-day operations, and reduces all forms of waste; and maintains awareness of information and operational technology developments to anticipate and support timely adoption of improvements.

6. Employee and Leadership Development. A sustainable utility recruits, develops, and retains a workforce that is competent, motivated, adaptive, and safety-focused; establishes a participatory, collaborative organization dedicated to continual learning, improvement, and innovation; ensures employee institutional knowledge is retained, transferred, and improved upon over time; emphasizes and invests in opportunities for professional and leadership development, taking into account the differing needs and expectations of a multigenerational workforce and for resource recovery operations; and establishes an integrated and well-coordinated senior leadership team.
7. Enterprise Resiliency. A sustainable utility ensures leadership and staff work together internally, and coordinate with external partners, to anticipate, respond to, and avoid problems; proactively identifies, assesses, establishes tolerance levels for, and effectively manages a full range of business risks (including interdependencies with other services and utilities, legal, regulatory, financial, environmental, safety, physical and cyber security, knowledge loss, talent, and natural disaster-related) consistent with industry trends and system reliability goals; and plans for and actively manages around business continuity.
8. Infrastructure Strategy and Performance. A sustainable utility understands the condition of and costs associated with critical infrastructure assets; plans infrastructure investments consistent with community needs, anticipated growth, system reliability goals, and relevant community priorities, building in a robust set of adaptation strategies (e.g., for changing weather patterns, customer base); maintains and enhances the condition of all assets over the long-term at the lowest possible life-cycle cost and acceptable risk consistent with customer, community, and regulator-supported service levels; and assures asset repair, rehabilitation, and replacement efforts are coordinated within the community to minimize disruptions and other negative consequences.
9. Community Sustainability. A sustainable utility takes an active leadership role in promoting and organizing community sustainability improvements through collaboration with local partners (e.g., transportation departments, electrical utilities, planning departments, economic development organizations, watershed and source water protection groups); manages operations, infrastructure, and investments to support the economic, environmental, and social health of its community; and integrates water resource management with other critical community infrastructure, social and economic development planning to support community-wide resilience, support for disadvantaged households, community sustainability, and livability.
10. Water Resource Sustainability. A sustainable utility ensures the availability and sustainable management of water for its community and watershed, including water resource recovery; understands its role in the complete water cycle; understands "fit for purpose" water reuse options; and integrates utility objectives and activities with other watershed managers and partners. A sustainable utility understands and plans for the potential for water resource variability (e.g., changing weather patterns, including extreme events, such as drought and flooding); utilizes as appropriate a full range of watershed investment and engagement strategies (e.g., Integrated Planning); engages in long-term integrated water resource management; and ensures that current and future customer, community, and ecological water-related needs are met.

Self-assessment of a utility's status with respect to these 10 attributes is designed to guide a utility through a 360-degree review of utility performance and to identify priorities for improvement within a specific community context. It is designed to help utilities protect their current infrastructure investment, motivate their workforce, and move from reaction mode to proactive mode to address relevant community-utility needs to provide affordable and sustainable services into the future (U.S. EPA et al. 2017).

24.5 APPLICATIONS

Three Illustrative Examples complement the material presented in this chapter on future water security issues for the United States.

Illustrative Example 24.1

The U.S. EPA has a stated *Clean Water and Drinking Water Infrastructure Sustainability Policy* (https://www.epa.gov/sites/production/files/2016-01/documents/clean-water-and-drinking-water-infrastructure-sustability-policy.pdf). The policy addresses sustainable practices at three interdependent levels. What are these three interdependent sustainability programs and how are they related to the EPA's goal for sustainable water infrastructure?

Solution

The policy addresses how more sustainable practices can be achieved through three interdependent program initiatives:

1. Sustainable Water Infrastructure Planning. For sustainable water infrastructure, flexible and comprehensive planning processes should be adopted to pursue water infrastructure investments that are cost effective over their life cycle, resource efficient, and consistent with community sustainability goals. This planning effort should result in infrastructure investments that reflect public health, water quality, and climate change considerations, the desired level of service, conservation of natural resources, alternative approaches such as natural or "green" systems, decentralized solutions, or other innovative approaches consistent with community values.
2. Sustainable Water Sector Systems. Water sector water supply and wastewater treatment systems should employ effective utility management practices (as described previously) to build and maintain the technical, financial, and managerial capacity and viability necessary to ensure long-term sustainability.
3. Sustainable Communities. Investments in various infrastructure sectors, such as housing, transportation, and water, should be coordinated to best align with a community's goals for livability and sustainable growth.

Each of these levels supports the others and support the EPA's overall goal to ensure that federal investments, policies, and actions support sustainable water infrastructure to best support existing communities, enhance economic competitiveness and promote affordable neighborhoods.

Illustrative Example 24.2

An important aspect of the *Effective Utility Management* (U.S. EPA et al. 2017) primer self-assessment component is developing measurable performance indicators for the 10 attributes presented as essential for sustainable water utilities. Appendix C in the *Effective Utility Management* primer provides example performance indicators to be used in the self-assessment. Pick 1 of the 10 attributes of a sustainable water utility and describe the performance indicators provided as examples for the self-assessment.

Solution

This is an open-ended question with variable answers based on the specific attribute selected for review. The discussion provided is an example of performance indicators suggested for the attribute community sustainability.

1. Watershed-Based Infrastructure Planning. This measure addresses utility efforts to consider watershed-based approaches when making management decisions affecting infrastructure planning and investment options. An example performance indicator might

include: *Does the utility employ alternative, watershed-based approaches to align infrastructure decisions with overall watershed goals and potentially reduce future infrastructure costs (yes/no)?*

2. Green Infrastructure. This measure addresses green infrastructure, which includes both the built and natural/unbuilt environment. Utilities may promote source water protection and conservation green infrastructure approaches in support of water conservation (e.g., per capita demand reduction) and water quality protection objectives. Examples of performance indicators might include: *Has the utility explored green infrastructure approaches and opportunities that are aligned with the utility's mandate, goals, and objectives and community interests (yes/no)? Does the utility have procedures that incorporate green infrastructure approaches and performance into new infrastructure investments (yes/no)?*
3. Greenhouse Gas Emissions. This measure will help drinking and wastewater utilities understand and reduce their individual contributions to area greenhouse gas emissions. Trends indicate that water utility emissions of these gases will likely be of interest to stakeholders. Examples of performance indicators might include: *Net (gross minus offsets) greenhouse gas emissions in tons of carbon dioxide (CO_2), nitrous oxide (N_2O), methane (CH_4), and, as applicable, hydrofluorocarbons (HFCs) and perfluorocarbons (PFCs) over time compared to baseline emissions. Percentage of utility energy demand met by renewable energy resources.*
4. Service Affordability. This measure addresses drinking water and wastewater service affordability, which centers on community members' ability to pay for water services. To the extent possible within its operating and regulatory contexts, the utility will want to consider and balance keeping water services affordable while ensuring the rates needed for long-term infrastructure and financial integrity. Examples of performance indicators might include: *Bill affordability (households for which rates may represent an unaffordable level) (percent): 100 × (number of households served for which average water bill is > "X" percent (often 2%–2.5%) of median household income ÷ total number of households served). Low-income billing assistance program coverage (percent): 100 × (number of customers enrolled in low-income billing assistance program ÷ number of customers who are eligible for enrollment in low-income billing assistance program).*
5. Community Economic Development. This measure assesses the extent to which utility operations play a role in local economic development (e.g., by attracting new employers to the area, enabling residential or commercial growth, or through job creation). Examples of performance indicators might include: *Change in tax base (dollars or percentage change) related to new water infrastructure. Number of jobs created by utility infrastructure investments: Internal to the utility; Contracted by the utility; or Through a new employer brought to the community as a result of utility infrastructure. Green infrastructure economic benefits: Crime reduction (percentage change); and Increase in local property values (percentage change).*

Illustrative Example 24.3

Water conservation is claimed to be a highly cost-effective approach to extending existing water supplies and increasing the security and sustainability of water sources. Provide documentation that this is indeed the case.

Solution

This is an open-ended question and can be answered by considering the reduction in water use by high-efficiency water fixtures, quantifying the water savings on an annual basis, and determining the value of the water saved annual in a specific local setting. The payback of water conservation estimates will be dependent on the value of water and the cost of implementing specific water conservation practices. Some practices, such as turning off a tap during the brushing of teeth, cost nothing and have an instant payback, others require the investment and installation of new fixtures that, depending on water costs, may have payback periods of 2–3 years. The U.S. EPA (1995) provides several examples of cost savings for conservation practices across the United States. One such example follows.

The Texas Water Development Board estimated that the use of water-efficient plumbing fixtures should save a typical four-member household 55,800 gallons of water and $627 in lower water and energy (i.e., water heating and pumping) costs per year, making the payback for plumbing fixture upgrades significantly less than 1 year. When aggregating these savings across the entire state, low-flow fixtures might reduce water use statewide by 805 million gallons per day (MGD) by the Year 2040, reducing the need to build new water and wastewater treatment plants by 15% or more than $68 million per year.

REFERENCES

Institute for Sustainable Infrastructure (ISI). 2019. Envision Version 3. Washington, DC: Institute for Sustainable Infrastructure. www.sustainableinfrastructure.org.

U.S. Department of Homeland Security and U.S. Environmental Protection Agency. 2015. *Water and Wastewater Systems Sector-Specific Plan*. Washington, DC: Office of Infrastructure Protection and the Office of Water. https://www.dhs.gov/sites/default/files/publications/nipp-ssp-water-2015-508.pdf.

U.S. Environmental Protection Agency (U.S. EPA). 1990. *Preventing Pollution through Efficient Water Use*. OW WH-556, OPPE PM-222. Washington, DC: Office of Water, Office of Pollution Prevention.

U.S. Environmental Protection Agency (U.S. EPA). 1995. *Cleaner Water Through Conservation*. EPA 841-B-95-002. Washington, DC: Office of Water. https://nepis.epa.gov/Exe/ZyPDF.cgi/20004MQA.PDF?Dockey=20004MQA.PDF.

U.S. Environmental Protection Agency, Association of Metropolitan Water Agencies, Association of Clean Water Administrators, Association of Safe Drinking Water Administrators, American Public Works Association, American Water Works Association, National Association of Clean Water Agencies, National Association of Water Companies, Water Environment Federation, Water Environment and Reuse Foundation, and Water Research Foundation. 2017. *Effective Utility Management. A Primer for Water and Wastewater Utilities*. Washington, DC: EPA Office of Wastewater Management. https://www.epa.gov/sites/production/files/2017-01/documents/eum_primer_final_508-january2017.pdf.

World Health Organization (WHO). 2017. *Progress on Drinking Water, Sanitation and Hygiene 2017. Update and SDG Baselines*. Geneva, Switzerland: World Health Organization (WHO) and the United Nations Children's Fund (UNICEF). https://apps.who.int/iris/bitstream/handle/10665/258617/9789241512893-eng.pdf;jsessionid=BFB5D93E1AC22CFE3A7EAB3C0F603BF9?sequence=1.

Appendix A: Units and Common Abbreviations

A.1 SI Multiples and Prefixes

Multiples and Submultiples		Prefixes	Symbols
1,000,000,000,000	10^{12}	tera	T
1,000,000,000	10^{9}	giga	G
1,000,000	10^{6}	mega	M
1,000	10^{3}	kilo	k
100	10^{2}	hecto	h
10	10^{1}	deka	da
Base unit 1	10^{0}		
0.1	10^{-1}	deci	d
0.01	10^{-2}	centi	c
0.001	10^{-3}	milli	m
0.000001	10^{-6}	micro	μ
0.000000001	10^{-9}	nano	n
0.000000000001	10^{-12}	pico	P
0.000000000000001	10^{-15}	femto	f
0.000000000000000001	10^{-18}	atto	a

A.2 Selected Common Abbreviations

Å, A	angstrom unit of length
abs	absolute
amb	ambient
app. MW, M	apparent molecular weight
atm, ATM	atmospheric
at. wt.	atomic weight
b.p.	boiling point
bbl	barrel
Btu	British thermal unit
cal	calorie
cg	centigram
cm	centimeter
cgs system	centimeter-gram-second system
conc.	concentrated, concentration
cc, cm^3	cubic centimeter
cu ft, ft^3	cubic feet
cfh	cubic feet per hour
cfm	cubic feet per minute
cfs	cubic feet per second
m^3, M^3	cubic meter
°	degree
°C	degree Celsius, degree Centigrade

(*Continued*)

A.2 (*Continued*) Selected Common Abbreviations

°F	degree Fahrenheit
°R	degree Rankine
ft	foot
ft-lb	foot pound
fpm	feet per minute
fps	feet per second
fps system	foot-pound-second system
f.p.	freezing point
gal	gallon
gr	grain
g, gm	gram
hr	hour
in	inch
kcal	kilocalorie
kg	kilogram
km	kilometer
liq	liquid
L	liter
log	logarithm (common)
ln	logarithm (natural)
m.p.	melting point
m, M	meter
μm	micrometer (micron)
mks system	meter-kilogram-second system
mph	miles per hour
mg	milligram
mL	milliliter
mm	millimeter
mμ	millimicron
min	minute
mol wt, MW, M	molecular weight
oz	ounce
ppb	parts per billion
pphm	parts per hundred million
ppm	parts per million
lb	pound
psi	pounds per square inch
psia	pounds per square inch absolute
psig	pounds per square inch gage
rpm	revolutions per minute
s	second
sp gr, SG	specific gravity
sp ht	specific heat
sp wt	specific weight
sq	square
scf	standard cubic foot
STP	standard temperature and pressure
temp	temperature
wt	weight

Appendix B: Conversion Constants

B.1 Conversion Constants

To Convert from	To	Multiply by
Length		
m	cm	100
m	mm	1,000
m	microns (μm)	10^6
m	angstroms (Å)	10^{10}
m	in	39.37
m	ft	3.281
m	mi	6.214×10^{-4}
ft	in	12
ft	m	0.3048
ft	cm	30.48
ft	mi	1.894×10^{-4}
Mass		
kg	g	1,000
kg	lb	2.205
kg	oz	35.24
kg	ton	2.268×10^{-4}
kg	grains	1.543×10^4
lb	oz	16
lb	ton	5×10^{-4}
lb	g	453.6
lb	kg	0.4536
lb	grains	7,000
Time		
s	min	0.01667
s	hr	2.78×10^{-4}
s	day	1.157×10^{-7}
s	week	1.653×10^{-6}
s	yr	3.171×10^{-8}
Force		
N	kg-m/s^2	1
N	Dynes	10^5
N	g-cm/s^2	10^5
N	lb_f	0.2248
N	lb • ft/s^2	7.233
lb_f	N	4.448
lb_f	dynes	4.448×10^5
lb_f	g-cm/s^2	4.448×10^5
lb_f	lb-ft/s^2	32.17

(Continued)

B.1 (*Continued*) Conversion Constants

To Convert from	To	Multiply by
Pressure		
atm	N/m^2 = Pa	1.013×10^5
atm	kPa	101.3
atm	bars	1.013
atm	$dynes/cm^2$	1.013×10^6
atm	lb_f/in^2 = psi	14.696
atm	mm Hg at 0°C (torr)	760
atm	in Hg at 0°C	29.92
atm	ft H_2O at 4°C	33.9
atm	in H_2O at 4°C	406.8
psi	Atm	6.80×10^{-2}
psi	mm Hg at 0°C (torr)	51.71
psi	in H_2O at 4°C	27.70
in H_2O at 4°C	atm	2.458×10^{-3}
in H_2O at 4°C	psi	0.0361
in H_2O at 4°C	mm Hg at 0°C (torr)	1.868
Volume		
m^3	L	1,000
m^3	cm^3 (cc, mL)	10^6
m^3	ft^3	35.31
m^3	gal (U.S.)	264.2
m^3	qt	1,057
ft^3	in^3	1,728
ft^3	gal (U.S.)	7.48
ft^3	m^3	0.02832
ft^3	L	28.32
Power		
W	J/s	1
W	cal/s	0.2390
W	ft-lb_f/s	0.7376
W	kW	10^{-3}
kW	Btu/s	0.949
kW	hp	1.341
hp	ft-lb_f/s	550
hp	kW	0.7457
hp	cal/s	178.2
hp	Btu/s	0.707
Concentration		
$\mu g/m^3$	lb/ft^3	6.243×10^{-11}
$\mu g/m^3$	lb/gal	8.346×10^{-12}
$\mu g/m^3$	gr/ft^3	4.370×10^{-7}
gr/ft^3	$\mu g/m^3$	2.288×10^6
gr/ft^3	g/m^3	2.288
lb/ft^3	$\mu g/m^3$	1.602×10^{10}
lb/ft^3	$\mu g/L$	1.602×10^7
lb/ft^3	lb/gal	7.48

(*Continued*)

B.1 (*Continued*) Conversion Constants

To Convert from	To	Multiply by
Viscosity		
P (poise)	g/cm-s	1
P	cP (centipoise)	100
P	kg/m-hr	360
P	lb/ft-s	6.72×10^{-2}
P	lb/ft-hr	241.9
P	lb/m-s	5.6×10^{-3}
lb/ft-s	P	14.88
lb/ft-s	g/cm-s	14.88
lb/ft-s	kg/m-hr	5.357×10^{3}
lb/ft-s	lb/ft-hr	3,600
Heat Capacity		
cal/g-°C	Btu/lb°F	1
cal/g-°C	kcal/kg-°C	1
cal/g-°C	cal/gmol-°C	Molecular Weight
cal/gmol-°C	Btu/lbmol-°F	1
J/g-°C	Btu/lb-°F	0.2389
Btu/lb-°F	cal/g-°C	1
Btu/lb-°F	J/g-°C	4.186
Btu/lb-°F	Btu/lbmol-°F	Molecular Weight
Energy		
J	N-m	1
J	erg	10^{7}
J	dyne-cm	10^{7}
J	kW-hr	2.778×10^{-7}
J	cal	0.2390
J	ft-lb_f	0.7376
J	Btu	9.486×10^{-4}
cal	J	4.186
cal	Btu	3.974×10^{-3}
cal	ft-lb_f	3.088
Btu	ft-lb_f	778
Btu	hp-hr	3.929×10^{-4}
Btu	cal	252
Btu	kW-hr	2.93×10^{-4}
ft-lb_f	cal	0.3239
ft-lb_f	J	1.356
ft-lb_f	Btu	1.285×10^{-3}

B.2 Common Engineering Conversion Factors

Length

1 ft = 12 in = 0.3048 m, 1 yard = 3 ft
1 mi = 5,280 ft = 1,609.344 m
1 nautical mile (nmi) = 6,076 ft

Mass

1 slug = 32.174 lb = 14.594 kg
1 lb = 0.4536 kg = 7,000 grains

Acceleration and Area

1 ft/s^2 = 0.3048 m/s^2
1 ft^2 = 0.092903 m^2

Mass Flow and Mass Flux

1 slug/s = 14.594 kg/s
1 lb/s = 0.4536 kg/s
1 kg·/m^2·s = 0.2046 lb/ft^2·s = 0.00636 slug/ ft^2·s

Pressure

1lb$_f$/ft^2 = 47.88 Pa, 1 torr = 1 mm Hg
1 psi = 144 psf, 1 bar = 10^5 Pa
1 atm = 2,116.2 psf = 14.696 psi = 101,325 Pa = 29.9 in Hg = 33.9 ft H_2O

Power

1 hp = 550 ft-lb$_f$/s = 745.7 W
1 ft-lb$_f$/s = 1.3558 W
1 Watt = 3.4123 Btu/h = 0.00134 hp

Specific weight

1 lb$_f$/ft^3 = 157.09 N/m^3

Viscosity

1 slug/(ft-s) = 47.88 kg/(m-s) = 478.8 poise (P)
1 P = 1 g/(cm-s) = 0.1 kg/(m-s) = 0.002088 slug/(ft-s)

Temperature Scale Readings

°F = (9/5)°C + 32
°C = (5/9) (°F–32)
°R = F + 459.69
K = °C + 273.16
°R = (1.8) K

Volume

1 ft^3 = 0.028317 m^3 = 7.481 gal
1 bbl = 42 U.S. gal
1 U.S. gal = 231 in^3 = 3.7853 L = 4 qt = 0.833 Imp. Gal
1 L = 0.001 m^3 = 0.035315 ft^3 = 0.2642 U.S. gal

Density

1 slug/ft^3 = 515.38 kg/m^3, 1 g/cm^3 = 1,000 kg/m^3
1 lb/ft^3 = 16.0185 kg/m^3, 1 lb/in^3 = 27.68 g/cm^3

Velocity

1 ft/s = 0.3048 m/s
1 knot = 1 nmi/hr = 1.6878 ft/s
1 mi/hr = 1.4666666 ft/s = 0.44704 m/s

Volume Flow

1gal/min = 0.02228 ft^3/s = 0.06309 L/s
1 million gal/day = 1.5472 ft^3/s = 0.04381 m^3/s

Force and Surface Tension

1 lb$_f$ = 4.448222 N = 16 oz
1 dyne = 1 g·cm/s^2 = 10^{-5} N
1 kg$_f$ = 2.2046 lb$_f$ = 9.80665 N
1 U.S. (short) ton = 2,000 lb$_f$
1 N = 0.2248 lb$_f$

Energy and Specific Energy

1 ft-lb$_f$ = 1.35582 J, 1 hp-hr = 2,544.5 Btu
1 Btu = 252 cal = 1,055.056 J = 778.17 ft-lb$_f$
1 cal = 4.185 J, 1 ft-lb$_f$/lb = 2.9890 J/kg

Heat Flux

1 W/m^2 = 0.3171 Btu/(hr-ft^2)

Kinematic Viscosity

1 ft^2/hr = 2.506 × 10^{-5} m^2/s
1 ft^2/s = 0.092903 m^2/s
1 stoke (st) = 1 cm^2/s = 0.0001 m^2/s = 0.001076 ft^2/s

Thermal Conductivity[a]

1 cal/(s-cm-°C) = 242 Btu/(hr-ft-°R)
1 Btu/(hr-ft-°R) = 1.7307 W/(m-K)

Specific Heat or Gas Constant[a]

1 (ft-lb$_f$)/(slug-°R) = 0.16723 (N-m)/(kg-K)
1 Btu/(lb-°R) = 4,186.8 J/(kg-K)

[a] Note that the intervals in absolute (Kelvin) and °C are equal. Also, Δ1°R = Δ1°F
Latent heat: 1 J/kg = 4.2995 × 10^{-4} Btu/lb = 10.76 lb$_f$-ft/slug = 0.3345 lb$_f$/lb
1 Btu/lb = 2,325.9 J/kg
Heat transfer coefficient: 1 Btu/(hr-ft-°F) = 5.6782 W/(m^2-°C)
Heat generation rate: 1 W/m^3 = 0.09665 Btu/(hr-ft^3)
Heat transfer per unit length: 1 W/m = 1.0403 Btu/(hr-ft)
Mass transfer coefficient: 1 m/s = 11.811 ft/hr, 1 lbmol/(hr-ft^2) = 0.0013562 kgmol/(s-m^2)

Index

Note: Page numbers in italic and bold refer to figures and tables, respectively.

D

E

H

I

K

L

M

P

Q

R

S

T

X

Y

Z